Solid State Chemistry

Solid State Chemistry

An Introduction
Fifth Edition

Elaine A. Moore

Lesley E. Smart

CRC Press is an imprint of the
Taylor & Francis Group, an **informa** business

Fifth edition published 2021
by CRC Press
2 Park Square, Milton Park, Abingdon, Oxon, OX14 4RN

and by CRC Press
6000 Broken Sound Parkway NW, Suite 300, Boca Raton, FL 33487-2742

First edition publishing by Chapman & Hall 1992
Fourth edition published by CRC Press 2012

CRC Press is an imprint of Informa UK Limited

British Library Cataloguing-in-Publication Data
A catalogue record for this book is available from the British Library

Library of Congress Cataloging-in-Publication Data

Names: Moore, Elaine (Elaine A.), author. | Smart, Lesley, author.
Title: Solid state chemistry : an introduction / Elaine A. Moore, Lesley E. Smart.
Description: Fifth edition. | Boca Raton : CRC Press, [2021] | Includes bibliographical references and index.
Identifiers: LCCN 2020013268 (print) | LCCN 2020013269 (ebook) | ISBN 9780367135720 (paperback) | ISBN 9780367135805 (hardback) | ISBN 9780429027284 (ebook)
Subjects: LCSH: Solid state chemistry.
Classification: LCC QD478 .S53 2021 (print) | LCC QD478 (ebook) | DDC 541/.0421--dc23
LC record available at https://lccn.loc.gov/2020013268
LC ebook record available at https://lccn.loc.gov/2020013269

ISBN: 978-0-367-135805 (hbk)
ISBN: 978-0-367-13572-0 (pbk)
ISBN: 978-0-429-027284 (ebk)

Typeset in Times
by Deanta Global Publishing Services, Chennai, India

Visit the companion website: www.crcpress.com/cw/moore

Dedication

This edition is dedicated to the memory of Lesley E. Smart (1947–2016), lead author on previous editions of this book.

Contents

Preface to the Fifth Edition

One of the great issues of the current time is sustainability. This edition, therefore, includes a new chapter introducing this subject and providing tools for sustainable approaches. written by an expert in the field, Mary Anne White. Solid-state devices, such as batteries and solar cells, can contribute to sustainability, but also, sustainable methods of synthesizing solid materials is a current important area of research. These are included in appropriate chapters. Solid-state chemistry research has moved on since the previous edition. I have taken this opportunity to add new topics and reorganise the content. Kröger–Vink and short Hermann–Mauguin notations have been added to the chapter on crystal structures. Cryogenic electron microscopy is now included with physical methods. Batteries have been taken out of the chapter on defects and given a chapter of their own together with fast-ion conductors. The section on MOFS has been expanded and one on COFs added. To do this and keep the book at an affordable size means that some topics have been dropped or given less space, including applications of defect solids, zeolites and superconductors, fuel cells, ALPOs, and metal oxide frameworks. Two chapters have been rewritten by new authors, Neil Allan and Liana Vella-Zarb, and I am very grateful for their help. I thank Dr. Simon Collinson for his expertise and helpful discussions with green chemistry aspects of this text

I would like to thank the production team at Taylor & Francis for their continued support and for providing reviewers for the first drafts, who provided many useful comments which I hope you will find have improved this book.

I wish to acknowledge the use of the Chemical Database Service at Daresbury (Fletcher, D. A., Meeking, R. F., and Parkin, D., 'The United Kingdom Chemical Database Service', *J. Chem. Inf. Comput. Sci.* **36**, 746–749, 1996), and the Inorganic Crystal Structure Database. (Berghoff, G. and Brown, I. D., 'The Inorganic Crystal Structure Database (ICSD)', in *Crystallographic Databases*, F. H. Allen et al., eds. Chester, International Union of Crystallography, 1987.) Also, WebLab Viewer Pro or Discovery Studio Visualiser from Dassault Systèmes Biovia Corp.

Elaine A. Moore

Preface to the Fourth Edition

Solid-state chemistry is still a rapidly advancing field, and the past few years have seen many new developments and discoveries in this field. This edition aims, as previously, not only to teach the basic science that underpins the subject, but also to direct the reader to the most modern techniques and to expanding and new areas of research. In order to do this, some of the subjects covered in the previous editions have either been briefly discussed or dropped altogether, to make way for the new. More space has been devoted to electron microscopy and to the techniques that are available with synchrotron radiation; there are new sections on nanomaterials and their synthesis and on computer modeling of solids, periodic mesoporous organosilicas (PMOs), metal organic frameworks (MOFs), metal oxide frameworks, metamaterials, and multiferroics. The section on molecular metals has been dropped and the discussion on conducting polymers has been added at the end of the expanded section on carbon nanoscience. Throughout, new scientific terms and concepts are highlighted in bold typeface where they are first introduced.

We thank our colleagues and families, who have supported us throughout this project, Professor Mark Weller for his very helpful review, and, in particular, Drs. Craig McFarlane and Mike Mortimer and Professor Peter Atkins, who have generously given their time and expertise in explaining and correcting certain points of science.

We wish to acknowledge the use of the Chemical Database Service at Daresbury (Fletcher, D. A., Meeking, R. F., and Parkin, D., 'The United Kingdom Chemical Database Service', *J. Chem. Inf. Comput. Sci.* **36**, 746–749, 1996), and the Inorganic Crystal Structure Database. (Berghoff, G. and Brown, I. D., 'The Inorganic Crystal Structure Database (ICSD),' in *Crystallographic Databases*, F. H. Allen et al., eds. Chester, International Union of Crystallography, 1987.) Also, WebLab Viewer Pro or Discovery Studio Visualiser from Accelerys Inc. was used to create the crystal structures.

Lesley E. Smart
Elaine A. Moore

Authors

Elaine A. Moore studied chemistry as an undergraduate at Oxford University and then stayed on to complete a DPhil in theoretical chemistry with Peter Atkins. After a two-year postdoctoral position at the University of Southampton, she joined the Open University in 1975, becoming a lecturer in chemistry in 1977, senior lecturer in 1998, and reader in 2004. She retired in 2017 and currently has an honorary position at the Open University. She has produced OU teaching texts in chemistry for courses at levels 1, 2, and 3 and written texts in astronomy at level 2 and physics at level 3. She is coauthor of *Metals and Life* (RSC Publishing, 2009) and of *Concepts in Transition Metal Chemistry* (RSC Publishing, 2010), which were part of a level 3 Open University course in inorganic chemistry and co-published with the Royal Society of Chemistry. She was team leader for the production and presentation of an Open University level 2 chemistry module delivered entirely online. She is a Fellow of the Royal Society of Chemistry and a Senior Fellow of the Higher Education Academy. She was co-chair for the successful Departmental submission of an Athena Swan bronze award.

Her research interests are in theoretical chemistry applied mainly to solid-state systems and is author or coauthor of over 50 papers in refereed scientific journals. A long-standing collaboration in this area led to her being invited to help run a series of postgraduate workshops on computational materials science hosted by the University of Khartoum.

Lesley E. Smart studied chemistry at Southampton University, United Kingdom, and after completing a PhD in Raman spectroscopy, she moved to a lectureship at the (then) Royal University of Malta. After returning to the United Kingdom, she took an SRC Fellowship to Bristol University to work on X-ray crystallography. From 1977 to 2009, she worked at the Open University chemistry department as a lecturer, senior lecturer, and Molecular Science Programme director, and held an honorary senior lectureship there until her death in 2016.

At the Open University, she was involved in the production of undergraduate courses in inorganic and physical chemistry and health sciences. She was the coordinating editor and an author of *The Molecular World* course, a series of eight books and DVDs co-published with the Royal Society of Chemistry, authoring two of these, *The Third Dimension* (RSC Publishing, 2002) and *Separation, Purification and Identification* (RSC Publishing, 2002). Her most recent books are *Alcohol and Human Health* (Oxford University Press, 2007) and *Concepts in Transition Metal Chemistry* (RSC Publishing, 2010). She has an entry in *Mothers in Science: 64 Ways to Have It All* (RSC Publishing, 2016; downloadable from the Royal Society website). She served on the Council of the Royal Society of Chemistry and as the chair of their Benevolent Fund.

Her research interests were in the characterisation of the solid state, and she authored publications on single-crystal Raman studies, X-ray crystallography, Zintl phases, pigments, and heterogeneous catalysis and fuel cells.

Contributors

Neil Allan
School of Chemistry
University of Bristol
Bristol, United Kingdom

Mary Anne White
Department of Chemistry
Dalhousie University
Halifax, Canada

Liana Vella-Zarb
Crystal Engineering Laboratories
Department of Chemistry
University of Malta
Msida, Malta

List of Units, Prefixes, and Constants

BASIC SI UNITS

Physical Quantity (and Symbol)	Name of SI Unit	Symbol for Unit
Length (l)	Metre	M
Mass (m)	Kilogram	Kg
Time (t)	Second	S
Electric current (I)	Ampere	A
Thermodynamic temperature (T)	Kelvin	K
Amount of substance (n)	Mole	Mol
Luminous intensity (I_v)	Candela	Cd

DERIVED SI UNITS

Physical Quantity (and Symbol)	Name of SI Unit	Symbol for SI Derived Unit and Definition of Unit
Frequency (v)	Hertz	Hz ($=s^{-1}$)
Energy (U), enthalpy (H)	Joule	J ($= kg\ m^2\ s^{-2}$)
Force	Newton	N ($= kg\ m\ s^{-2} = J\ m^{-1}$)
Power	Watt	W ($= kg\ m^2\ s^{-3} = J\ s^{-1}$)
Pressure (p)	Pascal	Pa ($= kg\ m^{-1}\ s^{-2} = N\ m^{-2} = J\ m^{-3}$)
Electric charge (Q)	Coulomb	C (= A s)
Electric potential difference (V)	Volt	V ($= kg\ m^2\ s^{-3}\ A^{-1} = J\ A^{-1}\ s^{-1}$)
Capacitance (c)	Farad	F ($= A^2\ s^4\ kg^{-1}\ m^{-2} = A\ s\ V^{-1} = A^2\ s^2\ J^{-1}$)
Resistance (R)	Ohm	O ($= V\ A^{-1}$)
Conductance (G)	Siemen	S ($= A\ V^{-1}$)
Magnetic flux density (B)	Tesla	T ($= V\ S\ m^{-2} = J\ C^{-1}\ s\ m^{-2}$)

SI PREFIXES

10^{-18}	10^{-15}	10^{-12}	10^{-9}	10^{-6}	10^{-3}	10^{-2}	10^{-1}	10^{3}	10^{6}	10^{9}	10^{12}	10^{15}	10^{18}
alto	femto	pico	nano	micro	milli	centi	deci	kilo	mega	giga	tera	peta	Exa
a	f	p	n	μ	m	c	d	k	M	G	T	P	E

FUNDAMENTAL CONSTANTS

Constant	Symbol	Value
Speed of light in a vacuum	c	2.997925×10^{8} m s^{-1}
Charge of a proton and charge of an electron	e	1.602189×10^{-19} C
Avogadro constant	N_A	6.022045×1023 mol^{-1}
Boltzmann constant	k	1.380662×10^{-23} J K^{-1} mo^{l-1}
Gas constant	$R = N_A k$	$8.31441 \times$ J K^{-1}
Faraday constant	$F = N_A e$	9.648456×10^{4} C mol^{-1}
Planck constant	h	6.626176×10^{-34} J s
	$\hbar = \frac{h}{2\pi}$	1.05457×10^{-34} J s
Vacuum permittivity	ε_0	8.854×10^{-12} F m^{-1}
Vacuum permeability	μ_0	$4\pi \times 10^{-7}$ J s^{2} C^{-2} m^{-1}
Bohr magneton	μ_B	9.27402×10^{-24} J T^{-1}
Electron g value	g_e	2.00232
Electron mass	m_e	9.1095×10^{-31} kg
Proton mass	m_p	1.6726×10^{-27} kg
Neutron mass	m_n	1.6749×10^{-27} kg

MISCELLANEOUS PHYSICAL QUANTITIES

Name of Physical Quantity	Symbol	SI Unit
Enthalpy	H	J
Entropy	S	J K^{-1}
Gibbs function	G	J
Standard change of molar enthalpy	$\Delta H_m^\ominus$	J mol^{-1}
Standard of molar entropy	$\Delta S_m^\ominus$	J K^{-1} mol^{-1}
Standard change of molar Gibbs function	$\Delta G_m^\ominus$	J mol^{-1}
Wave number	$\sigma = \frac{1}{\lambda}$	cm^{-1}
Atomic number	Z	Dimensionless
Conductivity	Σ	S m^{-1}
Molar bond dissociation energy	D_m	J mol^{-1}
Molar mass	$M = \left(\frac{m}{n}\right)$	kg mol^{-1}

THE GREEK ALPHABET

alpha	Α	α	nu	Ν	ν
beta	Β	β	xi	Ξ	ξ
gamma	Γ	γ	omicron	Ο	ο
delta	Δ	δ	pi	Π	π
epsilon	Ε	ε	rho	Ρ	ρ
zeta	Ζ	ζ	sigma	Σ	σ
eta	Η	η	tau	Τ	τ
theta	Θ	θ	upsilon	Υ	υ
iota	Ι	ι	phi	Φ	φ
kappa	Κ	κ	chi	Χ	χ
lambda	Λ	λ	psi	Ψ	ψ
mu	Μ	μ	omega	Ω	ω

1 An Introduction to Crystal Structures

The last few decades have seen much exciting research into solid-state chemistry. We have seen immense strides in the development of nanotechnology and solids contributing to the development of sustainable energy devices such as photovoltaic cells, fuel cells, and batteries, to mention but a few areas. It would be impossible to cover all the developments and topics in detail in an introductory text such as this, but we endeavour to give you a flavour of the excitement that some of the research has engendered and perhaps, more importantly, sufficient background with which to understand these developments and those that are yet to come.

Most substances, helium being a notable exception, if cooled sufficiently form a solid phase; the vast majority form one or more **crystalline** phases, where the atoms, molecules, or ions pack together to form a regular repeating array. This book is concerned mostly with the structures of metals, ionic solids, and extended covalent structures; structures that do not contain discrete molecules as such, but that comprise extended arrays of atoms or ions. We look at the structure of, and bonding in, these solids, how the properties of a solid depend on its structure, and how the properties can be modified by changes to the structure.

1.1 INTRODUCTION

To understand the solid state, we need to have some insight into the structure of simple crystals and the forces that hold them together, so it is here that we start this book. Crystal structures are usually determined by the technique of **X-ray crystallography**. This technique relies on the fact that the distances between the atoms in the crystals are of the same order of magnitude as the wavelength of X-rays (of the order of 1 Å or 100 pm): a crystal thus acts as a three-dimensional diffraction grating to a beam of X-rays. The resulting diffraction pattern can be interpreted to give the internal positions of the atoms in the crystal very precisely, thus defining interatomic distances and angles. (Some of the principles underlying this technique are discussed in Chapter 2, where we review the physical methods available for characterising solids.) Most of the structures discussed in this book would have been determined in this way.

1.2 LATTICES AND UNIT CELLS

Crystals are regular-shaped solid particles with flat shiny faces. In 1664, Robert Hooke first noted that the regularity of their external appearance is a reflection of a high degree of internal order. Crystals of the same substance, however, vary in shape considerably. In 1671, Nicolas Steno observed that this is not because their internal structure varies, but because some faces develop more than others. The angles between similar faces on different crystals of the same substance are always identical. The constancy of the interfacial angles reflects the internal order within the crystals. Each crystal is derived from a basic 'building block' that repeats over and over again, in all directions, in a perfectly regular way. This building block is known as the **unit cell**.

In order to talk about and compare the many thousands of crystal structures that are known, there has to be a way of defining and categorising the structures. This is achieved by defining the shape, symmetry, and also the size of each unit cell and the positions of the atoms within it.

1.2.1 Lattices

The simplest regular array is a line of evenly spaced objects such as that depicted by the commas in Figure 1.1a. There is a dot at the same place in each object: if we now remove the objects leaving the dots, we have a line of equally spaced dots of spacing a (Figure 1.1b). The line of dots is called the **lattice**, and by definition each **lattice point** (dot) has identical surroundings. This is the only example of a one-dimensional lattice and it can vary only in the spacing a. There are five possible two-dimensional lattices, and everyday examples of these can be seen all around in wallpapers and tiling.

1.2.2 One- and Two-Dimensional Unit Cells

The unit cell for the one-dimensional lattice in Figure 1.1a lies between the two vertical lines. If we took this unit cell and repeated it over and over again, we would reproduce the original array. Notice that it does not matter where in the structure we place the lattice points as long as they each have identical surroundings. In Figure 1.1c, we have moved the lattice points and the unit cell, but repeating this unit cell will still give the same array—we have simply moved the origin of the unit cell. There is never one unique unit cell that is 'correct'; there are always many that could be chosen and the choice depends both on convenience and convention. This is equally true in two and three dimensions.

The unit cells for the two-dimensional lattices are parallelograms with their corners at equivalent positions in the array, that is, the corners of a unit cell are

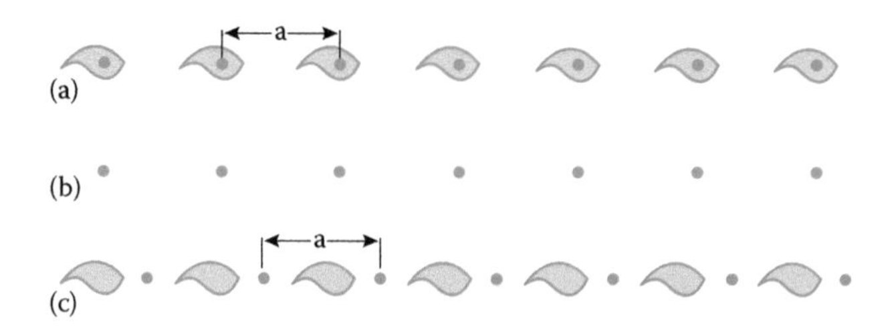

FIGURE 1.1 (a–c) A one-dimensional lattice and the choice of unit cells.

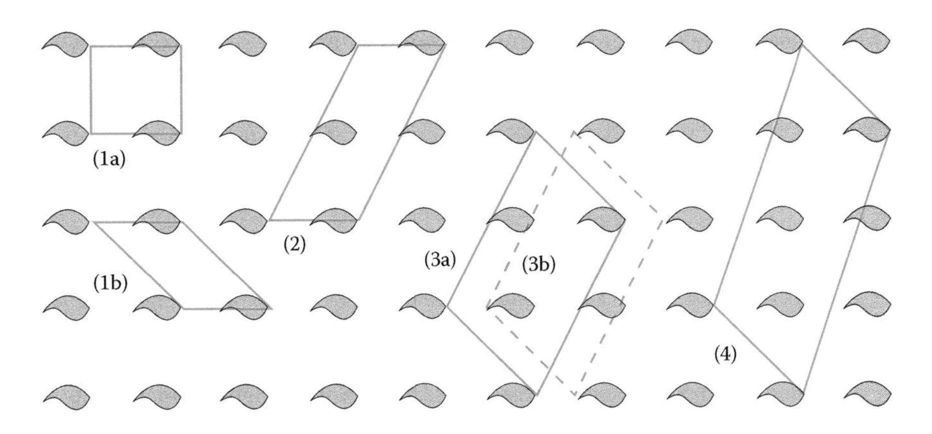

FIGURE 1.2 Choice of unit cell in a square two-dimensional lattice.

lattice points. In Figure 1.2, we show a square array with several different unit cells depicted. All of these, if repeated, would reproduce the array, but it is conventional to choose the smallest cell that fully represents the symmetry of the structure. Both unit cells (1a) and (1b) are of the same size, but clearly (1a) shows that the array is square, and this would be the conventional choice. Figure 1.3 demonstrates the same principles but for a centred rectangular array, where (a) would be the conventional choice because it includes information on the centring; the smaller unit cell (b) loses this information. It is always possible to define a noncentred oblique unit cell, but information about the symmetry of the lattice may be lost by doing so.

Unit cells such as (1a) and (1b) in Figure 1.2 and (b) in Figure 1.3 have a lattice point at each corner. However, they each contain a total of one lattice point because each lattice point is shared by four adjacent unit cells. They are known as **primitive unit cells** and are given the symbol **P**. The unit cell marked (a) in Figure 1.3 contains a total of two lattice points—one from the shared four corners and the other totally enclosed within the cell. This cell is said to be **centred** and is given the symbol **C**.

1.3 SYMMETRY

Before we take the discussion of crystalline structures any further, we need to discuss the concept of **symmetry**. This concept is extremely useful when it comes to describing the shapes of both individual molecules and regular repeating structures,

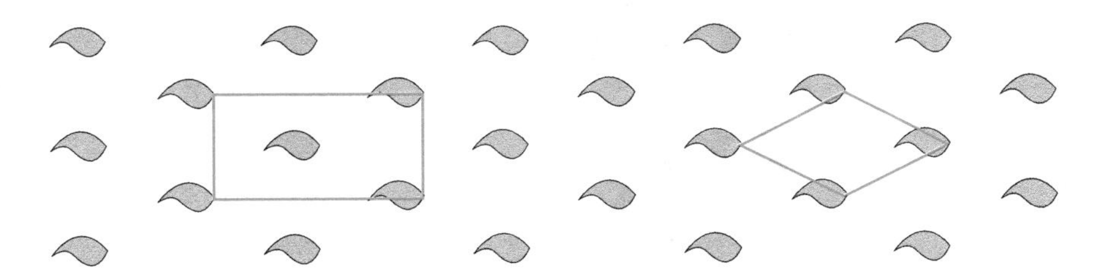

FIGURE 1.3 Choice of unit cell in a centred-rectangular lattice.

as it provides a way of describing similar features in different structures so that they become unifying features. The symmetry of objects in everyday life is something that we tend to take for granted and recognise easily without having to think about it. Take some simple examples illustrated in Figure 1.4. If we imagine a mirror dividing a spoon in half along the plane indicated, then we can see that one half of the spoon is a mirror image or reflection of the other half. Similarly with the paintbrush, only now there are two mirror planes at right angles which divide it. This symmetry feature is called a **plane of symmetry**.

Objects can also possess rotational symmetry. In Figure 1.4c, imagine an axle passing through the centre of the snowflake; in the same way as a wheel rotates about an axle, if the snowflake is rotated through one-sixth of a revolution, then the new position is indistinguishable from the old. Similarly, in Figure 1.4d, rotating the seven-sided UK 50p coin about a central axis by one-seventh of a revolution brings us to the same position as we started (ignoring the pattern on the surface). These axes are known as **axes of symmetry** or **rotational axes**. The symmetry axes and planes possessed by objects are examples of **symmetry elements**.

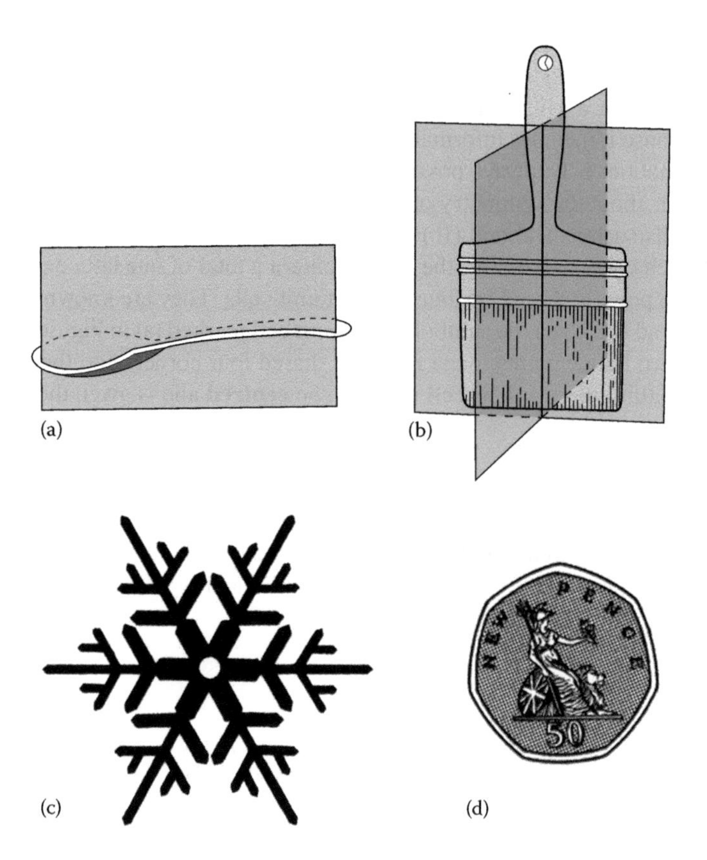

FIGURE 1.4 Common objects displaying symmetry: (a) a spoon, (b) a paintbrush, (c) a snowflake, (d) a British 50p coin.

1.3.1 Symmetry Notation

Two forms of symmetry notation are in common use. The **Schoenflies** notation is useful for describing the symmetry of individual molecules. The **Hermann–Mauguin** notation can be used to describe the symmetry of individual molecules; additionally, it is also used to describe the relationship between different molecules in space—their so-called **space symmetry**—and so it is the form most commonly met in the solid state. We shall therefore concentrate on the Hermann–Mauguin notation.

1.3.2 Axes of Symmetry

In the same way that we have previously seen for the snowflake and the 50p coin, molecules can possess rotational symmetry. Figure 1.5 illustrates this for several molecules.

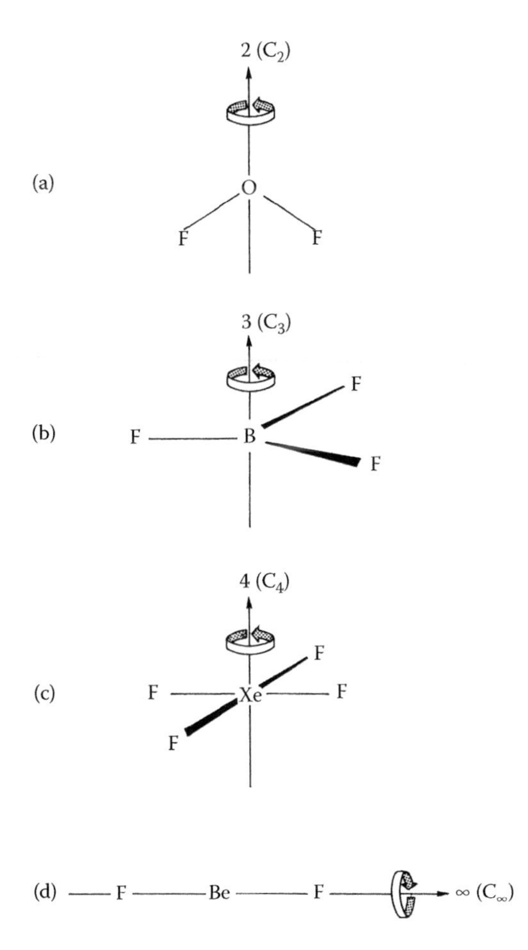

FIGURE 1.5 Axes of symmetry in molecules: (a) twofold axis in OF_2, (b) threefold axis in BF_3, (c) fourfold axis in XeF_4, and (d) ∞-fold axis in BeF_2.

In Figure 1.5a, the rotational axis is shown as a vertical line through the O atom in OF_2; rotation about this line by 180° in the direction of the arrow produces an identical-looking molecule. The axis of symmetry in this case is a twofold axis because we have to perform the rotation twice to return the molecule to its starting position.

Axes of symmetry are denoted by the symbol n, where n is the order of the axis. So the rotational axis of the OF_2 molecule is 2.

The BF_3 molecule in Figure 1.5b possesses a threefold axis of symmetry (3) because each one-third of a revolution leaves the molecule looking the same, and three turns bring the molecule back to its starting position.

In the same way, the XeF_4 molecule in Figure 1.5c has a fourfold axis (4) and four quarter-turns are necessary to bring it back to its starting position; this highest-order axis is then denoted as the **principal axis**. Notice that this molecule also has four 2 rotational axes at right angles to the 4 axis.

All linear molecules have an ∞ axis, which is illustrated for the BeF_2 molecule in Figure 1.5d; however small an angle of rotation, it always looks identical. The smallest angle of rotation possible is $1/\infty$, and so this principal axis is an infinite-order axis of symmetry.

1.3.3 Planes of Symmetry

Mirror planes occur in isolated molecules and in crystals such that everything on one side of the plane is a mirror image of the other. In a structure such a mirror plane is given the symbol m. Molecules may possess one or more **planes of symmetry**, and the diagrams in Figure 1.6 illustrate some examples. The planar OF_2 molecule has two planes of symmetry (Figure 1.6a): one is the plane of the molecule and the other is at right angles to the former. For all planar molecules, the plane of the molecule is a plane of symmetry. The diagram for BF_3 only show the planes of symmetry that

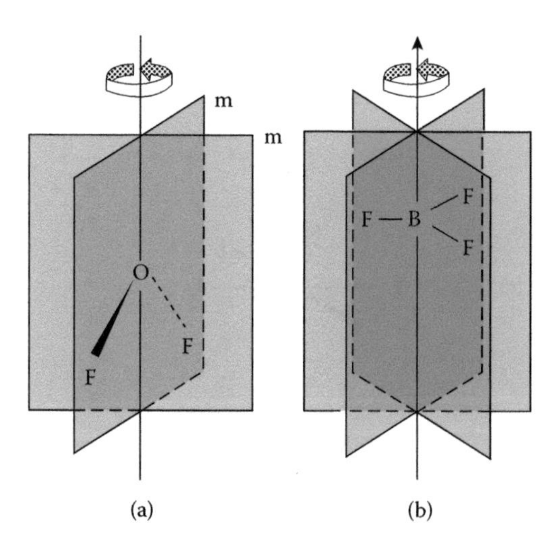

FIGURE 1.6 Planes of symmetry in molecules: (a) planes of symmetry in OF_2, (b) planes of symmetry in BF_3.

are perpendicular to the plane of the molecule. A symmetry plane which is parallel with the principal axis is known as vertical and one perpendicular to it horizontal. Vertical symmetry planes that do not pass through atoms are designated dihedral planes. In particular this is the case for CH_4 and other tetrahedral molecules.

1.3.4 Inversion

In this section, we show the third symmetry element called **inversion** through an **inversion centre** or **centre of symmetry**; it is given the symbol $\bar{1}$. In this element, you have to imagine a line drawn from any atom in the molecule, through the centre of symmetry and then continued for the same distance the other side; if, for every atom, this meets with an identical atom on the other side, then the molecule has a centre of symmetry. Of the molecules in Figure 1.5, BeF_2 has a centre of symmetry, while BF_3 and OF_2 do not. The benzene molecule (C_6H_6) possesses a centre of symmetry in the middle of the ring.

1.3.5 Inversion Axes and the Identity Element

The **inversion axis** is a combination of rotation and inversion and is given the symbol $\bar{n}$. The symmetry element consists of a rotation by $1/n$ of a revolution about the symmetry axis, followed by inversion through the centre of symmetry. An example of a 4-fold inversion axis is shown in Figure 1.7 for a tetrahedral molecule such as CF_4. The molecule is shown inside a cube as this makes it easier to see the symmetry elements. Anticlockwise rotation about the axis through 90° takes F_1 to the position shown as a dotted F, and inversion through the centre then takes this atom to the F_3 position.

The group of five symmetry elements is completed by the so-called **identity element**, which is no change—the equivalent of multiplying by 1.

1.3.6 Operations

Whereas the symmetry elements are essentially static properties of the object, actions such as rotating or reflecting a molecule are called **symmetry operations**, and the five symmetry elements have five symmetry operations associated with them.

The group of all the symmetry operations acting on a single object to describe the repetition of identical parts of the object while one point remains fixed is known as its **point group**. For the determination of the point groups of common molecules, we refer you to the standard texts on the subject (see the Further Reading section at the end of this book).

1.4 SYMMETRY IN CRYSTALS

The discussion so far has only shown the symmetry elements that belong to individual molecules. However, in the solid state, we are interested in regular arrays of atoms, ions, or molecules, and they too are related by these same symmetry elements. Figure 1.8 shows examples (not real) of how molecules could be arranged in

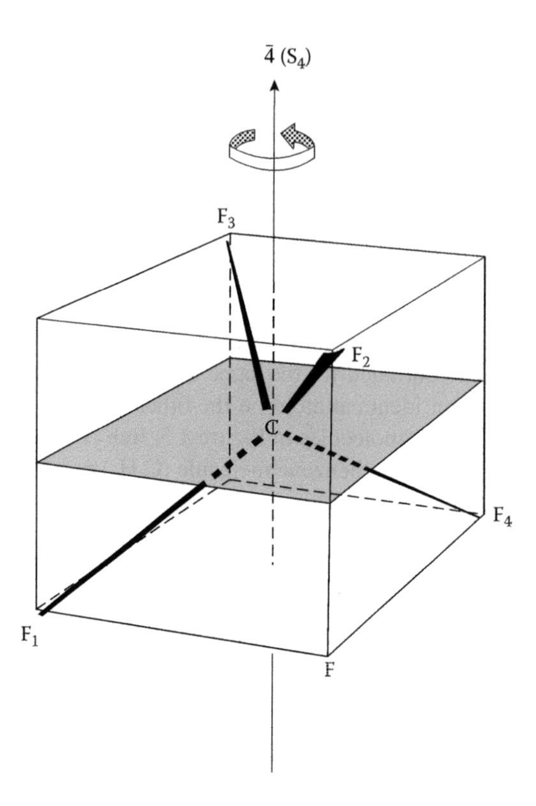

FIGURE 1.7 The $\bar{4}$ inversion (improper) axis of symmetry in the tetrahedral CF_4 molecule.

a crystal. In Figure 1.8a, two OF_2 molecules are related to one another by a plane of symmetry; in Figure 1.8b, three OF_2 molecules are related to one another by a threefold axis of symmetry; and in Figure 1.8c, two OF_2 molecules are related by a centre of inversion. Notice that in Figure 1.8b and 1.8c, the molecules are related in space by a symmetry element that they themselves do not possess; this is said to be their **site symmetry**.

1.4.1 Translational Symmetry Elements

The **glide plane** combines translation with reflection. Figure 1.9 shows an example of this symmetry element. The diagram shows part of a repeating three-dimensional structure projected onto the plane of the page; the circle represents a molecule or ion in the structure and there is a distance a between identical positions in the structure in the x direction (this follows the convention that a unit cell based on the x, y, and z axes has unit cell dimensions a, b, and c in those respective directions). The + sign next to the circle indicates that the molecule lies above the plane of the page in the z direction. The plane of symmetry is in the xz plane perpendicular to the page and is indicated by the dashed line. The symmetry element consists of reflection through this plane of symmetry, followed by translation. The translation can be either in the x or in the z direction (or along a diagonal) and the translation distance is half of

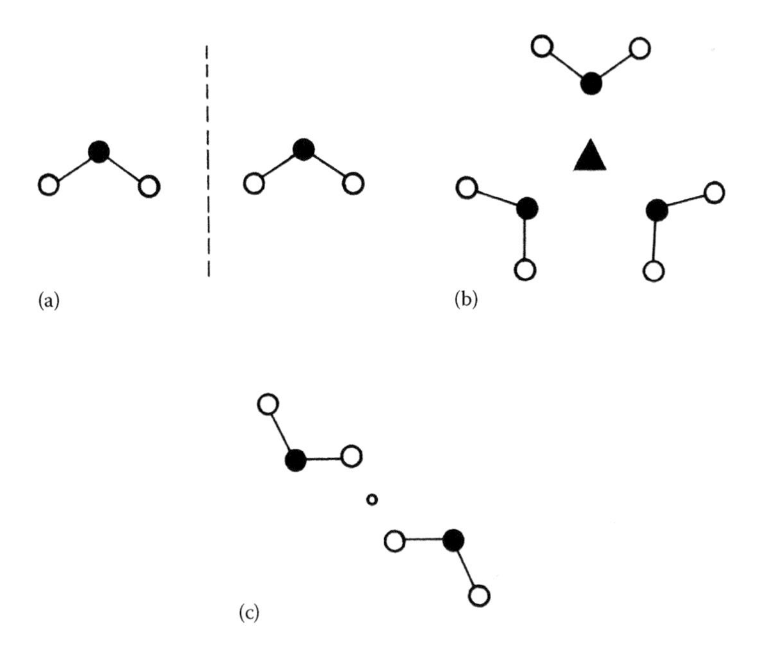

FIGURE 1.8 Symmetry in solids: (a) two OF_2 molecules related by a plane of symmetry, (b) three OF_2 molecules related by a threefold axis of symmetry, and (c) two OF_2 molecules related by a centre of inversion.

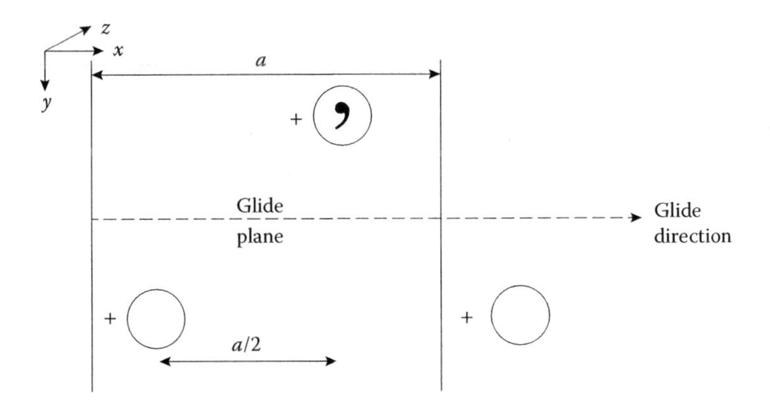

FIGURE 1.9 An *a* glide perpendicular to *b*.

the repeat distance in that direction. In the example illustrated, the translation takes place in the *x* direction; the repeat distance between identical molecules is therefore *a*, and the translation is by $a/2$. This symmetry element is called an **a glide**. If the translation had been along the *z* direction, the translation would have been by $c/2$ and is called a **c glide**. Translation along a diagonal gives an n glide. In centred cells, a more complex glide plane, a diamond glide plane, d, can exist. This is present in the structure of diamond.

You will notice two things about the molecule generated by the a glide: first, it still has a + sign against it, because the reflection in the plane leaves the z coordinate the same and, second, it now has a comma on it. Some molecules when reflected through a plane of symmetry are enantiomorphic, which means that they are not superimposable on their mirror image: the presence of the comma indicates that this molecule could be an enantiomorph.

The **screw axis** combines translation with rotation. Screw axes have the general symbol n_i, where n is the rotational order of the axis, that is, twofold, threefold, etc., and the translation distance is given by the ratio i/n. Figure 1.10 illustrates a 2_1 screw axis. In this example, the screw axis lies along the z direction and so the translation must also be in the z direction by $c/2$, where c is the repeat distance in the z direction. Notice that in this case, the molecule starts above the plane of the page (indicated by the + sign), but the effect of a twofold rotation is to take it below the plane of the page (– sign). Figure 1.11 probably illustrates this more clearly and also shows the different effects that the rotational and screw axes of the same order have on a repeating structure. Rotational and screw axes produce objects that are superimposable on the original. All other symmetry elements—glide plane, symmetry plane, inversion centre, and inversion axis—produce a mirror image of the original.

1.5 THREE-DIMENSIONAL LATTICES AND THEIR UNIT CELLS

There are seven lattice systems defined by their symmetry: triclinic, monoclinic, orthorhombic, tetragonal, rhombohedral, hexagonal, and cubic (Table 1.1).

As we saw in two dimensions, a lattice can always be defined by a primitive unit cell (P) with a lattice point at each corner; the same is true in three dimensions. However, in three dimensions, there are three other types of lattices with unit cells larger than the primitive ones; all four types are listed next and are illustrated in Figure 1.12.

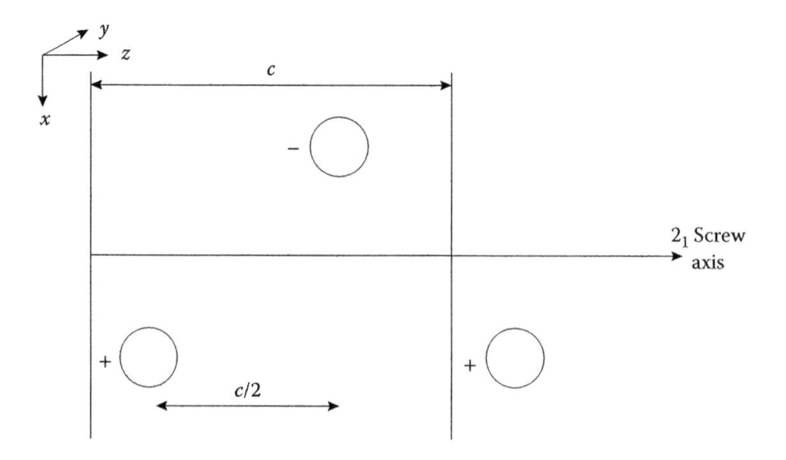

FIGURE 1.10 A 2_1 screw axis along z.

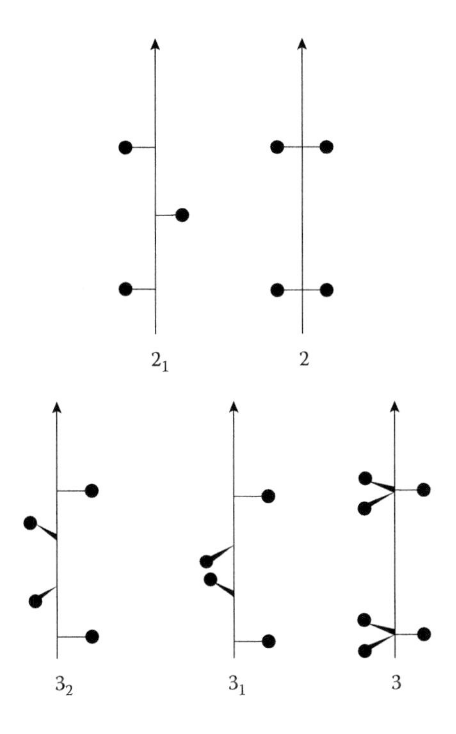

FIGURE 1.11 Comparison of the effects of twofold and threefold rotation axes and screw axes.

TABLE 1.1
Seven Lattice Systems

System	Unit cell	Minimum symmetry requirements
Triclinic	$\alpha \neq \beta \neq \gamma \neq 90°$ $a \neq b \neq c$	None
Monoclinic	$\alpha = \gamma = 90°$ $\beta \neq 90°$ $a \neq b \neq c$	One twofold axis or one symmetry plane
Orthorhombic	$\alpha = \beta = \gamma = 90°$ $a \neq b \neq c$	Any combination of three mutually perpendicular twofold axes or planes of symmetry
Tetragonal	$\alpha = \beta = \gamma = 90°$ $a = b \neq c$	One fourfold axis or one fourfold improper axis
Rhombohedral	$\alpha = \beta = \gamma \neq 90°$ $a = b = c$	One threefold axis
Hexagonal	$\alpha = \beta = 90°$ $\gamma = 120°$ $a = b \neq c$	One sixfold axis or one sixfold improper axis
Cubic	$\alpha = \beta = \gamma = 90°$ $a = b = c$	Four threefold axes at 109°28′ to each other

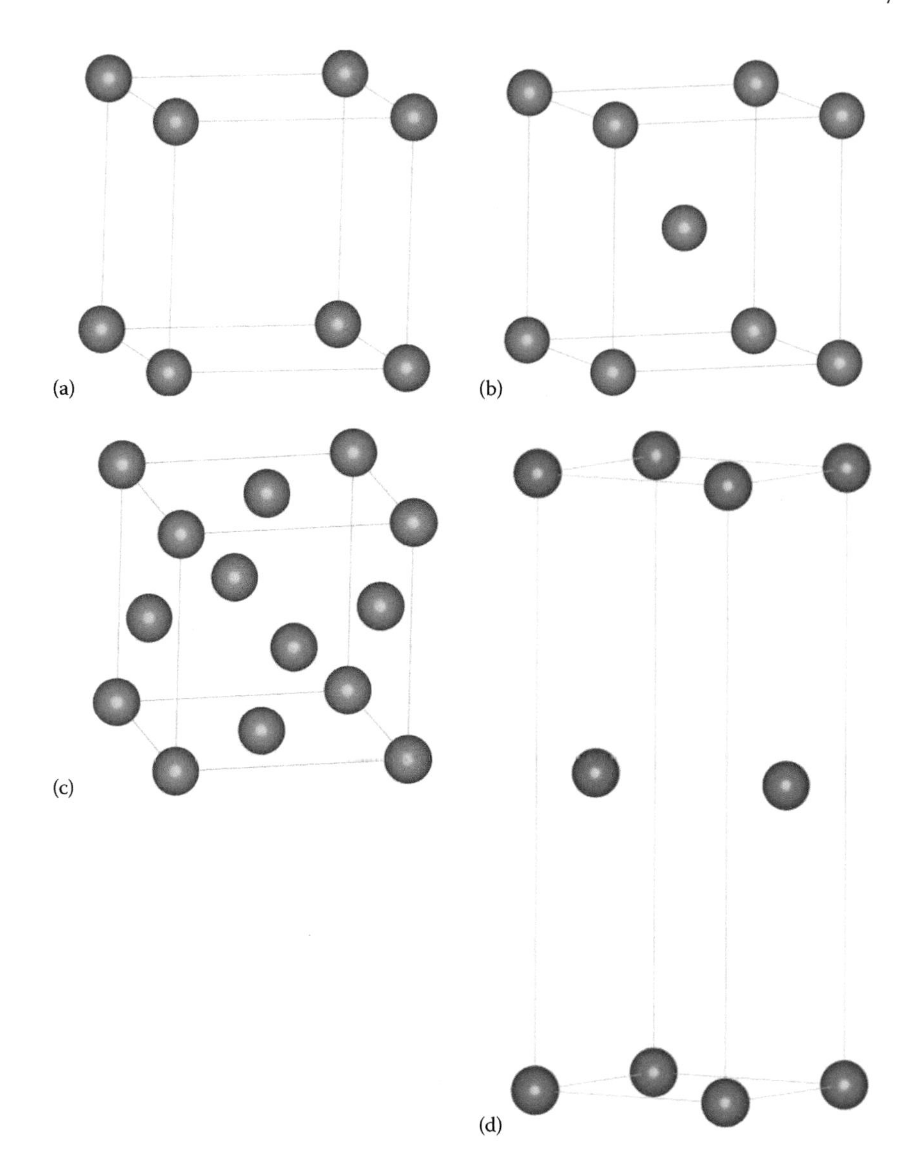

FIGURE 1.12 (a) Primitive (P), (b) body-centred (I), (c) face-centred (F), and (d) face-centred (A, B, or C) unit cells.

1. The **primitive** unit cell—symbol **P**—has a lattice point at each corner.
2. The **body-centred** unit cell—symbol **I**—has a lattice point at each corner and one at the centre of the cell.
3. The **face-centred** unit cell—symbol **F**—has a lattice point at each corner and one at the centre of each face.
4. The **face-centred** unit cell—symbol **A, B**, or **C**—has a lattice point at each corner, and one at the centres of each of a pair of opposite faces, for example, an A-centred cell has lattice points at the centres of the *bc* faces.

These lattice types are distributed among the lattice systems to give the 14 Bravais lattices listed in Table 1.2. Notice, for instance, that it is not possible to have an A-centred (or B- or C-centred) cubic unit cell; if only two of the six faces are centred, the unit cell necessarily loses its cubic symmetry.

The unit cell of a three-dimensional lattice is a parallelepiped defined by three distances a, b, and c, along the x, y, and z axes, respectively, and three angles α, β, and γ, where α is the angle between the y and z axes; β between x and z; and γ between x and y (Figure 1.13).

Because the unit cells are the basic building blocks of the crystals, they are space-filling, that is, they must pack together to fill all space. The unit cells of the seven lattice systems are illustrated in Figure 1.14. The seven lattice systems together with the shapes of their unit cells, as determined by minimum symmetry requirements, are detailed in Table 1.2.

It is instructive to note how much of a structure these various types of unit cell represent. We noted a difference between the centred and the primitive two-dimensional

TABLE 1.2
Bravais Lattices

Lattice system	Lattice types
Cubic	P, I, F
Tetragonal	P, I
Orthorhombic	P, C, I, F
Hexagonal	P
Rhombohedral	R[a]
Monoclinic	P, C
Triclinic	P

[a] The primitive unit cell of the rhombohedral system is normally given the symbol R.

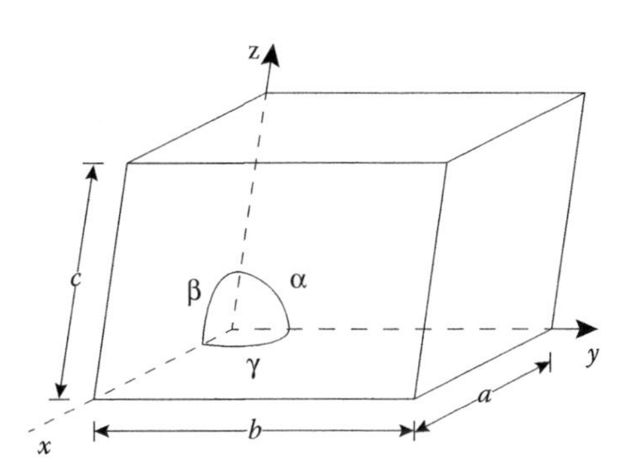

FIGURE 1.13 Definition of axes, unit cell dimensions, and angles for a general unit cell.

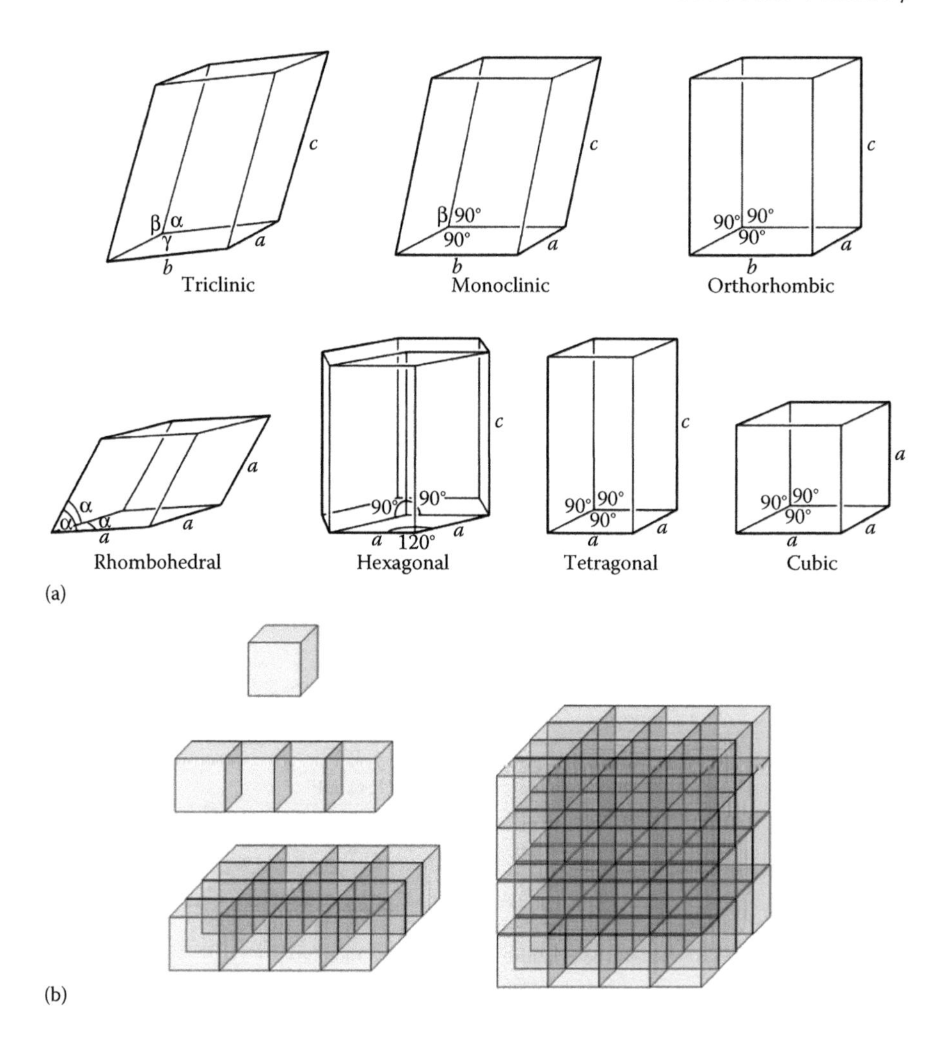

FIGURE 1.14 (a) Unit cells of the seven crystal systems. (b) Assemblies of cubic unit cells in one, two, and three dimensions.

unit cells, where the centred cell contains two lattice points whereas the primitive cell contains only one. We can work out similar occupancies for the three-dimensional case. The number of unit cells sharing a particular lattice point depends on its site. A corner site is shared by eight unit cells, an edge site by four, a face site by two, and a lattice point at the body centre is not shared by any other unit cell (Figure 1.15). Using these figures, we can work out the number of lattice points in each of the four types of cells in Figure 1.12. The results are listed in Table 1.3.

The symmetry of a crystal is a point group taken from a point at the centre of a perfect crystal and the set of symmetry operations leave this point fixed while moving each atom of the crystal to the position of an atom of the same kind. Only certain point groups are possible because of the constraint made by the fact that unit cells must be able to stack exactly with no spaces—thus, for instance, only one-,

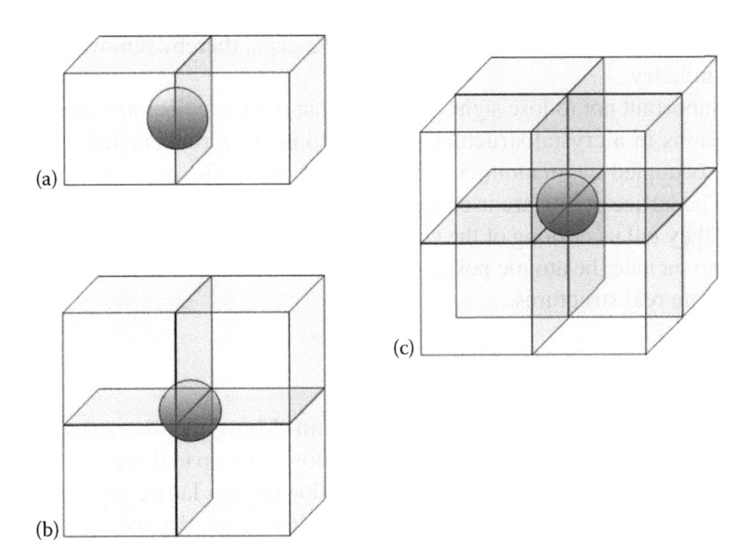

FIGURE 1.15 Unit cells showing a molecule on (a) a face, (b) an edge, and (c) a corner.

TABLE 1.3
Number of Lattice Points in the Four Types of Unit Cell

Name	Symbol	Number of lattice points in unit cell
Primitive	P	1
Body-centred	I	2
Face-centred	A or B or C	2
All face-centred	F	4

two-, three-, four-, and six-fold rotational axes are possible. Combining this with the planes of symmetry and centres of symmetry, we find that there are 32 crystallographic point groups compatible with the translational symmetry in three dimensions. If we combine the 32 crystal point groups with the 14 Bravais lattices, we find that there are 230 three-dimensional **space groups** that crystal structures can adopt, that is, 230 different space-filling patterns.

We note here that a class of crystals that breaks these rules and that can exhibit 5-, 8-, 10-, or 12-fold symmetry has been found, although we will not cover these further. These are **quasicrystals.** Dan Shechtman was awarded the 2011 Nobel Prize in Chemistry for the discovery of these structures. In 1982 he observed an electron diffraction pattern from a rapidly cooled Al–Mn alloy that showed a pattern of sharp spots which exhibited 10-fold symmetry—a symmetry that cannot be space-filling. The sharpness of the peaks showed that the crystal must contain long-range order. In this crystal, the ordered areas of fivefold symmetry cannot fit together to fill all space, so other atomic arrangements fit into the gaps, thereby removing the translational symmetry.

It is important not to lose sight of the fact that the lattice points represent equivalent positions in a crystal structure and not atoms. In a real crystal, a lattice point could be occupied by an atom, a complex ion, a molecule, or even a group of molecules. The lattice points are used to simplify the repeating patterns within a structure, but they tell us nothing of the chemistry or bonding within the crystal—for that we have to include the atomic positions; this we will do later in this chapter when we look at some real structures.

1.5.1 Space Group Labels

In the most common system (**short Hermann–Mauguin**) the space groups are labeled with the lattice type (P, F, C, I, or R) followed by up to three symbols describing the symmetry of the cell. The symbols following the lattice type give the symmetry of the crystal with respect to particular directions, depending on the crystal class. The symmetry elements are given symbols as shown in Tables 1.4 and 1.5.

All the space groups are documented in the International Tables for Crystallography (see the Further Reading section at the end of this book). We shall just look briefly at an example using a real unit cell to give you a feeling for the labeling system. Figure 1.16 shows the unit cell of caesium chloride, CsCl.

This belongs to the space group $Pm\bar{3}m$. P is the lattice type. You may have thought this was I. However in order to be body-centred, the atom at the centre must be the same element as those on the corners. As the atom at the body centre differs from those on the corners, this is a primitive unit cell, P. For a cubic lattice, three symbols describing the symmetry elements in three directions follow the lattice type. The first direction is the x, y, or z axis of the cell. The second is one of the body diagonals and the third is a face diagonal. Symmetry planes are chosen rather than symmetry axes where both occur. So first we have to consider symmetry with respect to the x,

TABLE 1.4
Symbols for Symmetry Elements in Hermann–Mauguin Space Group Labels

Identity	Centre of inversion	Mirror plane	Glide plane	Diagonal glide plane	Diamond glide plane
1	−1	m	a, b, or c	n	d

TABLE 1.5
Symbols for Symmetry Elements Involving Axes in Hermann–Mauguin Space Group Labels

Rotation axis	2, 3, 4, or 6
Screw axis	2_1, 3_1, 3_2 4_1, 4_2, 4_3, 6_1, 6_2, 6_3, 6_4, or 6_5

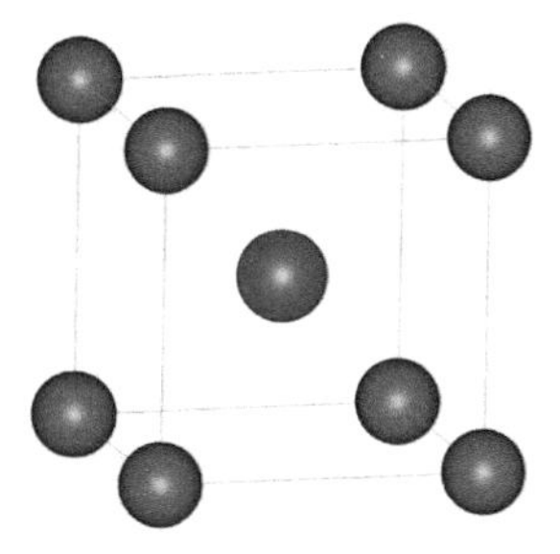

FIGURE 1.16 CsCl unit cell.

y, and *z* axes. There are planes of reflection in the unit cell at right angles to all these. Reflection or mirror planes are given the label m. Next consider symmetry elements with respect to the body diagonals. The body diagonal of the unit cell is a three-fold axis. This is common to all cubic systems and because the cell has a centre of inversion, it is written –3 or $\bar{3}$. This axis is not easy to spot and looking at the unit cell, you may have only spotted a 4-fold rotation axis. It is more easily seen if you tip the cube so that it is balanced on the origin with the (111) position vertically above it. Finally symmetry with respect to the face diagonals is added. Another reflection plane is included to give Pm-3m or Pm$\bar{3}$m.

1.5.2 Packing Diagrams

Drawing structures in three dimensions is not easy and so crystal structures are often represented by two-dimensional plans or projections of the unit cell contents—in much the same way as an architect makes building plans. These projections are called **packing diagrams** as they are particularly useful in small molecule structures for showing how the molecules pack together in the crystal and thus their intermolecular interactions.

The position of an atom or an ion in a unit cell is described by its **fractional coordinates**; these are simply the coordinates based on the unit cell axes (known as the **crystallographic axes**) and are expressed as fractions of the unit cell lengths. It has the simplicity of a universal system that enables unit cell positions to be compared from one structure to another regardless of a variation in the unit cell size.

To take a simple example, in a cubic unit cell with $a = 1000$ pm, an atom with an x coordinate of 500 pm has a fractional coordinate of $x/a = 500/1000 = 0.5$ in the x direction. Similarly, in the y and z directions, the fractional coordinates are given by y/b and z/c, respectively.

A packing diagram is shown in Figure 1.17 for CsCl. The projection is shown on the *yx* plane, that is, we are looking at the unit cell straight down the *z* axis. The *x*-fractional coordinate of any atoms/ions lying in the top or bottom face of the unit cell will be 0 or 1 (depending on where you take the origin), and it is conventional for this not to be marked on the diagram. Any *z* coordinate that is not 0 or 1 is marked in a convenient place on the diagram.

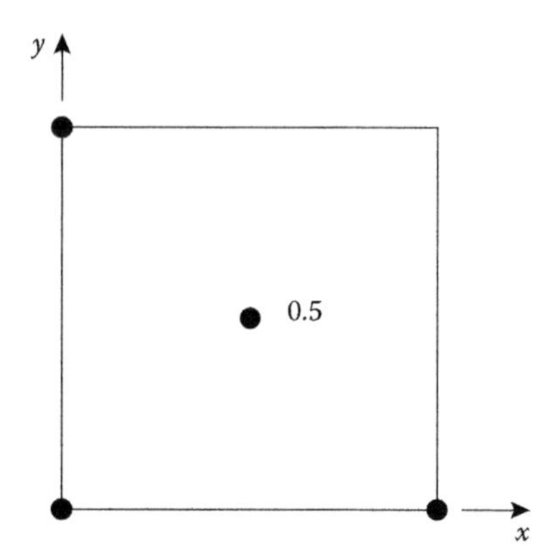

FIGURE 1.17 Packing diagram for CsCl.

1.6 CLOSE PACKING

Diagrams such as Figure 1.16 represent a crystal by a unit cell with ions, atoms, or molecules occupying positions within the cell. An alternative way of looking at crystal structures is to regard them as a collection of spheres packed together. In this view, the majority of metallic elements have a crystal structure in which the spheres are arranged to minimize the unoccupied space. This is known as **close packing.** Figure 1.18 shows two possible arrangements for a layer of such identical spheres. On squeezing the square layer in Figure 1.18a, the spheres would move to the positions in Figure 1.18b so that the layer takes up less space. The layer in Figure 1.18b—(layer A)—is said to be close-packed. The hollow spaces between the spheres are of two types, which have been marked with dots and crosses, respectively.

To build up a close-packed structure in three dimensions, we must now add a second layer—layer B. The spheres of the second layer can only sit in half of the hollows of the first layer, and layer B in Figure 1.19 (shown in blue) sits directly over the hollows marked with a cross (although it makes no difference which type we choose).

When we come to add a third layer, there are again two possible options as to where it can go. First, it could go directly over layer A, in the unmarked hollows; if we repeated this stacking sequence, we would build up the layers ABABABA… and so on. This is known as **hexagonal close packing (hcp)** (Figure 1.20a). In this structure, the hollows marked with a dot are never occupied by spheres, leaving very small channels through the layers (Figure 1.20b).

In the second option, the third layer could be placed over those hollows, as in Figure 1.19b, marked with a dot. This third layer, which we now label C, would not be directly over either A or B, and the stacking sequence when repeated would be ABCABCABC… and so on. This is known as **cubic close packing (ccp)** (Figure 1.21). Cubic close packing results in a face-centred cubic cell and so can also be labeled **fcc**. It is not obvious that Figure 1.21 shows a face-centred cell. This is because the planes illustrated are actually perpendicular to the body diagonal of the face-centred cell we described earlier. Figure 1.22 shows the close-packed layers as they appear in the fcc unit cell.

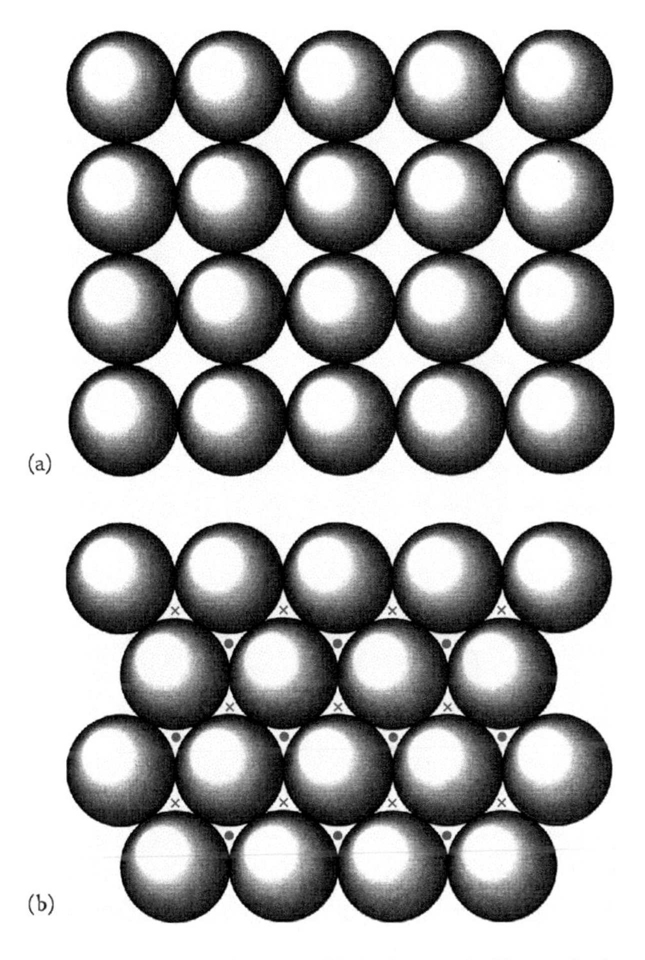

FIGURE 1.18 (a) A square array of spheres. (b) A close-packed layer of spheres.

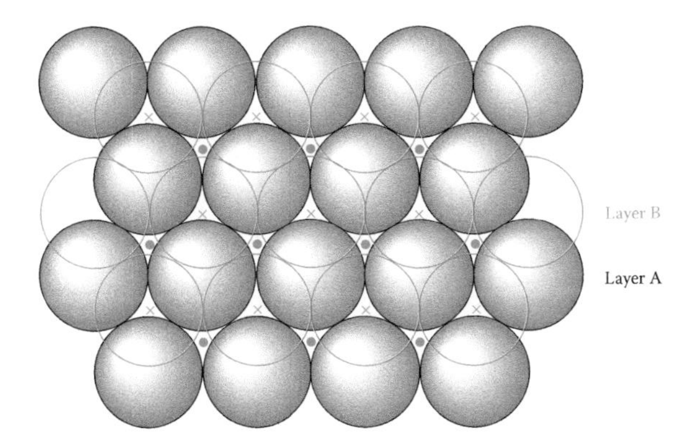

FIGURE 1.19 Two layers of close-packed spheres.

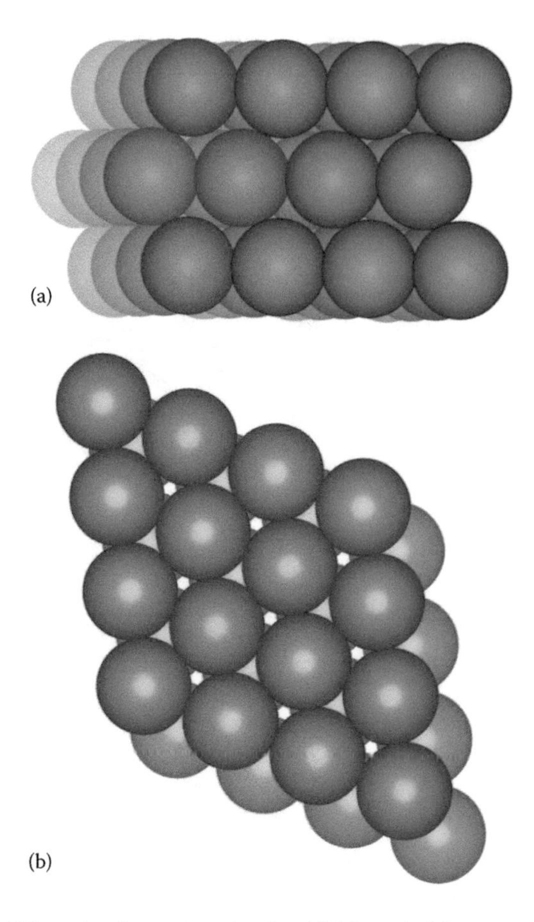

FIGURE 1.20 (a) Three *hcp* layers showing the ABAB… stacking sequence. (b) Three *hcp* layers showing the narrow channels through the layers.

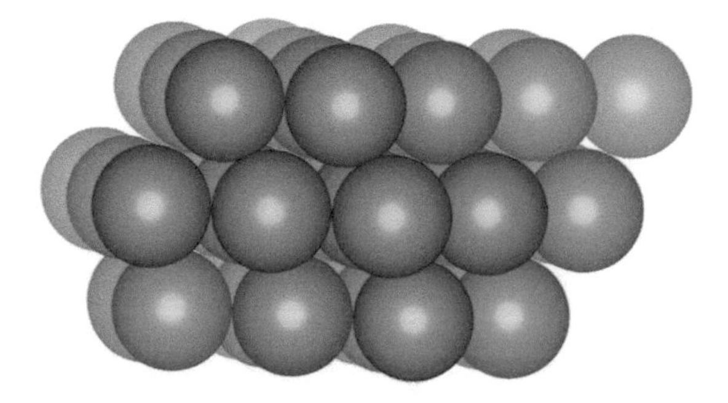

FIGURE 1.21 Three *ccp* layers.

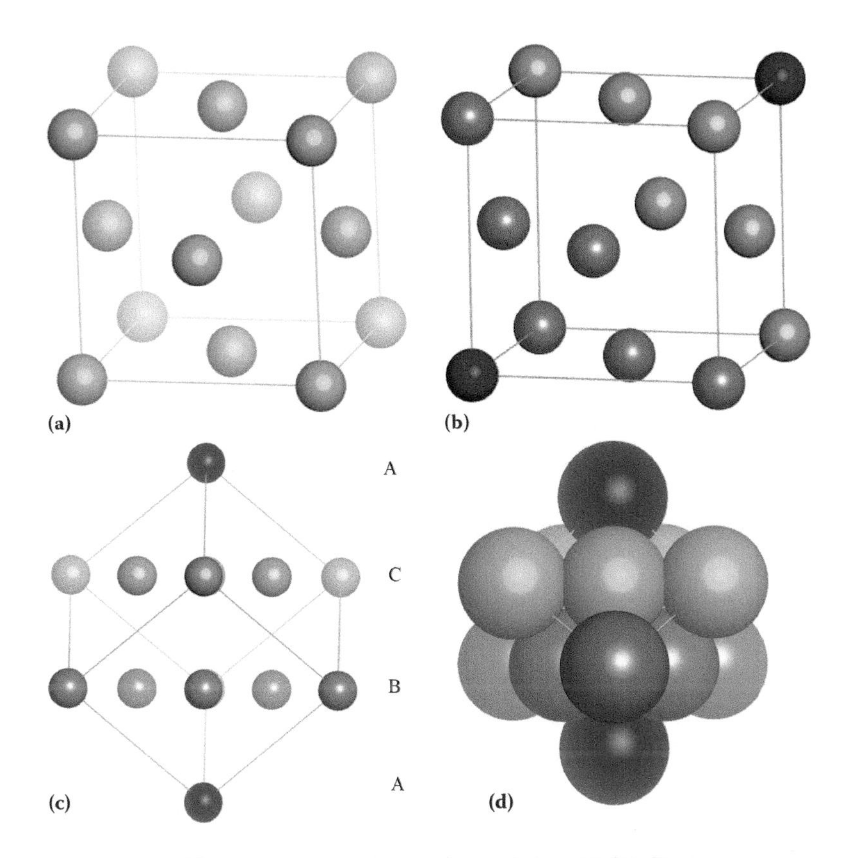

FIGURE 1.22 A face-centred cubic unit cell as conventionally drawn and rotated to show the *ccp* close-packed layers.

Close packing represents the most efficient use of space when packing identical spheres—74% of the total volume is occupied by the spheres: the **packing efficiency** is said to be 74%. Each sphere in the structure is surrounded by twelve equidistant neighbours—six in the same layer, three in the layer above, and three in the layer below: the **coordination number** of an atom in a close-packed structure is thus 12.

Another important feature of close-packed structures is the shape and number of the small amounts of space trapped in between the spheres. There are two different types of spaces within a close-packed structure: the first that we will consider is called an **octahedral hole**. Figure 1.23a shows two close-packed layers again, but with the octahedral holes shaded. Each of these holes is surrounded by six spheres, three in layer A and three in layer B; the centres of these spheres lie at the corners of an octahedron, hence the name (Figure 1.23b). If there are n spheres in the array, then there are also n octahedral holes.

Similarly, Figure 1.24a shows two close-packed layers, now with the second type of space, **tetrahedral holes**, shaded. Each of these holes is surrounded by four spheres with centres at the vertices of a tetrahedron (Figure 1.24b). If there are n spheres in the array, then there are $2n$ tetrahedral holes.

The octahedral holes in a close-packed structure are much bigger than the tetrahedral holes—they are surrounded by six atoms rather than four.

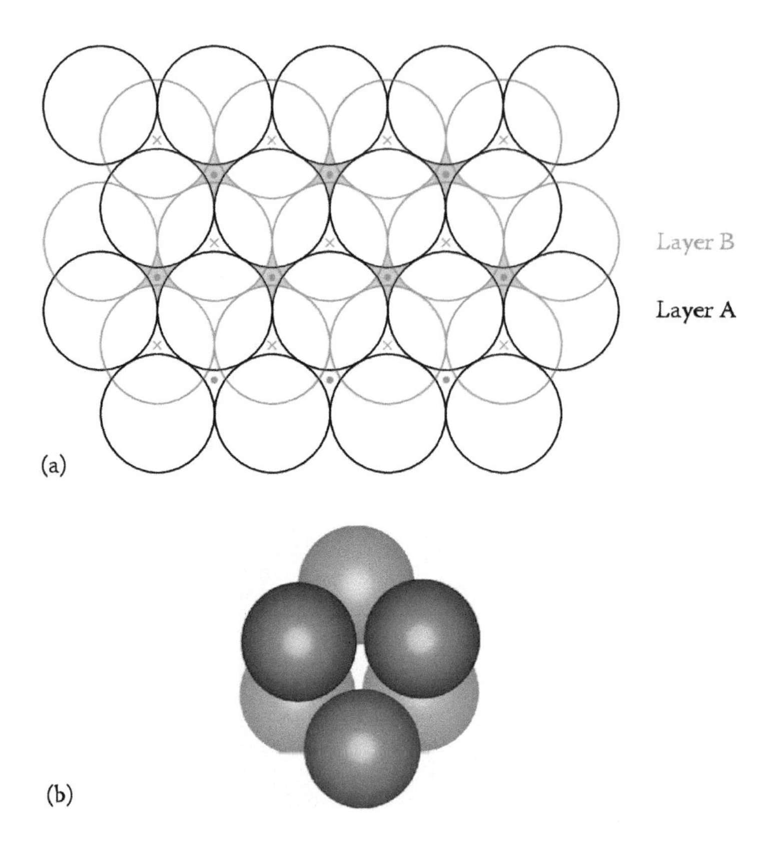

FIGURE 1.23 (a) Two layers of close-packed spheres with the enclosed octahedral holes shaded. (b) A computer representation of an octahedral hole.

There are, of course, innumerable stacking sequences possible when repeating close-packed layers; however, hexagonal close-packed and cubic close-packed structures are those of maximum simplicity and are those most commonly encountered in the crystal structures of the metallic elements. Only two other stacking sequences are found in perfect crystals of the elements—an ABAC repeat in La, Pr, Nd, and Am; and a nine-layer repeat ABACACBCB in Sm.

1.6.1 Body-Centred and Primitive Structures

Some metals do not adopt a close-packed structure, but have a slightly less efficient packing method: it is the **body-centred cubic (bcc) structure**, shown in Figure 1.12b. In this structure, an atom in the middle of a cube is surrounded by eight identical and equidistant atoms at the corners of the cube, the coordination number has dropped from 12 to 8, and the packing efficiency is now 68%, compared with 74% for close packing.

The simplest of the cubic structures is the **primitive cubic structure**. This is built by placing square layers, like the one shown in Figure 1.18a, directly on top of one another. Figure 1.25a illustrates this, and we can see in Figure 1.25b that each atom sits at the corner of a cube. The coordination number of an atom in this structure is six.

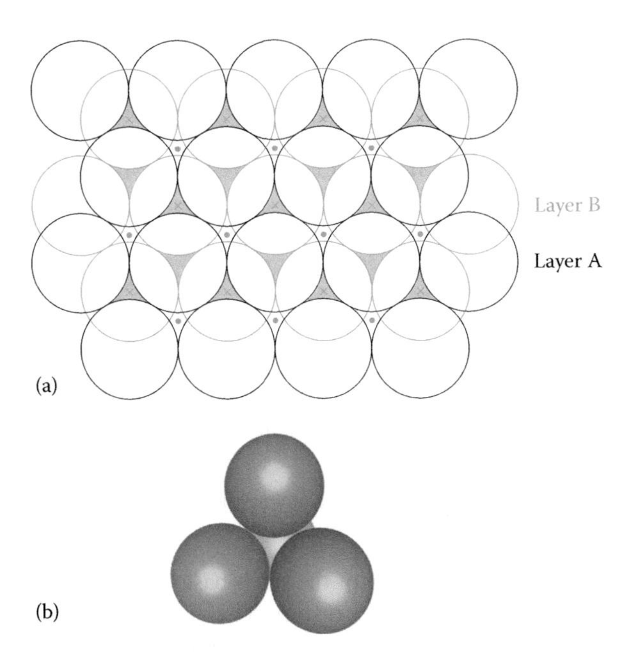

FIGURE 1.24 (a) Two layers of close-packed spheres with the tetrahedral holes shaded. (b) A computer representation of a tetrahedral hole.

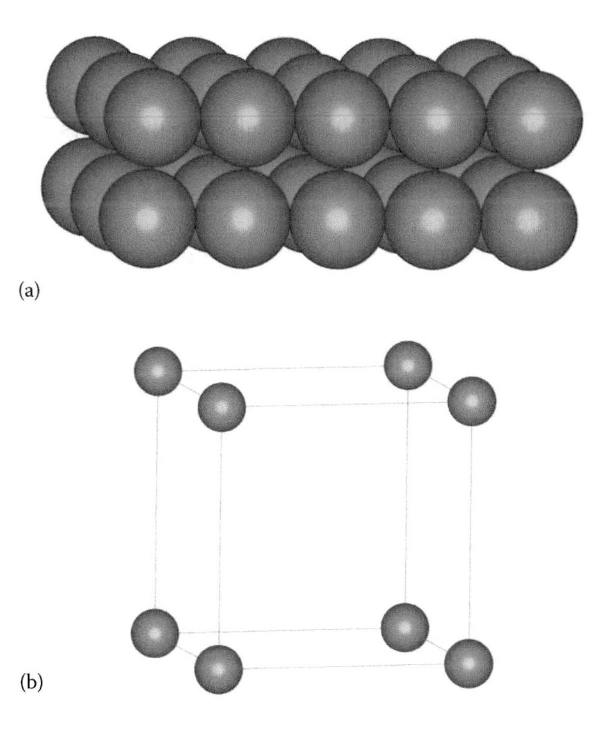

FIGURE 1.25 (a) Two layers of a primitive cubic array. (b) A cube of atoms from this array.

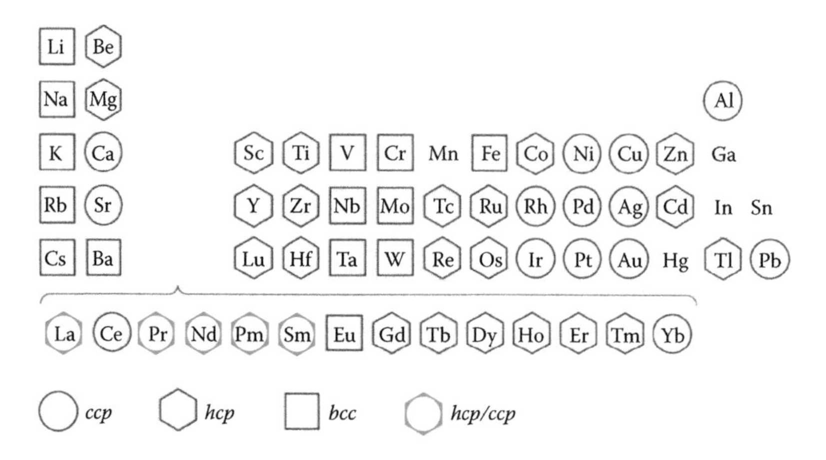

FIGURE 1.26 Occurrence of packing types among the metals.

The majority of metals have one of the three basic structures, *hcp*, *ccp*, or *bcc*; polonium alone adopts the primitive structure. The distribution of the packing types among the most stable forms of the metals at 298 K is shown in Figure 1.26. As we noted earlier, a very few metals have a mixed *hcp/ccp* structure of a more complex type. The structures of the actinoids tend to be rather complex and are not included.

1.7 CRYSTAL PLANES–MILLER INDICES

The crystal faces, both when they grow and when they are formed by cleavage, tend to be parallel either to the sides of the unit cell or to the planes in the crystals that contain a high density of atoms. It is useful to be able to refer to both the crystal faces and the planes in the crystals in some way—to give them a label—and this is usually done by using **Miller indices**.

First, we will describe how Miller indices are derived for lines in two-dimensional nets and then move on to look at planes in three-dimensional lattices. Figure 1.27 shows a rectangular net with several sets of lines, and a unit cell is marked on each set with the origin of each in the bottom left-hand corner corresponding to the directions of the x and y axes. A set of parallel lines are defined by two indices, h and k, where h and k are the number of parts into which a and b, the unit cell edges, are divided by the lines. Thus, the indices of a line hk are defined so that the line intercepts a at a/h and b at b/k. Start by finding a line next to the one passing through the origin. In the set of lines marked A, the line next to the one passing through the origin leaves a undivided ($a/1$), but divides b into two ($b/2$); both intercepts lie on the positive side of the origin, so in this case the indices of the set of lines hk are *12*, called the 'one-two' set. If the set of lines lies parallel to one of the axes, then there is no intercept and the index becomes zero. If the intercepted cell edge lies on the negative side of the origin, then the index is written with a bar on the top, for example, $\bar{2}$, called 'bar-two'. Notice that if we had selected the line on the other side of the origin in A, we would have indexed the lines as the $\overline{12}$; there is

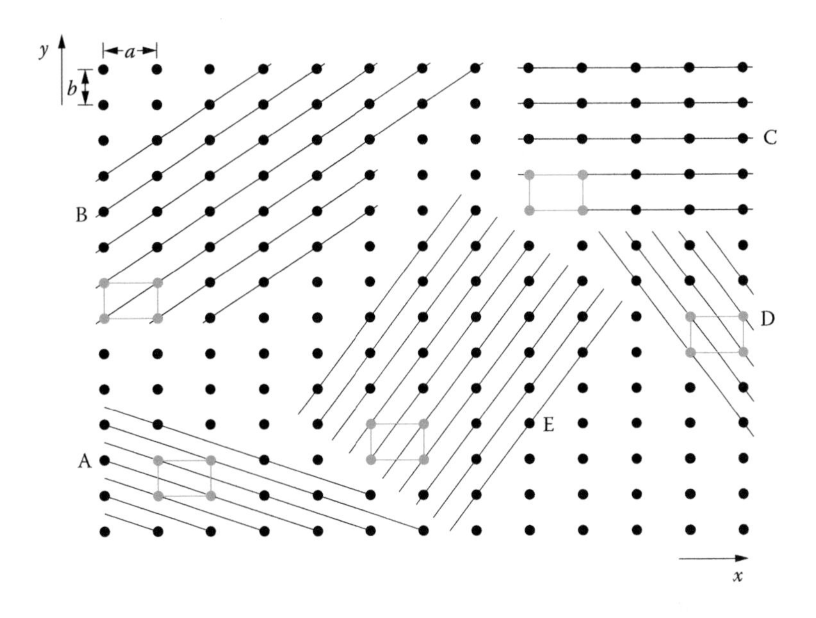

FIGURE 1.27 A rectangular net showing five sets of lines, A–E, with unit cells marked.

no difference between the two pairs of indices and always the *hk* and the $\overline{hk}$ lines are the same set of lines. Notice also, in Figure 1.27, that the lines with the lower indices are more widely spaced.

The Miller indices for planes in three-dimensional lattices are given by *hkl*, where *l* is now the index for the *z* axis. The principles are exactly the same. Thus, a plane is indexed *hkl* when it makes intercepts *a*/*h*, *b*/*k*, and *c*/*l* with the unit cell edges *a*, *b*, and *c*, respectively. Figure 1.28 shows some cubic lattices with various planes shaded. The positive directions of the axes are marked and these are orientated to conform with the conventional right-hand rule, as illustrated in Figure 1.29. In Figure 1.28a, the shaded planes lie parallel to *y* and *z*, but leave the unit cell edge *a* undivided; thus, the Miller indices of these planes are *100*. Again, take note that the *hkl* and $\overline{hkl}$ planes are the same.

1.7.1 Interplanar Spacings

It is sometimes useful to be able to calculate the perpendicular distance d_{hkl} between parallel planes (Miller indices *hkl*). When the axes are at right angles to one another (orthogonal), the geometry is simple and for an orthorhombic system where $a \neq b \neq c$ and $\alpha = \beta = \gamma = 90°$, this gives:

$$\frac{1}{d_{hkl^2}} = \frac{h^2}{a^2} + \frac{k^2}{b^2} + \frac{l^2}{c^2}$$

Other relationships are summarised in Table 1.6.

TABLE 1.6
d-Spacings in Different Crystal Systems

Lattice system	d_{hkl} as a function of Miller indices and lattice parameters
Cubic	$\frac{1}{d^2}=\frac{h^2+k^2+l^2}{a^2}$
Tetragonal	$\frac{1}{d^2}=\frac{h^2+k^2}{a^2}+\frac{l^2}{c^2}$
Orthorhombic	$\frac{1}{d^2}=\frac{h^2}{a^2}+\frac{k^2}{b^2}+\frac{l^2}{c^2}$
Hexagonal	$\frac{1}{d^2}=\frac{4}{3}\left(\frac{h^2+hk+k^2}{a^2}\right)+\frac{l^2}{c^2}$
Monoclinic	$\frac{1}{d^2}=\frac{1}{\sin^2\beta}\left(\frac{h^2}{a^2}+\frac{k^2\sin^2\beta}{b^2}+\frac{l^2}{c^2}-\frac{2hl\cos\beta}{ac}\right)$

1.8 CRYSTALLINE SOLIDS

We start this section by looking at the structures of some simple **ionic solids**.

An **ionic bond** is formed between two oppositely charged ions because of the electrostatic attraction between them. Ionic bonds are strong, but are also nondirectional; their strength decreases with the increasing separation of the ions. Ionic crystals are therefore composed of infinite arrays of ions that have packed together in such a way as to maximise the coulombic attraction between oppositely charged ions and to minimise the repulsions between ions of the same charge. We expect to find ionic compounds in the halides and oxides of Group 1 and 2 metals and it is with these crystal structures that this section begins.

In many structures, we find that the bonding is not purely ionic, but it possesses some degree of **covalency**: the electrons are shared between the two bonding atoms and are not merely transferred from one to the other. This is particularly true for the elements in the centre of the periodic table. This point is taken up in Section 1.8.4, where we discuss the size of ions and the limitations of the concept of ions as hard spheres.

Sections 1.8.5 and 1.8.6 look at the crystalline structures of covalently bonded species. In Section 1.8.5, extended covalent arrays are investigated, such as the structure of **diamond**—one of the forms of elemental carbon—where each atom forms strong covalent bonds to the surrounding atoms, forming an infinite three-dimensional network of localized bonds throughout the crystal. In Section 1.8.6, we look at molecular crystals that are formed from small, individual, covalently bonded molecules. These molecules are held together in the crystal by weak forces known collectively as **van der Waals forces**. These forces arise due to the interactions between the dipole moments in the molecules. Molecules that possess a permanent dipole can interact with one another (**dipole–dipole interaction**) and also with ions (**charge–dipole interaction**). Molecules that do not possess a dipole also interact with each other because 'transient dipoles' arise due to the movement of electrons, and these

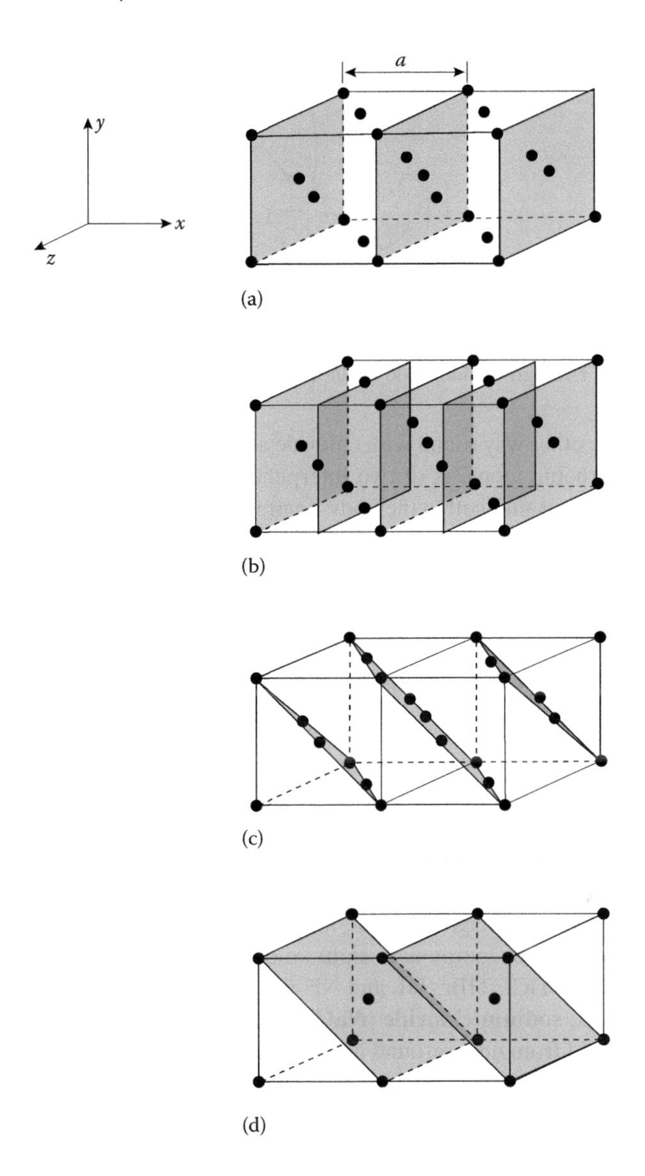

FIGURE 1.28 (a–c) Planes in a face-centred cubic lattice. (d) Planes in a body-centred cubic lattice (two unit cells are shown).

in turn induce dipoles in adjacent molecules. The net result is a weak attractive force known as the **London dispersion force**, which falls off very quickly with distance.

1.8.1 Ionic Solids with Formula MX

The cubic unit cell of the **caesium chloride structure (CsCl)** was shown in Figure 1.16, Section (1.5.1). It shows a caesium ion (Cs^+) at the centre of the cube, surrounded by eight chloride ions (Cl^-) at the corners. It could equally well have

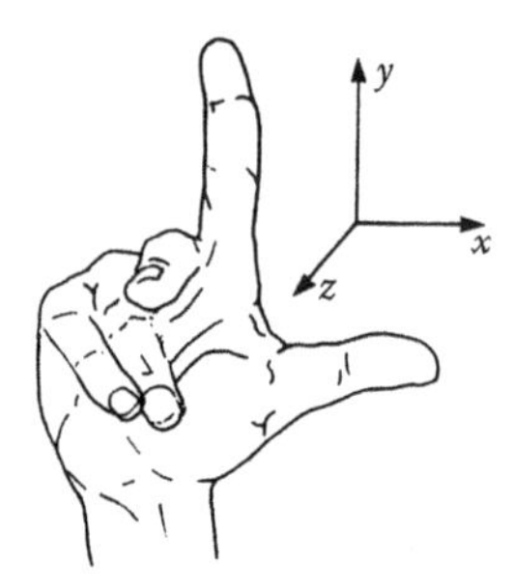

FIGURE 1.29 The right-handed rule for labeling axes.

been drawn the other way round with chloride at the centre and cesium at the corners because the structure consists of two interpenetrating primitive cubic arrays. Note the similarity of this unit cell to the body-centred cubic structure adopted by some of the elemental metals such as the Group 1 (alkali) metals. We noted earlier, however, that the caesium chloride structure is not body-centred cubic. This is because the environment of the caesium at the centre of the cell is not the same as the environment of the chlorides at the corners; a body-centred cell would have the same type of atom at both corners, that is, at (0, 0, 0) etc., *and* at the body centre $\left(\frac{1}{2},\frac{1}{2},\frac{1}{2}\right)$ because they are lattice points. The 8 atoms on the corners form a cubic primitive unit cell. In CsCl, each caesium is surrounded by eight chlorines at the corners of a cube and vice versa, so the coordination number of each type of atom is eight. The unit cell contains one formula unit of CsCl, with each of the eight corner chlorines being shared by eight unit cells.

Caesium is a large ion (ionic radii are discussed in detail in Section 1.8.4), so it is able to coordinate the eight chloride ions around it. Other compounds with large cations that can also accommodate eight anions and crystallise with this structure include CsBr, CsI, TlCl, TlBr, TlI, and NH_4Cl.

Common salt, **sodium chloride (NaCl)**, is also known as **rock salt**. It is mined all over the world from underground deposits left by the dried-up remains of ancient seas. A unit cell of the sodium chloride structure is illustrated in Figure 1.30. The unit cell is cubic and the structure consists of two interpenetrating face-centred arrays: one of Na^+ and the other of Cl^- ions. The space group of NaCl is Fm-3m. F refers to the face-centred cell formed by either array. Like CsCl, the structure has planes of symmetry perpendicular to *x*, *y*, *z*, a 3-fold axis along the body diagonal, and planes of symmetry perpendicular to the face diagonals. Figure 1.31 shows the body diagonal and the crystal viewed along this axis.

Each sodium ion is surrounded by six equidistant chloride ions situated at the corners of an octahedron and, in the same way, each chloride ion is surrounded by six sodium ions: we say that the coordination is 6: 6.

An alternative way of viewing this structure is to think of it as a cubic close-packed array of chloride ions with sodium ions filling all the octahedral holes. Filling all the octahedral holes gives a Na: Cl ratio of 1:1 with the structure as illustrated in Figure 1.30. Looking closely at Figure 1.28, you can see that one type of ion occupies

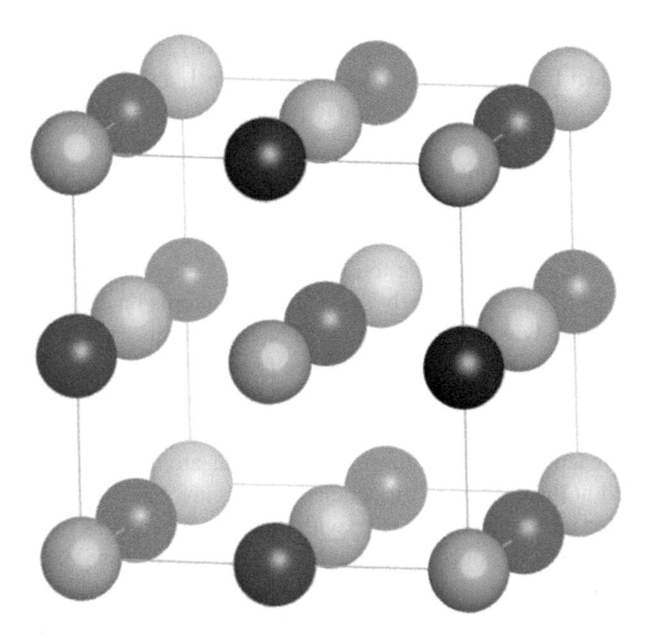

FIGURE 1.30 NaCl unit cell. Key: Na blue; Cl grey (or vice versa).

each vertex and the centre of each face, hence a face-centred cube. The close-packed layers are not apparent in this view as they lie at right angles to a cube diagonal (see Figure 1.31). Interpreting simple ionic structures in terms of the close packing of one of the ions with the other ion filling some or all of either the octahedral or the tetrahedral holes is extremely useful: it makes it particularly easy to see both the coordination geometry around a specific ion and the available spaces within a structure.

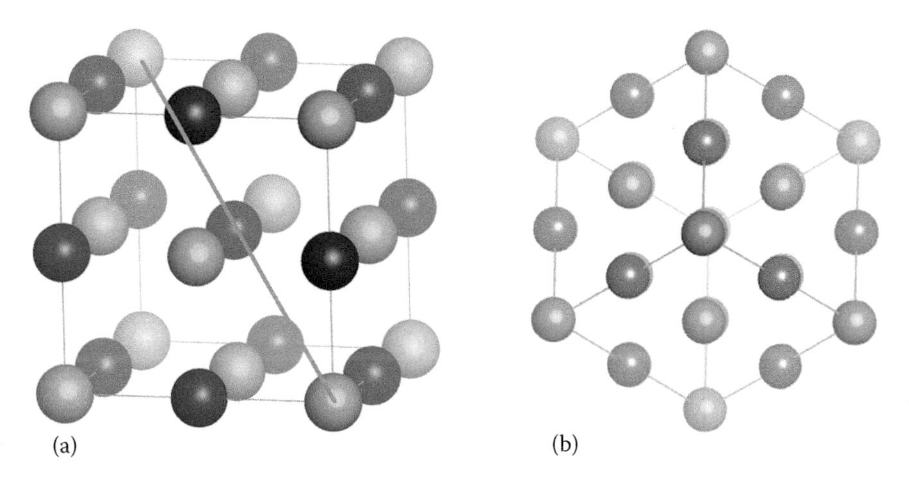

FIGURE 1.31 (a) The body diagonal of the NaCl unit cell. (b) The unit cell viewed along this axis.

The coordination number of sodium (and chlorine) is lower in this structure: as you might expect from its relative position in Group 1, a sodium ion is smaller than a caesium ion, so it is now only possible to pack six chlorides around it and not eight as in caesium chloride.

The sodium chloride unit cell contains four formula units of NaCl. There are 8 Na atoms on corners. These are each shared between 8 unit cells and so contribute 1/8 each to the unit cell. There are 6 on face centres which are shared between 2 unit cells and thus contribute 1/2. The total number of Na atoms in the unit cell is thus $8 \times 1/8 + 6 \times 1/2 = 1 + 3 = 4$. Similarly there are 12 Cl atoms on edges contributing 1/4 each and 1 Cl entirely within the unit cell. The total number is $12 \times 1/4 + 1$. So there are also 4 Cl atoms in the unit cell and hence four formula units.

Table 1.7 lists some of the compounds that adopt the NaCl structure.

Many of the structures described in this book can also be viewed as linked octahedra, where each octahedron consists of a metal atom surrounded by six other atoms situated at the corners of an octahedron (Figure 1.32a and 1.32b). These can also be depicted as viewed from above with contours marked, as in Figure 1.32c.

Octahedra can link together through corners, edges, and faces, as seen in Figure 1.33. The linking of octahedra by different methods effectively eliminates the atoms because some of the atoms are now shared between them: two MO_6 octahedra linked through a vertex have the formula M_2O_{11}; two MO_6 octahedra linked through an edge have the formula M_2O_{10}; two MO_6 octahedra linked through a face have the formula M_2O_9.

TABLE 1.7
Compounds with NaCl (Rock-Salt) Type of Crystal Structure

Most alkali halides, MX, and AgF, AgCl, AgBr
All alkali hydrides, MH
Monoxides, MO, of Mg, Ca, Sr, Ba
Monosulfides, MS, of Mg, Ca, Sr, Ba

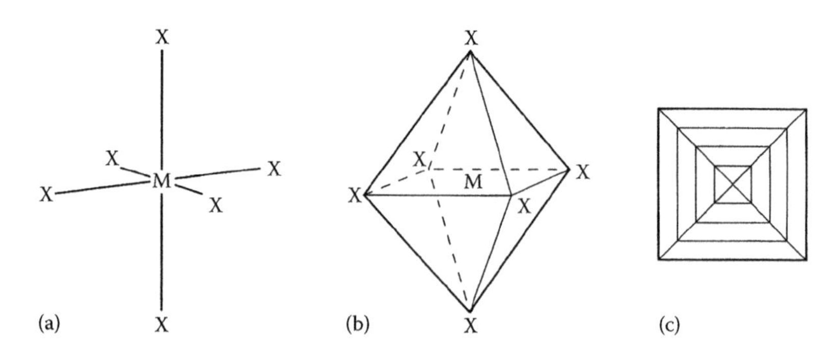

FIGURE 1.32 (a) An $[MX_6]$ octahedron, (b) a solid octahedron, and (c) a plan of an octahedron with contours.

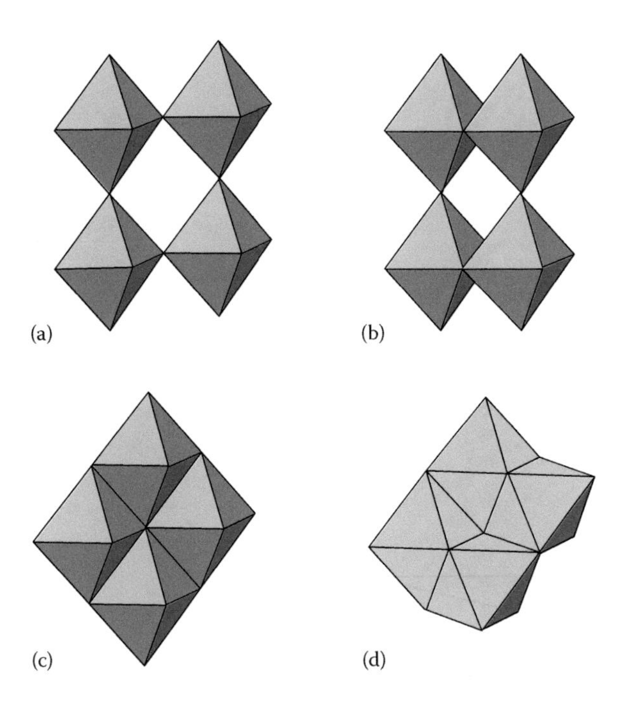

FIGURE 1.33 Conversion of (a) corner-shared MX_6 octahedra to (b) edge-shared octahedra, and (c) edge-shared octahedra to (d) face-shared octahedra.

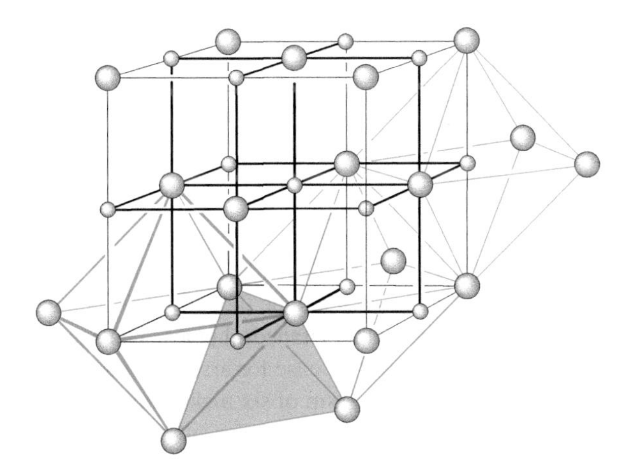

FIGURE 1.34 NaCl structure showing edge-sharing of octahedra and the enclosed tetrahedral space (shaded).

The NaCl structure can be described in terms of $NaCl_6$ octahedra sharing edges. An octahedron has 12 edges, and in the NaCl structure each one is shared by two octahedra. This is illustrated in Figure 1.34, which shows an NaCl unit cell with three $NaCl_6$ octahedra shown in outline and one of the resulting tetrahedral spaces depicted by shading.

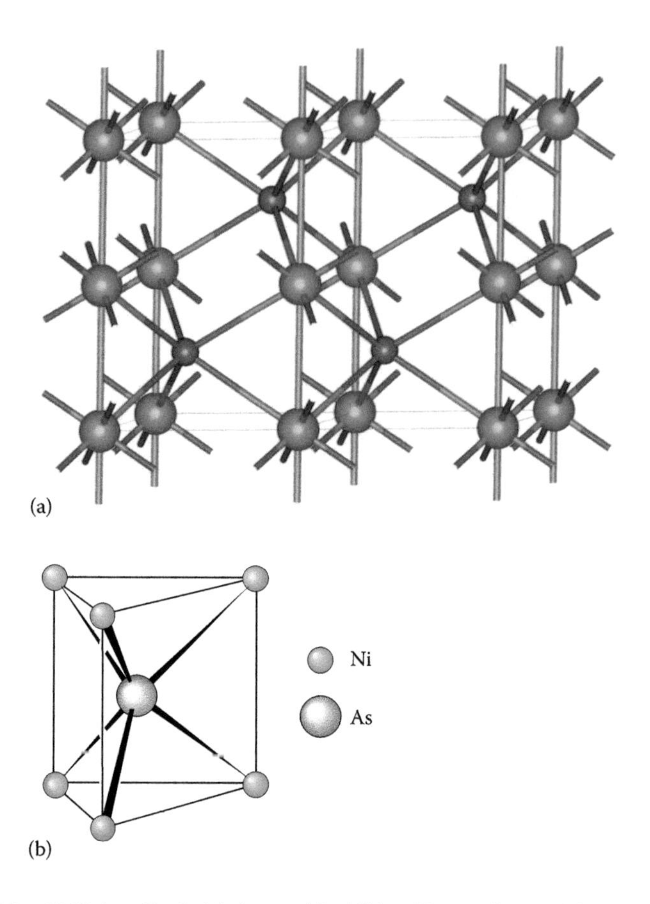

FIGURE 1.35 (a) Unit cell of nickel arsenide, NiAs. (For undistorted *hcp* $c/a = 1.633$, but this ratio is found to vary considerably.) Key: Ni, blue; As, grey. (b) The trigonal prismatic coordination of arsenic in NiAs.

The **nickel arsenide (NiAs) structure** is the equivalent of the sodium chloride structure in hexagonal close packing. It can be thought of as an *hcp* array of arsenic atoms with nickel atoms occupying the octahedral holes. The geometry of the nickel atoms is thus octahedral. This is not the case for arsenic, however, and each arsenic atom sits at the centre of a **trigonal prism** of six nickel atoms (Figure 1.35).

The unit cells of two structures, **zinc blende** (or **sphalerite**) and **wurtzite (ZnS)**, are shown in Figures 1.36 and 1.37, respectively. They are named after two different naturally occurring mineral forms of zinc sulfide (ZnS). Structures of the same element or compound that differ only in their atomic arrangements are termed **polymorphs**. Zinc blende is often contaminated by iron, making it very dark in colour and thus lending it the name of 'Black Jack'.

The zinc blende structure can be thought of as a *ccp* array of sulfide ions with zinc ions occupying every other tetrahedral hole in an ordered manner. Each zinc ion is thus tetrahedrally coordinated by four sulfides and vice versa. Notice that if all the atoms were identical, the structure would be exactly the same as that of diamond

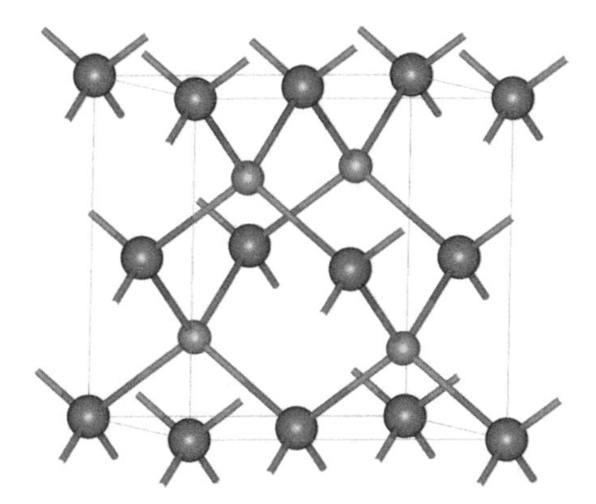

FIGURE 1.36 Crystal structure of zinc blende or sphalerite, ZnS. Key: Zn, blue; S, grey (or vice versa).

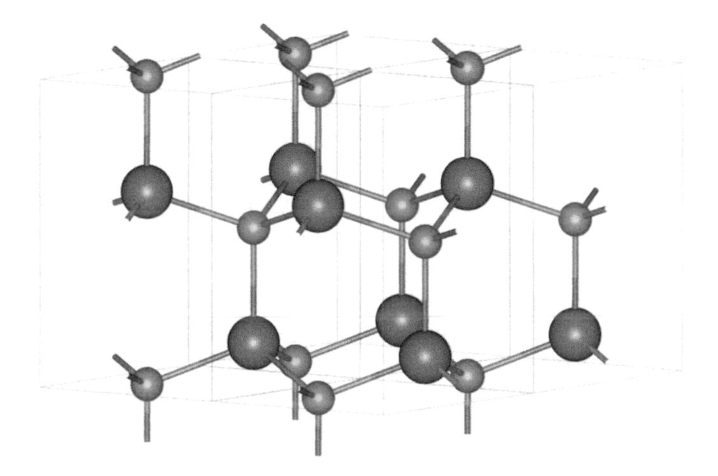

FIGURE 1.37 Crystal structure of wurtzite, ZnS. Key: Zn, blue; S, grey.

(see Section 1.8.5). Notice also that the atomic positions are equivalent, and we could equally well generate the structure by swopping the zinc and sulfur ions. Compounds adopting this structure include the copper halides and the Zn, Cd, and Hg sulfides.

The wurtzite structure is composed of an *hcp* array of sulfide ions with alternate tetrahedral holes occupied by zinc ions. Each zinc ion is tetrahedrally coordinated by four sulfide ions and vice versa. Compounds adopting the structure include BeO, ZnO, and NH_4F.

Notice how the coordination numbers of the structures we have looked at so far have changed. The coordination number for close packing, where all the atoms are identical, is 12. In the CsCl structure, it is eight; in NaCl, six; and in both the ZnS structures, four. As a general rule, the larger the cation, the more anions it can pack around itself (see Section 1.8.4).

1.8.2 Solids with General Formula MX_2

The **fluorite structure** is named after the mineral form of calcium fluoride (CaF_2), which is found in the United Kingdom in the famous Derbyshire 'Blue John' mines. The structure is illustrated in Figure 1.38. It can be thought of as related to a *ccp* array of calcium ions with fluorides occupying all the tetrahedral holes. There is a problem with this as a description, however, because calcium ions are rather smaller than fluoride ions and so, physically, fluoride ions would not be able to fit into the tetrahedral holes of a calcium ion array. Nevertheless, it gives an exact description of the relative positions of the ions in the structure. The diagram in Figure 1.38a clearly shows the fourfold tetrahedral coordination of the fluoride ions. Notice also that the larger octahedral holes are vacant in this structure—one of them is located at the body centre of the unit cell in Figure 1.38a surrounded by the six (blue) calcium ions.

This becomes a very important feature when we come to look at the movement of ions through crystal structures in Chapter 6.

By drawing small cubes with fluoride ions at each corner, as shown in Figure 1.38b, you can see that there is an eightfold cubic coordination of each calcium cation. Indeed, it is possible to move the origin and redraw the unit cell so that this feature can be seen more clearly, as shown in Figure 1.38c. The unit cell is now divided into eight smaller cubes called **octants**, with each alternate octant occupied by a calcium cation.

In the **antifluorite structure**, the positions of the cations and anions are merely reversed, and the description of the structure as cations occupying all the tetrahedral holes in a *ccp* array of anions becomes more physically realistic. In the example with the biggest anion and smallest cation, lithium telluride (Li_2Te), the tellurium ions are approximately close-packed (even though there is a considerable amount of covalent bonding). For the other compounds adopting this structure, such as the oxides and sulfides of alkali metals, M_2O and M_2S, the description accurately shows the relative positions of the atoms, but the anions could not be described as close-packed because they are not touching each other, and the cations are too big to fit in the tetrahedral holes. This means that the anion–anion distance is greater than for close packing.

These are the only structures where 8: 4 coordination is found. Many fast ion conductors are based on these structures (see Chapter 6).

Both of the **cadmium chloride and cadmium iodide (CdI_2) structures** are based on the close packing of the appropriate anions with half of the octahedral holes occupied by cations. In both structures, the cations occupy all the octahedral holes in every other close-packed anion layer, giving an overall layer structure with 6: 3 coordination. The cadmium chloride ($CdCl_2$) structure is based on a *ccp* array of chloride ions, whereas the cadmium iodide (CdI_2) structure is based on an *hcp* array of iodide ions. The cadmium iodide structure is shown in Figure 1.39; in (a) we can see that an iodide anion is surrounded by three cadmium cations on one side but by three iodides on the other, that is, it is not completely surrounded by ions of the opposite charge, as we would expect for an ionic structure. This is evidence that the bonding in some of these structures is not entirely ionic, as we have tended to imply so far. This is a point that is taken up again in more detail in Section 1.8.4.

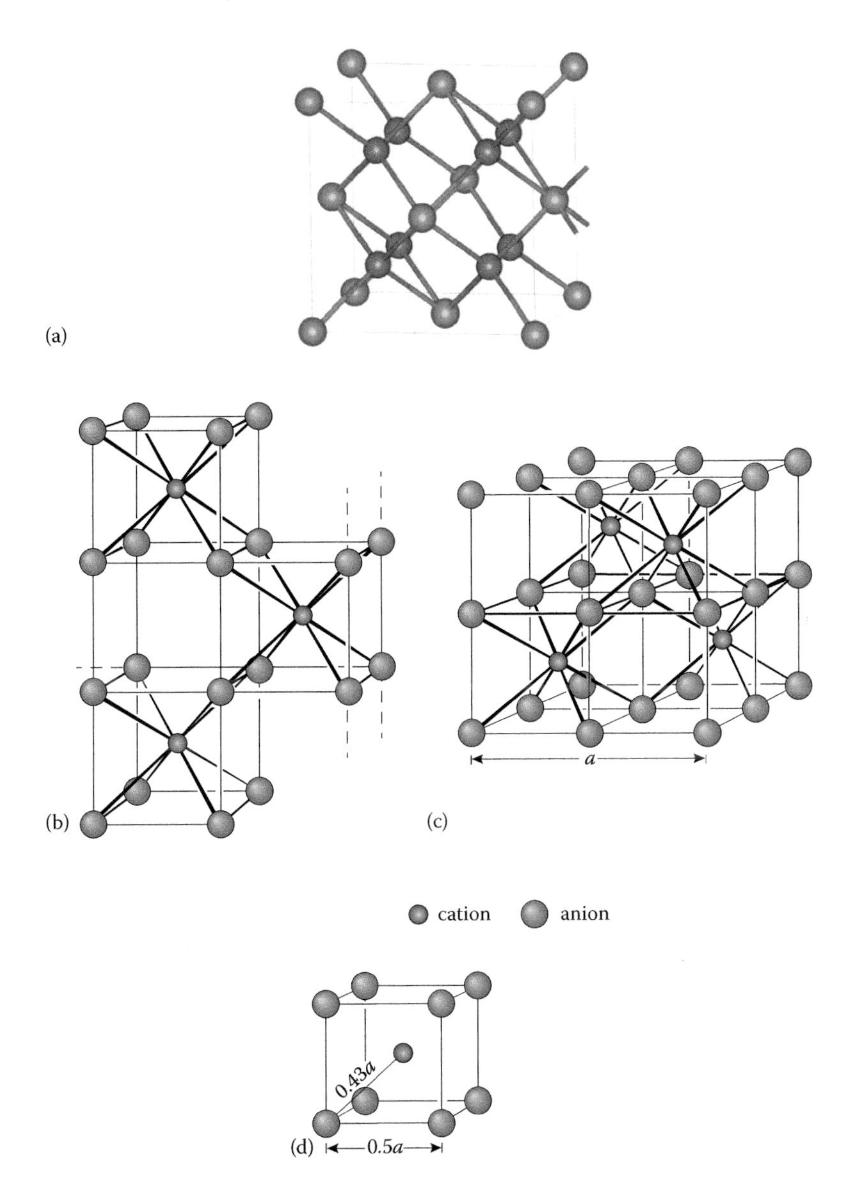

FIGURE 1.38 Crystal structure of fluorite, CaF_2. (a) Computer-generated unit cell as a *ccp* array of cations. Key: Ca, blue; F, grey. (b, c) The same structure redrawn as a primitive cubic array of anions. (d) Relationship of unit cell dimensions to the primitive anion cube (the octant).

The **rutile structure** is named after a mineral form of titanium dioxide (TiO_2). Rutile has a very high refractive index, scattering most of the visible light incident on it; therefore, it is the most widely used white pigment in paints and plastics. A unit cell is illustrated in Figure 1.40 The unit cell is tetragonal and the structure again demonstrates 6: 3 coordination but this time it is not based on close packing:

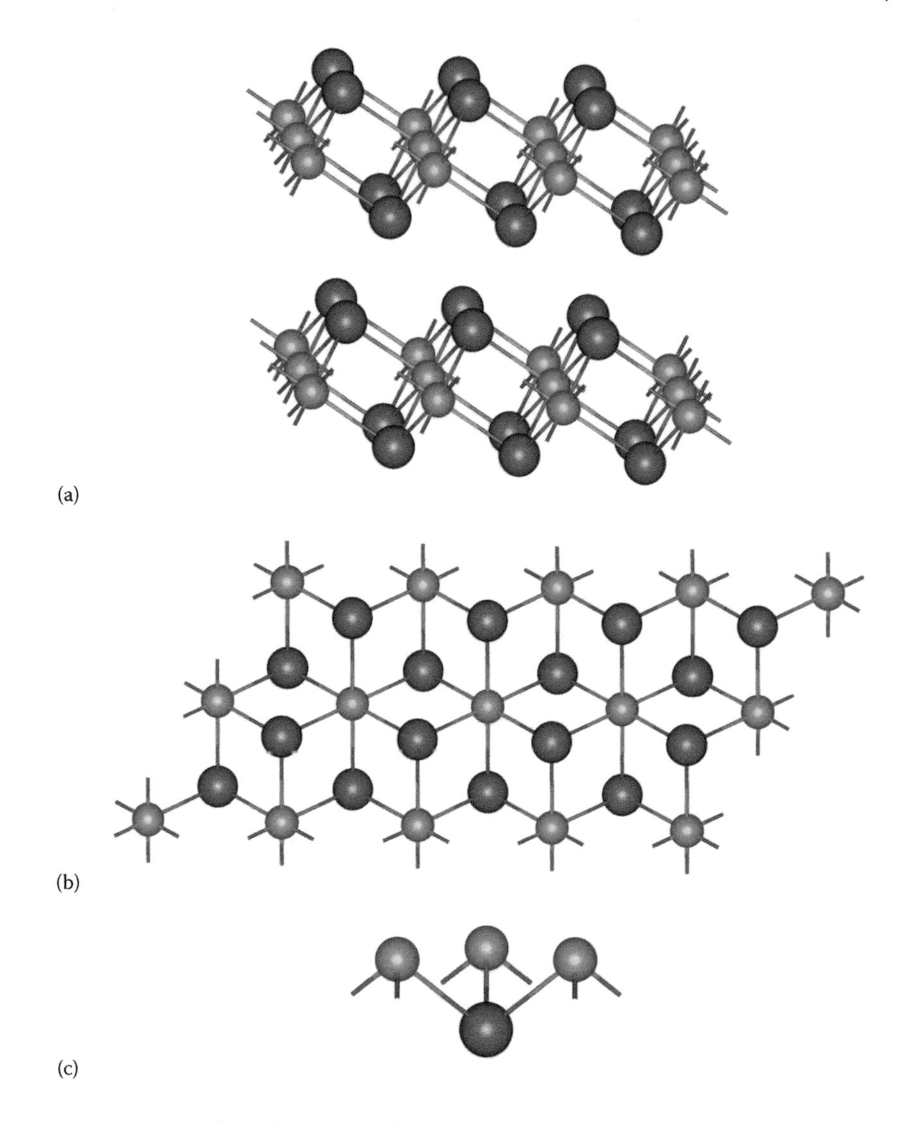

FIGURE 1.39 (a) Crystal structure of cadmium iodide, CdI_2. (b) Structure of the layers in CdI_2 and $CdCl_2$: the halogen atoms lie in planes above and below that of the metal atoms. (c) Coordination around one iodine atom in CdI_2. Key: Cd, blue; I, grey.

each titanium atom is coordinated by six oxygen atoms at the corners of a (slightly distorted) octahedron and each oxygen atom is surrounded by three planar titanium ions that lie at the corners of an (almost) equilateral triangle. It is not geometrically possible for the coordination around Ti to be a perfect octahedron and for the coordination around O to be a perfect equilateral triangle.

The structure can be viewed as chains of linked TiO_6 octahedra, where each octahedron shares a pair of opposite edges, and the chains are linked by sharing vertices (Figure 1.40b). Figure 1.40c shows a plan of the unit cell, which is facing down the chains of octahedra so that they are seen in projection.

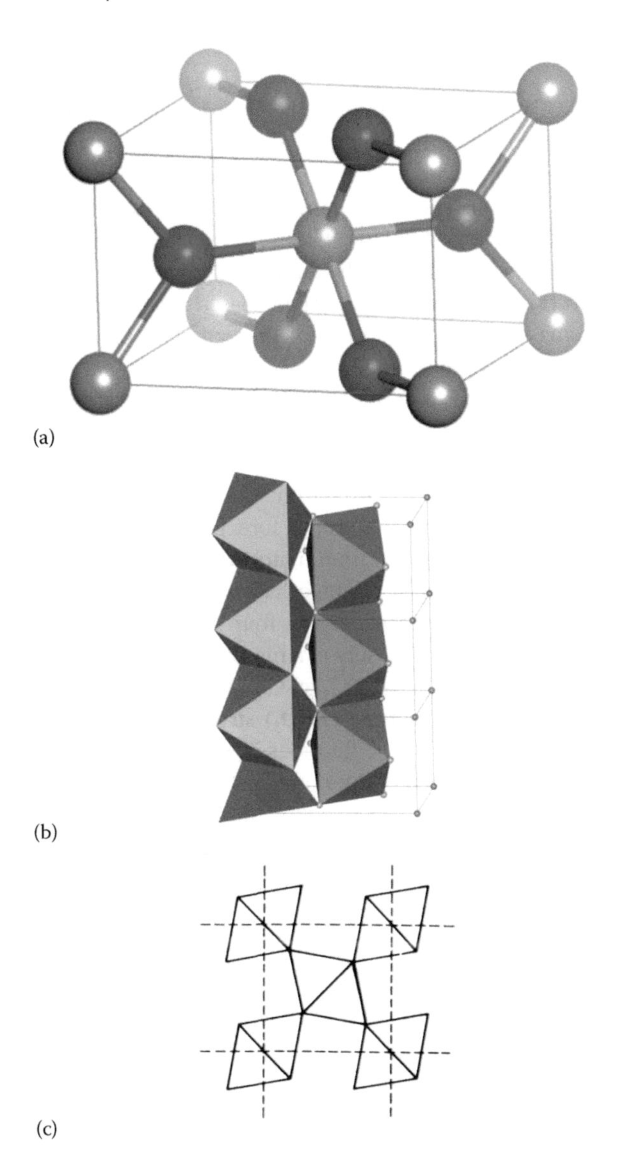

FIGURE 1.40 Crystal structure of rutile, TiO_2. (a) Unit cell, (b) parts of two chains of linked [TiO_6] octahedra, and (c) projection of structure on base of unit cell. Key: Ti, blue; O, grey.

Occasionally, the **antirutile structure** is encountered where the metals and the nonmetals have swapped. The **β-cristobalite structure** is named after a mineral form of silicon dioxide (SiO_2). The silicon atoms are in the same positions as both the zinc and the sulfur ions in zinc blende (or the carbon atoms in diamond, which we look at in Section 1.8.5), and each pair of silicon atoms is bonded to an oxygen atom midway between them. The only metal halide adopting this structure is beryllium fluoride (BeF_2), which is characterised by 4: 2 coordination, with Be tetrahedrally coordinated by F.

1.8.3 Other Important Crystal Structures

As the valency of the metal increases, the bonding in these simple binary compounds becomes more covalent and the highly symmetrical structures characteristic of the simple ionic compounds occur far less frequently, with molecular and layer structures being common. There are many thousands of inorganic crystal structures; here, we briefly describe just a few of those that are commonly encountered and those that occur in later chapters of this book.

The **bismuth triiodide structure (BiI_3)** is based on an *hcp* array of iodides with the bismuth atoms occupying one third of the octahedral holes. Alternate pairs of layers have two thirds of the octahedral sites occupied.

The mineral **corundum (α-Al_2O_3)** is the basis for ruby and sapphire gemstones, with their colour depending on the impurities (see Chapter 8). It is very hard—second only to diamond on the Mohs scale. This structure can be described as an *hcp* array of oxygen atoms with two thirds of the octahedral holes occupied by aluminium atoms (Figure 1.41). As we have seen before, geometrical constraints dictate that octahedral coordination of the aluminium atoms precludes tetrahedral coordination of the oxygen atoms. However, it is suggested that this structure is adopted in preference to other possible ones because the four aluminium atoms surrounding an oxygen atom approximate most closely to a regular tetrahedron. The structure is also adopted by Ti_2O_3, V_2O_3, Cr_2O_3, α-Fe_2O_3, α-Ga_2O_3, and Rh_2O_3.

The **rhenium trioxide structure (ReO_3)** structure (also called the aluminium fluoride structure) is adopted by the fluorides of Al, Sc, Fe, Co, Rh, and Pd and also by the oxides WO_3 (at high temperature) and ReO_3 (see Section 5.8.1). The structure consists of ReO_6 octahedra linked together through *each* corner to give a highly symmetrical three-dimensional network with cubic symmetry. Part of the structure is shown in Figure 1.42a, the linking of the octahedra in Figure 1.42b, and the unit cell in Figure 1.42c.

Three important **mixed oxide structures** are **spinel**, **perovskite**, and **ilmenite**. The **spinels** have the general formula AB_2O_4, taking their name from the mineral spinel $MgAl_2O_4$, where, generally, A is a divalent ion (A^{2+}) and B is trivalent

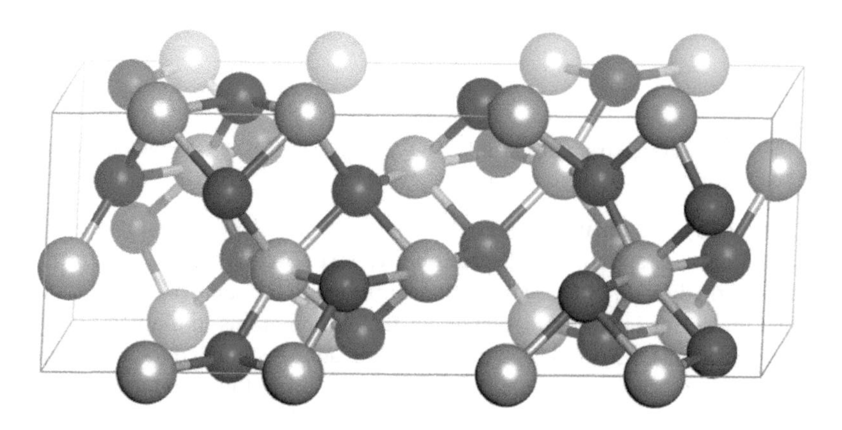

FIGURE 1.41 Crystal structure of corundum.

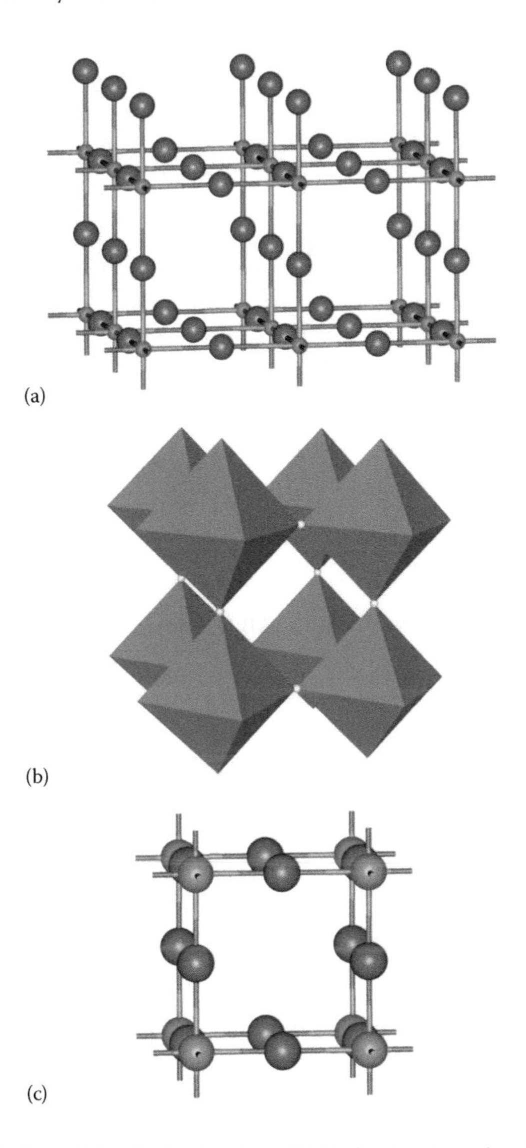

FIGURE 1.42 (a) Part of the ReO_3 structure, (b) ReO_3 structure showing the linking of $[ReO_6]$ octahedra, and (c) unit cell. Key: Re, blue; O, grey.

(B^{3+}). The structure can be thought of as being based on a cubic close-packed array of oxide ions, with A^{2+} ions occupying tetrahedral holes and B^{3+} ions occupying octahedral holes. A spinel crystal containing nAB_2O_4 formula units has $8n$ tetrahedral holes and $4n$ octahedral holes, so, accordingly, one eighth of the tetrahedral holes are occupied by the A^{2+} ions and one half of the octahedral holes by the B^{3+} ions. The face-centred unit cell is illustrated in Figure 1.43. The A ions are tetrahedrally coordinated, whereas the B ions occupy octahedral sites. Spinels with this structure include compounds of the formula MAl_2O_4, where M is Mg, Fe, Co, Ni, Mn, or Zn.

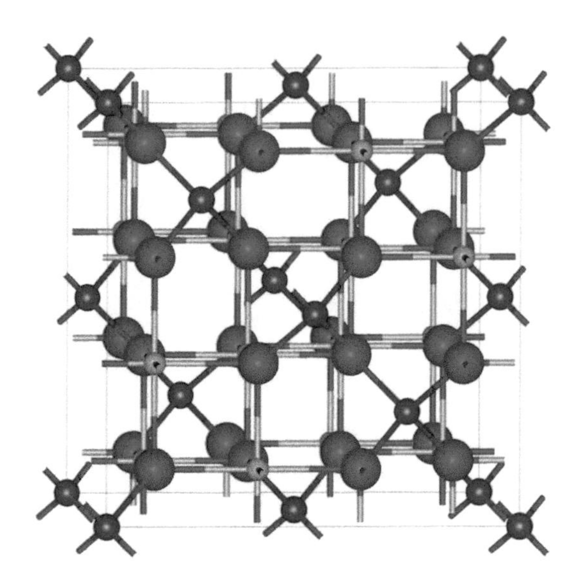

FIGURE 1.43 Spinel structure, $CuAl_2O_4$ (AB_2O_4). Key: Cu, blue; Al, mauve; O, red.

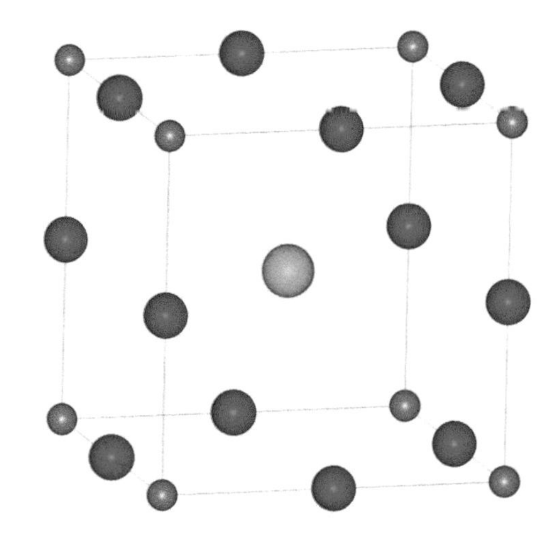

FIGURE 1.44 Perovskite structure of compounds ABX_3, such as $CaTiO_3$. Key: Ca, green; Ti, silver; O, red.

When compounds of the general formula AB_2O_4 adopt the **inverse-spinel structure**, the formula is better written as $B(AB)O_4$, because this indicates that half of the B^{3+} ions now occupy tetrahedral sites, and the remaining half, together with the A^{2+} ions, occupy the octahedral sites. Examples of inverse spinels include magnetite, Fe_3O_4 (see Section 9.7), $Fe(MgFe)O_4$, and $Fe(ZnFe)O_4$.

The **perovskite structure** is named after the mineral $CaTiO_3$. A unit cell is shown in Figure 1.44: this unit cell is termed the A-type, because if we take the general formula ABX_3 for the perovskites, then in this cell, the A atom is at the centre. The central Ca (A) atom is coordinated to 8 Ti atoms (B) at the corners and to 12 oxygens (X) at the midpoints of the cell edges. The structure can be usefully described in other ways. First, perovskite has the same octahedral framework as ReO_3 based on BX_6 octahedra with an A atom added in at the centre of the cell (see Figure 1.42b). Secondly, it can be thought of as a *ccp* array of A and X atoms, with the B atoms occupying the octahedral holes (compare with the unit cell of NaCl in Figure 1.30 to check this). Compounds adopting this structure include $SrTiO_3$, $SrZrO_3$, $SrHfO_3$, $SrSnO_3$, and $BaSnO_3$. When the B atom is much smaller than the A atom, it can be displaced within the cell giving rise to a ferroelectric effect (Chapter 9). The structures of the high-temperature cuprate superconductors are based on this structure (see Chapter 10).

The **ilmenite structure** is adopted by oxides of the formula ABO_3, where A and B are similar in size and their charges add up to a total charge of +6. The structure is named after the mineral $Fe^{II}Ti^{IV}O_3$, and is very similar to the corundum structure described above—an *hcp* array of oxygen atoms—but now there are two different cations present, occupying two thirds of the octahedral holes.

The structures related to close packing are summarised in Table 1.8.

TABLE 1.8
Structures Related to Close-Packed Arrangements of Anions

Formula	Cation: anion coordination	Type and number of holes occupied	Cubic close packing	Hexagonal close packing
MX	6: 6	All octahedral	Sodium chloride: NaCl, FeO, MnS, TiC	Nickel arsenide: NiAs, FeS, NiS
	4: 4	Half tetrahedral—every alternate site occupied	Zinc blende: ZnS, CuCl, γ-AgI	Wurtzite: ZnS, β-AgI
MX_2	8: 4	All tetrahedral	Fluorite: CaF_2, ThO_2, ZrO_2, CeO_2	None
	6: 3	Half octahedral; alternate layers have fully occupied sites	Cadmium chloride: $CdCl_2$	Cadmium iodide: CdI_2, TiS_2
MX_3	6: 2	One-third octahedral; alternate pairs of layers have two thirds of the octahedral sites occupied		Bismuth iodide: BiI_3, $FeCl_3$, $TiCl_3$, VCl_3
M_2X_3	6: 4	Two-thirds octahedral		Corundum: α-Al_2O_3, α-Fe_2O_3, V_2O_3, Ti_2O_3, α-Cr_2O3
ABO_3		Two-thirds octahedral		Ilmenite: $FeTiO_3$
AB_2O_4		One-eighth tetrahedral and one-half octahedral	Spinel: $MgAl_2O_4$ inverse spinel: $MgFe_2O_4$, Fe_3O_4	Olivine: Mg_2SiO_4

1.8.4 Ionic Radii

We know from quantum mechanics that atoms and ions do not have precisely defined radii. However, from the foregoing discussion of ionic crystal structures, we have seen that ions pack together in an extremely regular fashion in crystals, and that their atomic positions and thus, their interatomic distances can be measured very accurately. It is a very useful concept, therefore, particularly for those structures based on close packing, to think of ions as hard spheres, each with a particular radius.

If we take a series of alkali metal halides, all with the rock-salt structure, if we replace one metal ion with another, say sodium with potassium, we would expect the metal–halide internuclear distance to change by the same amount each time if the concept of an ion as a hard sphere with a particular radius holds true. Table 1.9 shows the results of this procedure for a range of alkali halides, with the change in internuclear distance on swopping one ion for another in bold.

From Table 1.9, we can see that the change in internuclear distance upon changing the ion is not constant, and also that the variation is not great. This provides us with an experimental basis that it is not unreasonable to approximate the ions as having fixed radii, despite the picture being not precisely true. We would expect a model based on hard spheres to be only an approximation, because atoms and ions are squashable entities, they are polarisable and their shape and size will be affected by their environment. Nevertheless, it is a useful concept to develop a bit further as it enables us to describe some of the ionic crystal structures in a simple pictorial way.

There have been many suggestions as to how individual ionic radii can be assigned, and the literature contains several different sets of values, each set being named after the person(s) who originated the method of determining the radii. We briefly describe some of these methods before listing the values most commonly used at present. It is most important to remember that you must not mix radii from more than one set of values, because, even though the values vary considerably from one set to another, each set is internally consistent; that is, if you add together two radii from one set of values, you will obtain an approximately correct internuclear distance as determined from the crystal structure.

The internuclear distances can be determined by X-ray crystallography, and in order to obtain values for individual ionic radii from these, the value of one radius

TABLE 1.9

Interatomic Distances of Some Alkali Halides, r_{M-X} (pm)

	F^-		Cl^-		Br^-		I^-
Li^+	201	**56**	257	**18**	275	**27**	302
	30		**24**		**23**		**21**
Na^+	231	**50**	281	**17**	298	**25**	323
	35		**33**		**31**		**30**
K^+	266	**48**	314	**15**	329	**24**	353
	16		**14**		**14**		**13**
Rb^+	282	**46**	328	**15**	343	**23**	366

needs to be fixed by some method. Originally, in 1920, Landé suggested that in the alkali halide with the largest anion and smallest cation—LiI—we can assume that the iodide ions must be in contact with each other and that the tiny Li^+ ion is inside the octahedral hole. As the Li-I distance is known, it is then a matter of simple geometry to determine the iodide radius. Once the iodide radius is known, and assuming it stays constant, then the radii of the metal cations can be found from the structures of the other metal iodides. Bragg and Goldschmidt later extended the list of ionic radii using similar methods. It is very difficult to come up with a consistent set of values for the ionic radii because, of course, the ions are not hard spheres, they are somewhat elastic and their radii are affected by their environment, such as the nature of the oppositely charged ligand and the coordination number. Pauling proposed a theoretical method of calculating the radii from the internuclear distances, and he produced a set of values that are internally consistent and also show the expected trends in the periodic table. Pauling's method was to take a series of alkali halides with isoelectronic cations and anions and assume that they are in contact with each other; if we assume that each radius is inversely proportional to the effective nuclear charge felt by the outer electrons of the ion, the radius for each ion can be calculated from the internuclear distance. Divalent ions undergo additional compression in a lattice and compensation has to be made for this effect in calculating their radii. With some refinements, this method gave a consistent set of values that were widely used for many years; they are usually known as **effective ionic radii**.

It is also possible to determine accurate electron density maps for the ionic crystal structures using X-ray crystallography. Such a map is shown for NaCl and LiF in Figure 1.45. The electron density contours fall to a minimum—although, note, not to zero—in between the nuclei, and it has been suggested that this minimum position should be taken as the radius position for each ion. These experimentally determined ionic radii are often called **crystal radii**; the values are somewhat different from the

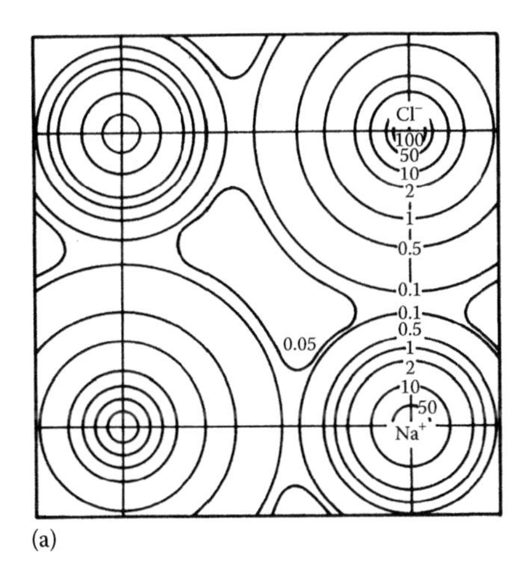

(a)

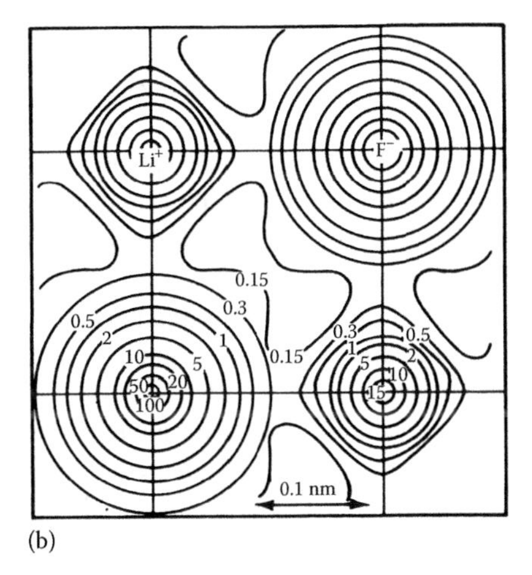

(b)

FIGURE 1.45 Electron density maps for (a) NaCl and (b) LiF.

older sets and tend to make the anions smaller and the cations bigger than previously calculated. Shannon and Prewitt have compiled the most comprehensive set of radii, using data from almost a thousand crystal structure determinations and based on the conventional values of 126 and 119 pm for the radii of the $O^{2-}$and F^- ions, respectively. These values differ by a constant factor of 14 pm from the older traditional values, but it is generally accepted that they correspond more closely to the actual physical sizes of ions in a crystal. A selection of these data is shown in Table 1.10.

Several important trends in the sizes of ions can be noted from the data in Table 1.10:

1. The radii of ions within a group of the periodic table, such as the alkali metals, increase with atomic number, *Z*: as you go down the group, there are more electrons and the outer ones are further from the nucleus.
2. In a series of isoelectronic cations, such as Na^+, Mg^{2+}, and Al^{3+}, the radii decrease rapidly with increasing positive charges. The number of electrons is constant, but the nuclear charge increases pulling the electrons in, and the radii decrease.
3. For pairs of isoelectronic anions, for example, F^- and O^{2-}, the radii increase with increasing charges because the more highly charged ion has a smaller nuclear charge.
4. For elements with more than one oxidation state, for example, Ti^{2+} and Ti^{3+}, the radii decrease as the oxidation state increases: in this case, the nuclear charge stays the same but the number of electrons that it acts on decreases.
5. As you move across the periodic table, for a series of similar ions, such as the first row transition metal divalent ions (M^{2+}), there is an overall decrease in radii. This is due to an increase in the nuclear charge across the table because electrons in the same shell do not screen the nucleus from each other very well. A similar effect is seen for the M^{3+} ions of the lanthanoids and this is known as the **lanthanoid contraction**.
6. For transition metals, the spin state affects the ionic radii.
7. The crystal radii increase with an increase in the coordination number—see examples Cu^+ and Zn^{2+} in Table 1.10.

The picture of ions as hard spheres works best for fluorides and oxides, both of which are small, fairly uncompressible ions. As the ions get larger, they are more easily compressed the electron cloud is more easily distorted—and they are said to be more **polarisable**.

When we were discussing particular crystal structures in the previous section, we noted that a larger cation, such as Cs^+, was able to pack eight chloride ions around itself, whereas the smaller Na^+ could accommodate only six. If we continue to think of ions as hard spheres, for a particular structure, as the ratio of the cation and anion radii changes, there will come a point when the cation is so small that it will no longer be in contact with the anions. The lattice would not be stable in this state because the negative charges would be too close together for comfort and we would predict that the structure would change to one of lower coordination, allowing the anions to move further apart. If the ions are hard spheres, using simple geometry, it is possible

TABLE 1.10

Crystal Radii/pm

						H											He
						126(− 1)											
Li	Be											B	C	N	O	F	Ne
90(+1)	59(+2)											41 (+3)	30(+44)	132(−3)[a]	126(−2)	119(−1)	—
														30(+3)			
Na	Mg											Al	Si	P	S	Cl	Ar
116(+1)	86(+2)											68(+3)	54(+4)	58(+3)	170(−2)	167(−1)	—
K	Ca	Sc	Ti	V	Cr	Mn	Fe	Co	Ni	Cu	Zn	Ga	Ge	As	Se	Br	Kr
152(+1)	114(+2)	89(+3)	100 (+2)	93(+2)	87/94)+2)[b]	81/97 (+2)[b]	75/921+2)[b]	79/89(+2)[b]	83(+2)	91(+1)	88(+2)	76 (+3)	87 (+2)	72(+3)	184(−2)	182(−1)	—
			81 (+3) 75 (+4)	78(+3)	76(+3)	72/79(+3)[b]	69/79(+3)[b]	69/75(+3)[b]	70/741+3)[b]	87(+2) 68(+3)			67(+4)				
Rb	Sr	Y	Zr							Ag	Cd	In	Sn	Sb	Te	I	Xe
166(+1)	132(+2)	104(+3)	8 6 (+4)							129(+1)	109(+2)	94(+3)	136(+2)[c]	90(+3)	207(−2)	206(−1)	62(+8)
													83 (+4)				
Cs	Ba	Lu	Hf							Au	Hg	Tl	Pb	Bi	Po	At	Rn
181 (+1)	149(+2)	100 (+3)	85 (+4)							151+1)	116(+2)	103(+3)	133(+2)	117(+3)	—	—	—

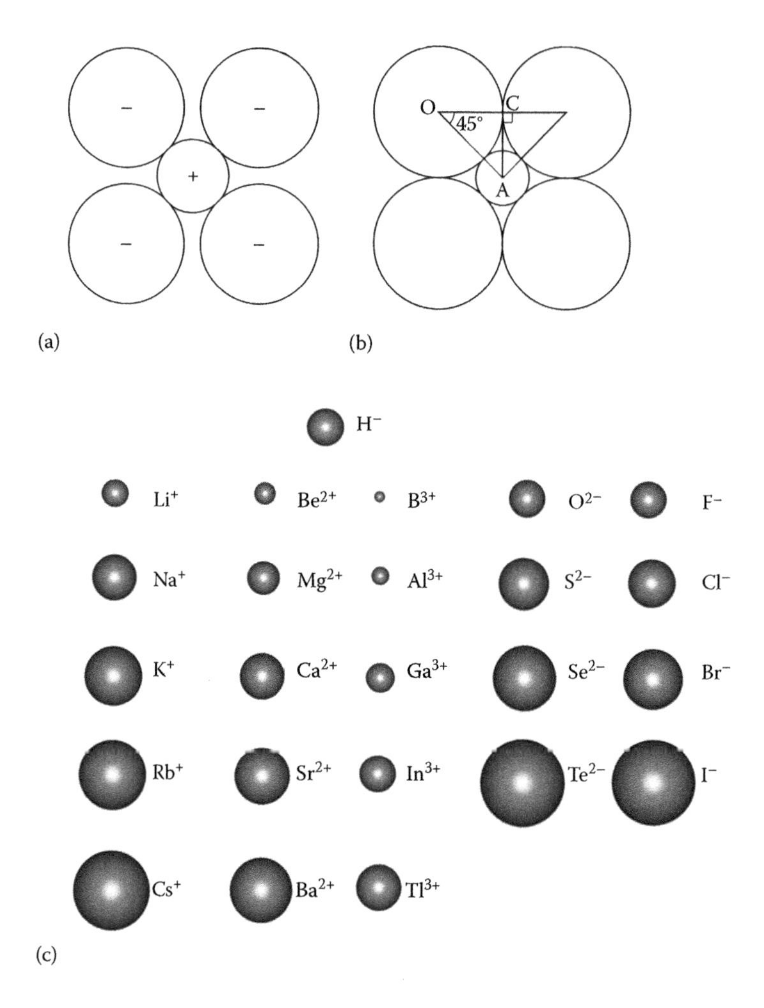

FIGURE 1.46 (a) Anions packed around a cation on a horizontal plane, (b) anion–anion contact on a horizontal plane through an octahedron, and (c) relative sizes of the typical element ions.

to quantify the radius ratio (i.e., $r_{cation}/r_{anion} = r_+/r_-$) at which this happens, and this is illustrated for the octahedral case in Figure 1.46.

Taking a plane through the centre of an octahedrally coordinated metal cation, the stable situation is shown in Figure 1.46a, and the limiting case for stability, when the anions are touching each other, is shown in Figure 1.46b. The anion radius (r_-) in Figure 1.46b is OC, and the cation radius (r_+) is (OA – OC). From the geometry of the right-angled triangle, we can see that cos 45° = OC/OA = 0.707. The radius ratio (r_+/r_-) is given by (OA – OC)/OC = ((OA/OC) – 1) = (1.414 – 1) = 0.414. Using similar calculations, it is possible to calculate the limiting ratios for the other geometries: these are summarised in Table 1.11.

On this basis, we would expect to be able to use the ratio of ionic radii to predict possible crystal structures for a given substance. But does it work? Unfortunately,

TABLE 1.11
Limiting Radius Ratios for Different Coordination Numbers

Coordination number	Geometry	Limiting radius ratio	Possible structures
		0.225	
4	Tetrahedral		Wurtzite, zinc blende
		0.414	
6	Octahedral		Rock salt, rutile
		0.732	
8	Cubic		Cesium chloride, fluorite
		1.00	

only about 50% of the structures are correctly predicted. This is because the model is too simplistic—ions are not hard spheres, but are polarised under the influence of other ions. In larger ions, the valence electrons are further away from the nucleus and also shielded from its influence by the inner core electrons, so the electron cloud is more easily distorted. The ability of an ion to distort an electron cloud—its polarising power—is greater for small ions with high charge, and the distortion of the electron cloud means that the bonding between two such ions becomes more directional in character. This means that in these cases the bonding involved is rarely truly ionic but frequently involves at least some degree of covalency. The higher the formal charge on a metal ion, the greater is the proportion of covalent bonding between the metal and its ligands. The higher the degree of covalency, the less likely is the concept of ionic radii and radius ratios to work. It also seems that there is little energy difference between the six-coordinate and eight-coordinate structures and that the six-coordinate structure is usually preferred. In fact, eight-coordinate structures are rarely found; for instance there are no eight-coordinate oxides. The preference for the six-coordinate rock-salt structure is thought to be due to the covalent bonding contribution: in this structure, the three orthogonal p orbitals lie in the same direction as the vectors joining the cation to the surrounding six anions, so they are well placed for good overlap of the orbitals necessary for σ bonding to take place. The potential overlap of the p orbitals in the caesium chloride structure is less favourable.

1.8.5 Extended Covalent Arrays

In the last section, we noted that many 'ionic' compounds in fact possess some degree of covalency in their bonding. As the formal charge on an ion becomes greater, we expect the degree of covalency to increase, and so we would generally expect compounds of elements in the centre of the periodic table to be covalently bonded. Indeed, some of these elements themselves are covalently bonded solids at room temperature. Examples include elements such as Group 3, boron; Group 14, carbon, silicon, and germanium; Group 15, phosphorus and arsenic; and Group 16, selenium and tellurium, which form **extended covalent arrays** in their crystal structures.

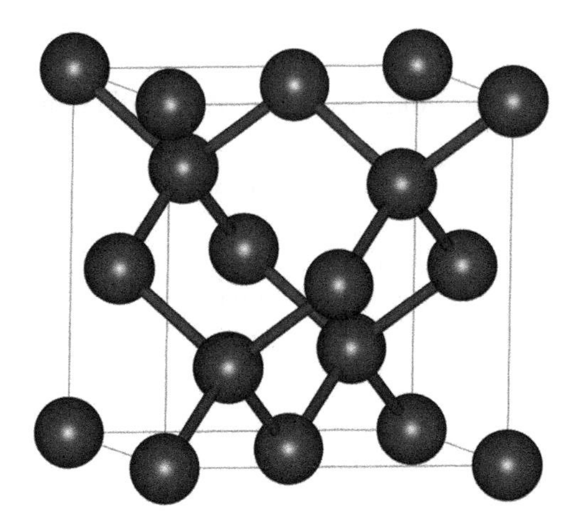

FIGURE 1.47 A unit cell of the diamond structure.

Take, for instance, one of the forms of carbon—**diamond**. Diamond has a cubic crystal structure with an F-centred lattice (Figure 1.47); the positions of the atomic centres are the same as in the zinc blende structure, with carbon now occupying both the zinc and the sulfur positions. Each carbon is equivalent and is tetrahedrally coordinated to four others, forming a covalently bonded **giant molecule** throughout the crystal. The carbon–carbon distances are all identical (154 pm). It is interesting to note how the different types of bonding have affected the coordination number, as here we have identical atoms, but now restricted to four because this is the maximum number of covalent bonds that carbon can form. In the case of a metallic element, such as magnesium forming a crystal, the structure is close-packed with each atom 12-coordinated (bonding in metals is discussed in Chapter 4). The covalent bonds in diamond are strong, and the rigid three-dimensional network of atoms makes diamond the hardest substance known; it also has a high melting temperature (m.t. 3773 K). Silicon carbide (SiC), known as carborundum, also has this structure with silicon atoms and carbon atoms alternating throughout; it, too, is very hard and is used for polishing and grinding.

1.8.6 Bonding in Crystals

Although bonding in solids will be dealt with in detail in Chapter 4, at this point it is convenient to summarise the different types of bonding that we have met so far in crystal structures. In Section 1.6, we considered the structures of metallic crystals held together by metallic bonding. In Section 1.8, we looked at structures such as NaCl and CsCl, which have ionic bonding, and later saw the influence of covalent bonding in the layer structures of $CdCl_2$ and CdI_2.

A simple picture of **metallic bonding** is that metals consist of a regular array of metal cations surrounded by a 'sea' of electrons. These electrons occupy the space between the cations, binding them together, but they are able to move under the influence of an external field, thus accounting for the electrical conductivity of metals. A more sophisticated picture will be given in Chapter 4.

An **ionic bond** forms between two oppositely charged ions due to the electrostatic attraction between them. The attractive force (F) is given by **Coulomb's law**, $F \propto \left[(q_1 q_2)/r^2\right]$ where q_1 and q_2 are the charges on the two ions and r is the distance between them. A similar but opposite repulsive force is experienced by two ions of the same charge. Ionic bonds are strong and nondirectional; the energy of the interaction is given by force × distance and is inversely proportional to the separation of the charges, r. Ionic forces are effective over large distances compared with other bonding interactions. Ions pack together in regular arrays in ionic crystals, in such a way as to maximise coulombic attraction and minimise repulsions.

In **covalent bonding**, the electrons are shared between two atoms, resulting in a build-up of electron density between the atoms. Covalent bonds are strong and directional. In a covalent bond, electronegative elements such as oxygen and nitrogen attract an unequal share of the bonding electrons, such that one end of the bond acquires a partial negative charge (δ–) and the other end acquires a partial positive charge (δ+). The separation of the negative and positive charge creates an **electric dipole**, and the molecule can align itself in an electric field. Such molecules are said to be **polar** and possess a permanent **dipole moment**. The partial electric charges on polar molecules can attract one another in a **dipole–dipole interaction**. The dipole–dipole interaction is about 100 times weaker than ionic interactions and falls off quickly with distance, as a function of $1/r^3$.

Polar molecules can also interact with ions in a **charge–dipole interaction**, which is about 10–20 times weaker than ion–ion interactions and which decreases with distance as $1/r^2$.

Even if molecules do not possess a permanent dipole moment, weak forces can exist between them. The movement of the valence electrons creates 'transient dipoles', and these, in turn, induce dipole moments in the adjacent molecules. The transient dipole in one molecule can be attracted to the transient dipole in a neighbouring molecule, and the result is a weak, short-range attractive force, known as the **London dispersion force**. These dispersion forces drop off rapidly with distance, decreasing as a function of $1/r^6$.

The weak nonbonded interactions that occur between molecules are often collectively referred to as **van der Waals forces**.

In one special case, polar interactions are strong enough for them to be exceptionally important in dictating the structure of the solid and liquid phases. Where hydrogen is bonded to a very electronegative element, such as oxygen or fluorine, there is a partial negative charge (δ–) on the electronegative element and an equal and opposite charge (δ+) on the hydrogen. The positively charged $H^{\delta+}$ can also be attracted to the partial negative charge on a neighbouring molecule, forming a weak bond, known as a **hydrogen bond**, O–H-----O, pulling the three atoms almost into a straight line. A network of alternating weak and strong bonds is built up, for example, as can be seen in water (H_2O) and in hydrogen fluoride (HF). The longer, weaker hydrogen bonds can be thought of as dipole–dipole interactions and are particularly important in biological systems and in any crystals that contain water.

1.8.7 Atomic Radii

An atom in a covalently bonded molecule can be assigned a **covalent radius** (***r***$_c$) and also a nonbonded radius, known as the **van der Waals radius**.

Covalent radii are calculated from half the interatomic distance between two singly bonded, like atoms. For diatomic molecules such as F_2, this is no problem, but for other elements, like carbon, which do not have a diatomic molecule, an average value is calculated from a range of compounds that contain a C–C single bond.

The van der Waals radius is defined as a nonbonded distance of closest approach and is calculated from the smallest interatomic distances in crystal structures that are considered to be not bonded to one another. Again, these are average values compiled from many crystal structures. If the sum of the van der Waals radii of two adjacent atoms in a structure is greater than the measured distance between them, then it is supposed that there is some bonding between them. Table 1.12 gives the covalent and the van der Waals radii for the typical elements.

1.8.8 Molecular Structures

Finally, we consider crystal structures that do not contain any extended arrays of atoms—the many crystals that contain small, discrete, covalently bonded molecules that are held together only by weak forces.

Examples of molecular crystals are found throughout organic, organometallic, and inorganic chemistry. The crystals are characterised by low melting and boiling temperatures. We will look at just two examples, carbon dioxide and water (ice), both familiar, small, covalently bonded molecules.

When gaseous carbon dioxide (CO_2) is cooled sufficiently, it forms a molecular crystalline solid, as illustrated in Figure 1.48. Notice that the unit cell contains clearly

TABLE 1.12
Single-Bond Covalent Radii and van der Waals Radii (in Parentheses) for the Typical Elements/pm

Group 1	Group 2	Group 3	Group 14	Group 15	Group 16	Group 17	Group 18
							He
							— (140)
Li	Be	B	C	N	O	F	Ne
135	90	80	77 (170)	74 (155)	73 (152)	71 (147)	— (154)
Na	Mg	Al	Si	P	S	Cl	Ar
154	130	125	117 (210)	110 (180)	104 (180)	99 (175)	— (188)
K	Ca	Ga	Ge	As	Se	Br	Kr
200	174	126	122	121 (185)	117 (190)	114 (185)	— (202)
Rb	Sr	In	Sn	Sb	Te	I	Xe
211	192	141	137	141	137 (206)	133 (198)	— (216)
Cs	Ba	Tl	Pb	Bi	Po	At	Rn
225	198	171	175	170	140	—	—

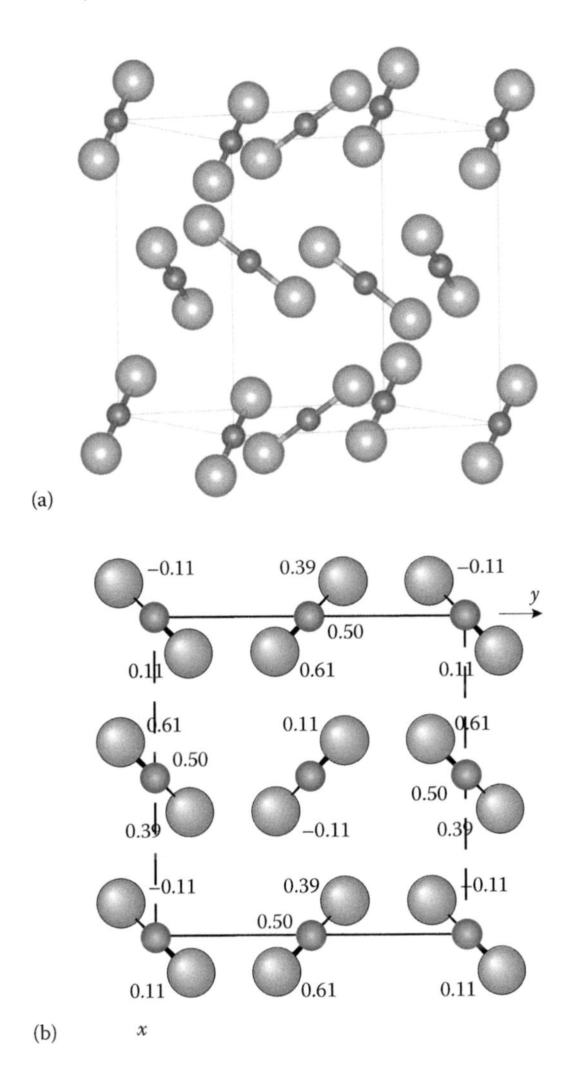

FIGURE 1.48 **(a)** Crystal structure of CO_2. (b) A packing diagram of the unit cell of CO_2 projected onto the *xy* plane. The heights of the atoms are expressed as fractional coordinates of *c*. Key: C, blue; O, grey.

discernible CO_2 molecules, which are covalently bonded, and are held together in the crystal by weak van der Waals forces.

The structure of one form of ice (crystalline water) is shown in Figure 1.49. Each H_2O molecule is tetrahedrally surrounded by four others. The crystal structure is held together by the hydrogen bonds formed between a hydrogen atom of one water molecule and the oxygen of the adjacent molecule, forming a three-dimensional arrangement throughout the crystal. This open hydrogen-bonded network of water molecules makes ice less dense than water, so that it floats on the surface of water.

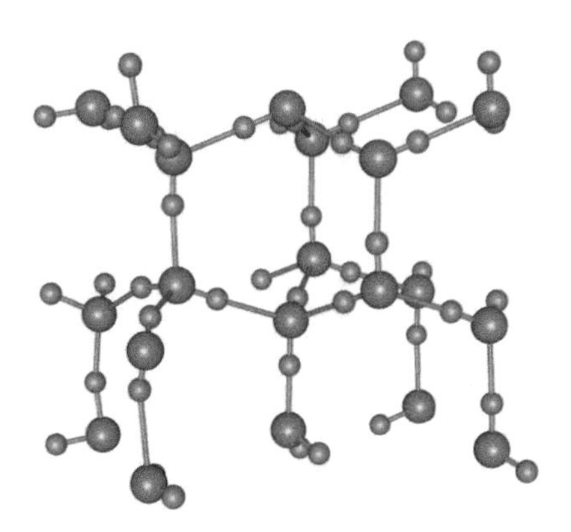

FIGURE 1.49 Crystal structure of ice. Key: H, blue; O, grey.

TABLE 1.13
Classification of Crystal Structures

Type	Structural unit	Bonding	Characteristics	Examples
Ionic	Cations and anions	Electrostatic, nondirectional	Hard, brittle, crystals of high m.t.; moderate insulators; melts are conducting	Alkali metal halides
Extended covalent array	Atoms	Mainly covalent	Strong hard crystals of high m.t.; insulators	Diamond, silica
Molecular	Molecules	Mainly covalent between atoms in molecules, van der Waals or hydrogen bonding between molecules	Soft crystals of low m.t. and large coefficient of expansion; insulators	Ice, organic compounds
Metallic	Metal atoms	Band model (see Chapter 4)	Single crystals are soft; strength depends on structural defects and grain; good conductors; m.t. vary but tend to be high	Iron, aluminum, sodium

A summary of the various types of crystalline solids is given in Table 1.13, relating the type of structure to its physical properties. It is important to realise that this only gives a broad overview and is intended as a guide only: not every crystal will fall exactly into one of these categories.

1.9 LATTICE ENERGY

The **lattice energy** (*L*) of an ionic crystal is a measure of the strength of the bonding and is defined as the standard enthalpy change when one mole of the solid is formed from the gaseous ions, for example, for NaCl, $L = \Delta H_{\mathrm{m}}^{\ominus}$ for the reaction in Equation 1.1 = −787 kJ mol^{-1}; an exothermic process:

$$\mathrm{Na^{+}}(\mathrm{g}) + \mathrm{Cl^{-}}(\mathrm{g}) = \mathrm{NaCl}(\mathrm{s}) \tag{1.1}$$

Because of the difficulty in producing gaseous ions, it is not possible to measure lattice energies directly, but they can be determined experimentally using a thermochemical **Born–Haber cycle**, or they can be calculated from electrostatics, as we will see in Section 1.9.2.

1.9.1 BORN–HABER CYCLE

A Born–Haber cycle is the application of **Hess's law** to the enthalpy of formation of an ionic solid at 298 K. Hess's law states that the enthalpy of a reaction is the same whether the reaction takes place in one or in several steps. A Born–Haber cycle for a metal chloride (MCl) is shown in Figure 1.51; the metal chloride is formed from the constituent elements in their standard state in the equation at the bottom and by the clockwise series of steps above. From Hess's law, the sum of the enthalpy changes for each step around the cycle can be equated with the standard enthalpy of formation, and we get

$$\Delta H_{\mathrm{f}}^{\ominus}(\mathrm{MCl,s}) = \Delta H_{\mathrm{atm}}^{\ominus}(\mathrm{M,s}) + I_1(\mathrm{M}) + \frac{1}{2}D(\mathrm{Cl-Cl}) - E(\mathrm{Cl}) + L(\mathrm{MCl,s}). \tag{1.2}$$

The terms in the Born–Haber cycle are defined in Table 1.14 together with some sample data. By rearranging Equation 1.2, we can write an expression for the lattice energy in terms of the other quantities, which can then be calculated if the values for these are known. Notice that the way in which we have defined lattice energy gives **negative** (exothermic) values; you may find Equation 1.1 written the other way round in some texts, in which case they will quote positive lattice energies. Notice also that electron affinity is defined as the heat evolved when an electron is added to an atom; as an enthalpy change refers to the heat absorbed, the electron affinity and the enthalpy change for that process have opposite signs (Figure 1.50).

Cycles such as this can be constructed for other compounds such as oxides, MO, sulfides, MS, higher-valent metal halides, MX_n, etc. The difficulty in these cycles usually comes in the determination of values for the electron affinity (*E*). In the case of an oxide, it is necessary to know the double electron affinity for oxygen (the negative of the enthalpy change of the following reaction):

$$2\mathrm{e^{-}}(\mathrm{g}) + \mathrm{O}(\mathrm{g}) = \mathrm{O^{2-}}(\mathrm{g}) \tag{1.3}$$

TABLE 1.14
Terms in the Born–Haber Cycle

Term	Definition of the reaction to which the term applies	NaCl/kJ mol^{-1}	AgCl/kJ mol^{-1}
$\Delta H^{\ominus}_{atm}(M)$	Standard enthalpy of atomization of metal M: M(s) = M(g)	107.8	284.6
$I_1(M)$	First ionization energy of metal M: $M(g) = M^+(g) + e^-(g)$	494	732
$\frac{1}{2}D(Cl-Cl) - E(Cl)$	Half the dissociation energy of $Cl_2 : \frac{1}{2}Cl_2(g) = Cl(g)$	122	122
$-E(Cl)$	Minus the electron affinity of chlorine is the enthalpy change for: $Cl(g) + e^-(g) = Cl^-(g)$	−349	−349
$L(MCl,s)$	Lattice energy of MCl(s): $M^+(g) + Cl^-(g) = MCl(s)$		
$\Delta H^{\ominus}_f(MCl, s)$	Standard enthalpy of formation of MCl(s): $M(s) + \frac{1}{2}Cl_2(g) = MCl(s)$	−411.1	−127.1

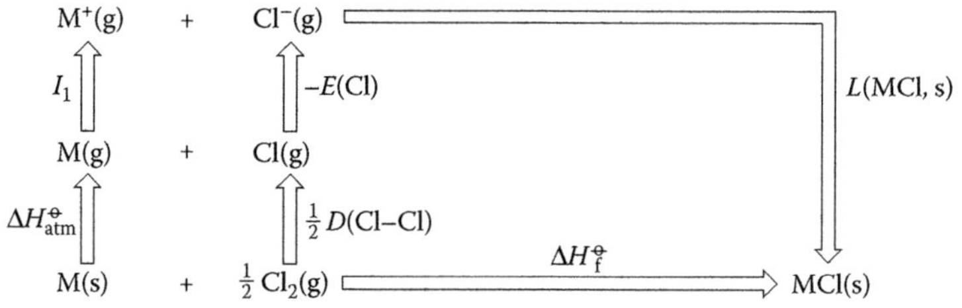

FIGURE 1.50 The Born–Haber cycle for a metal chloride, MCl.

This can be broken down into two stages:

$$e^-(g) + O(g) = O^-(g) \tag{1.4}$$

and

$$e^-(g) + O^-(g) = O^{2-}(g). \tag{1.5}$$

Proton affinities can be found in a similar way: proton affinity is defined as the enthalpy change of the reaction shown in Equation 1.6, where a proton is lost by the species, A:

$$AH^+(g) = A(g) + H^+(g) \tag{1.6}$$

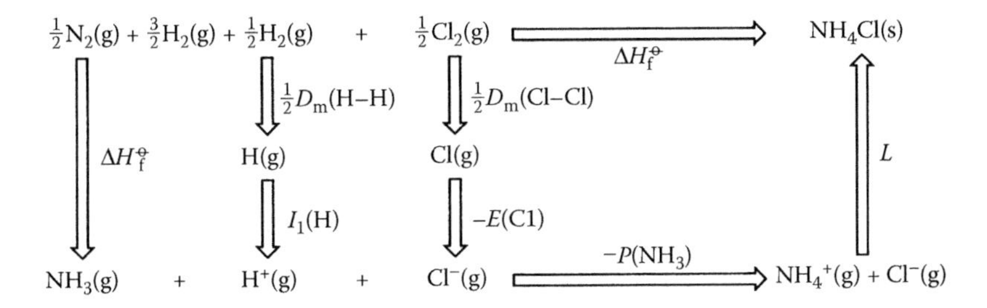

FIGURE 1.51 Thermochemical cycle for the calculation of the proton affinity of ammonia.

Again, this value can be obtained from a suitable thermochemical cycle, provided the lattice energy is known. Take as an example the formation of the ammonium ion, $NH_4^+(g)$:

$$NH_3(g) + H^+(g) = NH_4^+(g) \tag{1.7}$$

The enthalpy change of the reaction in Equation 1.7 is minus the proton affinity of ammonia, $-P(NH_3,g)$. This could be calculated from the thermochemical cycle shown in Figure 1.51, provided the lattice energy of ammonium chloride is known.

Interestingly, it was arguments and calculations of this sort that led Neil Bartlett to the discovery of the first noble gas compound, $XePtF_6$. Bartlett had prepared a new complex, O_2PtF_6, which, by analogy with the diffraction pattern of $KPtF_6$, he formulated as containing the dioxygenyl cation, $\left[O_2^+\right]\left[PtF_6^-\right]$. He realised that the ionisation energies of oxygen and xenon are very similar and that although the radius of the Xe^+ ion is slightly different, because the PtF_6^- anion is very large, the lattice energy of $\left[Xe^+\right]\left[PtF_6^-\right]$ should be very similar to that of the dioxygenyl complex and, therefore, should exist. Accordingly, he mixed xenon and PtF_6 and obtained the orange-yellow solid of xenon hexafluoroplatinate—the first noble gas compound. (Although, in fact, the compound turned out not to have the structure that Bartlett predicted because at room temperature $XePtF_6$ reacts with another molecule of PtF_6 to give a product containing $[XeF]^+[PtF_6]^-$ and $[PtF_5]^-$.)

It is impossible to determine the enthalpy of reaction for Equation 1.5 experimentally, so this value can only be found if the lattice energy is known—a Catch-22 situation! To overcome problems such as this, methods of calculating (rather than measuring) the lattice energy have been devised and they are described in the next section.

1.9.2 Calculating Lattice Energies

For an ionic crystal of known structure, it should be a simple matter to calculate the energy released on bringing together the ions to form the crystal, using the equations of simple electrostatics. The energy of an ion pair, M^+ and X^- (assuming they are point charges), separated by distance *r*, is given by **Coulomb's law**:

$$E = -\frac{e^2}{4\pi\varepsilon_0 r} \tag{1.8}$$

and where the *magnitudes* of the charges on the ions are Z_+ and Z_-:

$$E = -\frac{Z_+ Z_- e^2}{4\pi\varepsilon_0 r} \tag{1.9}$$

(e is the electronic charge, 1.6×10^{-19} C, and ε_0 is the permittivity of a vacuum, 8.854×10^{-12} F m^{-1}).

The energy due to coulombic interactions in a crystal is calculated for a particular structure by summing all the ion-pair interactions, thereby producing an infinite series. The series includes terms due to the attraction of the opposite charges on cations and anions and also repulsion terms due to cation–cation and anion–anion interactions. Some of these interactions are shown in Figure 1.52 for the NaCl structure: the Na^+ ion at the centre is surrounded by 6 Cl^- ions as next neighbours at a distance r, then by 12 Na^+ cations at distance $\sqrt{2r}$, then by 8 Cl^- anions at distance $\sqrt{3r}$, followed by a further 6 cations at $2r$ and so on.

The coulombic energy of interaction is given by the summation of all these interactions:

$$E_c = -\frac{e^2}{4\pi\varepsilon_0 r}\left(6 - \frac{12}{\sqrt{2}} + \frac{8}{\sqrt{3}} - \frac{6}{2} + \frac{24}{\sqrt{5}} \cdots\right)$$

or

$$E_C = -\frac{e^2}{4\pi\varepsilon_0 r}\left(\frac{6}{\sqrt{1}} - \frac{12}{\sqrt{2}} + \frac{8}{\sqrt{3}} - \frac{6}{\sqrt{4}} + \frac{24}{\sqrt{5}} \cdots\right) \tag{1.10}$$

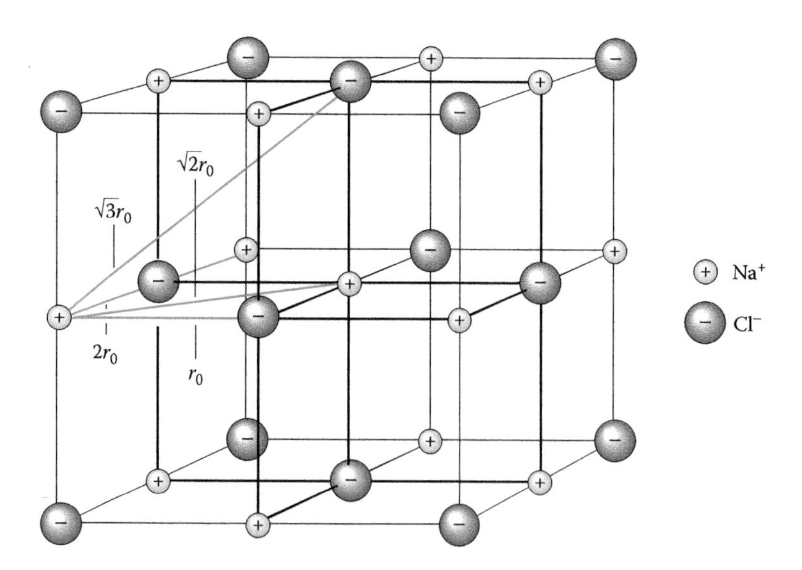

FIGURE 1.52 Sodium chloride structure showing internuclear distances.

TABLE 1.15
Madelung Constants for Some Common Ionic Lattices

Structure	Madelung constant, A	Number of ions in formula unit, v	$\frac{A}{v}$	Coordination
Cesium chloride (CsCl)	1.763	2	0.882	8: 8
Sodium chloride (NaCl)	1.748	2	0.874	6: 6
Fluorite (CaF_2)	2.519	3	0.840	8: 4
Zinc blende (ZnS)	1.638	2	0.819	4: 4
Wurtzite (ZnS)	1.641	2	0.821	4: 4
Corundum (Al_2O_3)	4.172	5	0.835	6: 4
Rutile (TiO_2)	2.408	3	0.803	6: 3

The term inside the brackets is known as the **Madelung constant (*A*)**, in this case, for the NaCl structure. The series is slow to converge; nevertheless, the values of the Madelung constant have been computed, not only for NaCl but also for most of the simple ionic structures. For one mole of NaCl, we can write

$$E_C = -\frac{N_A A e^2}{4\pi\varepsilon_0 r} \tag{1.11}$$

where N_A is the **Avogadro number**, 6.022×10^{23} mol^{-1} (note that the expression is multiplied by N_A and not by $2N_A$, even though there are N_A cations and N_A anions present; this avoids counting every interaction twice!). The value of the Madelung constant is dependent only on the geometry of the lattice and not on its dimensions; values for various structures are given in Table 1.15.

Ions, of course, are not point charges but consist of positively charged nuclei surrounded by electron clouds. At small distances, these electron clouds repel each other, and this too needs to be taken into account when calculating the lattice energy of the crystal. At large distances, the repulsion energy is negligible; however, as the ions approach one another closely, it increases very rapidly. Max Born suggested that the form of this repulsive interaction could be expressed by

$$E_R = \frac{B}{r^n} \tag{1.12}$$

where B is a constant and $\boldsymbol{n}$ (known as the **Born exponent**) is large and also a constant.

We can now write an expression for the lattice energy in terms of the energies of the interactions that we have considered:

$$L = E_C + E_R = -\frac{N_A A Z_+ Z_- e^2}{4\pi\varepsilon_0 r} + \frac{B}{r^n} \tag{1.13}$$

TABLE 1.16

Constants Used to Calculate *n*

Ion Type	Constant
[He]	5
[Ne]	7
[Ar]	9
[Kr]	10
[Xe]	12

The lattice energy will be a minimum when the crystal is at equilibrium, that is, when the internuclear distance is at the equilibrium value of r_0. If we minimise the lattice energy, we get

$$L = \frac{N_A A Z_+ Z_- e^2}{4\pi\varepsilon_0 r_0}\left(1 - \frac{1}{n}\right) \tag{1.14}$$

This is known as the Born–Landé equation: the values of r and n can be obtained from X-ray crystallography and from compressibility measurements, respectively. The other terms in the equation are well-known constants, and when values for these are substituted, we get

$$L / \text{kJ mol}^{-1} = -\frac{1.389 \times 10^5 \, A Z_+ Z_-}{r_0 (\text{pm})}\left(1 - \frac{1}{n}\right) \tag{1.15}$$

If the unit of r_0 is picometere, then the unit of L will be kilojoules per mole.

Pauling showed that the values of n can be approximated with reasonable accuracy for compounds of ions with noble gas configurations, by averaging empirical constants for each ion. The values of these constants are given in Table 1.16; for example, n for rubidium chloride (RbCl) is 9.5 (average of 9 and 10) and for strontium chloride (SrC12) it is 9.33 (average of 9, 9, and 10).

Notice the dramatic effect that the charge on the ions has on the value of the lattice energy. A structure containing one doubly charged ion has a factor of 2 in the equation ($Z_+Z_- = 2$), whereas one containing two doubly charged ions is multiplied by a factor of 4 ($Z_+Z_- = 4$). Structures containing multiply charged ions tend, therefore, to have much larger (numerically) lattice energies.

The Russian chemist A. F. Kapustinskii noted that if the Madelung constants (A) for a number of structures are divided by the number of ions in one formula unit of the structure (v), the resulting values are almost constant (see Table 1.15), varying only between approximately 0.88 and 0.80, and this led to the idea that it would be possible to set up a general lattice energy equation that could be applied to any crystal regardless of its structure. We can now set up a general equation and use the resulting equation to calculate the lattice energy of an unknown structure. First, replace the Madelung constant (A) in the Born–Landé equation (Equation 1.15) with the value from the NaCl structure (0.874v) and r_0 by $(r_+ + r_-)$, where r_+ and r_- are the cation and anion radii for six-coordination, giving

$$L / \text{kJ mol}^{-1} = -\frac{1.214 \times 10^5 \, \nu Z_+ Z_-}{r_+ + r_- (\text{pm})}\left(1 - \frac{1}{n}\right) \tag{1.16}$$

If n is assigned an average value of 9, we arrive at

$$L / \text{kJ mol}^{-1} = -\frac{1.079 \times 10^5 \, \nu Z_+ Z_-}{r_+ + r_- (\text{pm})} \tag{1.17}$$

Equations 1.14 and 1.15 are known as the Kapustinskii equations.

1.9.2.1 Computer Modeling

Rather than simply using formulas like these, we can use a computer package to calculate lattice energies. Modern computer packages for solids use a different method for calculating the Coulomb interaction. The **Ewald method** divides the summation into near and far contributions. For the near contribution, a charge distribution taking the form of a Gaussian function, $\exp[-r^2/a^2]$, where a determines the width or spread of the Gaussian, is added around each point charge. These distributions have a net charge equal and opposite to that of the point charges. Gaussian functions are easy to deal with computationally and a can be chosen so that the charge distribution drops off rapidly. This ensures that beyond the width of the Gaussian, other point charges see a net charge of zero and hence no interaction. This effectively reduces the Coulomb interaction to a short-range interaction between the point charge and the charge distribution. The interaction between the Gaussian charge distributions forms the far contribution. It can be calculated and subtracted from the near contribution.

In these calculations it is common to use an expression due to Buckingham for the repulsive interaction, $E = A\exp\left(\frac{-r}{\rho}\right) - \frac{C}{r^6}$, where r is the distance from the centre of the ion and A, ρ, and C are adjustable parameters. To account for the polarisability of the ions a simple but effective model is used in which the ion is divided into a core and a shell. The shell can expand and contract around the core under the influence of a simple harmonic force, $F = -k(r - r_0)^2$, where r is the radius of the shell, k is a force constant (known as a spring constant), and r_0 is the equilibrium bond length. The smaller the force constant, the easier it is for the shell to alter its size. It is also possible to allow the shell to move off-centre to mimic deformation of the ion. Polyatomic ions can be modeled by assuming the atoms of the ion are covalently bonded to each other. Such methods reproduce the lattice energy with a high degree of accuracy.

Computational methods are frequently used for crystals with many atoms in the unit cell such as zeolites (Chapter 7) and also to calculate the energy of defects in crystals (Chapter 5) and the paths followed by ions moving through a solid (Chapter 6).

1.10 SUMMARY

In this opening chapter, we have introduced many of the principles and ideas that lie behind a discussion of the crystalline solid state.

1. Crystalline solids can be viewed as repeating units, unit cells, forming a lattice.
2. For 3-dimensional lattices there are 7 lattice systems defined by their symmetry which can pack together with no spaces.
3. There are 4 ways of arranging lattice points within a unit cell, the lattice types, primitive, body-centred, face-centred (all faces), and face-centred (a pair of opposite faces).
4. Combining the lattice systems with the possible lattice types for each system gives 14 Bravais lattices.
5. The unit cell can belong to one of 32 crystallographic symmetry point groups. Combining these with the Bravais lattices produces 230 space groups.
6. Crystal structures can be imaged in a number of ways:
   - a 3-d diagram of a unit cell with ions, atoms, molecules on the lattice points.
   - a 2-d projection of the unit cell; packing diagrams.
   - a close-packed array of hard spheres with occupation of tetrahedral and/or octahedral holes where needed.
   - linked octahedra or tetraheda.
7. Taking the approximation of ionic crystals as arrays of spheres, several methods have been developed to estimate the radii of different ions. While each method will produce a set of radii that will give internuclear distances to a good approximation, the values of individual radii vary and values from different sets cannot be used together.
8. The lattice energy of an ionic crystal is the energy released when the crystal forms from gaseous ions. This is not directly measurable. A value can be obtained from a thermodynamic cycle (Born–Haber cycle) when measured values of the other quantities in the cycle are available. It is also possible to calculate lattice energies using one of several approximate formulas that have been developed or a computer program designed for this purpose.
9. Covalent bonding is present in many structures and can give rise to lower-than-expected coordination numbers and/or layer structures.
10. When only covalent bonding is present, we tend to see a rather different type of crystal structure as there are interactions that act in particular directions only.

QUESTIONS

1. Does the CF_4 molecule in Figure 1.7 possess a centre of inversion? What other rotation axis is coincident with the inversion axis, $\bar{4}$?

2. The unit cell of $Sr_2Mn_2O_5F$ was predicted to have parameters a = 383.0 pm, b = 381.6 pm, c = 816.6 pm, $\alpha = \beta = \gamma = 90°$. Which lattice system does the unit cell belong to?
3. Chromium(III) oxide crystallizes with a hexagonal unit cell. What are the angles, α, β, and γ of its unit cell?
4. How many formula units, CaF_2, are there in the fluorite unit cell shown in Figure 1.38c?
5. Assign a space group label to the structure of the ideal perovskite, Figure 1.44.
6. Using Figure 1.30, 1.36, 1.44, and 1.42c together with models if necessary, draw packing diagrams for (a) NaCl, (b) ZnS (zinc blende), (c) perovskite, and (d) ReO_3.
7. Figure 1.44 shows an A-type perovskite unit cell (that is, one with A (in ABO_3) at the centre). A unit cell with B at the centre is equally valid and is known as a B-type perovskite cell (Figure 1.53). Draw a packing diagram for the B-type cell.
8. Calculate the packing efficiency of the simple cubic primitive cell. Assume the atoms in this cell are hard spheres that just touch.
9. Index the sets of lines in Figure 1.27 marked B, C, D, and E.
10. Index the sets of planes in Figure 1.28 b–d.
11. Estimate a value for the radius of the iodide ion. The distance between the lithium and iodine nuclei in lithium iodide is 300 pm.
12. Calculate a radius for F^- from the data in Table 1.9 for NaI and NaF. Repeat the calculation using RbI and RbF.
13. Using simple geometric arguments show that the limiting radius for an ion in cubic (8-fold) coordination is 0.732*r*.
14. Calculate the value of the Madelung constant for the structure in Figure 1.54. All bond lengths are equal and all bond angles are 90°. Assume that there

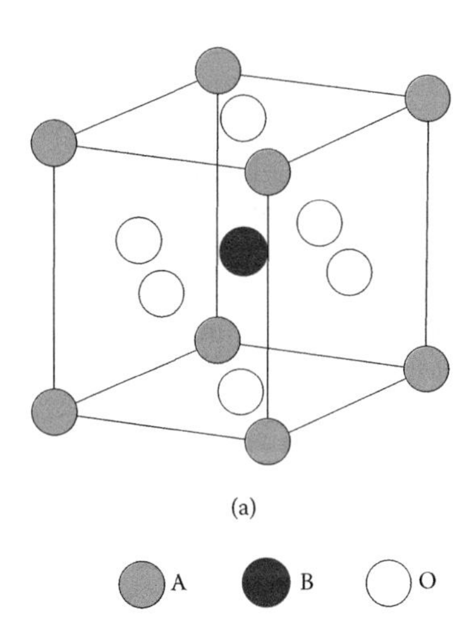

FIGURE 1.53 B-type perovskite cell.

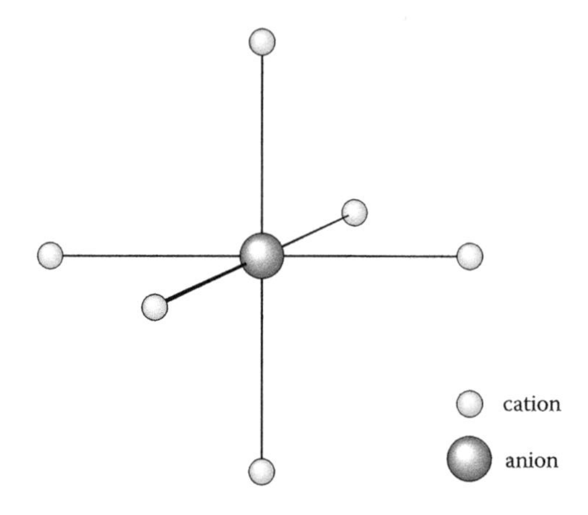

FIGURE 1.54 Structure for Question 14. All bond lengths are equal and bond angles are 90°.

TABLE 1.17
Values of the Born–Haber Cycle Terms for KCl/kJ mol^{-1}

Term	Value
$\Delta H^{\ominus}_{atm}$	89.1
I_1	418
$\frac{1}{2}D(\text{Cl}-\text{Cl})-$	122
–E(Cl)	–349
$\Delta H^{\ominus}_{f}(\text{KCl, s})$	–436.7

are no ions other than those shown in Figure 1.54 and that the charges on the cation and anion are +1 and –1, respectively.

15. Calculate a value for the lattice energy of potassium chloride using Equation 1.15. Compare this with the value that you calculate from the thermodynamic data in Table 1.17.
16. Calculate a value for the electron affinity of oxygen for two electrons. Take magnesium oxide (MgO) as a model and devise a suitable cycle. Use the data given in Table 1.18.
17. Calculate a value for the proton affinity of ammonia using the cycle in Figure 1.51 and the data in Table 1.19.

TABLE 1.18
Values of the Born–Haber Cycle Terms for MgO/kJ mol^{-1}

Term	Value
$\Delta H^{\ominus}_{atm}(Mg, s)$	147.7
$I_1(Mg)$	736
$I_2(Mg)$	1452
$\frac{1}{2}D(O-O)$	249
$\Delta H^{\ominus}_{f}(MgO, s)$	–601.7
$E(O)$	141

TABLE 1.19
Values of the Born–Haber Cycle Terms for NH_4Cl/kJ mol^{-1}

Term	Value
$\Delta H^{\ominus}_{f}(NH_3)$	–46.0
$\Delta H^{\ominus}_{f}(NH_4Cl_{,s})$	–314.4
$\frac{1}{2}D(H-H)$	218
$\frac{1}{2}D(Cl-Cl)$	122
$I(H)$	1314
$E(Cl)$	349
$r_+(NH_4^+)$	151 pm

2 Physical Methods for Characterizing Solids

Liana Vella-Zarb

2.1 INTRODUCTION

There is a vast array of physical methods for investigating the structures of solids, each technique with its own strengths and weaknesses—some techniques are able to investigate the local coordination around a particular atom or its electronic properties, whereas others are suited to elucidating the long-range order of the structure. Others investigate just the surface structure. No book could do justice to all the techniques on offer, so here we describe just some of the more commonly available or recently featured techniques, and try to show what information can be gleaned from each one, and its limitations. We start with X-ray diffraction by powders and single crystals. Single crystal X-ray diffraction is used to determine atomic positions precisely and therefore the bond lengths (to a few tens of picometres*) and bond angles of molecules within the unit cell. It gives an overall, average picture of a long-range ordered structure, but is less suited to giving information on the structural positions of defects, dopants, and non-stoichiometric regions. It is often very difficult to grow single crystals, but most solids can be made as a crystalline powder. Powder X-ray diffraction is probably the most commonly employed technique in solid-state inorganic chemistry and has many uses from analysis and assessing phase purity, to determining structure. Single crystal X-ray diffraction techniques have provided us with the structures upon which the interpretation of most powder data is based. Structure determination using X-ray techniques is not only suited for minerals or inorganic solids. The vast improvements in modern X-ray sources (rotating anodes, microfocus tubes, synchrotron radiation) and detectors have also enabled the collection of sufficiently good data from very small crystals, and have extended their application to compounds having very large molecules. Thus, structure determination of macromolecules such as proteins and viruses is possible. Following the sections on diffraction techniques, Sections 2.7–2.14 cover other commonly-used techniques for characterising solids. Examples of such established techniques include thermal analysis or electron microscopy, but we will also briefly discuss more recent techniques which experienced significant technical improvement over the past years, e.g.

* Different physical techniques commonly use a variety of units, often nanometres, picometres or Ångstroms, Å (1 Å = 0.1 nm or 100 pm). Although we have tried to use SI units as far as possible throughout this book, you will find Ångstroms also used, particularly in this chapter, where it is common in the literature of the subject.

cryo-electron microscopy (for which Dubochet, Frank, and Henderson were awarded the Nobel Prize in Chemistry in 2017) or submicron X-ray microscopy/tomography.

2.2 X-RAY DIFFRACTION

2.2.1 Generation of X-Rays

The discovery of X-rays was made by a German physicist, Wilhelm Röntgen, in 1895, for which he was awarded the first Nobel Prize in Physics in 1901. The benefits of his discovery in terms of medical diagnosis and treatment and in investigating molecular and atomic structure are immeasurable, and yet Röntgen was a man of such integrity that he refused to make any financial gain out of his discovery, believing that scientific research should be made freely available.

An electrically heated filament, usually tungsten, emits electrons, which are accelerated by a high potential difference (20–50 kV) and allowed to strike a metal target or anode which is water cooled (Figure 2.1a). The anode emits a continuous spectrum of 'white' X-radiation but superimposed on this are sharp, intense X-ray peaks (K_α, K_β) as shown in Figure 2.1b. The frequencies of the K_α and K_β lines are characteristic of the anode metal; the target metals most commonly used in X-ray crystallographic studies are copper and molybdenum, which have K_α lines at 154.18 pm and 71.07 pm respectively. These lines occur because the bombarding electrons knock out electrons from the innermost K shell ($n=1$) and this in turn creates vacancies which are filled by electrons descending from the shells above. The decrease in energy appears as radiation; electrons descending from the L shell ($n=2$) give the K_α lines and electrons from the M shell ($n=3$) give the K_β lines. (These lines are actually very closely spaced doublets—$K_{\alpha 1}$, $K_{\alpha 2}$ and $K_{\beta 1}$, $K_{\beta 2}$—which are usually not resolved.) As the atomic number, Z, of the target increases, the lines shift to a shorter wavelength.

Normally, in X-ray diffraction, monochromatic radiation (single wavelength or a very narrow range of wavelengths) is required. Usually the K_α line is selected and the K_β line is filtered out by using a filter made of a thin metal foil of the element adjacent (Z–1) in the periodic table; thus nickel effectively filters out the K_β line of copper, and niobium is used for molybdenum. A drawback to using filters is the very high background (from the white X-ray spectrum) and the fact that the X-ray beam is still not purely monochromatic and unpolarised. A monochromatic beam of higher quality, lower background, and of pure $K_{\alpha 1}$ radiation can be obtained by reflecting the beam from a plane in a single crystal, normally made of graphite, silicon, or germanium (the reasons why this works will become obvious after you have read the next section).

Although the actual X-ray diffraction technique did not change, recent developments in the construction of highly sensitive area detectors now enable time-resolved X-ray diffraction, as well as the possibility to collect data from very small, weakly diffracting samples, such as macromolecules (proteins), and molecular frameworks, amongst others. This led to the requirement of stronger, more focused X-ray beams. Thus, apart from the 'standard' sealed X-ray tube (Figure 2.1), other types of, laboratory X-ray tubes are available, such as rotating anode, microfocus, or metal-jet-anode X-ray tubes. These novel types of X-ray tubes still follow the same principle for the

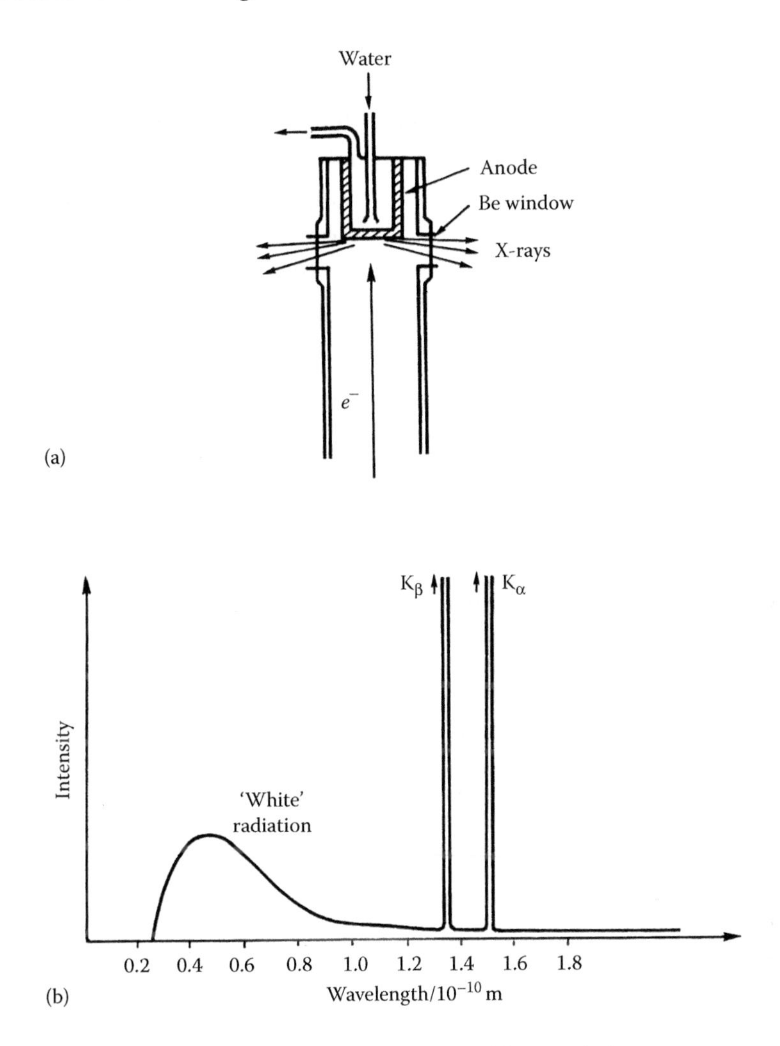

FIGURE 2.1 (a) Section through an X-ray tube. (b) An X-ray emission spectrum.

production of X-rays. The rotating anode and metal-jet anode increase the brightness or brilliance of the X-ray beam, and thus the density of the emitted X-rays. In this manner higher power can be used and a more efficient cooling of the anode is possible by either (a) rotating the anode or (b) using a liquid metal jet, e.g. gallium. A microfocus tube focuses the X-ray beam to a smaller diameter by different optics, which then results in a brighter source, while using less power. Table 2.1 summarizes the various features of different laboratory X-ray tubes. Compared to a normal X-ray tube as explained above, the brilliance of the X-ray beam can be increased by a factor of 20 when using sealed microfocus tubes and by up to 260 when using the newest-generation metal-jet-anode tubes. Costs and maintenance do increase considerably when using these more advanced sources. Thus, laboratories often still choose the less expensive normal sealed X-ray tube.

TABLE 2.1
Approximate X-Ray Beam Brilliance for the Main Types of In-House Laboratory Sources (from *Acta Crystallogr. D* 69, 1283–1288)

System	Power (W)	Actual spot on anode (μm)	Brilliance (photons s^{-1} mm^{-2} $mrad^{-1}$)
Standard sealed tube	2000	1000×1000	0.1×10^9
Standard rotating-anode	3000	3000×300	0.6×10^9
Microfocus sealed tube	50	150×30	2.0×10^9
Metal-jet	200	20×20	26×10^9

If X-rays of the highest brilliance and purity are needed, this cannot be achieved in a standard laboratory setting, and a different X-ray source is required: the radiation emitted in the storage ring of a particle accelerator, also called synchrotron radiation. Synchrotrons are particle accelerators in which electrons are forced into a circular pathway at high speed. The electrons emit a spectrum of intense polychromatic (white) radiation ranging from soft UV to hard X-rays over a very tightly defined angle tangential to the ring. By the combination of the correct filters, monochromators, and mirrors, radiation of a precise wavelength can be used for different analytical methods, amongst them X-ray diffraction. Synchrotron radiation offers a brilliance which can reach a factor of 10^{12} compared to a normal sealed laboratory X-ray tube. More details about synchrotrons can be found in Section 2.10.1.

2.2.2 Diffraction of X-Rays

By 1912, the nature of X-rays—whether they were particles or waves—was still unresolved; a demonstration of X-ray diffraction effects was needed to prove their wave nature. This was eventually achieved by Max von Laue using a crystal of copper sulfate as the **diffraction grating**, work which earned him the Nobel Prize for Physics in 1914. Crystalline solids consist of regular arrays of atoms, ions, or molecules with interatomic spacings of the order of 100 pm. For diffraction to take place, the wavelength of the incident light has to be of the same order of magnitude as the spacings of the grating. Because of the periodic nature of the internal structure, it is possible for crystals to act as a three-dimensional diffraction grating to light of a suitable wavelength: a Laue photograph is shown in Figure 2.2.

This discovery was immediately noted by W. H. and W. L. Bragg (father and son), and they started experiments on using X-ray crystal diffraction as a means of structure determination. In 1913 they first determined the crystal structure of NaCl, and they went on to solve many structures including those of KCl, ZnS, CaF_2, $CaCO_3$, and diamond. W. L. (Lawrence) Bragg noted that X-ray diffraction behaves like 'reflection' from the planes of atoms within the crystal and that only at specific orientations of the crystal with respect to the source and detector are X-rays 'reflected' from the planes. It is not like the reflection of light from a mirror, as this requires that the angle of incidence equals the angle of reflection, and this is possible for all

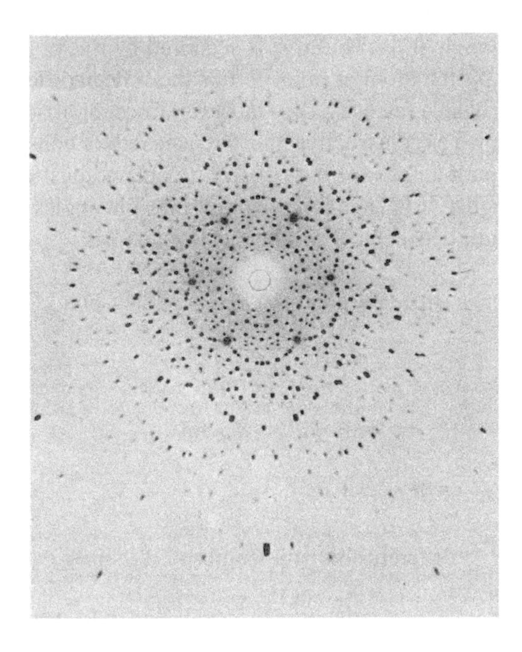

FIGURE 2.2 X-ray diffraction by a crystal of beryl using the Laue method. (From Moore, W. J., *Physical Chemistry*, 5th ed. 1972. Reprinted with courtesy of Eastman Kodak.)

angles. With X-ray diffraction, reflection only occurs when the conditions for **constructive interference** are fulfilled.

Figure 2.3 illustrates the **Bragg condition** for the reflection of X-rays by a crystal. The array of black points in the diagram represents a section through a crystal and the lines joining the dots mark a set of parallel planes with Miller indices *hkl* and **interplanar spacing** d_{hkl}. A parallel beam of monochromatic X-rays ADI is incident

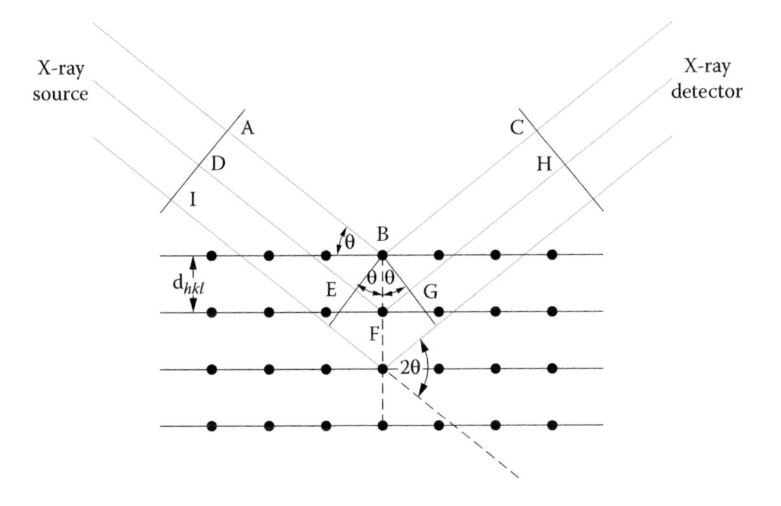

FIGURE 2.3 Bragg reflection from a set of crystal planes with a spacing d_{hkl}.

to the planes at an angle θ_{hkl}. The ray A is scattered by the atom at point B and the ray D is scattered by the atom at point F. For the reflected beams to emerge as a single beam of reasonable intensity, they must reinforce, or arrive in phase with, one another. This is known as constructive interference, and for constructive interference to take place, the path lengths of the interfering beams must differ by an integral number of wavelengths. If BE and BG are drawn at right angles to the beam, the difference in path length between the two beams is given by:

$$\text{difference in path length} = \text{EF} + \text{FG}$$

but,

$$\text{EF} = \text{FG} = d_{hkl} \sin \theta_{hkl}$$

so,

$$\text{difference in path length} = 2d_{hkl} \sin \theta_{hkl} \tag{2.1}$$

This must be equal to an integral number, n, of wavelengths. If the wavelength of the X-rays is θ, then

$$n\lambda = 2d_{hkl} \sin \theta_{hkl} \tag{2.2}$$

This is known as the **Bragg equation**, and it relates the spacing between the crystal planes, d_{hkl}, to the particular Bragg angle, θ_{hkl}, at which reflections from these planes are observed (mostly the subscript hkl is dropped from the Bragg angle θ without any ambiguity as the angle is unique for each set of planes).

When $n=1$, the reflections are called first order, and when $n=2$ the reflections are second order, and so on. However, the Bragg equation for a second-order reflection from a set of planes hkl is

$$2\lambda = 2d_{hkl} \sin \theta$$

which can be rewritten as

$$\lambda = 2\frac{d_{hkl}}{2} \sin \theta \tag{2.3}$$

Equation 2.3 represents a first-order reflection from a set of planes with interplanar spacing $d_{hkl}/2$. The set of planes with interplanar spacing $d_{hkl}/2$ has Miller indices $2h\ 2k\ 2l$. So, the second-order reflection from hkl is indistinguishable from the first-order reflection from $2h\ 2k\ 2l$, and the Bragg equation may be written more simply as

$$\lambda = 2d_{hkl} \sin \theta \tag{2.4}$$

2.3 SINGLE CRYSTAL X-RAY DIFFRACTION

From a single crystal it is possible to measure the position and intensity of the *hkl* reflections accurately and from this data determine not only the unit cell dimensions and space group, but also the precise atomic positions in three-dimensional space. In most cases this can be done with speed and accuracy, and it is one of the most powerful structural techniques available to a chemist.

2.3.1 THE IMPORTANCE OF INTENSITIES

So far we have only discussed the effects of using crystals as three-dimensional diffraction gratings for X-rays. But you may have wondered why one goes to all this trouble. If we want to magnify an object to see its structure in more detail then why not use a lens system as in a microscope or a camera? Here a lens system focuses the light which is scattered from an object (which if left alone would form a diffraction pattern) and forms an image. Why not use a lens to focus X-rays and avoid all the complications? The problem is that there is no suitable way in which X-rays can be focused, and so the effect of the lens has to be simulated by a mathematical calculation on the information contained in the diffracted beams. Much of this information is contained in the intensity of each beam, but as always, there is a snag! The recording methods do not record all of the information in the beam because they only record the intensity and are insensitive to the phase. Intensity is proportional to the square of the amplitude of the wave, and the phase information is lost. Unfortunately it is this information which derives from the atomic positions in a structure. When light is focused by a lens, this information is retained.

So far we have seen that if we measure the Bragg angle of the reflections and successfully assign them to the correct set of lattice planes (Miller indices), then we get information on the size of the unit cell (*via* the Bragg equation 2.4) and, if it possesses any translational symmetry elements, also on the symmetry. The assignment of the correct indices/lattice planes for each reflection observed in the pattern and thus, the unit cell parameters of the crystal structure, is called indexing and will be discussed in Section 2.4. In addition we have seen that the intensity of each reflection is different and this too can be measured. In early photographic work, the relative intensities of the spots on the film were assessed by eye with reference to a standard, and later a scanning microdensitometer was used. In modern diffractometers intensities are no longer collected individually (by orienting the crystal subsequently in all possible positions which fulfil the Bragg equation) with a scintillation counter. Instead, the diffracted beam is intercepted by a two-dimensional detector, usually either a CCD (charge-coupled device), an image plate (replacing a photographic film) or a hybrid pixel detector. The big advantage of two-dimensional detectors is the simple fact that a great number of intensities can be captured at once. It is a true, electronic, alternative to the photographic film. In all cases the Bragg angle, intensity, and profile (standard deviation) of each intensity (reflection) can be recorded electronically. The variety and pros and cons of the currently available two-dimensional detectors lie outside the scope of this book, but for the interested reader some books and reviews are suggested in the Further Reading section for this chapter at the end of this book.

The interaction which takes place between X-rays and a crystal involves the electrons in the crystal: the more electrons an atom possesses, the more strongly it will scatter the X-rays. The effectiveness of an atom in scattering X-rays is called the **scattering factor** (or **form factor**), and is given the symbol f_0. The scattering factor depends not only on the atomic number, but also on the Bragg angle θ and the wavelength of the X-radiation: as the Bragg angle increases the scattering power drops. The decrease in scattering power with angle is due to the finite size of an atom; the electrons are distributed around the nucleus and as θ increases, the X-rays scattered by an electron in one part of the atom are increasingly out of phase with those scattered in a different part of the electron cloud (see Figure 2.14a).

Why are intensities important? A simple example demonstrates this clearly. We know that the heavier an atom is, the better it is at scattering X-rays. On the face of it, we might think that the planes containing the heavier atoms will give the more intense reflections. While this is true, the overall picture is more complicated than that because there are interactions with the reflected beams from other planes to take account of, and these may produce destructive interference. Consider the diffraction patterns produced by NaCl and KCl crystals, both of which have the same structure (Figure 2.4). The structure can be thought of as two interlocking *ccp* arrays of Na^+ and Cl^- ions. The unit cell shown in the figure has close-packed layers of Cl^- ions which lie parallel to a body diagonal, with indices *111*. Lying exactly halfway in between the *111* layers, and parallel to them, are close-packed layers of Na^+ ions; this means that a reflection from the Cl^- close-packed layers is exactly out of phase with that from the equivalent Na^+ layers. Because a chloride ion has 18 electrons it scatters the X-rays more strongly than a sodium ion with 10 electrons, the reflections partially cancel, and the intensity of the *111* reflection will be fairly weak. The *222* layers contain the close-packed layers of both Na^+ and Cl^-, and this will be a strong reflection because the reflected waves will now reinforce one another. When we come to look at the equivalent situation in KCl, the reflection from the *111* layers containing K^+ ions is exactly out of phase with the reflection from the Cl^- close-packed layers. But K^+ and Cl^- are isoelectronic, and so their scattering factors for X-rays are virtually

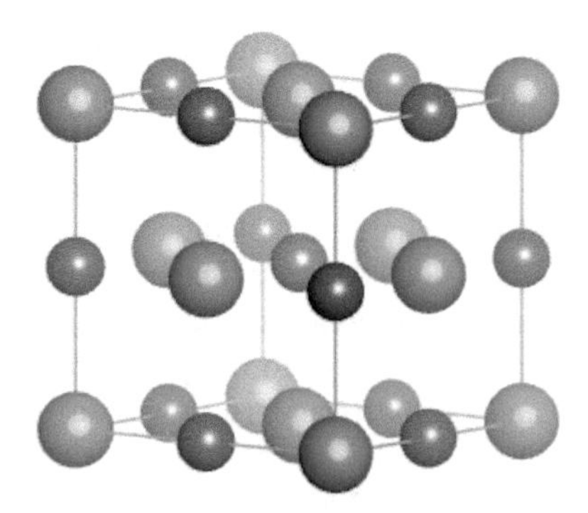

FIGURE 2.4 NaCl unit cell showing the close-packed *111* and *222* planes.

identical and the net effect is that the two reflections cancel out, and the *111* reflection appears to be absent. Similarly this means that the first observed reflection in the diffraction pattern from KCl is the *200* and it would be very easy to mistake this for the 100 reflection from a primitive cubic cell with a unit cell length half that of the real face-centered cell.

The resultant of the waves scattered by all the atoms in the unit cell, in the direction of the *hkl* reflection, is called the **structure factor**, F_{hkl}, and is dependent on both the position of each atom and its scattering factor. It is given by the general expression for *j* atoms in a unit cell:

$$F_{hkl} = \sum_{j} f_j e^{2\pi i\left(hx_j + ky_j + lz_j\right)} \tag{2.5}$$

where f_j is the scattering factor of the *j*th atom and $x_j\, y_j\, z_j$ are its fractional coordinates. Series such as this can also be expressed in terms of sines and cosines, more obviously reflecting the periodic nature of the wave; they are known as **Fourier series**. In a crystal which has a centre of symmetry and *n* unique atoms in the unit cell (the unique set of atoms is known as the **asymmetric unit**) the expression in 2.5 simplifies to

$$F_{hkl} = 2\sum_{n} f_n \cos 2\pi\left(hx_n + ky_n + lz_n\right) \tag{2.6}$$

The electron density distribution within a crystal can be expressed in a similar way as a three-dimensional Fourier series:

$$\rho(x, y, z) = \frac{1}{V}\sum_{h}\sum_{k}\sum_{l} F_{hkl} e - 2\pi i\left(hx + ky + lz\right) \tag{2.7}$$

where $\rho(x,y,z)$ is the electron density at a position $x\ y\ z$ in the unit cell and V is the volume of the unit cell. Notice the similarity between the expressions in equations 2.5 and 2.7. In mathematical terms the electron density is said to be the **Fourier transform** of the structure factors and vice versa. This relationship means that if the structure factors are known, then it is possible to calculate the electron density distribution in the unit cell, and thus the atomic positions.

The intensities of the *hkl* reflections, I_{hkl}, are measured as described above and form the data set for a particular crystal. The intensity of a reflection is proportional to the square of the structure factor:

$$I_{hkl} \propto F_{hkl}^2 \tag{2.8}$$

Taking the square root of the intensity gives a value for the magnitude of the structure factor (mathematically this is known as the modulus of the structure factor denoted by vertical bars either side).

$$\left|F_{hkl}\right| \propto \sqrt{I_{hkl}} \tag{2.9}$$

Because the diffracted intensity is proportional to the square of the modulus of the structure factor, $|F_{hkl}|^2$, the intensities of the *hkl* and $\overline{hkl}$ reflections are equal. This is known as **Friedel's law**, and the *hkl* and reflections as a Friedel pair. The law only holds true when there is no absorption.

Before this information can be used, the data set has to undergo some routine corrections, a process known as **data reduction**. The **Lorentz correction** relates to the geometry of the collection mode; the **polarization correction**, ***p***, allows for the fact that the non-polarized X-ray beam may become partly polarized on reflection, and an **absorption correction** is often applied to data, particularly for inorganic structures, because the heavier atoms absorb some of the X-ray beam, rather than just scattering it. Corrections can also be made for **anomalous dispersion**, which affects the scattering power of an atom when the wavelength of the incident X-ray is close to its absorption edge. These corrections are applied to the scattering factor, f_0, of the atom.

The structure factor (and thus the intensity of a reflection) is dependent on both the position of each atom and its scattering factor. The structure factor can be calculated, therefore, from a knowledge of the types of atoms and their positions using equations such as 2.5 or 2.6. It is a great challenge in X-ray crystallography that we need to be able to do the reverse of this calculation—we have the measured magnitudes of the structure factors, and from them we want to calculate the atomic positions. But there is the snag, which we mentioned earlier, known as **the phase problem**. When we take the square root of the intensity we only obtain the modulus of the structure factor, and so we only know its magnitude and not its sign. The phase information is unfortunately lost and we need it in order to calculate the electron density distribution and thus the atomic positions.

2.3.2 Solving Single Crystal Structures

It would seem to be an unresolvable problem—to calculate the structure factors we need the atomic positions and to find the atomic positions we need both the amplitude and the phase of the resultant waves, and we only have the amplitude. Fortunately, many scientists over the years have worked at finding ways around this problem, and have been extremely successful, to the extent that for many systems the solving of the crystal structure has become a routine and fast procedure.

Single crystal X-ray diffraction data is nowadays collected using a computer-controlled diffractometer which measures the Bragg angle θ and the intensity I for each *hkl* reflection. Modern diffractometers employ two-dimensional detectors (see Section 2.3.1), so that a great number of reflections can be collected and measured at the same time. A full data set, i.e. the intensities of all *hkl* reflections, which can be thousands of reflections, can be accumulated in hours rather than the days or weeks of earlier times when the intensity of each reflection had to be collected individually using a scintillation counter. A full data set is not usually collected, as to reduce collection time, crystallographers only collect a unique set of data which is symmetry related to all the rest—the higher the symmetry of the crystal, the smaller the set needed. It is important, however, to collect all the data, including the Friedel pairs, if the properties of anomalous dispersion are to be exploited in determining the absolute configuration of a structure.

To summarize what we know about a structure:

- The size and shape of the unit cell are determined, usually from short preliminary scanning routines carried out directly on the diffractometer.
- The reflections are indexed, and from the systematic absences the Bravais lattice and the translational symmetry elements of the structure are determined: this information often leads to the correct choice of space group, or narrows the possibilities down to a choice of only a few.
- The intensities of the indexed reflections (or a symmetry-related subset) are measured and stored as a data file.
- Correction factors are applied to the raw intensity data.
- Finally, the square roots of the corrected data are taken to give a set of observed structure factors. These are known as $\boldsymbol{F}_{\mathbf{obs}}$ or $\boldsymbol{F}_{\mathbf{o}}$.
- In order to calculate the electron density distribution in the unit cell, we need to know not only the magnitudes of the structure factors, but also their phase.

Crystal structures are solved by creating a set of trial phases for the structure factors. Until around 10 years ago, basically two main methods existed for doing this. The first is known as the **Patterson** method and since it relies on the presence of at least one (but not many) heavy atoms in the unit cell it is useful for solving many inorganic molecular structures. The second is via **direct methods** and these are best used for structures in which the atoms have similar scattering properties. Direct methods calculate mathematical probabilities for phase values and hence enable the construction of an electron density map of the unit cell; theoreticians have produced packages of accessible computer programs for solving and refining structures.

More recently, so-called **dual space methods** were developed in which alternate (random) phase refinement in real and reciprocal space takes place. Different algorithms were developed (e.g. charge flipping or intrinsic phasing) and these are being used very successfully for almost every type of molecular compound.

Programs which use combinations of the above algorithms and approaches together with the finding of the correct space group at once are now also used very often. This is mainly possible now because of the increased calculation power of multi-core processors in normal PCs and laptops.

Once the atoms in a structure have been located, a calculated set of structure factors, $\boldsymbol{F}_{\mathbf{calc}}$ or $\boldsymbol{F}_{\mathbf{c}}$, is determined for comparison with the F_{obs} magnitudes, and the positions of the atoms are refined using **least-squares methods**, for which standard computer programs are available. In practice, atoms vibrate about their equilibrium positions; this is often called **thermal motion**, although it depends not only on the temperature, but also on the mass of the atom and the strengths of the bonds keeping it in place. The higher the temperature, the greater the amplitude of vibration and the larger the volume over which the electron density is spread, thus causing the scattering power of the atom to fall more quickly. Part of the refinement procedure is to allow the electron density of each atom to refine in a sphere around the nucleus.

Structure determinations usually quote an adjustable parameter known as the **isotropic displacement parameter, *B*** (also called the **isotropic temperature factor**). The electron density of each atom can also be refined within an ellipsoid around its nucleus, when an **anisotropic displacement parameter** correction is applied which has six adjustable parameters.

The residual index, or ***R* factor**, gives a measure of the difference between the observed and calculated structure factors and therefore of how well the structure has refined. It is defined as

$$R = \frac{\sum \left| \left(|F_o| - |F_c| \right) \right|}{\sum |F_o|} \tag{2.10}$$

and is used to give a guide to the correctness and precision of a structure. In general, the lower the R value, the better the structure determination. R values should be used with caution as it is not unknown for structures to have a low R value and still be wrong, although fortunately this does not happen often. There are no hard-and-fast rules for the expected value of R, and interpreting them is very much a matter of experience. It is usually taken as a rule of thumb for small molecule structures that a correct structure for a reasonable quality data set would refine to well below an R of 0.1; anything above should be viewed with some degree of suspicion. That said, most structures nowadays, if collected from good-quality crystals on modern diffractometers, would usually refine to below R 0.05 and often to below R 0.03.

A good structure determination, as well as having a low R value, will also have low standard deviations on both the atomic positions and the bond lengths calculated from these positions. This is probably a more reliable guide to the quality of the refinement.

When a single crystal of a solid can be produced, X-ray diffraction provides an accurate, definitive structure, with bond lengths determined to tenths of a picometre. In recent years, the technique has been transformed from a very slow method reserved only for the most special structures, to a method of almost routine analysis: with modern machines, suites of computer programs and fast computers are used to solve several crystal structures per week.

It is now standard academic practice to deposit final published structures and their data in a crystallographic database, of which there are several: the Cambridge Structural Database, CSD (for small organic and organometallic molecules); the Inorganic Crystal Structure Database, ICSD, the Crystallographic Open Database, COD; CRYSTMET for metals and alloys; the Protein Data Bank, PDB; and the Nucleic Acid Database, NDB. These databases check the deposited structure and its data through their own software to ensure internal consistency.

2.3.3 High-Energy X-Ray Diffraction

Solving the structure of solids that have limited long-range order has long been a problem, for instance in glasses. With the recent upsurge in the preparation of new nanocrystalline materials there is an increased incentive to be able to solve such structures. A nanoparticle is usually considered to have at least one dimension less than 100 nm (nanostructures are discussed in Chapter 11). In solids the limited order on a nano-scale can be due to several reasons: it can be composed of nanoparticles such as buckytubes, or sheets; there may be small nanometre-sized ordered areas within larger structures; mesoporous materials have nanometre-size holes (Chapter 7); and nanocomposites are composed of intimate mixtures of nanoparticles. Later sections in this chapter deal with electron microscopy and scanning probe techniques which map the surface layers of materials, but determining the short-range 3-D structure in such materials has presented a real challenge that has only recently begun to be solved.

Scientists have turned to X-ray diffraction using synchrotron sources or neutron diffraction (see next section). Synchrotrons produce monochromatic beams of very-high-energy (and thus short-wavelength) X-rays (see Sections 2.2 and 2.9 for more detail) which collect over smaller scattering angles and give higher resolution.

As we have seen, the planes of atoms in a perfectly ordered crystal reflect an X-ray beam giving a sharp spot—the Bragg reflection. In a solid that has only a small volume of ordered structure, the Bragg reflections will be fewer in number, weaker, and much less sharp, and will be superimposed on a diffuse background from the scattering of the rest of the structure.

High-energy X-ray diffraction experimental data are collected in much the same way as for powder diffraction (see Section 2.4), but usually over longer periods of time, and other factors which contribute to the intensity must be corrected for, such as background scatter, polarization, absorption, and the Compton effect (ionization of an electron from the sample due to inelastic scattering).

Using a technique called **atomic pair distribution function, PDF**, it is possible to analyse such patterns, and determine interatomic distances and coordination numbers within the ordered areas. PDF is based on all the atomic scattering, both sharp and diffuse, and calculates the probability of finding atomic pairs at a particular distance. As this does not identify particular atoms, computer models are then fitted to the data to determine the structure.

2.4 POWDER DIFFRACTION

2.4.1 Powder Diffraction Patterns

A finely ground crystalline powder contains a very large number of small crystals, known as crystallites, which are oriented randomly to one another. If such a sample is placed in the path of a monochromatic X-ray beam, diffraction will occur from planes in those crystallites which happen to be oriented at the correct angle to fulfil the Bragg condition. The diffracted beams make an angle of 2θ with the incident beam. Because the crystallites can be oriented in all directions while still maintaining the Bragg condition, the reflections lie on the surface of cones whose semi-apex angles are equal to the deflection angle 2θ (Figure 2.5a). In the **Debye–Scherrer** photographic method, a strip of film was wrapped around the inside of an X-ray camera (Figure 2.5b) with a hole to allow the collimated incident beam through and a

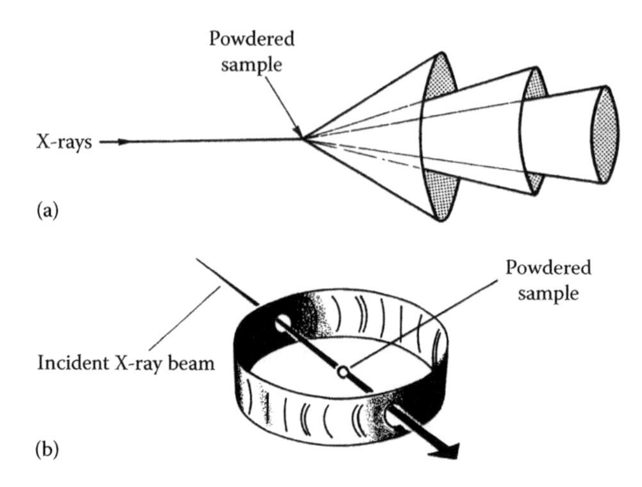

FIGURE 2.5 (a) Cones produced by a powder diffraction experiment. (b) Experimental arrangement for a Debye–Scherrer photograph.

beam-stop to absorb the undiffracted beam. The sample was rotated to bring as many planes as possible into the range that satisfies the diffraction condition, and the cones were recorded as arcs on the film. Using the radius of the camera and the distance along the film from the centre, the Bragg angle 2θ, and thus the d_{hkl} spacing for each reflection could be calculated.

Even though the geometry of this setup remained basically unchanged, the collection of powder diffraction patterns is now almost always performed by automatic diffractometers (Figure 2.6a), using a scintillation, silicon strip, CCD detector, or an imaging plate to record the angle and the intensity of the diffracted beams, which are plotted as intensity against 2θ (Figure 2.6b).

2D powder diffraction can provide very useful information about the crystallinity, size distribution, and preferred orientation of particles in a powder. For example, plate- or needle-like crystals will not orient themselves completely randomly, to different intensities of certain reflections dependent on how the sample is oriented. Apart from Debye–Scherrer geometry, where reflections are collected by transmitting the X-ray beam through the sample (e.g. inside a spinning capillary), diffraction patterns can also be collected by reflecting the X-ray beam on to the sample (Bragg–Brentano Geometry). The data, both position and intensity, are readily measured and stored on a computer for analysis.

The difficulty in the powder method lies in deciding which planes are responsible for each reflection; this is known as 'indexing the reflections', i.e. assigning the correct *hkl* index to each reflection. Although this is often possible for simple compounds in high-symmetry systems (as explained in the following sections), it is quite challenging to do for many larger and/or less symmetrical systems.

2.4.2 Absences Due to Lattice Centring

First, consider a primitive cubic system. From Equation 2.11, we see that the planes giving rise to the reflection with the smallest Bragg angle will have the largest d_{hkl}

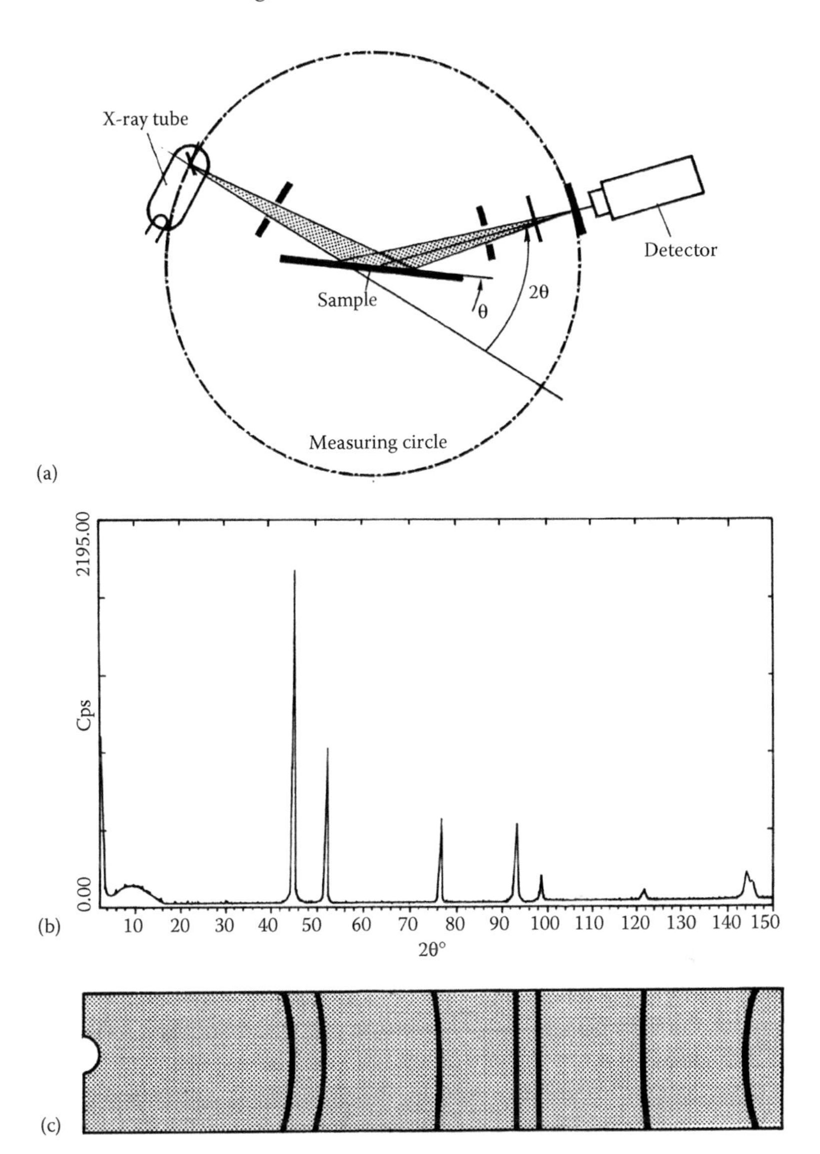

FIGURE 2.6 (a) Diagram of a powder diffractometer. (b) A powder diffraction pattern for Ni powder compared with (c) the Debye–Scherrer photograph of Ni powder.

spacing. In a primitive cubic system the *100* planes have the largest separation and thus give rise to this reflection, and as $a=b=c$ in a cubic system, the *010* and the *001* also reflect at this position. For the cubic system with a unit cell dimension a, the spacing of the reflecting planes is given by Equation 2.11

$$d_{hkl} = \frac{a}{\sqrt{\left(h^2 + k^2 + l^2\right)}} \tag{2.11}$$

Combining this with the Bragg equation (Equation 2.4) gives

$$\lambda = \frac{2a \sin \theta_{hkl}}{\sqrt{\left(h^2 + k^2 + l^2\right)}}$$

and rearranging gives

$$\sin^2 \theta_{hkl} = \frac{\lambda^2}{4a^2}\left(h^2 + k^2 + l^2\right) \tag{2.12}$$

For the primitive cubic class all integral values of the indices h, k, and l are possible. Table 2.2 shows the values of hkl in order of increasing value of $(h^2+k^2+l^2)$ and therefore of increasing sin θ values.

One value in the sequence, 7, is missing: this is because there are no possible integral values for $(h^2+k^2+l^2)=7$. There are also other higher missing values where $(h^2+k^2+l^2)$ cannot be an integer, 15, 23, 28, etc., but note that this is only an arithmetical phenomenon and is nothing to do with the structure.

Taking Equation 2.12, if we plot the intensity of diffraction of the powder pattern of a primitive cubic system against $\sin^2\theta_{hkl}$ we would get six equi-spaced lines with the 7th, 15th, 23rd, etc., missing. Consequently it is easy to identify a primitive cubic system and by inspection to assign indices to each of the reflections.

The cubic unit cell dimension a can be determined from any of the indexed reflections using Equation 2.12. The experimental error in measuring the Bragg angle is constant for all angles, so to minimize error, either the reflection with the largest Bragg angle is chosen, or more usually, a least-squares refinement to all the data is used.

The pattern of observed lines for the two other cubic crystal systems, body-centred and face-centred, is rather different from that of the primitive system. The differences arise because the centring leads to destructive interference for some reflections and these extra missing reflections are known as **systematic absences**.

Consider the *200* planes shown shaded in the F face-centred cubic unit cells shown in Figure 2.7; if a is the cell dimension, they have a spacing $a/2$. Figure 2.8 shows the reflections from four consecutive planes in this structure. The reflection from the *200* planes is exactly out of phase with the *100* reflection. Throughout the crystal there are equal numbers of the two types of planes with the result that complete destructive interference occurs and no *100* reflection is observed. Examining reflections from all the planes for the F face-centered system in this way, we find that in order for a reflection to be observed, the indices must be either all odd or all even.

TABLE 2.2
Values of $(h^2+k^2+l^2)$

Hkl	*100*	*110*	*111*	*200*	*210*	*211*	*220*	*300=221*
$(h^2+k^2+l^2)$	1	2	3	4	5	6	8	9

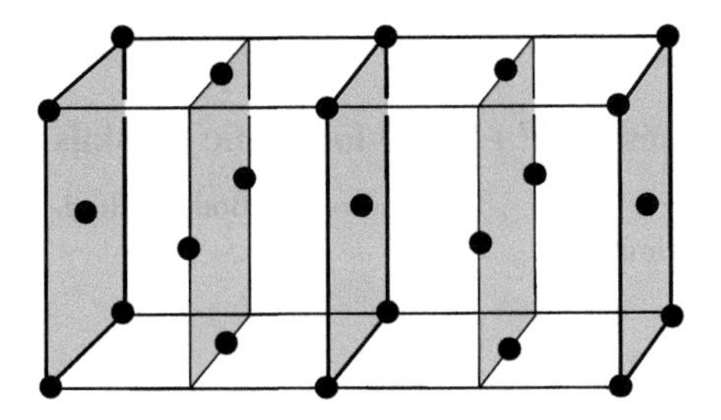

FIGURE 2.7 Two F-centered unit cells showing the *200* planes shaded.

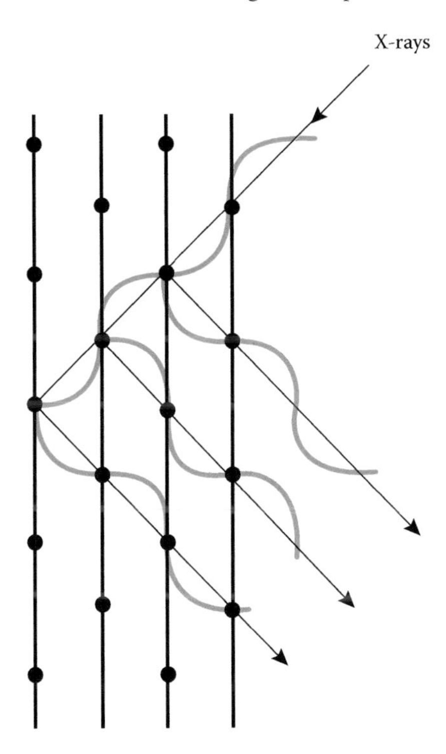

FIGURE 2.8 The *100* reflection from an F-centered cubic lattice.

A similar procedure for the body-centered cubic system finds that for reflections to be observed the sum of the indices must be even.

It is possible to characterize the type of Bravais lattice present by the pattern of systematic absences. Although our discussion has centered on cubic crystals, these absences apply to all crystal systems, not only to cubic, and are summarized in Table 2.4 at the end of the next section. The allowed values of $h^2+k^2+l^2$ are listed in Table 2.3 for each of the cubic lattices.

Using this information and Equation 2.12, we can see that if the observed $\sin^2\theta$ values for a pattern are in the ratio 1: 2: 3: 4: 5: 6: 8, …, then the unit cell is likely to be primitive cubic, and the common factor is $\frac{\lambda^2}{4a^2}$.

TABLE 2.3
Allowed Values of $(h^2+k^2+l^2)$ for Cubic Crystals

Forbidden numbers	Primitive, P	Face-centred, F	Body-centred, I	Corresponding *hkl* values
	1			100
	2		2	110
	3	3		111
	4	4	4	200
	5			210
	6		6	211
7				–
	8	8	8	220
	9			221, 300
	10		10	310
	11	11		311
	12	12	12	222
	13			320
	14		14	321
15				–
	16	16	16	400

A face-centered cubic unit cell can also be recognized: if the first two lines have a common factor, A, then dividing all the observed $\sin^2\theta$ values by A gives a series of numbers, 3, 4, 8, 11, 12, 16, …, and A is equal to $\frac{\lambda^2}{4a^2}$.

A body-centered cubic system gives the values of $\sin^2\theta$ in the ratio 1:2:3:4:5:6:7:8 with the values 7 and 15 apparently not missing, but now the common factor is $\frac{2\lambda^2}{4a^2}$.

2.4.3 Systematic Absences Due to Screw Axes and Glide Planes

The presence of translational symmetry elements in a crystal structure can be detected because they each lead to a set of systematic absences in the *hkl* reflections. Figures 1.10 and 1.11 showed how a two-fold screw (2_1) along z introduces a plane of atoms exactly halfway between the *001* planes: reflections from these planes will destructively interfere with reflections from the *001* planes and the *001* reflection will be absent, as will any reflection for which l is odd. The effect of a glide plane (Figure 1.9) is to introduce a plane of atoms halfway along the unit cell in the direction of the glide. For an a glide perpendicular to b, the *10l* reflection will be absent, and in general the *h0l* reflections will only be present when h is even. Systematic absences are summarized in Table 2.4.

There are now fairly powerful computer programmes for indexing in existence, and powder diffraction patterns can be indexed fairly readily for the high-symmetry

TABLE 2.4

Systematic Absences Due to Translational Symmetry Elements

Symmetry element		Affected reflection	Condition for reflection to be present
Primitive lattice	P	hkl	None
Body-centered lattice	I	hkl	$h+k+l$=even
Face-centered lattice	A	hkl	$k+l$=even
	B		$h+l$=even
	C		$h+k$=even
Face-centered lattice	F		$h\ k\ l$ all odd or all even
2-fold screw, 2_1 along 4-fold screw, 4_2 along 6-fold screw, 6_3 along	a	$h00$	h=even
3-fold screw, 3_1, 3_2 along 6-fold screw, 6_2, 6_4 along	c	$00l$	l divisible by 3
4-fold screw 4_1, 4_3 along	a	$h00$	h divisible by 4
6-fold screw, 6_1, 6_5 along	c	$00l$	l divisible by 6
glide plane perpendicular to	b	$h0l$	
translation $\frac{a}{2}$ (a glide)			h=even
translation $\frac{c}{2}$ (c glide)			l=even
$\frac{b}{2}+\frac{c}{2}$ (n glide)			$h+l$=even
$\frac{b}{4}+\frac{c}{4}$ (d glide)			$h+l$ divisible by 4

crystal classes such as cubic, tetragonal, and hexagonal. For the other systems, the pattern often consists of a large number of overlapping lines, and indexing can be much more difficult.

From the cubic unit cell dimension a, we can calculate the volume of the unit cell, V. If the density, ρ, of the crystals is known, then the mass of the contents of the unit cell, M, can also be calculated

$$\rho = \frac{M}{V} \tag{2.13}$$

From a knowledge of the molecular mass, the number of molecules, Z, in the unit cell can be calculated. Examples of these calculations are in the questions at the end of this chapter.

The density of crystals can be determined by preparing a mixture of liquids (in which the crystals are insoluble!) such that the crystals neither float nor sink: the crystals then have the same density as the liquid. The density of the liquid can be determined in the traditional way using a density bottle.

2.4.4 Uses of Powder X-Ray Diffraction

2.4.4.1 Identification of Unknowns and Phase Purity

Powder diffraction is a very powerful technique that can be quite challenging because as the structures become more complex, the number of reflections increases with decreasing symmetry present in the crystal structure. Overlap becomes a serious problem and it is sometimes difficult to index and measure the intensities of the reflections. It is usefully used as a fingerprint method for detecting the presence of a known compound or phase in a product. This is made possible by the existence of a huge library of powder diffraction patterns which are held and regularly updated by the International Centre for Diffraction Data, ICDD. Other databases exist, e.g. Pauling file project database, some of them are free of charge (e.g. Crystallographic Open Database). These files are available via download. When the powder diffraction pattern of a sample has been recorded along with both the d_{hkl} spacings and intensity of the lines, these can be matched against the patterns of known compounds in the libraries.

The identification of compounds using powder diffraction is useful for qualitative analysis, such as mixtures of small crystals in geological samples. It also gives a rough check of the purity of a sample, determining which compounds are present and in what proportions—but note that powder diffraction does not detect amorphous products or impurities of less than about 5 percent.

Powder diffraction provides information about the crystal structure of a compound: (a) the positions of the reflections that are a result of the parameters and symmetry of the unit cell, and (b) the intensity pattern of the reflections caused by the 3D arrangement of atoms or molecules within the unit cell. Powder patterns can confirm whether two similar compounds, where one metal substitutes for another for instance, have an isomorphous structure or whether a compound with the same stoichiometry or composition has formed a different crystal structure because atoms or molecules are arranged differently (polymorphism).

2.4.4.2 Crystallite Size

As the crystallite size decreases, the width of the diffraction peak increases. To either side of the Bragg angle, the diffracted beam will interfere destructively and we expect to see a sharp peak. However, the destructive interference is the result of the summation of all the diffracted beams, and close to the Bragg angle it takes diffraction from very many planes to produce complete destructive interference. In small crystallites there are not enough planes to produce complete destructive interference, and so we see a broadened peak.

The Debye–Scherrer formula enables the thickness of a crystallite to be calculated from the peak widths:

$$T = \frac{C\lambda}{B\cos\theta} = \frac{C^2}{\left(B_M^2 - B_S^2\right)^{\frac{1}{2}}\cos\theta} \tag{2.14}$$

where T is the crystallite thickness, λ the wavelength of the X-rays (T and λ have the same units), θ the Bragg angle, and B is the full-width at half-maximum, FWHM, of the peak (radians) corrected for instrumental broadening. (B_M and B_S are the FWHMs of the sample and of a standard respectively. A highly crystalline sample with a diffraction peak in a similar position to the sample is chosen and this gives a measure of the broadening due to instrumental effects.)

This method is particularly useful for plate-like crystals with distinctive shear planes (e.g. the *111*), as measuring the peak width of this reflection gives the thickness of the crystallites perpendicular to these planes.

It is a common feature of solid-state reactions that reaction mixtures become more crystalline on heating, as is evidenced by the X-ray diffraction patterns with sharper signals.

2.4.4.3 Following Reactions and Phase Diagrams

Powder X-ray diffraction is also a useful method for following the progress of a solid-state reaction and determining mechanisms, and also for determining phase diagrams (Figure 2.9). By collecting an X-ray pattern at regular intervals as the sample is heated on a special stage in the diffractometer, evolving phases can be seen as new peaks which appear in the pattern, with a corresponding decrease in the peaks due to the starting material(s). Figure 2.10 follows the phase transition of ferrosilicon from the non-stoichiometric alpha phase (Fe_xSi_2, $x=0.77$–0.87) to the stoichiometric beta phase $FeSi_2$, when it is kept at 600°C for a period of time. In Figure 2.11 we see

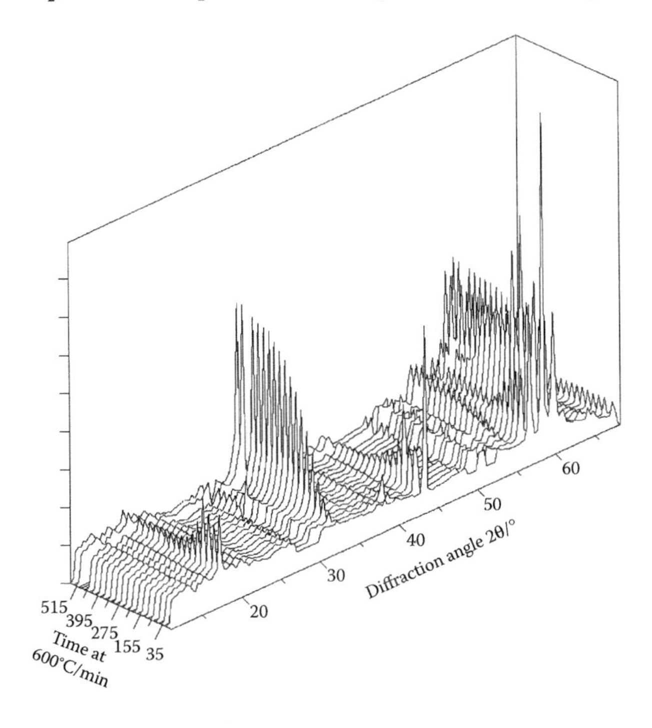

FIGURE 2.9 Powder XRD patterns show the phase changes in ferrosilicon with time, when heated at 600°C. (From Professor Frank J. Berry. With permission.)

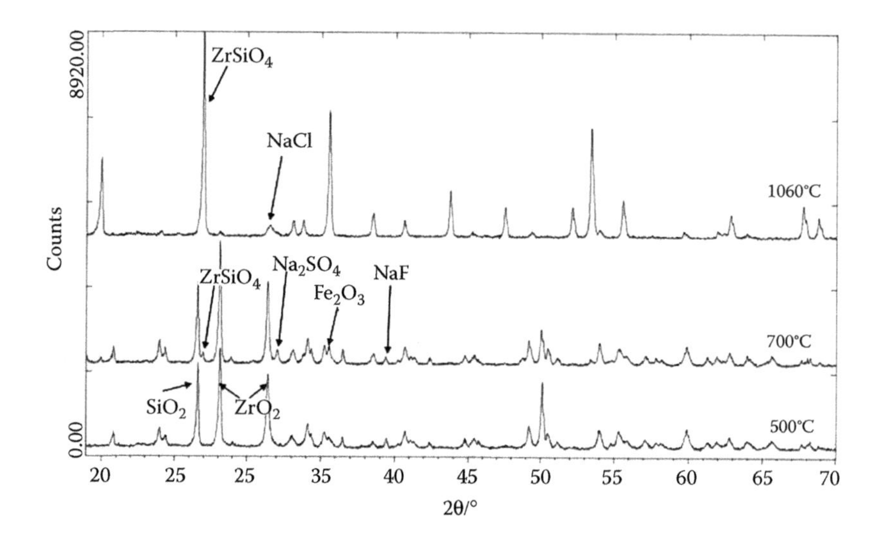

FIGURE 2.10 The phase evolution of iron-doped zircon ($ZrSiO_4$), from zirconia (ZrO_2), silica (SiO_2), ferrous sulfate ($FeSO_4$), and sodium halide mineralisers.

three powder diffraction patterns collected at different temperatures for a solid-state preparation of an Fe-doped zircon from powdered zirconia and silica according to the equation: $ZrO_2 + SiO_2 - ZrSiO_4$. Sodium halides are used to bring the reaction temperature down, and ferrrous sulfate was the source of iron; as the temperature of the reaction mixture is increased, the peaks due to zirconia and silica decrease while those of zircon increase, until at 1060°C, this is the major component. During the reaction, peaks due to intermediates such as Na_2SO_4 and Fe_2O_3 are also seen to evolve.

A careful comparison of the intensities of particular peaks using standards, enable not only the different phases to be identified but also the proportions of different phases to be determined so that a phase diagram can be constructed. In this way it is possible to study samples and phase changes under extreme conditions at high temperatures and pressures.

2.4.4.4 Structure Determination and the Rietveld Method

In a high-symmetry crystal system, there are very few peaks in the powder pattern, and they are often well resolved and well separated. It is then possible to measure their position and intensity with accuracy, and by the methods we described earlier, index the reflections and solve the structure. For larger and less symmetrical structures there are far more reflections which overlap considerably, rendering the process of extraction of intensities, and subsequently crystal structure determination, a little more laborious.

A method known as **Rietveld analysis** has been developed for refining crystal structures from powder diffraction data. The Rietveld method involves an interpretation of not only the line position but also the line intensities, and because there is so much overlap of the reflections in the powder patterns, the method developed by Rietveld involves analyzing the overall line profiles. Rietveld formulated a method of assigning each peak a Voigt profile (basically a convolution between a Gaussian and Lorentz profile) shape and then allowing the signals to overlap so that an overall line profile could be calculated. The method was originally developed for neutron diffraction. The actual structure solution (see Section 2.3.2) can be achieved via various methods or approaches. Whereas about 20 years ago, structure solution from powder data was only achievable if a good structure model from known structures was known, nowadays dual space methods such as charge flipping or simulated annealing provided a quantum leap in the field. Crystal structures of organic and inorganic molecules with molecular weights near 1000 Daltons have been solved and refined successfully. Following the finding of a trial structure a powder diffraction profile is calculated from it and then compared with the measured profiles. The trial structure can then be gradually modified by changing the atomic positions, and refined until a best-fit match with the measured pattern is obtained. The validity of the structure obtained is assessed by an R factor, and also by a difference plot of the two patterns (which should be as close to a flat line as possible). The approach is very similar to one described for structure solution from single crystal data (Section 2.3.2). The Rietveld method tends to work best if a good trial structure is found, or a similar structure is already known, for instance if the unknown structure is a slight modification of a known structure, with perhaps one metal changed for another (Figure 2.11).

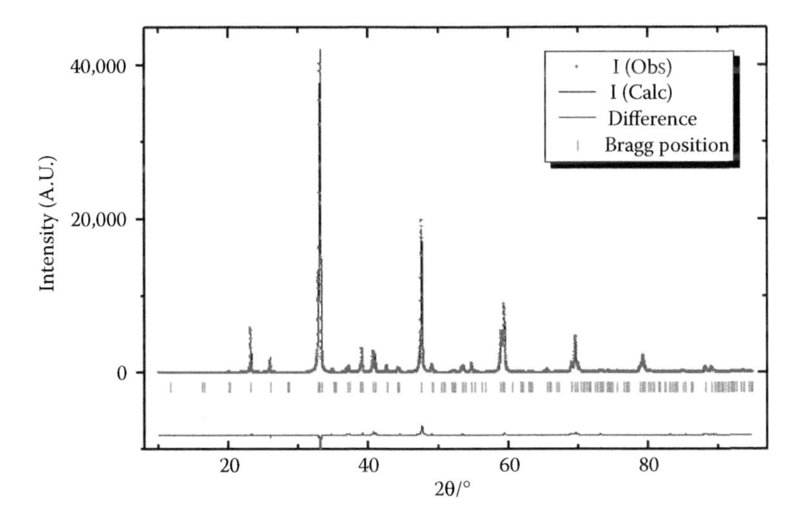

FIGURE 2.11 Rietveld analysis of perovskite with partial substitution of Ti with Ca. (Republished with permission of the Royal Society of Chemistry from Simulated annealing prediction of the crystal structure of ternary inorganic compounds using symmetry restrictions', Reinaudi, L. et al., *J. Am. Chem. Soc., Dalton Transactions,* 4258–4262, 2000. permission conveyed through Copyright Clearance Center, Inc.)

2.5 NEUTRON DIFFRACTION

Neutron diffraction, the elastic scattering of neutrons from a solid, can also be used for crystallographic structural determination. The technique was first developed in 1945 at Oak Ridge by Wollan, and he was soon joined by Schull in 1946. It wasn't until 1994 that Schull received his Nobel Prize for this work (Wollan having died), notably the longest gap between doing the work and getting the award! The vast majority of crystal structures published in the literature have been solved using X-ray diffraction; neutrons are much less commonly used because there are very few sources of neutrons available, whereas X-ray diffractometers can be housed in any laboratory. In addition, because the flux of a monochromatic source of neutrons is small, this necessitates the use of large single crystals and long counting times for the experiment in order to get sufficient intensity. Crystals typically have needed to be at least 1 mm in each direction, and it can be extremely difficult if not impossible to grow such large, perfect crystals, and most experiments are powder diffraction, or as we saw in the preceding section, of nanoparticles. However, high-energy neutron sources are now available, such as those at the Institut Laue-Langevin Grenoble, France, the Rutherford Appleton Laboratory, UK, the Oak Ridge National Laboratory, TN, and the need for these large single crystals in neutron studies is receding. Neutron diffraction does have different properties which can lend certain advantages: it can yield very precise lattice information, determine the position of light atoms, especially hydrogen, and map structural magnetic data.

The de Broglie relationship states that any beam of moving particles will display wave properties according to the formula

$$\lambda = \frac{h}{p} \tag{2.15}$$

where λ is the wavelength, p is the momentum of the particles ($p = mv$, mass × velocity), and h is Planck's constant. Neutrons are released in atomic fission processes from a uranium target, when they have very high velocities and very small wavelengths. The neutrons generated in a nuclear reactor can be slowed using heavy water. This results in a wavelength of about 0.1 nm (1 Å) therefore making them suitable for structural diffraction experiments. The neutrons generated have a range of wavelengths, and a monochromatic beam is formed using reflection from a plane of a single-crystal monochromator at a fixed angle (according to Bragg's law). Structural studies need a high flux of neutrons and this usually means that the only appropriate source is a high-flux nuclear reactor such as at Oak Ridge in the United States and Grenoble in France.

Alternative spallation sources are also available, such as the Rutherford laboratory in the UK, where the neutrons are produced by bombarding metal targets with high-energy protons. The diffraction experiments we have seen so far are set up with X-rays of a single wavelength, λ, so that in order to collect all the diffracted beams, the Bragg angle θ is varied (Bragg equation $n\lambda = 2d\sin\theta$). With the spallation source, the whole moderated beam with all its different wavelengths is used at a fixed angle and the diffraction pattern recorded as the function of the time of flight of the neutrons (if we substitute $v = D/t$ (velocity = distance × time) in the de Broglie relationship we see that the wavelength of the neutrons is proportional to t: $\lambda = \left(\frac{ht}{Dm}\right)$). Because this method uses all of the beam, it has the advantage of greater intensity.

The difference between X-ray and neutron diffraction techniques lies in the scattering process: X-rays are scattered by the electron cloud around the nucleus, whereas neutrons are scattered by the nucleus. The scattering factor for X-rays thus increases linearly with the number of electrons in the atom, so that heavy atoms (large atomic number Z) are much more effective at scattering than light atoms (small Z), but because of the size of the atoms relative to the wavelength of the X-rays, the scattering from different parts of the cloud is not always in phase, so the scattering factor decreases with sin $\theta//\lambda$ due to the destructive interference (Figure 2.12a). Because the nucleus is very small, neutron scattering factors do not decrease with sin θ/λ, and because nuclei are similar in size, they are all fairly similar in value and that of hydrogen is anomalously large due to the nuclear spin. The fact that the neutron scattering factors are almost invariant with sin θ/λ, means that the intensity of the data does not drop at high angles of θ as is the case with X-ray patterns, and so a neutron powder pattern tends to yield considerably more data. Neutron scattering factors are also affected randomly by **resonance scattering**, when the neutron is absorbed by the nucleus and released later. This means that neutron scattering factors cannot be predicted but have to be determined experimentally and they vary for different atoms and indeed for different isotopes (Figure 2.12b).

Note that because of the different scattering mechanisms, bond lengths determined by X-ray and neutron studies will be slightly different. The neutron determination will give the true distance between the nuclei, whereas X-ray values are distorted by the size of the electron cloud and so are shorter.

The excellent high-angle data obtained from neutron diffraction means that atomic positions are determined very precisely and so structures are solved to much higher resolution. The Rietveld profile analysis is used to refine the structures of crystalline powders—in fact this technique originated with neutron powder diffraction.

2.5.1 Uses of Neutron Diffraction

The fact that neutron scattering factors are similar for all elements means that light atoms scatter neutrons as effectively as heavy atoms and can therefore be located in the crystal structure; for example the X-ray scattering factors for deuterium and tungsten are 1 and 74 respectively, whereas the equivalent neutron values are 0.667 and 0.486. This property is particularly useful for locating hydrogen atoms in a structure, which can sometimes be difficult to do in an X-ray determination, especially if the hydrogen atoms are in the presence of a heavy metal atom. Accordingly, many neutron studies in the literature have been done with the express purpose of locating hydrogen atoms, or of exploring hydrogen bonding.

Neutrons are not absorbed by the crystals, so they are also useful for studying systems containing heavy atoms that absorb X-rays very strongly.

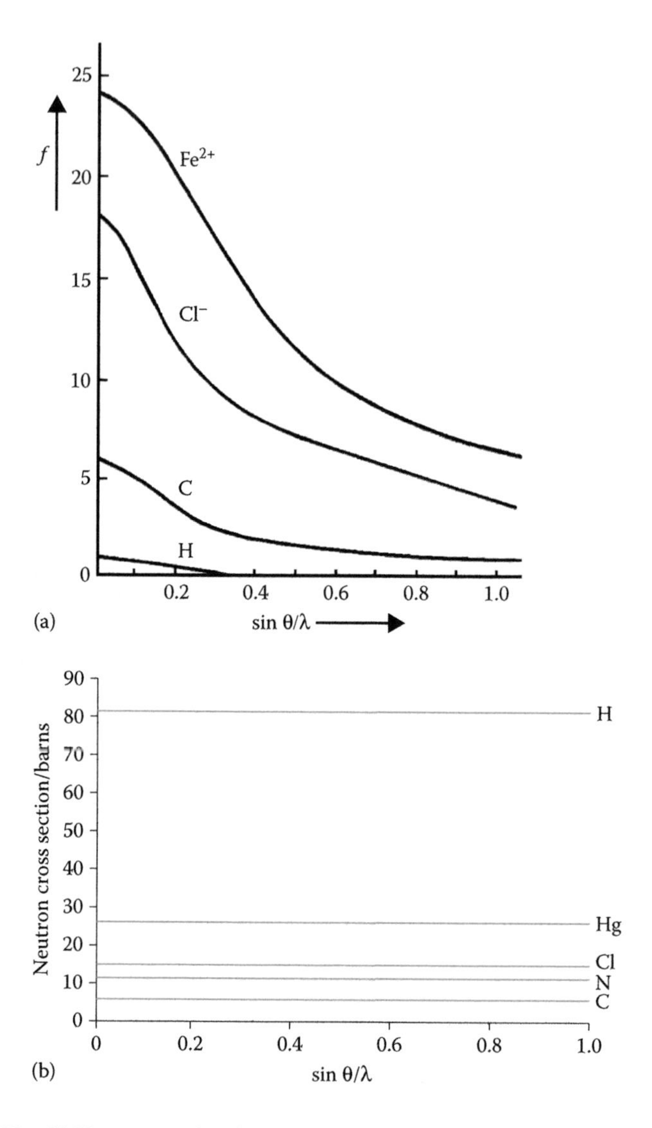

FIGURE 2.12 (a) X-ray scattering factors for hydrogen, carbon, chloride, and ferrous ions and (b) the neutron scattering cross-sections for several elements, as a function of sin θ/λ.

Atoms near each other in the periodic table have very similar X-ray scattering factors and cannot always be distinguished in an X-ray structure determination: oxygen and fluorine for instance, or similar metals in an alloy. A neutron structure determination may be able to identify atoms with similar atomic numbers.

As well as the scattering of neutrons by the nuclei, there is additional magnetic scattering of the neutrons from paramagnetic atoms. This arises because a neutron has spin and so possesses a magnetic moment which can interact with the magnetic moment of an atom. The atomic magnetic moment is due to the alignment of the electron spins, and so this interaction, like the scattering of X-rays, decreases with increasing Bragg angle due to the size of the electron cloud. As you will see later in Chapter 9, the magnetic moments of a paramagnetic crystal are arranged randomly, but in ferromagnetic, ferrimagnetic, and antiferromagnetic substances, the atomic magnetic moments are arranged in an ordered fashion. In ferromagnetic substances the magnetic moments are arranged so that they all point in the same direction and so reinforce one another; in antiferromagnetic substances the magnetic moments are ordered so that they completely cancel one another out, and in ferrimagnetic substances the ordering leads to a partial cancellation of magnetic moments. Magnetic scattering of a polarized beam of neutrons from these ordered magnetic moments gives rise to magnetic Bragg peaks. For instance, the structure of NiO, as determined by X-ray diffraction, is the same as that of NaCl. In the neutron study, however, below 120 K, extra peaks appear due to the magnetic interactions; these give a magnetic unit cell which has a cell length *twice* that of the standard cell. This arises (Figure 2.13) because the alternate close-packed layers of Ni atoms have their magnetic moments aligned in opposing directions, giving rise to antiferromagnetic behaviour.

Neutron data can yield such accurate lattice constants that the differences for a metal under stress can be mapped. This is used to analyse stresses in aero and car components.

Increasing attention is now being given to determining the structures of nanocrystalline materials.

2.6 X-RAY MICROSCOPY/X-RAY COMPUTED TOMOGRAPHY

The actual technique of X-ray microscopy (XRM), more often called X-ray Computed Tomography (CT), has existed for a long time. You may know this technique from medical imaging when patients need to undergo a 'CT scan' in order to obtain X-ray

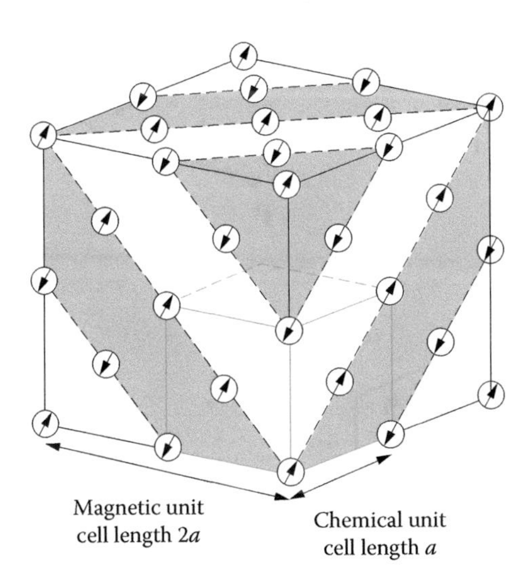

FIGURE 2.13 The magnetic ordering in NiO. Only the Ni planes are shown, and alternate close-packed layers have opposing magnetic moments. Note that the magnetic unit cell length is double that of the normal unit cell.

3D images. Hundreds of 2D X-ray images are taken from either the whole body or parts of it and these 'slices' are then recombined by a computer to create a 3D image. Developments during the past 15 years lead to the recognition of X-ray CT as an important analytical technique in Materials Science to gain 3D images with resolution in the micron and sub-micron/nano range (micro-CT, µCT, or sub-micron-CT, nano-CT). This is mostly due to the availability of stronger and more focused X-ray sources (e.g. microfocus X-ray tubes), more advanced optics and fast and sensitive 2D area detectors.

µCT and nanoCT in materials science allows 3D imaging of the entire microstructure of materials, in most cases non-destructively, and with spatial resolution which can meet and surpass that of optical microscopy. In contrast to optical or electron microscopy XRM is not a surface analytical technique. It truly provides a 3D image of the whole specimen. No lenses out of glass can be used when using X-rays, thus the X-ray beam hits the sample directly and dependent on the attenuation inside the material, radiographs are taken. In principle this functions just like taking an X-ray image in a clinic, only with a much better resolution and in three dimensions.

The basic setup of any XRM and the four various geometries applied in instruments is explained in Figure 2.14. In any case, however, radiographs of the sample from different angles have to be taken and recombined using a powerful computer. The approach is very similar to what needs to be done for single crystal X-ray diffraction. The major difference is that in the latter, diffraction images are taken instead of attenuation images or radiographs. The resolution of these X-ray images depends strongly on the size and coherence of the X-ray beam, the focusing optics and the sensitivity and resolution of the detector. The biggest challenge, thus, is to

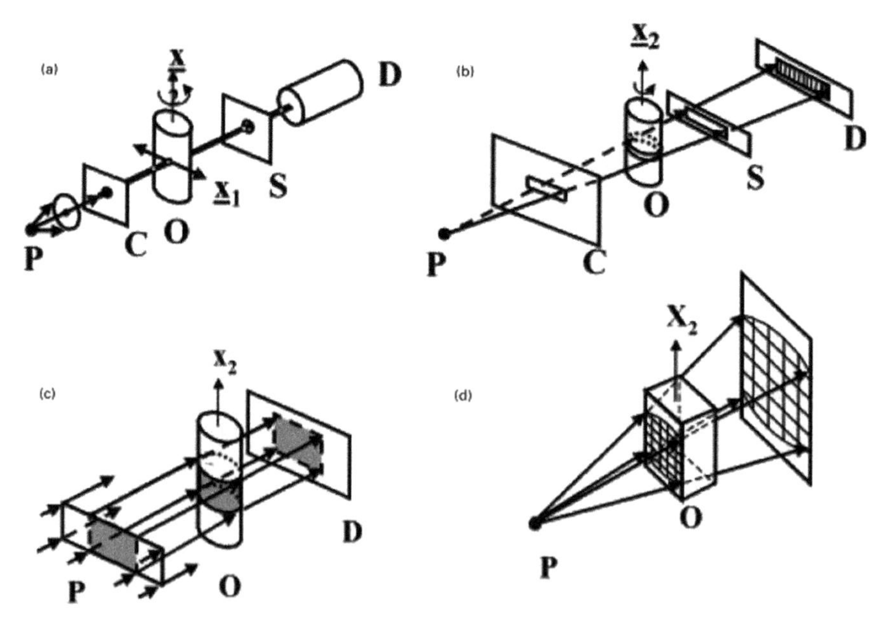

3 **Illustration of four experimental approaches to X-ray microtomography data collection: *a* pencil, *b* fan, *c* parallel, and *d* cone beam methods. P is X-ray source, C is collimator, O is object being imaged, x_2 is axis about which sample is rotated to produce different views required for reconstruction, S is slit, and D is detector**

FIGURE 2.14 Schematic diagram of an X-ray micro computer tomograph/microscope.

achieve high resolution, possibly independent of the size of the sample or object. Resolution in this case needs to be expressed three-dimensionally as spatial resolution. In two-dimensional imaging, this is best expressed with the actual size of the smallest increment used to produce the image, the pixel size. In three dimensions the smallest increment is called a voxel.

μCT achieves voxel sizes of around 50 μm whereas sub-micron or nanoCT can reach voxel sizes clearly below 1 μm. Spatial resolution strongly depends on the size of the sample to be analysed. Ideally, the sample should remain within the field of view (FOV) when rotated. The interplay between FOV and voxel size results in the true spatial resolution. The larger the sample, the larger the FOV needed and the larger the voxel size. Many manufactures achieve higher spatial resolution independent of the size of the specimen by e.g. using improved optical arrangements (additional use of lenses, different focusing optics amongst others). Different parameters are used to clarify this, such as RaaD (radiation at a specific distance), or simply the minimum achievable voxel. Values can easily differ by a factor of 10 or 100! With newest generation micro X-ray tomographs or microscopes 3D X-ray images of an object can be taken with a resolution just below 1 μm and a specimen size ranging between a few microns and ca. 30 cm.

The applications of this powerful method are broadening constantly and they include structural elucidation of composites, process control, non-invasive metrology, consumer electronics, performance prediction and failure, analysis of historical artefacts, transport in porous media, and stress analysis, in situ analysis of macrostructure of a material whilst heating or during gas–solid reactions.

Manufacturers are constantly adding new features to increase the range of applications (e.g. X-ray diffraction on micro- or nanoparticles, in situ chambers, microdiffraction). New techniques are developed annually with an aim to achieving better resolution and contrast, even in the nanometer region through the development of novel detectors or combined use of CCD detectors and optical microscopes. Figure 2.15 shows a typical setup of such a system.

Figures 2.16a–e provide five examples of the different areas and applications μCT can offer.

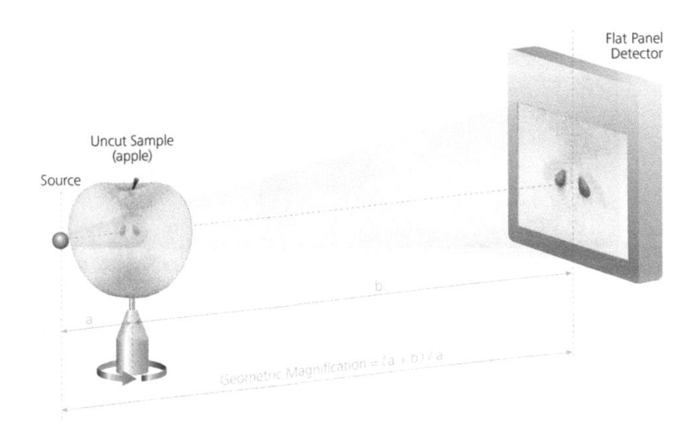

FIGURE 2.15 Conventional microCT architecture: sample must be close to the source to achieve resolution. (Courtesy of Carl Zeiss X-ray Microscopy, Inc.)

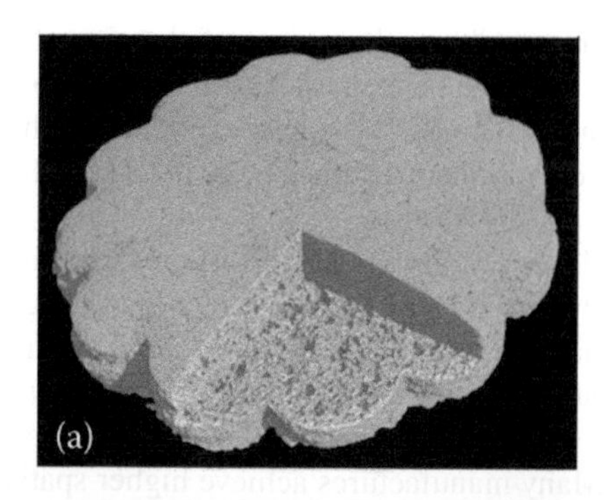

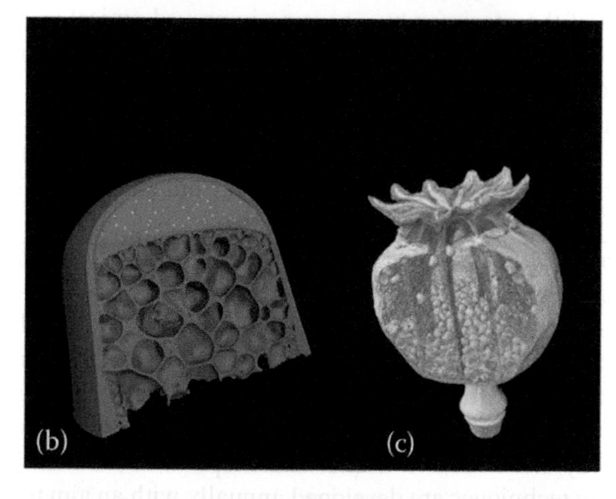

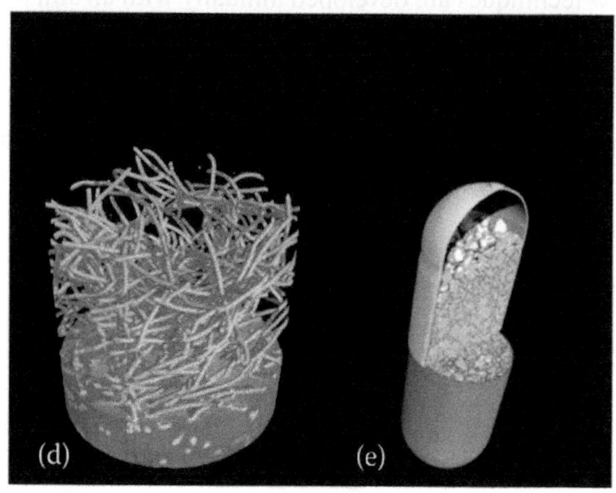

FIGURE 2.16 X-ray tomography of (a) Shortcake biscuit, (b) aerated chocolate, (c) poppy seed pod non-destructively sectioned to show the seeds inside, (d) wires non-destructively segmented from a 50-mm-diameter reinforced concrete core, (e) paracetamol capsule scanned to reveal particulate components. (μCT Images courtesy of The Hounsfield Facility, University of Nottingham.)

2.7 ELECTRON MICROSCOPY

Electron microscopy is widely used in the characterisation of solids to study structure, morphology, and crystallite size, to examine defects, and to determine the distribution of elements. The basic principle underlying an electron microscope is similar to that of an optical microscope only a beam of accelerated electrons is used rather than visible light. Because of the wave—particle duality of electrons they behave like electromagnetic radiation and, at high energies, have very short wavelengths, λ. The electron wavelength imposes a lower limit on the resolution of the and can be readily calculated: after acceleration through a voltage V the kinetic energy of the electron is $\frac{1}{2}mv^2 = \mathrm{eV}$, we can write $v = \frac{p}{m}$ where p is the momentum of the electron, so that $\frac{p^2}{2m} = \mathrm{eV}$ where e and m are the charge and mass of the electron respectively. Using the de Broglie relation $p = \frac{h}{\lambda}$ from Equation 2.14, where h is Planck's constant, gives $\frac{h^2}{\lambda^2} = \mathrm{eV}$. Rearranging: $\lambda = \frac{h}{\sqrt{2\,\mathrm{meV}}}$ and inserting values for h, m, and e gives

$$\lambda = 1.23(V)^{-1/2}\ \mathrm{nm}. \tag{2.15}$$

(This neglects relativistic corrections, which become significant as the kinetic energy approaches the electron rest mass energy of 511 keV.) For an accelerating potential V of 100 kV, λ=0.0123 nm (0.123 Å).

As you will see in the next pages, the lower limit of resolution of even the best electron microscopes is currently well below the actual resolutions achieved, unlike optical microscopes where resolutions close to their limit are obtainable. This reflects the fact that optical lenses are superior to their electromagnetic counterparts.

The electron beam is produced by heating a tungsten filament, a lanthanum hexaboride, LaB_6, crystal, or from a field emission gun, FEG, which uses a cathode of either tungsten or zirconium oxide. The beam is focused by magnetic coil magnets in a high vacuum (the vacuum prevents interaction of the beam with any extraneous particles) to a fine spot. Detection can be by scintillation counter, film, CCD, or monolithic active pixel sensors, also called CMOS (complementary metal–oxide–semiconductor) detectors. Through the most recent development of Cryo Electron Microscopy (Cryo EM), for which Dubochet, Frank, and Henderson were awarded the Nobel Prize in Chemistry 2017, resolution down to sub-nanometer levels can be achieved in the best instruments. The 3D structure of molecules (proteins, enzymes) of a molecular weight larger than 50,000 g/mol can be elucidated and thus, Cryo EM has become a true alternative to protein crystallography, mostly because there is no need to crystallise the compound.

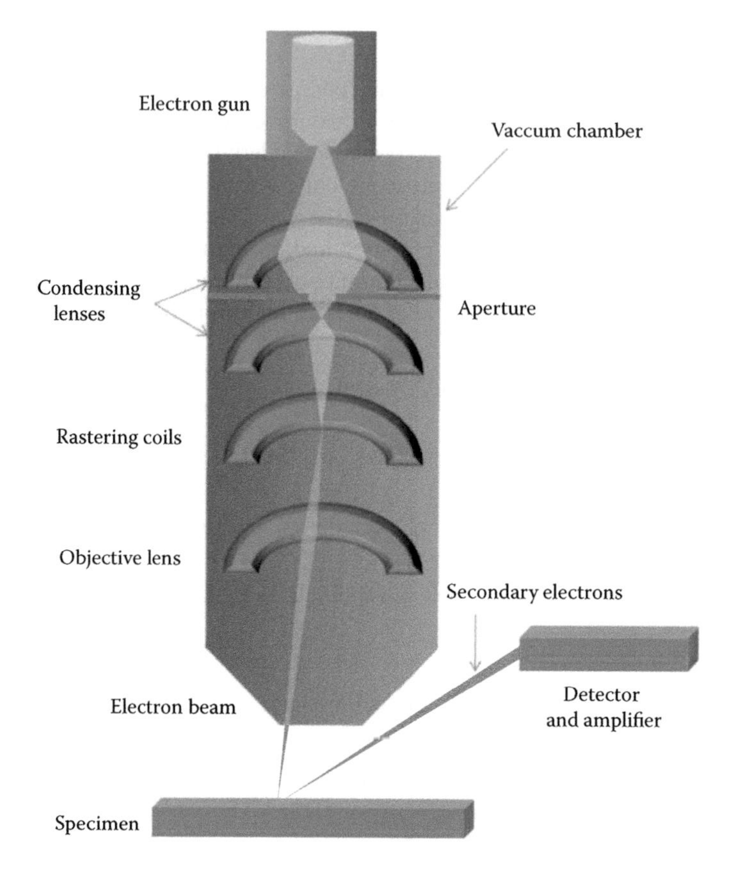

FIGURE 2.17 Schematic diagram of an SEM instrument.

2.7.1 Scanning Electron Microscopy, SEM

In this technique high-energy monochromatic electrons, generated at a potential of 1–30 kV, are formed into a finely focused beam which is rastered across the surface of the sample. Figure 2.17 shows a schematic diagram of the instrument. Electrons are generated at high voltage and are focused by a condenser lens; the high-angle electrons are excluded by an aperture, and the beam is then further focused by a second condenser lens. A set of coils sweeps the beam across the surface of the sample in a grid, and an objective lens finally focuses the beam on to the sample. The electrons impinge upon on the atoms within an interaction volume close to the surface, and signals produced from these atoms include back-scattered electrons (the reflection of electrons by elastic scattering), emitted secondary electrons from inelastic scattering, and characteristic X-rays (due to the production of secondary electrons) amongst others. An SEM image of the structure is mainly due to the low-energy secondary electrons: those produced deep within the sample will be absorbed, but those closer to the surface escape and can be collected and detected. These secondary electrons are accelerated toward a scintillation detector, amplified by a photomultiplier, and then displayed; the number of electrons determines the brightness of the image.

The back-scattered electrons have higher energy, and when these escape from the surface they are not deflected onto the image detector. However, these electrons can also be collected, and as the higher atomic number elements back-scatter the electrons more strongly, this information, together with the characteristic X-rays emitted from the inelastic scattering, enables both qualitative and quantitative elemental analysis of the surface.

Detection of the different signals can produce maps of the surface topography of samples such as catalysts, minerals, and polymers. It is also useful for looking at particle size, crystal morphology, magnetic domains, and surface defects (Figure 2.18), for performing elemental analysis as well as for investigating other properties such as electrical conductivity. A large depth of field gives a 3-D appearance to the image of the surface. A wide range of magnification can be used, anything from $\times 10$ to $\times 500{,}000$, and features down to sizes of about 1 nm and less can be resolved with the best instruments.

The samples must be conducting, at least at the surface, and nonconducting samples may need to be coated with gold, graphite, or similar; electrical earthing prevents charge building up on the surface.

The resolution of SEM relies on the size of the focused beam and on the lens system, and the two must be balanced carefully to give best results: the higher the current in the condenser lens, the smaller the spot, and thus the higher the resolution as it is more sensitive to smaller features on the surface, but this is at the expense of fewer electrons reaching the detector. The technique is not usually good enough to resolve individual atoms, as can the transmission electron microscope which uses higher-energy (and therefore shorter-wavelength) electrons (see next section). SEM is widely available in labs, and usually achieves a resolution from 0.5 to <20 nm depending on the instrument.

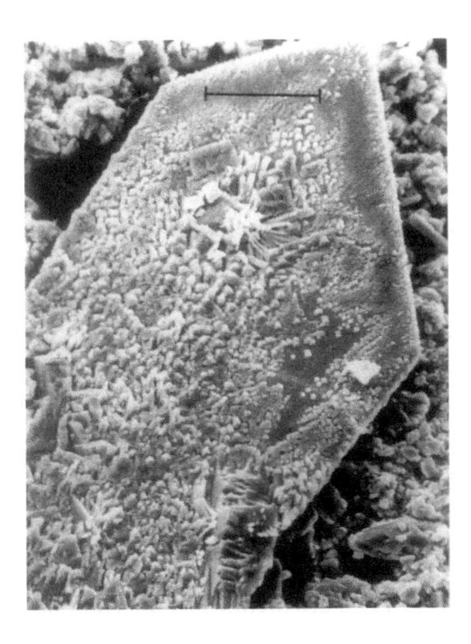

FIGURE 2.18 SEM showing crystals of $VSbO_4$ growing out of β-Sb_2O_4 following reaction with V_2O_5 (bar $= 40\ \mu m$). (From Professor Frank J. Berry. With permission.)

2.7.2 Transmission Electron Microscopy, TEM

TEM provides significantly higher resolution than a conventional microscope using visible light, and also than SEM, due to the very-high-energy electrons generated; resolution of less than 0.1 nm (<1 Å) at magnifications of 50 million and more have been achieved. TEM produces a direct image of a solid at high magnification from the transmission of electrons through an ultra-thin sample, using an electron gun and an assembly of electromagnetic lenses, all under high vacuum. A very thin sample, hundreds of nanometres thick at most, is prepared and subjected to a high energy, high intensity beam of electrons which are usually generated using either a tungsten filament or a lanthanum hexaboride crystal and subjected to a potential of 100–300 kV; after acceleration, the electrons are formed into a beam by two condenser lenses and then focused on to the sample by an objective lens. Additional complex magnetic lenses (quadrupole, hexapole, and octupole) can be incorporated to correct distortions in the beam.

Some of the electrons strike the atoms on the surface and are deflected (the deflected electrons may be elastically or inelastically scattered) whereas those which pass through the sample are detected and form a 2-D projection: the emergent beam is expanded by projector lenses onto a detector such as a fluorescent screen, film, or CCD. The instrument can be operated to select either the direct beam (bright-field image) or the diffracted beam (dark-field image).

Most commonly, when the primary beam is collected after passing through the sample, the image is seen as a dark feature on a bright background (**bright-field mode**); areas of the sample where more electrons are absorbed by the higher-atomic-number elements or thicker parts of the sample are seen as dark, and effectively a 2-D projection of the sample is obtained (Figure 2.19).

In the **dark-field mode**, however, a detector can be placed in front of the sample to collect the coherent back-diffracted electrons, the Bragg reflections. An aperture excludes the primary electron beam, and by tilting the specimen, particular Bragg reflections can be selected and the image formed is of a dark background with bright areas from the selected reflection(s). This method is particularly useful for studying the position and type of lattice defects (see Chapter 5).

High resolution instruments (sometimes called **High Resolution Transmission Electron Microscopy, HRTEM**) use methods of obtaining the phase of the electron beam as well as the amplitude. A large aperture allows the incident beam which passes directly through the sample and the scattered beams to pass to a detector. The coherent elastically scattered beams interfere and form an image (so called phase contrast) which can be interpreted to give atomic position information.

Because the electrons pass through the sample, TEM/HRTEM images the bulk structure, and so can detect crystal defects such as phase boundaries, shear planes, and so on. In the highest resolution instruments, resolution of less than 0.1 nm (<1 Å) is achievable, allowing individual atoms and defects to be imaged (Figure 2.20). By combining views of a crystal taken at different angles, a 3-D crystal structure can even be obtained.

The resolution limits of TEM are dictated by aberrations in the lens system which affect the focusing of the beam. There are three main problems: **astigmatism**,

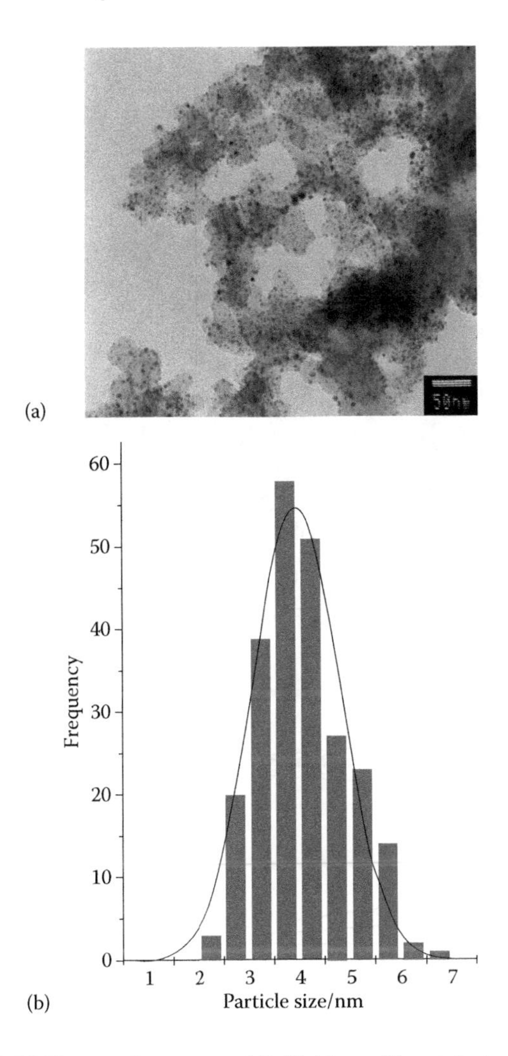

FIGURE 2.19 (a) TEM image of a supported Pt/Cr bimetallic catalyst on C. (b) Analysis of the metal particle sizes of this catalyst.

chromatic aberration, C_c, and **spherical aberration**, C_s. Astigmatism is due to the cross-section of the beam being elliptical rather than circular. Chromatic aberration occurs because electrons are not quite monochromatic but have a range of energies and thus also wavelengths. Spherical aberration is shown diagrammatically in Figure 2.21, and occurs because, as the electrons pass through the condensing lenses, the outer electrons of the beam which are diffracted through larger angles are brought to a focus at a different point on the axis to those in the centre, which are diffracted through small angles, creating a focal disc rather than a spot. Modern advances are now enabling fine corrections to be made for aberrations (see Section 2.7.4) such that resolution below 0.05 nm (<0.5 Å) has been achieved.

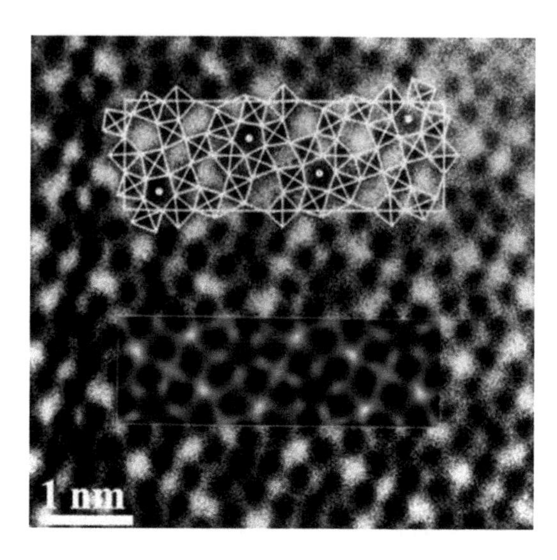

FIGURE 2.20 HRTEM image of $Nb_4W_{13}O_{47}$ tungsten bronze along *011*. The insets show the structural model and a simulation. (From Dr. Frank Krumeich. With permission.)

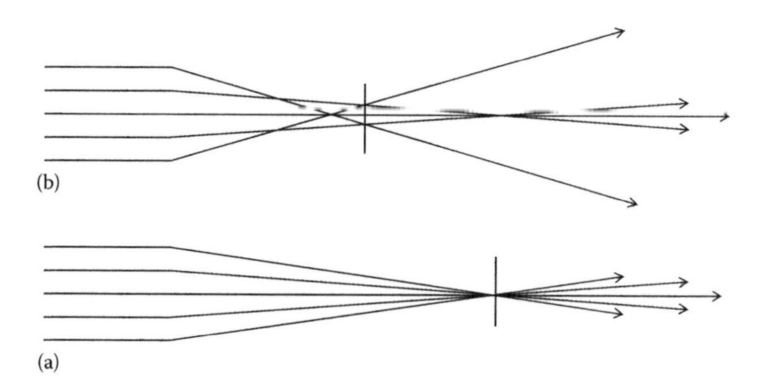

FIGURE 2.21 Diagram showing (a) electrons condensed to a single focal point and (b) electrons focused to a disc by a lens with spherical aberration.

2.7.3 Cryogenic Electron Microscopy (Cryo EM)

Cryogenic electron microscopy, or Cryo EM, is transmission electron microscopy at low temperatures. Up to 20 years ago, this simply meant freezing solids to lower temperatures in order to decrease the signal-to-noise ratio. Recently, however, this term is mainly used to describe a technique which enables the structure determination of large biological molecules such as proteins, viruses, or enzymes by the use of electron microscopy without the need of crystallisation. The development of Cryo EM by J. Dubochet, J. Frank, and R. Henderson was awarded with the Nobel Prize in Chemistry in 2017. The discovery is groundbreaking in the field of structural chemistry, simply because structural elucidation on atomic scale or resolution had only been possible by carrying out diffraction experiments on crystalline samples. Macromolecules such as proteins or enzymes are very difficult to crystallise and thus, often the structure determination of some of these compounds was not possible because single crystals of sufficient size and quality could not be grown.

Cryo EM is usually carried out by embedding the macromolecules in vitrified water, i.e. water which is solidified as an amorphous glass. Crystalline water (ice) makes it impossible as it would diffract the electron beam. As all electron microscopy has to be carried out under vacuum, it is crucial for the sample to be able to withstand such drastic atmospheric conditions. Embedding the sample in vitrified glass protects the compound from evaporation. Sample preparation consists of dropping an aqueous solution or suspension of the macromolecule on to a very fine grid. This is then immersed carefully in liquid ethane (–190°C). The temperature is of great importance here as water starts crystallising at temperatures above –140°C.

In order to obtain a well-resolved 3D image of the molecule multiple images need to be taken and processed with a powerful computer/software.

2.7.4 Energy Dispersive X-Ray Analysis, EDX (EDAX)

As we saw in Section 2.2.1, an electron beam incident on an element gives rise to the emission of characteristic X-rays. When an electron in an inner shell is excited and ejected from the shell, an electron hole is created; an electron from a higher energy shell drops down to fill the hole, releasing the excess energy in the form of an X-ray. In the electron microscope electrons which are scattered inelastically, the secondary electrons, cause different elements present to emit characteristic X-rays. These can be separated by a silicon–lithium detector placed within the objective lens system, and each signal is collected, amplified, and corrected for absorption and other effects, to give both qualitative and quantitative analysis of the elements present (for elements of atomic number greater than 11), a technique known as **EDX or EDAX, Energy Dispersive Analysis of X-rays** (Figure 2.22).

Newer detectors such as the silicon drift detector, SDD, can offer advantages over the usual Si(Li) variety as they collect over a larger angle, leading to shorter collection times and better count rate, and they also do not necessitate cooling with liquid nitrogen.

2.7.5 Scanning Transmission Electron Microscopy, STEM

A **scanning transmission electron microscope, STEM**, combines the scanning ability of SEM with the high resolution achieved in TEM. To achieve this the electron beam is finely focused down to a very small spot (<1 nm–10 nm) which is then rastered across the sample. The transmitted electrons are collected to provide a high resolution image of the inner structure and surface of the sample. As in TEM, the secondary and elastically scattered electrons, as well as the characteristic X-ray spectrum can be collected to give chemical structure and identification data. As it is a transmission technique, the sample must be prepared very carefully and should be ultra-thin. Because these instruments are designed to give very high resolution

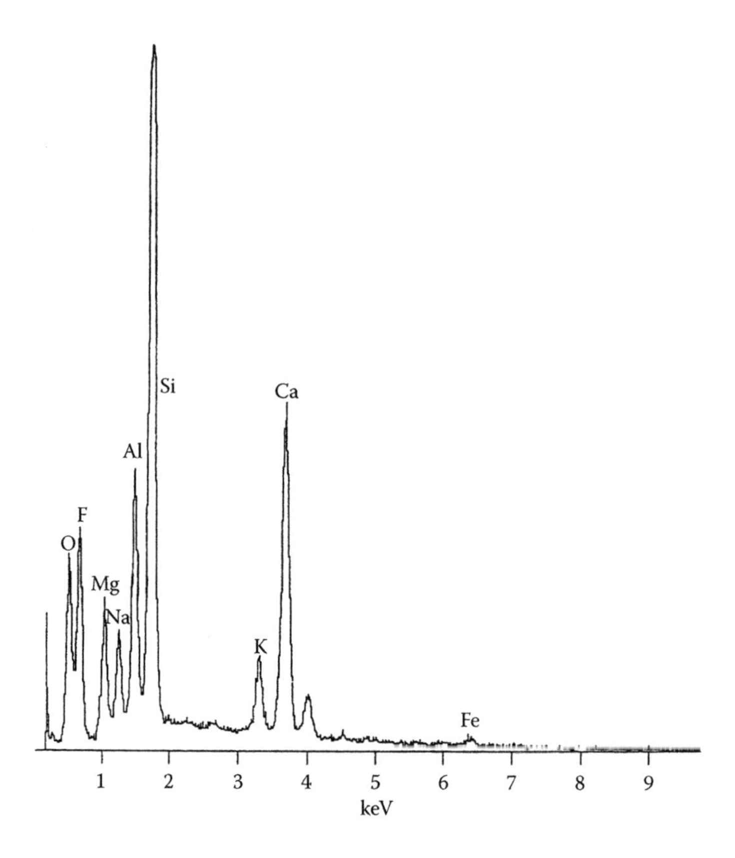

FIGURE 2.22 EDAX analysis of a glaze.

images, they have to be operated under very careful conditions which are thermally stable, vibration free, and radiation free—definitely no slamming of doors!

The high-angle scattered electrons in STEM also give rise to a very useful form of imaging, **high-angle annular dark-field imaging, HAADF** or **Z-contrast imaging**. The Bragg equation, $n\lambda = 2d \sin\theta$, tells us that the high energy, short-wavelength electrons used in these instruments will give very small Bragg angles for the coherently scattered electrons. However, high-angle scattered electrons are also produced which have been deflected by close interaction with the nuclei; in this case, each atom acts as an independent scatterer and there is no interference between these electrons, they are incoherent. The intensity of each incoherent scattering can simply be summed over all the atoms and is very sensitive to atomic number. These high-angle electrons can be collected by an annular dark-field detector, which collects electrons which pass through an annulus (ring) around the primary beam, allowing the low-angle Bragg scattered electrons to pass through (Figure 2.23). The intensity of each position in the image thus obtained is representative of the average atomic number of the column of atoms in that position in the sample, giving detailed information on the position of different types of atoms in the sample.

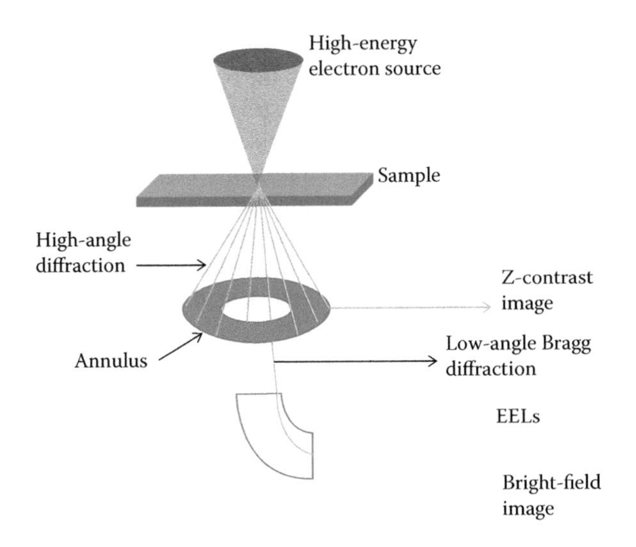

FIGURE 2.23 Schematic diagram of high-angle annular dark-field imaging.

2.7.6 Electron Energy Loss Spectroscopy, EELS

The schematic diagram of a STEM instrument shown in Figure 2.25(a) indicates another form of analysis of the transmitted beam of electrons through a thin sample, that of **electron energy loss spectroscopy, EELS**. In a primary beam an electron subjected to a 300 kV potential will have a kinetic energy of 300 keV; during inelastic collisions, some of this energy may be transferred to other electrons. The kinetic energy distribution of the transmitted electrons can be measured and characteristic losses observed and interpreted. Like EDX, this technique often adds another another source of data for the electron microscopist.

The inelastically scattered incident electrons lose energy during collisions due to interactions with both inner shell, valence, and conduction electrons and also due to the excitation of **phonons** (crystal lattice vibrations). Thus the electron beam can excite **plasmons** (quantized oscillations of the electron gas in a conducting solid), electronic transitions between the bands in the solid (see Chapter 4), and inner shell ionization.

The ionization of inner shell electrons is particularly useful for identifying different elements in the solid. During ionization a critical threshold energy, E_c, must be transferred from the incident electron to eject another electron from its energy level. The incident electron then travels on through the spectrometer to the detector with a reduced kinetic energy, and the difference between its initial and final energy is measured. More than the critical energy may be given to the ejected electron $E > E_c$. The intensity of the absorption decreases with increasing energy loss giving a sharp discontinuity in the spectrum which corresponds to the characteristic energy of the inner shell and is thus characteristic of the element; this is known as the **absorption edge**. Fine structure of the absorptions gives information about band structure, useful for the accurate measurement of band gaps on semiconductors. Measurement of the absorption intensities gives information on the concentration of the elements. The critical ionization energy is sensitive to the chemical environment of the atom, and so information about the bonding in the structure may be obtained from EELS.

In order of increasing energy loss, the processes of inelastic electron scattering taking place are: phonons, 0.02 eV; intraband transitions 5–25 eV; plasmons 5–25 eV; ionization 10–1000 eV. The spectra thus naturally divide into two regions with a low-loss region with energy losses due to plasmons and interband transitions up to ~50 eV and a high-loss region above 50 eV, due to ionisation from inner shells.

The technique is particularly useful for the detection of light elements.

2.7.7 superSTEM

In order to improve the resolution of STEM, it is necessary not only to have very high energy electrons and an ultra-fine spot, but to get rid of all the problems of instability and aberrations in the beam, and this was the aim of building the superSTEM instruments. The first **superSTEM** instruments in Europe were built at the Daresbury, UK facility. This site provides a stable geological basis of sandstone, and the instruments occupy their own building with separate foundations, separate electrical supply, carefully controlled temperature and air conditioning, and shielding from external sources of radiation and sound—everything that might cause a problem.

The ability to correct spherical aberration, C_s (Figure 2.21) in the beam has progressed immensely in recent years, thanks to the work of Krivanek, Dellby, and Brown at Cambridge University, UK. The approach has been to introduce negative spherical aberration which then cancels out the positive aberration. They have achieved this using a series of complex magnetic lenses (quadrupole and octupole) along the direction of travel of the beam, so that all rays focus at a common point. The final specifications of the instrument are impressive with current stability of $\leq$1 ppm and an improved resolution of about a factor of three to less than 0.1 nm (<1 Å) (Figure 2.23).

2.8 SCANNING PROBE MICROSCOPY, SPM

The characterisation of solid surfaces is often carried out using a range of techniques known collectively as **scanning probe microscopy, SPM**. These techniques produce direct images of surfaces down to atomic-level features. The common feature of all the techniques is the monitoring of the interaction of a very fine tip with the surface while it is moved across the surface in a very fine grid; information about the surface is gathered by the way the interaction changes with position. The pioneer scanning probe technique was **scanning tunnelling microscopy, STM** which enables conducting and semiconducting surfaces to be imaged at atomic resolution. It was invented in the early 1980s by Gerd Binnig and Heinrich Rohrer at IBM in Switzerland, and for which they were awarded the 1986 Nobel Prize in Physics. Looking for ways to study nonconducting surfaces, Binnig, Quate, and Gerber went on to develop the **atomic force microscope, AFM,** several years later. Since then a plethora of other microscopy techniques, which probe particular properties of surfaces at the atomic level, have been developed: for instance, near-field scanning optical microscopy, NSOM, uses a sub-wavelength-sized aperture rather than a lens to direct the light on to the sample and spatial resolutions of 20 nm and less can now be achieved; photoscanning microscopy, PSTM, tunnels photons using an optical tip; magnetic force microscopy, MFM, is used to investigate magnetic structure; electric force microscopy, EFM, can be used to look at electronic properties; others enable the mapping of surface friction, capacitance, adhesion, and Young's modulus. SPMs can also be used to investigate adsorbed molecules, to manipulate and move molecules, and to obtain the spectra of single molecules on the surface. More recently STM resolution has been increased to such an extent that the movement of molecules and atoms on the surface can be tracked, with exciting prospects for the study of diffusion, catalysis, and chemical reactions.

2.8.1 Scanning Tunnelling Microscopy, STM

In scanning tunnelling microscopy, STM, a sharp metal wire tip is brought sufficiently close (<1 nm) to the surface of a conducting or semiconducting solid sample, that electrons can move through the vacuum between the two. When a small voltage is applied to the tip, creating a potential difference, electrons from the sample surface are attracted to the positive tip (Figure 2.24). However, an energy barrier prevents the electrons from leaving the sample. The energy needed to eject an electron from the highest occupied level in the solid, the Fermi level, out into a surrounding vacuum, is known as the **work function**, φ. However, if the tip and surface are sufficiently close together, quantum mechanical effects come into play, their electron–wave functions can overlap, and there is a finite probability that the electron can be found on the other side of the barrier. This is known as quantum mechanical **tunnelling**, and produces a very small tunnelling current in the range of pico- to nanoamperes. The magnitude of the current is very sensitive to the size of the gap, changing by a factor of 10 when the distance changes by 0.1 nm.

The metal tip is scanned backwards and forwards across the solid, point-by-point and line-by-line (rastering), and the steep variation of the tunnelling current with distance gives an image of the atoms on the surface (Figure 2.24). The image is usually formed by keeping a constant tunnelling current and measuring the distance, thus creating contours of constant density of states on the surface (see Chapter 4).

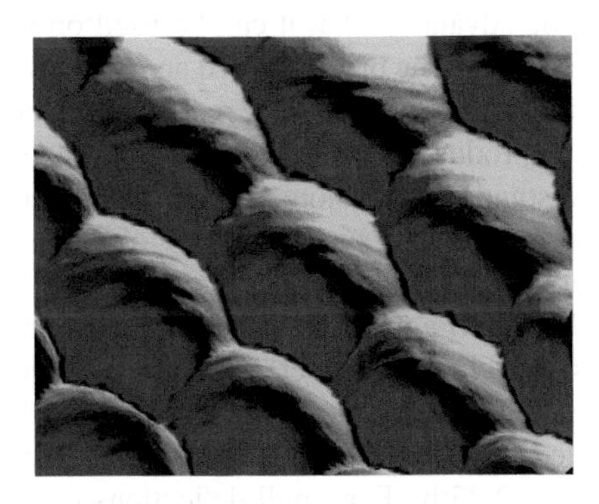

FIGURE 2.24 The Pt (*111*) face clearly showing the close-packed array of atoms.

The distance from tip to sample is typically 0.4–0.7 nm, and this very fine control of the movement of the tip is managed by the expansion and contraction of piezoelectric crystals (see Chapter 9) using feedback electronics to adjust the height. PZT, lead zirconium titanate, $PbZr_{1-x}Ti_xO_3$, piezoelectric crystals are typically used, moving by about 1 nm with 1 mV. The values of the adjustments are recorded, defining a grid of values which is then displayed as a greyscale image that can be artificially coloured by height, curvature, etc., to enhance particular features of interest. A faster scan of flatter surfaces can be achieved by maintaining a constant height, giving an image of the current changes and thus a charge density map of the surface.

By changing the sign of the potential, the tunnelling direction reverses, and thus STM can map either occupied or unoccupied density of states. Chemical information can be obtained by altering the voltage so that electrons from different types of energy levels tunnel across. The map thus shows features due to both the topography and the electronic structure, down to the level of showing the positions of individual atoms or defects. As an STM image maps either the density of states or the charge density at the surface, the images have to be manipulated and modelled to finally yield the structure, contours, and atomic positions on the surface.

The image resolution relies on the diameter of the scanning metal tip, so it is vital to be able to produce sharp tips with just a few atoms on the end; surprisingly, even cutting a wire at an angle with scissors can achieve this. The most commonly used tips are platinum–iridium, which are produced by mechanical shearing, and tungsten, produced by electrochemical etching. Most recently carbon nanotube tips have been introduced. To get good results, as well as working with very sharp tips, it is also important to work with very clean, stable surfaces in a vibration-free environment.

STM can also be used to manipulate the atoms on a surface—this and nanolithography will be discussed later in Chapter 11.

The atomic force microscope, AFM, is now widely available in labs, and it is to that technique that we turn next.

2.9 ATOMIC FORCE MICROSCOPY, AFM

Atomic force microscopy, AFM, is based on the detection of very small (in the order of nanonewtons) forces between a sharp tip (of the order of nanometres) and atoms on a surface and has the advantage that it can be used on nonconducting surfaces and under ambient conditions. The tip is scanned across the surface at subnanometre distances, and deflections due to attraction or repulsion by the underlying atoms are detected. The technique produces atomic-scale maps of the surface, and its use has been vital to the understanding of nanotechnology (Chapter 11). It can be used to determine the different forces acting on the surfaces of solids, such as van der Waals, chemical bonding, magnetic, capillary, friction, and so on. If the arm is brought very close to a surface, (~1–10 nm) a small attractive force is caused by van der Waals forces, but when it is brought even closer short range coulombic repulsion forces between the electron clouds of the atoms on the tip and surface begin to dominate and the arm is deflected upwards (Figure 2.25(a)). The resulting interatomic force curve is shown in Figure 2.25(b). For small deflections of the arm about the equilibrium position, the curve is almost linear and **Hooke's law,** which tells us that a

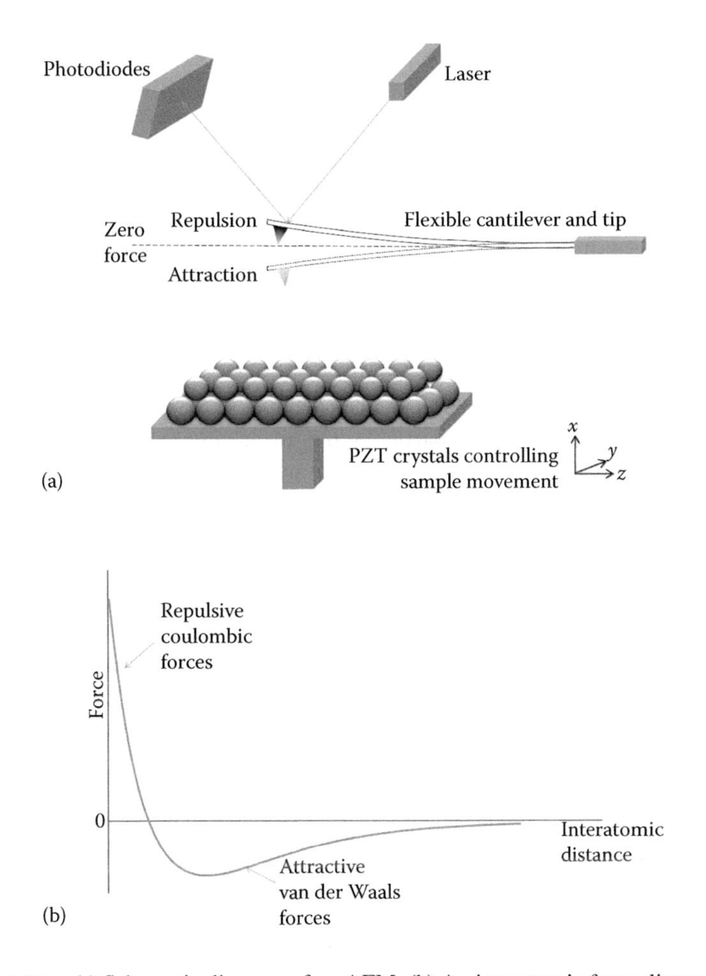

FIGURE 2.25 (a) Schematic diagram of an AFM. (b) An interatomic force distance curve.

movement, d, is directly proportional to the force, F, experienced, can be used to determine the repulsive force: $F=-kd$ where k is the force constant or spring constant (stiffness) of the arm. The deflection of the arm is detected by bouncing a laser beam from the arm to a detector consisting of a photodiode array.

The tip and its supporting cantilever arm are typically made of silicon or silicon nitride, Si_3N_4. Instead of scanning the surface at constant height, running the risk of collision and damage to the tip, a constant force between tip and surface is maintained via a feedback loop. The fine control of the movement is achieved through piezoelectric PZT crystals, which allow sub-nanometer adjustment in all three directions, thus enabling a 3-D image of the surface to be created.

An AFM can be operated in two main ways, static and dynamic. In the static **contact mode** the tip is dragged across the surface so close as to be in the region of repulsive forces, thus preventing the tip being attracted down on to the surface; to avoid damage to the tip it must be kept away from the surface and so thinner more flexible cantilevers are used in this mode. The deflection of the tip is measured and a feedback loop maintains a constant force as the tip is moved over the surface. By measuring the twist of the cantilever as well as deflection, the changes in friction of the surface can also be mapped.

In the **non-contact** dynamic mode a stiff cantilever is oscillated just above its resonance frequency of (usually) a few nanometres, and is kept further away from the surface where the attractive forces dominate. The oscillation frequency is modified by changes in the surface as the tip is scanned across. Once again a feedback loop maintains either a constant oscillation frequency or a constant amplitude. By measuring the deflection at each point on a grid, a 3-D map of the surface can be generated. This mode is used for softer surfaces as it reduces drag on the tip, and usually operates under ultra-high vacuum.

One of the problems in looking at surfaces under ambient conditions is the contamination of the surface and in particular the surface liquid (mainly water), in which the tip can get stuck. The **tapping** dynamic mode was developed to overcome this and here the cantilever is oscillated slightly below its resonant frequency, and closer to the surface so that it is alternately attracted and repelled. Stiff cantilevers tend to be used in this mode as the attraction by any surface liquid layer can cause the tip to stick. By operating in this way, less damage is caused to both the tip and the surface, and with it even adsorbed single molecules can be scanned.

Recent developments in AFM have achieved remarkable results. The atoms on the surface of a Sn and Pb alloy deposited on a Si surface have been chemically identified by very accurately measuring the dependence of the force on the distance between the tip and individual atoms and then using modelling to compare with previously measured fingerprints for each different component. Every atom in the chemical structure of a pentacene molecule (five fused benzene rings) has also been resolved using non-contact AFM. This amazing resolution has been achieved by creating a probe that has a single CO molecule attached to the end, and using a stiff cantilever with resonance amplitudes down to 0.02 nm. Images of water on a surface have been obtained using tips treated with a hydrophobic compound.

Further applications of AFM in nanoscience can be seen in Chapter 11.

2.10 X-RAY ABSORPTION SPECTROSCOPY, XAS

2.10.1 Extended X-Ray Absorption Fine Structure, EXAFS

EXAFS is used for the determination of short-range order in non-crystalline materials, such as gases, liquids, amorphous powders, and nanocrystalline materials, where conventional diffraction techniques are of little use. In high energy accelerators, electrons are injected into an electron storage ring, captured, and accelerated around this circular path by a series of magnets. When the electrons are accelerated to kinetic energies above the MeV range, they are travelling very close to the speed of light and they emit **synchrotron radiation** (Figure 2.26). There are just over fifty synchrotron sources in the world of which only four, so-called 'third-generation', are the most powerful: ESRF at Grenoble, France, and PETRA-III, Hamburg, Germany operate at 6 GeV; APS, at Argonne, IL, 7 GeV; SPring-8 in Japan at 8 GeV; these huge storage rings have circumferences of up to 2300 m respectively. The UK facility

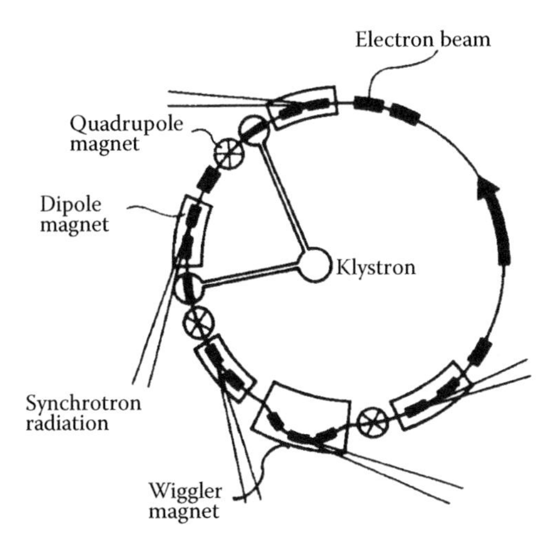

FIGURE 2.26 Storage ring.

DIAMOND Light Source has a ring of 561.6 m circumference and operates at 3 GeV. The X-radiation generated ranges from soft X-rays to hard X-rays (~300 eV–300 keV, 40 Å–0.04 Å). Unlike X-radiation from a conventional generator, synchrotron radiation is of uniform intensity across a broad band of wavelengths and several orders of magnitude (10^4–10^6) higher in intensity (Figure 2.27). The shortest X-ray wavelengths emerge as almost fully collimated, polarized beams.

In an EXAFS experiment the X-radiation is absorbed by a bound electron in a core shell (the K, L, or M shell) and ejected as a photoelectron. If one measures the absorption coefficient of the sample as a function of the X-ray frequency, a sharp rise or **absorption edge** is observed at the shell threshold energy (Figure 2.28) when the electron is ionised. Each element has its own characteristic shell energies, and this makes it possible to study one type of atom in the presence of many others, by **tuning** the X-ray energy to a particular absorption edge. The appropriate frequency X-radiation from the continuous synchrotron radiation is selected by using the Bragg reflection from a single plane of a carefully cut crystal such as Si *220*; often two crystals are used as is shown in the schematic diagram of a double crystal monochromator in Figure 2.29. By changing the Bragg angle of reflection, the frequency of the X-rays selected may be changed, and thus the absorption edges of a wide range of elements can be studied.

The waves of the ejected photoelectron from the shell can be thought of as a spherical wave emanating from the nucleus of the absorbing atom; this encounters neighbouring atoms and is partially scattered by them producing a phase shift (Figure 2.30). Depending on the phase shift experienced by the electron, the reflected waves can then interfere constructively or destructively with the outgoing wave, producing a net interference pattern at the nucleus of the original atom. Absorption by the original atom is now modified, and the effect is seen as sinusoidal oscillations or **fine structure** superimposed on the absorption edge (Figure 2.30) extending out to several hundred eV after the edge. The extent to which the outgoing wave is reflected

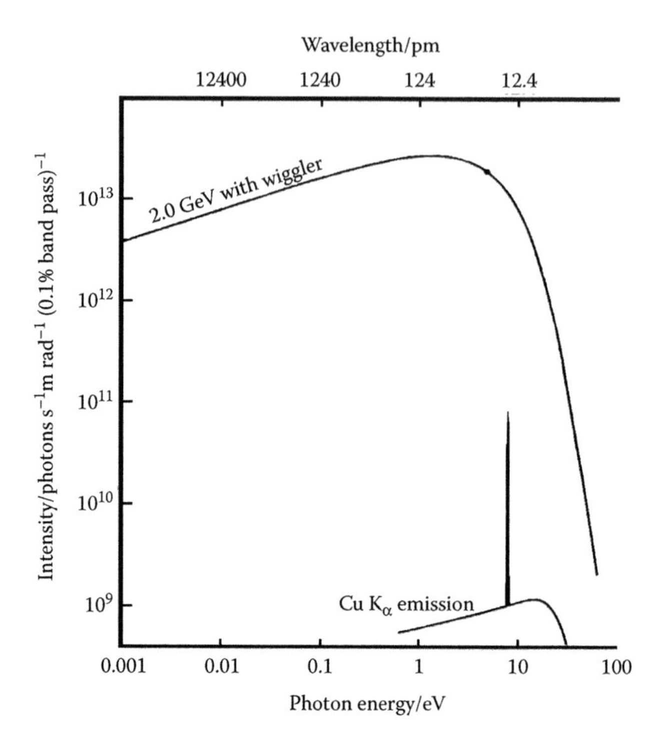

FIGURE 2.27 Synchrotron radiation.

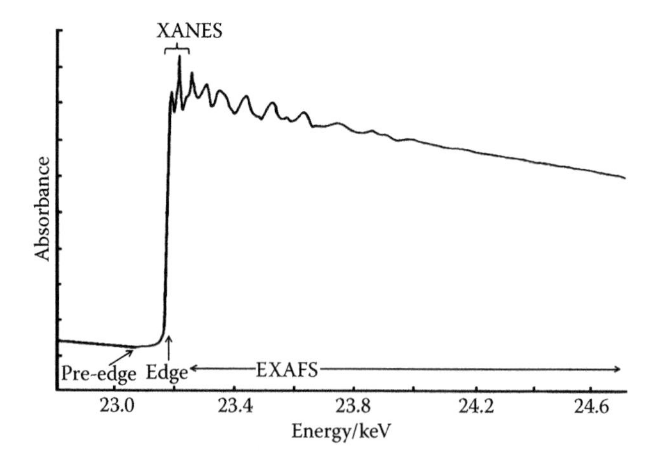

FIGURE 2.28 Rh absorption edge showing the pre-edge XANES and EXAFS regions.

by a neighbouring atom, and so the intensity of the reflected wave, is partly dependent on the scattering factor of that atom. The interference pattern making up the EXAFS thus depends on the number and the type of neighbouring atoms, and their distance from the absorbing atom.

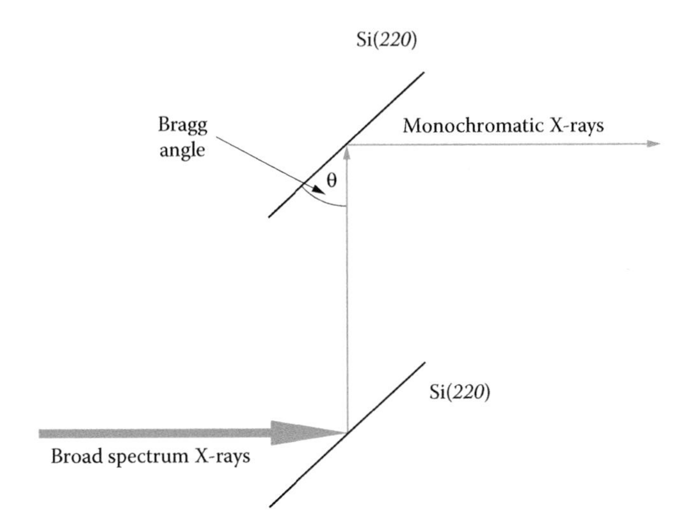

FIGURE 2.29 Bragg reflections from a double-crystal monochromator. From the Bragg equation, $n\lambda = 2d \sin \theta$, d for the planes of the crystal stays constant, so changing the angle changes the wavelength of the X-rays reflected. Two crystals are used to make the exit beam parallel to the entrance beam. Curved crystals focus the X-rays.

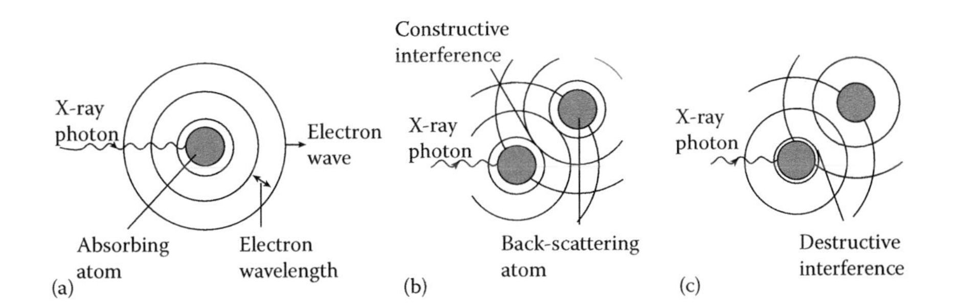

FIGURE 2.30 The EXAFS process: (a) the photoelectron is ejected by X-ray absorption, (b) the outgoing photoelectron wave (solid line) is backscattered constructively by the surrounding atoms (dashed line), and (c) destructive interference takes place between the outgoing and the backscattered wave.

The EXAFS function is obtained from the X-ray absorption spectrum by subtracting the absorption due to the free atom. A Fourier transform of the EXAFS data gives a **radial distribution function** which shows the distribution of the neighbouring atoms as a function of internuclear distance from the absorbing atom. The absorbing atom is surrounded by shells of neighbours, known as **coordination shells**. Finally the radial distribution function is fitted to a series of trial structural models until a structure which best fits the data is obtained, and the data is refined as a series of coordination shells surrounding the absorbing atom. The final structure will have refined number and types of atoms, and their distance from the absorbing atom. It is difficult to differentiate atoms of similar atomic number, and it is important to note that EXAFS only provides data on distance: there is no angular information. Depending on the quality of the data obtained, in favourable cases several coordination shells out to about 6 Å (0.6 nm) can be refined, with an accuracy of about 1 pm. The determination of the number of nearest neighbours is less precise.

In the example shown in Figure 2.31 a clay (a layered double hydroxide, LDH) was intercalated with a transition metal complex, $(NH_4)_2MnBr_4$. The EXAFS data in (a) shows the Mn K-edge EXAFS of the pure complex and one can note a coordination sphere of four Br atoms at a distance of 0.249 nm (2.49 Å), corresponding well to the tetrahedral coordination found in the X-ray crystal structure. However, after intercalation, the complex reacts with the layers in the clay, and the coordination

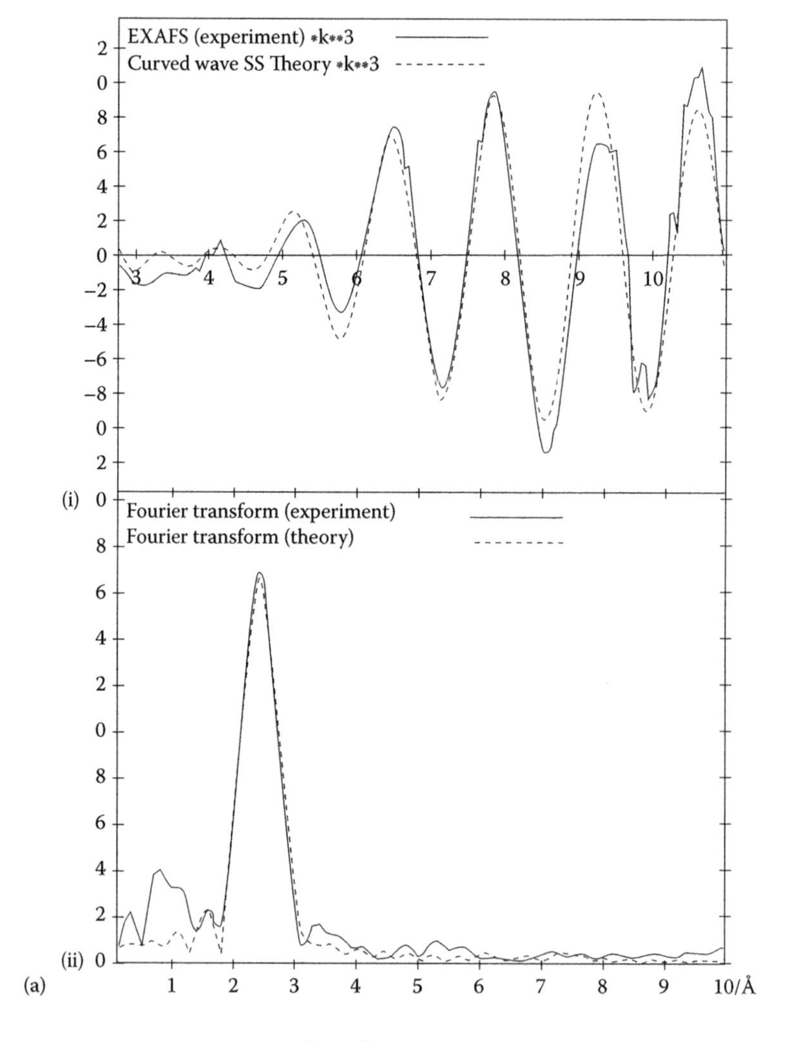

FIGURE 2.31 EXAFS data for (a) $\left(NH_4^+\right)_2 MnBr_4^{2-}$ and (b) the intercalation of $MnBr_4^{2-}$ in a layered double hydroxide clay: (i) extracted EXAFS data and (ii) the radial distribution function: solid line, experimental; dotted line, calculated.

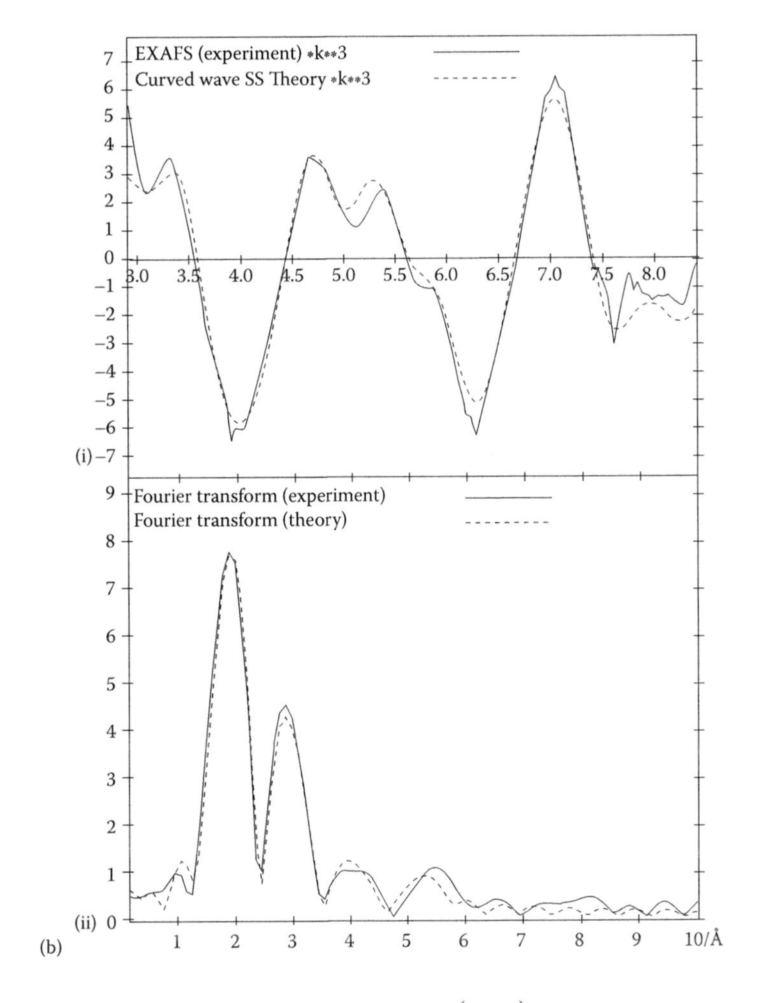

FIGURE 2.31 (CONTINUED) EXAFS data for (a) $(NH_4^+)_2$ $MnBr_4^{2-}$ and (b) the intercalation of $MnBr_4^{2-}$ in a layered double hydroxide clay: (i) extracted EXAFS data and (ii) the radial distribution function: solid line, experimental; dotted line, calculated.

changes to distorted octahedral where Mn is now su.rrounded by four O atoms at a distance of 0.192 nm (1.92 Å) and two Br atoms at a distance of 0.225 nm (2.25 Å).

We saw earlier (Figure 2.11) that powder X-ray diffraction can be used to follow phase changes over time with heating; Figure 2.32 shows the corresponding iron K-edge EXAFS analysis for the same ferrosilicon sample. As the alpha ferrosilicon changes into the beta phase, a shell of eight silicon atoms at about 2.34 Å is found to surround Fe in both forms, but the beta form is found to have only two Fe atoms coordinated to Fe 0.298 nm (2.97 Å), compared with 3.4 Fe atoms at 0.268 nm (2.68 Å) in the alpha phase.

X-rays are very penetrating, so EXAFS, like X-ray crystallography, is used to examine the structure of the bulk of a solid. It has the disadvantage that it only provides information on interatomic distances, but has the considerable advantage

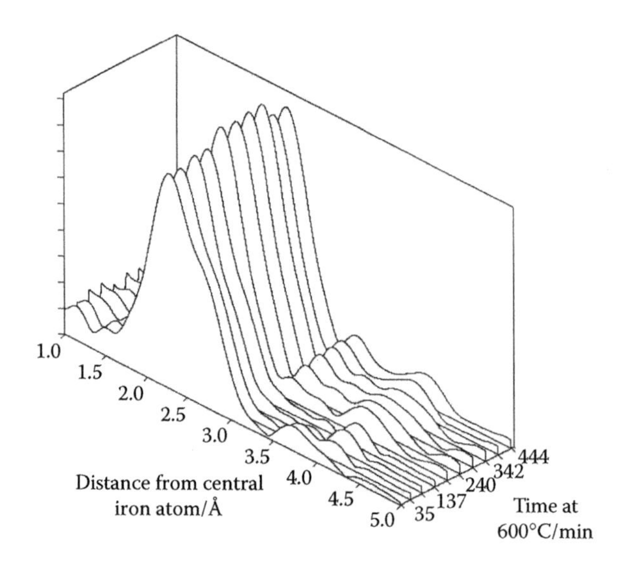

FIGURE 2.32 EXAFS patterns as a function of time, showing the phase evolution of β-ferrosilicon from the α-form when heated at 600°C. (From Professor Frank J. Berry. With permission.)

that it is not restricted to crystalline samples, and can be used on amorphous solids, glasses, and liquids. Additionally, through the use of different absorption edges, it can be used to investigate the coordination around more than one type of atom in the sample.

2.10.2 X-Ray Absorption Near-Edge Structure, XANES, and Near-Edge X-Ray Absorption Fine Structure, NEXAFS

These two techniques are in principle synonymous and refer to the regions of an X-ray absorption spectrum up to about 150 eV either side of an absorption edge; the spectrum contains the pre-edge, the position of the edge itself, and the XANES fine structure just after the edge (Figure 2.30). The acronym NEXAFS is now used almost exclusively to refer to the spectra of the light elements up to C, N, and O. The fine structure close to either side of the edge (Figure 2.30) is very dependent on the nature of the chemical environment of the absorbing atom, and the position of the edge itself depends on oxidation state.

The precise position of the absorption edge shifts to higher energy with the increasing chemical oxidation state of the absorbing atom. An edge may shift by up to 5 eV with each increment in oxidation number. The shifts and heights of the edge can be used to determine the oxidation state of the absorbing metal. It is often used as a fingerprinting method using the spectra of known standards, and is thus useful for determining the ratio of oxidation states in mixed oxides, composites, or impure samples.

The fine structure in the pre-edge region is due to both the oxidation state of the metal and to the coordination chemistry surrounding it: notice the dramatic changes in the pre-edge region in Figure 2.31 as both oxidation state and coordination around Mn change. The peaks here, before complete ionisation occurs at the absorption edge, are due to the excitation of electrons into vacant energy levels of the metal. We know from ligand-field theory just how much the pattern of molecular orbital energy levels changes when we move from octahedral coordination to distorted octahedral, and this is very different again from tetrahedral coordination. The number of electrons in the highest-occupied molecular orbitals varies with the oxidation state of the metal: for instance, Mn(II) is d^5 and Mn(IV) d^3. Both these factors therefore lead to the observation of a different pattern of absorption for each compound.

The changes in the shape of the immediate post-edge structure are due to the multiple scattering events (when the photoelectron is scattered sequentially by more than one atom) which affect these low-kinetic energy photoelectrons. Multiple scattering must also be dependent on atomic position and thus on the coordination and bonding around the absorbing atom.

XANES is not straightforward to interpret but has some advantages over EXAFS. It is a stronger signal and so can be carried out at lower concentration or sample size and can be used to determine the nature of contaminants or defects; it is not very temperature sensitive, and it can therefore be recorded at room temperature. In addition it can provide information on the bonding and molecular orbitals present, or on the band structure and density of available electronic states (see Chapter 4). It can also be used to follow the evolution of species during a reaction, and to follow the kinetics.

2.11 X-RAY PHOTOELECTRON SPECTROSCOPY (XPS)

IUPAC named this technique ESCA (Electron Spectroscopy for Chemical Analysis); however, XPS is the more commonly accepted name used by the larger part of the scientific community. XPS is a surface-sensitive quantitative spectroscopic technique used to identify elements, their concentrations, and their chemical state within the sample. The sample is bombarded with a monochromatic beam of soft X-rays (1–3 keV) which causes the emission of photoelectrons (photoemission) from both core and valence levels of surface atoms into the vacuum. The principle behind this phenomenon is that of Einstein's photoelectric effect, which states that when a solid is exposed to X-rays of energy $h\nu$, electrons in orbitals with binding energies less than $h\nu$ can be ejected, and subsequently detected in a spectrometer with a kinetic energy given by:

$$KE_i = hv - BE_i - \varphi_{sp} \tag{2.16}$$

where BE_i is the binding energy of an electron in an orbital i measured with respect to the Fermi level of the sample (the zero energy level reference in the solid), φ_{sp} being a correction for the work function of the spectrometer (typically 3 ± 5 eV). Usually X-rays emitted from Mg or Al targets are used ($K\alpha$ = 1253.6 eV and 1486.6 eV, respectively). The actual spectrum shows the number of emitted electrons as a function of their binding (kinetic) energy. The lines in the spectrum are due to core level ionization, Auger emission, and valence level emissions superimposed on the background of inelastically scattered electrons.

Thus, one of the core elements of every XPS instrument is an electron energy analyser. It measures the energy spectrum of the electrons emitted from the solid. Usually, a concentric hemispherical analyzer (CHA) is used in which electrons follow a distinct pathway, only passing from the entrance slit to the exit slit if they have the correct kinetic energy. This energy is determined by the potential difference maintained between the two hemispheres and their radii (typically 100–200 mm), which in turn makes it possible to exclude electrons of different energies (they will not reach the exit slit and therefore will not be counted by the detector). XPS operates in a constant pass energy mode, where the potentials on the CHA are set to a constant value to transmit electrons of a single energy, called the pass energy. This can be varied to higher or lower values to (a) maximize throughput for survey scans (higher values) or (b) to enhance resolution when detailed scans are acquired (lower values). The actual spectrum is produced by a scanned electron lens fitted to the entrance aperture. Larger operating distances between the sample and the analyzer and a constant energy resolution ($\Delta E/E_{pass}$) are achieved and maintained across the spectrum. Certain additional lens components or voltages can alter/improve the physical or electronic definition of the analysis area. Spatial resolutions of ca. 2–10 μm can be achieved in this manner. Position-sensitive detectors allow for higher electron counting statistics and are now almost exclusively used.

Figure 2.33 shows the main components of an XPS system and a typical spectrum.

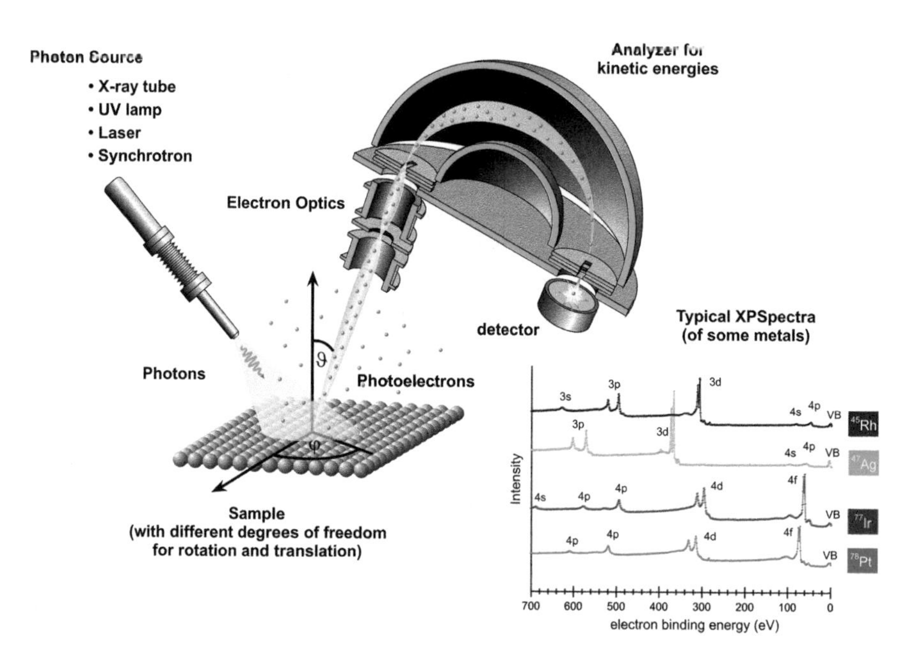

FIGURE 2.33 Schematic drawing of a typical XPS setup with photon source (X-rays, UV light, laser, or synchrotron radiation), a sample manipulation with different linear and rotational degrees of freedom, electron optics, an energy dispersive analyzer, and a detector. XP spectra (intensity vs. binding energy) are shown for four different metals with their element specific distribution of core level photoemission. (Courtesy of Frank Müller.)

XPS is an extremely powerful analytical technique for samples which can withstand the necessary ultra-high vacuum conditions necessary to carry out the experiment. The most important features are:

a) Quantitative analysis (down to ppm range) to identify every element with an atomic number of 3 (boron) or higher in the outermost 10–50 Å of a surface.
b) Determination of the oxidation or chemical state of elements in a compound.
c) If clean single crystalline surfaces are prepared (under ultra-high vacuum) XPS can be employed, in conjunction with other related methods, to probe properties such as:
   - band structure
   - electronic states
   - adsorption phenomena
   - initial stages of compound formation

Further reading about this powerful method is highly recommended as a more detailed description of it would unfortunately be out of the scope of this textbook.

2.12 SOLID-STATE NUCLEAR MAGNETIC RESONANCE SPECTROSCOPY

An NMR nucleus must have a nuclear spin, I, and thus a nuclear magnetic moment. The size of the magnetic moment depends both on I which can take values of 0, $\frac{1}{2}$, 1, $\frac{3}{2}$, 2, $\frac{5}{2}$, …, and on its magnetogyric ratio, γ. In the presence of a magnetic field Zeeman splitting occurs and $2I+1$ non-degenerate spin states are formed between which $2I$ allowed transitions can take place between adjacent energy levels. An RF signal at the NMR frequency causes transitions between the spin states to give the NMR spectrum.

For many years solid-state NMR spectra gave broad resonances which were difficult to interpret. A magnetic dipole interacts with all the other dipoles in its immediate environment. In solution NMR spectroscopy of nuclei with spin $I=½$, the dipolar interactions and anisotropic effects are averaged out by the constant molecular motion, but this is not so in the solid state, and in consequence the NMR spectra of their solids tend to be very broad and uninterpretable. In addition, because of the lack of movement in a solid, the spin-lattice relaxation times can be long, giving low signal-to-noise ratio.

Methods have been developed to give more sharply resolved spectra, such that **magnetic dipolar interactions** can be removed by the application of a high-power decoupling field at the resonance frequency. The chemical shift of a particular atom varies with the orientation of the molecule to the field. In a polycrystalline solid this gives a range of values, an effect known as the **chemical shielding anisotropy**, which broadens the resonance. In MAS a polycrystalline sample is formed into a cylinder and then placed in a magnetic field inclined at an angle β to the field. If the sample is then rotated around its cylindrical axis, a time averaging of the anisotropic interactions takes place. As long as the spinning is fast enough, each individual contribution to the resultant NMR signal depends on a term ($3\cos^2\beta - 1$). If this expression is set to equal to zero it can be solved to find the angle—the so-called magic angle—and this angle can be manually set. The term becomes zero when $3\cos^2\beta = 1$, or $\cos\beta = (1/3)^{1/2}$, i.e. $\beta = 54° 44'$. As long as the spinning is fast enough, the contributions from all crystallites in the sample are the same and just a single line is observed.

Spinning the sample in this way improves the resolution in chemical shift of the spectra and this sometimes lends its name to the technique as **magic angle spinning spectroscopy (MAS NMR)**. The spinning speed has to be greater than the frequency spread of the signal; if it is less, as may be the case for very broad bands, then a set of so-called 'spinning side-bands' are observed, and care is needed in assigning the central resonance. Spinning speeds of over 100 kHz are now being achieved. The technique for assigning the central resonance is to collect the spectra at two different spinning speeds, when the spinning side bands move but the central resonance stays fixed. MAS NMR is sometimes used as an umbrella term to imply the application of any or all of these techniques in obtaining a solid-state NMR spectrum.

Some isotopes have long spin-lattice relaxation times in solids, which give rise to poor signal-to-noise ratios. In order to increase the signal from these isotopes, the sensitivity can be improved by using a technique known as **cross-polarization**, where a complex pulse sequence transfers polarization from an abundant spin nucleus to the dilute spin thereby enhancing the intensity of its signal, improving signal-to-noise.

Advances in superconducting magnets, which can now provide over 20 T, and improved techniques for eliminating the broadening are now giving much higher resolution spectra and rendering the technique more accessible. The improved resolution can allow J couplings to be measured and this has meant the introduction of two-dimensional NMR methods. In some sophisticated experiments internuclear distances can be measured.

The ability to use density functional theory, DFT, methods to calculate chemical shifts and quadrupolar coupling constants now also means that correlations from structure to parameter can be made.

High-resolution spectra can be measured for many different NMR isotopes. Solid-state NMR has proved very successful in elucidating zeolite structures. Zeolites are three-dimensional framework silicate structures where many of the silicon sites are occupied by aluminium (Chapter 7). Because Al and Si are next to each other in the periodic table they have similar X-ray atomic scattering factors, and consequently are virtually indistinguishable on the basis of X-ray crystallographic data. It is possible to build up a picture of the overall shape of the framework with accurate atomic positions but difficult to decide which atom is Si and which is Al.

^{29}Si has a nuclear spin $I = 1/2$ and so gives sharp spectral lines with no quadrupole broadening or asymmetry; the sensitivity is quite high and ^{29}Si has a natural abundance of 4.7%. Pioneering work using solid-state NMR on zeolites was carried by Lippmaa and Engelhardt in the late 1970s. They showed that up to five peaks could be observed for the ^{29}Si spectra of various zeolites and that these corresponded to the five different Si environments that can exist. Each Si is coordinated by four oxygen atoms, but each oxygen can then be attached either to a Si or to an Al atom giving the five possibilities: $Si(OAl)_4$, $Si(OAl)_3(OSi)$, $Si(OAl)_2(OSi)_2$, $Si(OAl)(OSi)_3$, and $Si(OSi)_4$. They also showed, most importantly, that characteristic ranges of these shifts could be assigned to each coordination type—the more Al linkages, the more positive the shift. These ranges could then be used in further structural investigations of other zeolites (Figure 2.34). A solid-state NMR spectrum of the zeolite, analcite, is shown in Figure 2.35. Analcite has all five possible environments. Even with this information it is still an extremely complicated procedure to decide where each linkage occurs in the structure.

Most of the NMR nuclei do not have $I = ½$, but are quadrupolar nuclei, with a nuclear spin of $I > ½$ and a nuclear charge distribution that is not spherically symmetric: the distorted spheres of charge may be prolate (elongated) or oblate (flattened) spheroids. These nuclei present much more of a challenge in terms of signal broadening as the quadrupole interacts strongly with local electric field gradients caused by any non-symmetrical local chemical coordination around the nucleus. This provides an efficient spin-lattice relaxation mechanism for the magnetic energy levels, so giving rise to relatively short spin relaxation times in the solid state. The size of the coupling, ω_Q, depends on both the particular nucleus and its site symmetry and both first-order ($\propto \omega_Q$) and second-order ($\propto \omega_Q^2$) interactions affect the magnetic energy levels (Figure 2.36). In a highly symmetrical environment, such as octahedral or tetrahedral, the lines will be much sharper. The first-order interaction can be mostly suppressed by spinning at the magic angle. Methods of eliminating the broadening due to the second-order interaction have now been found, but they are not straightforward. The **double rotation method, DOR,** spins the sample

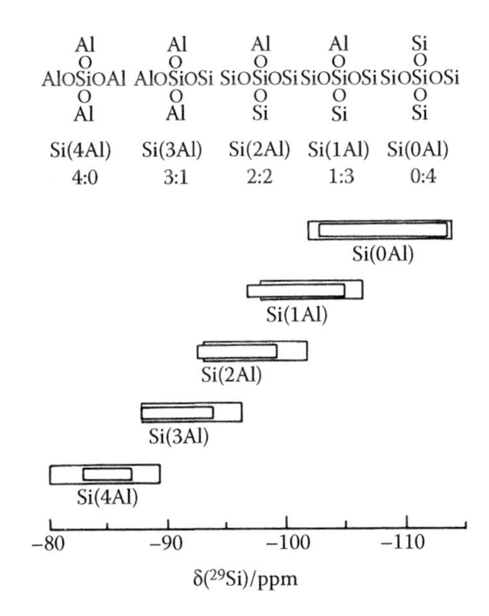

FIGURE 2.34 The five possible local environments of a silicon atom together with their characteristic chemical shift ranges. The inner boxes represent the ^{29}Si shift ranges suggested in the earlier literature. The outer boxes represent the extended ^{29}Si shift ranges, which are more unusual.

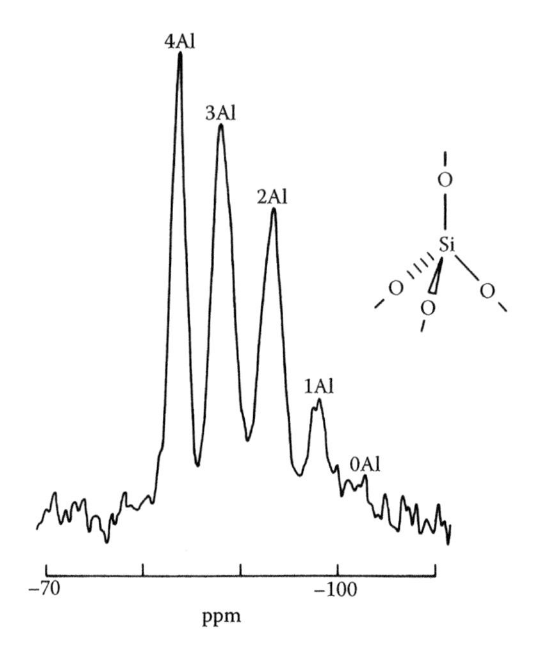

FIGURE 2.35 ^{29}Si solid-state NMR spectrum at 79.6 MHz of analcite, showing five absorptions characteristic of the five possible permutations of Si and Al atoms attached at the corners of the SiO_4 tetrahedron as indicated.

E

Zeeman interaction

Quadrupolar interaction—first-order $\propto \omega_Q$

Quadrupolar interaction—second order $\propto \omega_Q^2$

3/2

1/2

–1/2

–3/2

Central transition

FIGURE 2.36 Diagram showing the splitting of spin states when $I=3/2$.

simultaneously at two angles—the magic angle of 54.74° and 30.56° to suppress both interactions. This method uses a 'rotor-in-a-rotor' design which is technologically quite difficult. **Dynamic angle spinning, DAS,** spins sequentially at two different angles. More recently technically simpler techniques have been introduced known as **multiple-quantum magic-angle spinning, MQMAS,** first applied in 1995, and **satellite transition magic-angle spinning, STMAS,** introduced in 2000.

^{27}Al has a 100% natural abundance and a nuclear spin $I = \frac{5}{2}$, resulting in a strong resonance which is broadened and rendered asymmetric by the second-order quadrupolar effects. However, interpreting the ^{27}Al solid-state NMR spectra of minerals and zeolites is now achievable; for instance it has been used to determine the three aluminium positions in kyanite, Al_2SiO_5. Diagnostically it can also be used to distinguish different types of aluminium coordination: octahedrally coordinated $[Al(H_2O)_6]^{3+}$, which is frequently trapped as a cation in the pores of zeolites and gives a peak at about 0 ppm ($[Al(H_2O)_6]^{3+}$(aq) is used as the reference); tetrahedral $Al(OSi)_4$, which gives rise to a single resonance with characteristic Al chemical shift values for individual zeolites in the range 50–65 ppm; and $AlCl_4^-$, which may be present as a residue from the preparative process and which has a resonance at about 100 ppm.

2.13 THERMAL ANALYSIS

Thermal analysis methods investigate the properties of solids as a function of a change in temperature. They are useful for investigating phase changes, polymorphism, thermodynamics of reactions, decomposition, loss and uptake of solvent or gases, as well as for constructing phase diagrams. Thermal analysis has experienced a little renaissance in the field of solid-state chemistry during the past decades. The coupling of thermal analysis with various spectroscopic and spectrometric methods added the possibility to follow reactions in situ and to quantitatively analyse/identify the developed gases during temperature changes.

2.13.1 Differential Thermal Analysis, DTA

A phase change produces either an absorption or an evolution of heat. The sample is placed in one chamber, and a reference material, usually a solid that will not change phase over the temperature range of the experiment, in the other. Both chambers are heated at a controlled uniform rate in a furnace, and the difference in temperature between the two is monitored and recorded against time. Any reaction in the sample will show as a peak in the plot of differential temperature; exothermic reactions give an increase in temperature, and endothermic a decrease, so the peaks appear in opposite directions. Figure 2.37(a) shows three exotherms in the DTA of KNO_3, due to (i) a phase change from tetragonal to trigonal at 129°C, (ii) melting at 334°C, and (iii) decomposition above 550°C.

2.13.2 Thermogravimetric Analysis, TGA

In this experiment the weight of a sample is monitored as a function of time as the temperature is increased at a controlled uniform rate. The loss of solvent of crystallization, or volatiles such as oxygen, carbon dioxide, or decomposition products which are in gas form at various temperatures shows up as a weight loss. Oxidation or adsorption of gas shows up as a weight gain. The TGA plot for KNO_3 in Figure 2.41(b) shows a small weight loss up to about 550°C, probably due to the loss of adsorbed water, followed by a dramatic weight loss when the sample decomposes.

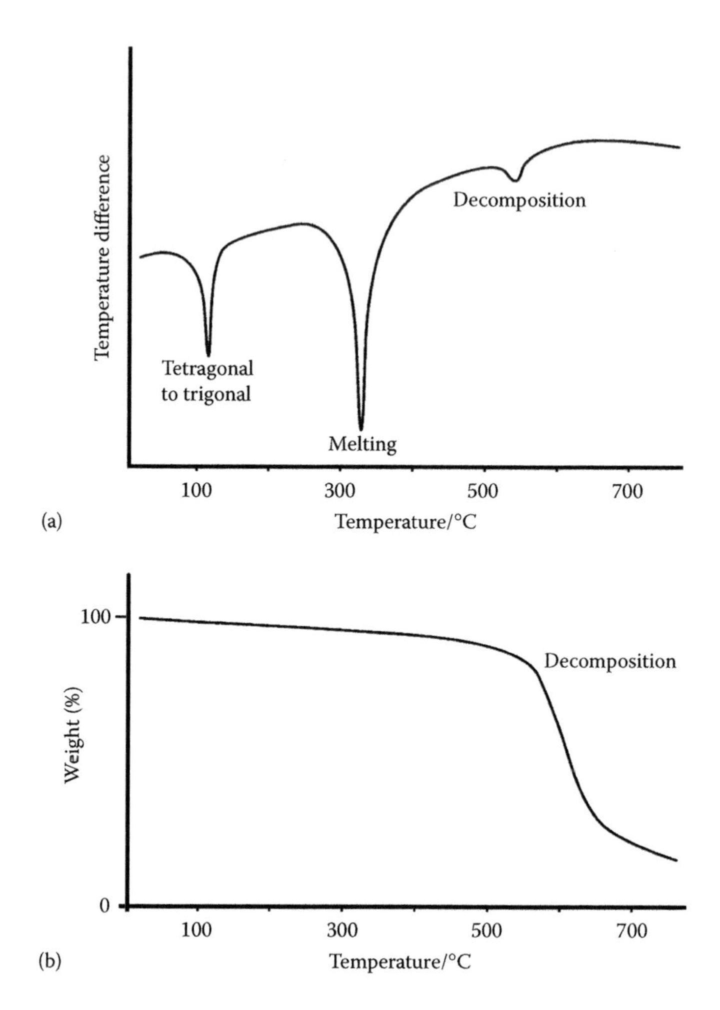

FIGURE 2.37 (a) The DTA and (b) the TGA trace for KNO_3.

2.13.3 Differential Scanning Calorimetry, DSC

Differential Scanning Calorimetry, DSC, measures the amount of heat released by a sample as the temperature is increased or decreased at a controlled uniform rate, and so it can be used to investigate chemical reactions as well as phase changes or the transformation into a different crystalline form. Thus, heats of reaction for phase changes can also be measured.

2.13.4 Simultaneous Thermal Analysis, STA, and Coupling with Spectroscopic or Spectrometric Methods

Devices which can analyse a sample in terms of the amount of heat released or adsorbed and the change in weight over a range of temperature is called STA, as it uses DSC and TGA simultaneously. The technique can further be enhanced by coupling the instrument with an evolved gas analyser which detects and, sometimes quantitatively, analyses all the gaseous compounds formed at a specific temperature and time. Methods coupled with STA are commonly FT-IR (Fourier-Transform Infra-Red spectroscopy), mass spectrometry, or gas chromatography. In this manner, detailed information is obtained not only from the sample itself; but the amount and identity of any evolved gas is also analysed at any specific time and temperature. The amount of information gained from such measurements can be immensely useful as now, every signal appearing in the TGA or DSC can also be interpreted better. Phase changes can be distinguished from chemical reactions or the formation of a different crystalline phase. If a loss in mass is recorded at a specific temperature, the identity, amount, and exact time when the evolved gas or gas mixture was released can be determined.

2.14 TEMPERATURE PROGRAMMED REDUCTION, TPR

Temperature programmed reduction measures the reaction of hydrogen with a sample at various temperatures. The results are interpreted in terms of the different species present in the sample and their amenability to reduction, and so can give information on the presence of different oxidation states, or the effect of a dopant in a lattice. It is useful for measuring the temperature necessary for the complete reduction of a catalyst and is commonly used to investigate the interaction of a metal catalyst with its support, or the effect of a promoter on a metal catalyst.

The sample is heated over time in a furnace under a flowing gas mixture, typically 10% H_2 in N_2. Hydrogen has a high electrical conductivity, so a decrease in hydrogen concentration is marked by a decrease in conductivity of the gas mixture; this change is measured by a thermal conductivity cell (katharometer) and is recorded either against time or temperature.

The peaks in a TPR profile (Figure 2.38) measure the temperature at which various reduction steps take place. The amount of hydrogen used can be determined

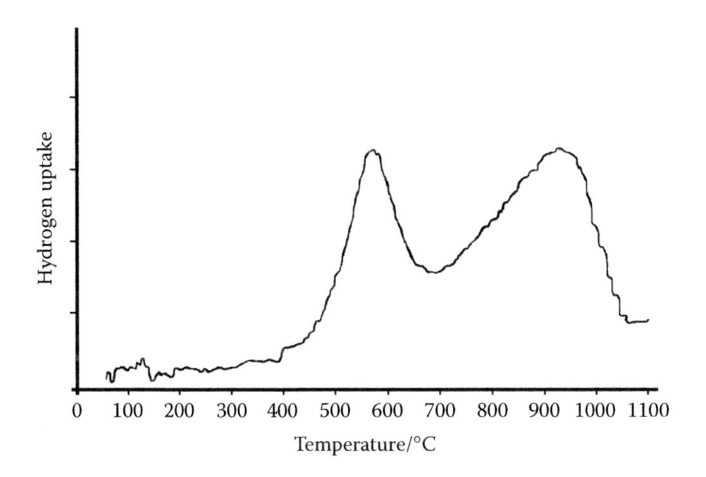

FIGURE 2.38 TPR trace for α-Fe_2O_3.

from the area under the peak. In the example shown, we see two peaks, indicating that the reduction of Fe_2O_3 takes place in two stages, first to Fe_3O_4 and then to metallic Fe.

Other temperature-programmed techniques include Temperature Programmed Oxidation and Sulfidation, TPO and TPS, for investigating oxidation and sulfidation behaviour, and Temperature Programmed Desorption, TPD (also called Thermal Desorption Spectroscopy, TDS), which analyses gases desorbed from the surface of a solid or a catalyst on heating.

2.15 OTHER TECHNIQUES

In an introductory general text such as this, it is impossible to do justice to the plethora of physical techniques which are currently available to the solid-state chemist. We have therefore concentrated on those which are most widely available either in the laboratory or in national facilities. For other techniques one will need to refer to more specialized texts. We have not covered vibrational spectroscopy (IR and Raman), as good treatments of this subject are to be found in most undergraduate physical chemistry textbooks. Our coverage is of necessity fairly brief, and much more detail is to be found in the texts listed in the Further Reading section at the end of this book.

2.16 SUMMARY

Table 2.5 summarises the characterisation methods which we have discussed in this chapter together with their applications.

TABLE 2.5
Comparison of the Different Characterisation Methods Described in This Chapter

Name	Abbreviation	Source	Applications	Max. resolution	Comments
X-ray Diffraction	XRD (PXRD, SXRD)	X-ray tube/synchrotron	3D (crystal) structure determination/identification	pm, bond distances with accuracy < 0.01 Å	
X-ray Microscopy/ Tomography	XRM (μCT)	X-ray tube/synchrotron	3D imaging	μm	
Scanning Electron Microscopy	SEM	Electron beam (1–30kV)	3D surface/appearance	nm	
Transmission Electron Microscopy	TEM	Electron beam (100–300kV)	Imaging though very thin sample	0.1 nm, 1 Å	Can be used for both 2D imaging and diffraction
Cryogenic Electron Microscopy	Cryo EM	Electron beam (100–300kV)	Structure determination of large molecules, i.e. proteins	0.1 n,, 1 Å	sample embedded in vitrified water at −190°C
Energy Dispersive X-ray Analysis	EDX (EDAX)	Electron beam (100–300kV)	Qualitative and quantitative elemental analysis	For elements with atomic number > 11	Inelastic scattering cause element specific emission of X-rays
Electron Energy Loss Spectroscopy	EELS	Electron beam (100–300kV)	Qualitative and quantitative elemental and band gap analysis	Especially good for analysis of light elements	Measurement of absorption energies through inelastic scattering
Scanning Tunnelling Microscopy	STM	Electrons tunnelling between metal tip and surface (d < 1 nm)	Surface imaging of conducting and semiconducting materials	nm	
Atomic Force Microscopy	AFM	Coulomb repulsion of metal tip on surface at d < 1 nm	Surface imaging of nonconducting materials	nm	Non-contact mode: stiff cantilever is oscillated just above its resonance frequency

(Continued)

TABLE 2.5 (CONTINUED)

Comparison of the Different Characterisation Methods Described in This Chapter

Name	Abbreviation	Source	Applications	Max. resolution	Comments
X-ray Absorption Spectroscopy	XAS (EXAFS, XANES, NEXAFS)	Synchrotron radiation in X-ray frequency range	Determination of short-range order in non-crystalline materials	Interatomic distances, coordination environment of crystalline and non-crystalline samples	Study of absorption edges and its fine structure of specific elements
X-ray Photoelectron Spectroscopy	XPS (ESCA)	Soft X-rays (1–3keV)	Quantitative analysis of surface elements and their chemical state	ppm, for elements with atomic number > 3	Analyses emitted photoelectrons
Solid-State Nuclear Magnetic Resonance	ssNMR	Magnetic field, RF signal	Qualitative analysis, identification, bond distances	Å range, bond distances	
Thermogravimetric Analysis	TGA	Balance, thermocouple, electric heating	Analysis of decomposition or solvent loss temperature, in situ reaction monitoring	Balance resolution 0.1 µg	Weight change as a function of time during heating/cooling
Differential Thermal Analysis	DTA	Thermocouple	Melting, crystallisation, polymerisation point, determination of phase changes	Temp. resolution 0.001 K	Monitors heat change compared to a reference material as a function of time
Differential Scanning Calorimetry	DSC	Thermocouple	Melting, crystallisation, polymerisation point, determination of phase changes	See above	Measures the amount of heat released while temperature is changed

QUESTIONS

1. What are the spacings of the *100*, *110*, and *111* planes in a cubic crystal system of unit cell dimension a? In what sequence would you expect to find these reflections in a powder diffraction photograph?
2. What is the sequence of the following reflections in a primitive cubic crystal: *220*, *300*, and *211*?
3. Nickel crystallizes in a cubic crystal system. The first reflection in the powder pattern of nickel is the *111*. What is the Bravais lattice?
4. The $\sin^2\theta$ values for Cs_2TeBr_6 are listed in Table 2.6 for the observed reflections. To which cubic class does it belong? Calculate its unit-cell length assuming Cu-K_α radiation of wavelength 154.2 pm.
5. X-ray powder data for NaCl is listed in Table 2.7. Determine the Bravais lattice, assuming that it is cubic.

TABLE 2.6
$\sin^2\theta$ Values for Cs_2TeBr_6

$\sin^2\theta$
.0149
.0199
.0399
.0547
.0597
.0799
.0947

TABLE 2.7
θ Values for NaCl

θ_{hkl}
13° 41′
15° 51′
22° 44′
26° 56′
28° 14′
33° 7′
36° 32′
37° 39′
42° 0′
45° 13′
50° 36′
53° 54′
55° 2′
59° 45′

6. Use the data given in Question 5 to calculate a unit-cell length for the NaCl unit cell.
7. If the unit cell length of NaCl is $a = 563.1$ pm and the density of NaCl is measured to be 2.17×10^3 kg m^{-3}; calculate Z, the number of formula units in the unit cell. (The atomic masses of sodium and chlorine are 22.99 and 35.45 respectively.)
8. A powder diffraction pattern establishes that silver crystallizes in a face-centered cubic unit cell. The *111* reflection is observed at $\theta = 19.1°$ using Cu–K_α radiation. Determine the unit cell length, a.
9. Calcium oxide crystallizes with a face-centered cubic lattice, $a = 481$ pm and a density $\rho = 3.35 \times 10^3$ kg m^{-3}: calculate a value for Z. (Atomic masses of Ca and O are 40.08 and 15.999 respectively.)
10. Thorium diselenide, $ThSe_2$, crystallizes in the orthorhombic system with $a = 442.0$ pm, $b = 761.0$ pm, $c = 906.4$ pm and a density $\rho = 8.5 \times 10^3$ kg m^{-3}: calculate Z. (Atomic masses of Th and Se are 232.03 and 78.96 respectively.)
11. Cu crystallizes with a cubic close-packed structure. The Bragg angles of the first two reflections in the powder pattern collected using Cu-K_α radiation are 21.6° and 25.15°. Calculate the unit cell length a, and estimate a radius for the Cu atom.
12. Arrange the following atoms in order of their ability to scatter X-rays: Na, Co, Cd, H, Tl, Pt, Cl, F, O.
13. In a cubic crystal we observe the *111* and *222* reflections, but not *001*. What is the Bravais lattice?
14. A student was asked to index a powder diffraction pattern for a material that was meant to be cubic. The student's answer, however, gave a triclinic cell. If the sample is diffracted using Cu $K_{\alpha 1}$ radiation ($\lambda = 1.54056$ Å), the following peaks are obtained along 2θ: 18.52°, 25.80°, 32.16°, 38.01°, 42.13°, 45.98°, 54.23°, 57.91°, 61.09°, and 64.53°. Index the data, showing any working clearly, assign the Miller indices, and deduce the crystal system. What are the lattice parameters of the material?
15. Interpret the EXAFS radial distribution function for $Co(CO)_4$ shown in Figure 2.39.

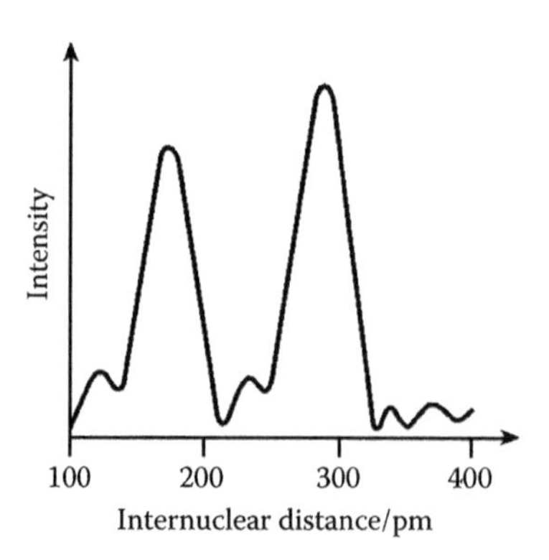

FIGURE 2.39 Co edge EXAFS radial distribution function for $Co(CO)_4$.

16. Figure 2.40 shows the ^{79}Si MAS NMR spectrum of the zeolite faujasite. Use this figure to determine which Si environments are most likely to be present.
17. Figure 2.41 shows the ^{29}Si spectrum of the same sample of faujasite as Figure 2.40, but after treatment with $SiCl_4$ and washing with water. What has happened to it?
18. A sample of faujasite was treated with $SiCl_4$ and four ^{27}Al MAS NMR spectra were taken at various stages afterwards (Figure 2.42). Describe carefully what has happened during the process.
19. A TGA trace for 25 mg of a hydrate of manganese oxalate, $MnC_2O_4 \cdot H_2O$ showed a weight loss of 20 mg at 100°C. What was the composition of the hydrate? A further weight loss occurs at 250°C, but then a weight *gain* at 900°C. What processes might be taking place?
20. Figure 2.43 shows the DTA and TGA traces for ferrous sulfate heptahydrate. Describe what processes are taking place.
21. What is the standard deviation of a bond length of 2.620(1)?

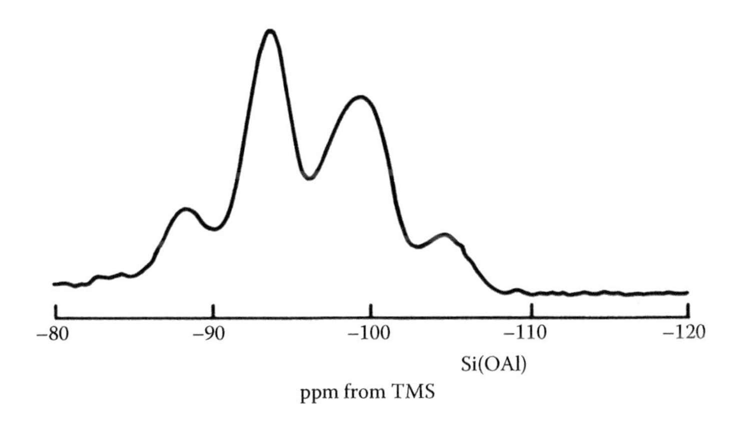

FIGURE 2.40 ^{29}Si MAS NMR spectrum at 79.6 MHz, of faujasite with Si/Al = 2.61 (zeolite-Y).

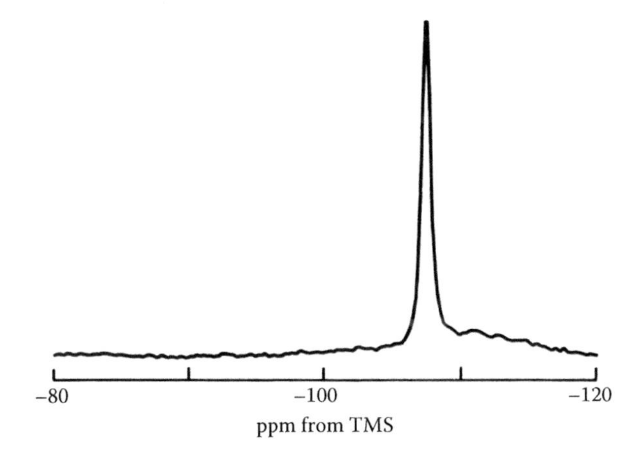

FIGURE 2.41 ^{29}Si MAS NMR spectrum at 79.6 MHz, of faujasite with Si/Al = 2.61 after successive dealumination with $SiCl_4$ and washings.

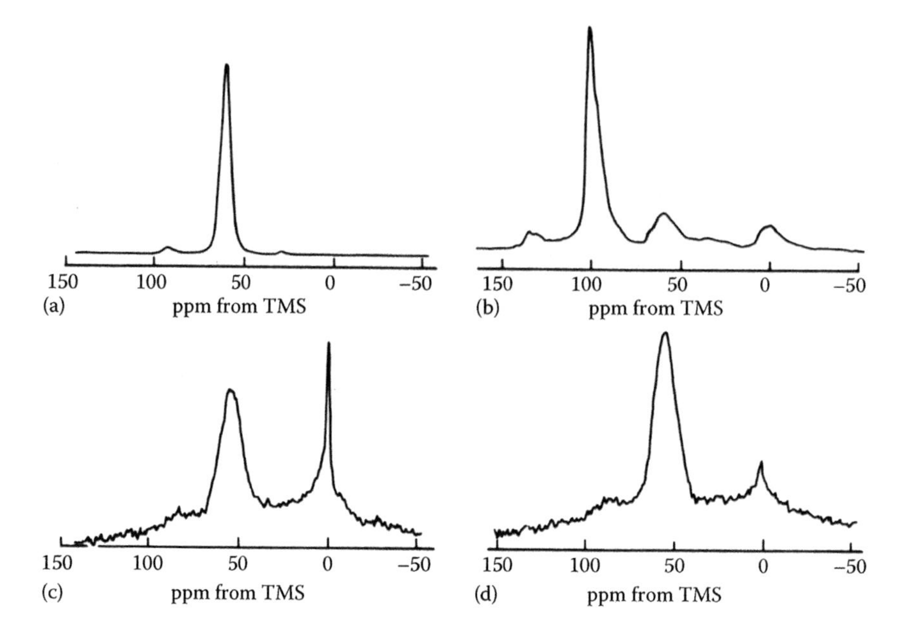

FIGURE 2.42 ^{27}Al MAS NMR spectra at 104.2 MHz obtained on faujasite samples at various stages of the $SiCl_4$ dealumination procedure. (a) Starting faujasite sample. (b) Intact sample after reaction with $SiCl_4$ before washing. (c) Sample (b) after washing with distilled water. (d) After several washings.

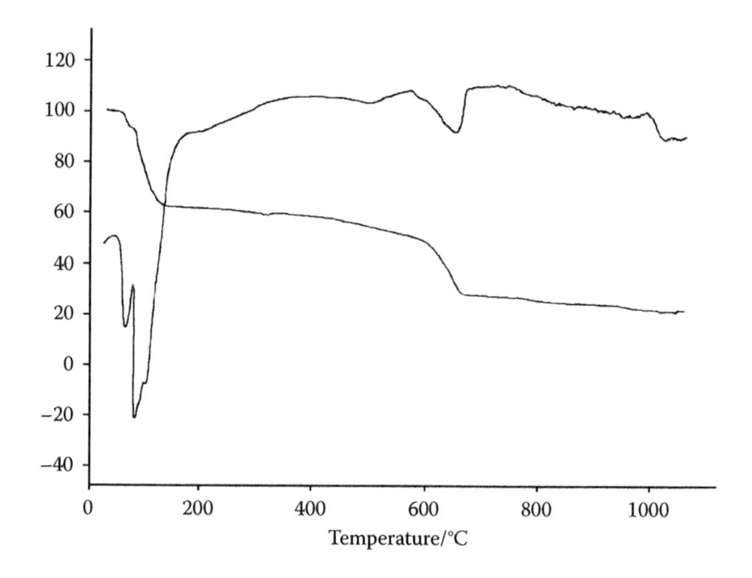

FIGURE 2.43 DTA and TGA traces for ferrous sulfate heptahydrate.

3 Synthesis of Solids

3.1 INTRODUCTION

Interest in the properties of solids and in the development of new materials has given rise to the development of a huge variety of methods for preparing them. The method chosen for any solid will depend not only on the composition of the solid but also on the form it is required in for its proposed use. For example, silica glass for fibre optics needs to be much freer of impurities than silica glass used for manufacturing laboratory equipment. Some methods may be particularly useful for producing solids in forms that are not stable under normal conditions, for example, the synthesis of diamond employs high pressures. Other methods may be chosen because they favour the formation of unusual oxidation states, for example, the preparation of chromium dioxide by the hydrothermal method, or because they promote the production of fine powders or, by contrast, large single crystals. For industrial use, a method that does not employ high temperatures could be favoured because of the ensuing energy savings.

In the preparation of solids, usually, care has to be taken to use stoichiometric quantities pure starting materials and to ensure that the reaction has been completed, because it is usually not possible to purify a solid once it has formed.

We do not have space here to discuss all the ingenious syntheses that have been employed, so we shall concentrate on those that are commonly used, with a few examples of the techniques used for solids with particularly interesting properties. The preparation of organic solid-state compounds and polymers is not covered as, generally, it involves organic synthesis techniques, which is a whole field in itself, and is covered in many organic textbooks.

It is difficult to impose a logical order on such a diverse subject. This chapter starts by considering the most basic and most commonly used method of preparing solids, **the ceramic method**: this grand title disguises the fact that it simply means grinding up the reactant solids and heating them hard until they react! We then go on to look at refinements of this method and ways of improving the uniformity of the reaction mixture and reducing the reaction temperature. The following sections on microwave heating and combustion synthesis show alternative methods of inducing solid-state reactions. Later sections concentrate on less well-known methods of preparing inorganic solids, such as using high pressures and gas-phase reactions. We also consider some methods used for the production of particularly pure solids, which are important in the semiconductor industry and for the preparation of single crystals.

3.2 HIGH-TEMPERATURE CERAMIC METHODS

3.2.1 Direct Heating of Solids

Ceramic method: The simplest and most common way of preparing solids is the ceramic method, which consists of heating together two nonvolatile solids, which react to form the required product. This method is widely used both industrially and in the laboratory and can be used to synthesise a whole range of materials, such as mixed metal oxides, sulfides, nitrides, aluminosilicates, and many others—the first high-temperature superconductors were made by a ceramic method. To take a simple example, we can consider the formation of zircon ($ZrSiO_4$), which is used in the ceramics industry as the basis of high-temperature pigments to colour the glazes on bathroom porcelain. It is made by the direct reaction of zirconia (ZrO_2) and silica (SiO_2) at 1300°C:

$$ZrO_2(s) + SiO_2(s) = ZrSiO_4(s).$$

The procedure is to take accurately weighed, stoichiometric amounts of the binary oxides, using the purest of solids, grind them in a pestle and mortar to give a uniform small particle size, and then heat in a furnace for several hours in an alumina crucible (Figure 3.1).

Despite its widespread use, the simple ceramic method has several disadvantages. High temperatures are generally required, typically between 500°C and 2000°C, and this requires a large input of energy. This is because the coordination numbers in these ionic binary compounds are high, varying from 4 to 12, depending on the size and charge of the ion, and it takes a lot of energy to overcome the lattice energy so that a cation can leave its position in the lattice and diffuse to a different site. In addition, the phase or compound desired may be unstable or decompose at such high temperatures. Reactions such as this can be very slow; increasing the temperature speeds up the reaction as it increases the diffusion rate of the ions. In general, the solids are not raised to their melting temperatures, so the reaction occurs in the solid state. Solid-state reactions can only take place at the interface of two solids, and once the surface layer has reacted, the reaction continues as the reactants diffuse from the bulk to the interface. Raising the temperature enables the reaction at the interface and the diffusion through the solid to be faster than at room temperatures; a rule of thumb suggests that a temperature of about two-thirds of the melting temperature of the solids gives a reasonable reaction time. Even so, diffusion is often the limiting step. Because of this, it is important that the starting materials are ground to give a small particle size and are very well mixed to maximise the surface contact area and minimize the distance that the reactants have to diffuse.

To achieve a homogeneous mix of small particles, it is necessary to be very thorough in grinding the reactants. The number of crystallite faces in direct contact with one another can also be improved by pelletising the mixed powders in a hydraulic press. Commonly, the reaction mixture is removed during the heating process and is reground to bring fresh surfaces in contact, thereby speeding up the reaction. Nevertheless, the reaction time is measured in hours; the preparation of the ternary

FIGURE 3.1 The basic apparatus for the ceramic method: (a) pestles and mortars for fine grinding; (b) a selection of porcelain, alumina, and platinum crucibles; and (c) a furnace.

oxide $CuFe_2O_4$ from CuO and Fe_2O_3, for example, takes 23 hours. The product is often not homogeneous in composition because as the reaction proceeds, a layer of the ternary oxide is produced at the interface of the two crystals, so ions now need to diffuse through this before they react. It is usual to take the initial product, grind it again, and reheat several times before a phasepure product is obtained. Usually trial and error has to be used to find out the best reaction conditions, with samples tested by powder X-ray diffraction to determine the phase purity.

Solid-state reactions up to 2000°C are usually carried out in furnaces that use resistance heating: the resistance of a metal element results in electrical energy being converted to heat, as in an electric fire. This is a common method of heating up to 2000°C; however, above this temperature, other methods have to be employed. Higher temperatures in samples can be achieved by directing an electric arc at the reaction mixture (to 3000°C), and for very high temperatures, a carbon dioxide laser (with output in the infrared) can give temperatures up to 4000°C.

Containers must be used for the reactions that can withstand both high temperatures and are also sufficiently inert not to react themselves; suitable crucibles are commonly made of silica (to 1400°C), alumina (to 1900°C), zirconia (to 2000°C), or magnesia (to 2400°C), but metals such as platinum (melting temperature 1700°C) and tantalum (melting temperature 2980°C) and graphite linings are used for some reactions.

If any of the reactant oxides are nonstoichiometric, volatile, hygroscopic, or air sensitive, then it may be possible to use carbonates instead. If this simple method of heating in the open atmosphere is no longer appropriate, a sealed tube method will be needed.

Sealed tube methods: Evacuated tubes are used when the products or reactants are sensitive to air or water or are volatile. An example of the use of this method is the preparation of samarium sulfide. In this case, sulfur has a low boiling temperature (444°C) and an evacuated tube is necessary to prevent it from boiling off and being lost from the reaction vessel.

The preparation of samarium sulfide (SmS) is of interest because it contains samarium (a lanthanoid element) in an unusual oxidation state of +2 rather than the more common state of +3. Samarium metal in powder form and powdered sulfur are mixed together in stoichiometric proportions and heated to around 730°C in an evacuated silica tube. (Depending on the temperature of the reaction, Pyrex or silica is the common choice for these reaction tubes, as both are fairly inert, and can be sealed onto a Pyrex vacuum system for easy handling.) The product from the initial heating is then homogenised and heated again, this time to around 2000°C in a tantalum tube (sealed by welding) by passing an electric current through the tube, the resistance of the tantalum providing the heating.

A high pressure of oxygen may be used to prepare an oxide of high-oxidation state or hydrogen for a reduction reaction. An inert atmosphere may be used to prevent oxidation or decomposition.

The pressures obtained in sealed reaction tubes can be very high, and it is not unknown for tubes to explode, however carefully they are made; it is thus very important to seal them carefully and to take safety precautions such as surrounding the tube with a protective metal container and using safety screens.

Special atmospheres: The preparation of some compounds must be carried out under a special atmosphere, but not necessarily at high pressures: an inert gas such as argon may be used to prevent oxidation to a higher-oxidation state; an atmosphere of oxygen can be used to encourage the formation of high-oxidation states; or, conversely, a hydrogen atmosphere can be used to produce a low-oxidation state. Experiments of this type are usually carried out with the reactant solids in a small boat placed in a tube in a horizontal-tube furnace. The gas is passed over the reactants for a time to expel all the air from the apparatus, and then it flows over the reactants during the heating and cooling cycles, exiting through a bubbler to maintain a positive pressure and prevent the ingress of air by back diffusion (Figure 3.2).

To obtain more homogeneous products and better reaction rates, methods involving particles smaller than those that can be obtained by grinding have been introduced. In a polycrystalline mixture of solid reactants, we might expect the particle size to be of the order of 10 μm; even careful and persistent grinding will only reduce the particle size to around 0.1 μm. Diffusion during a ceramic reaction therefore

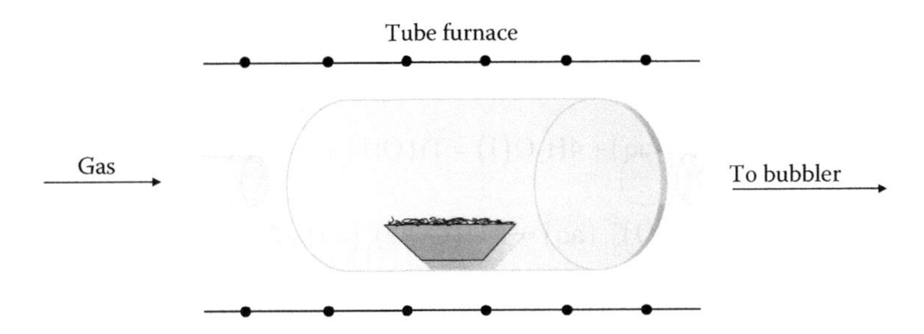

FIGURE 3.2 Reacting solids under a special gas atmosphere.

takes place across anywhere between 100 and 10,000 unit cells. Various ingenious methods, some physical and some chemical, have been pioneered to bring the components of the reaction either into more intimate contact or into contact at an atomic level, thereby reducing this diffusion path; in doing so, the reactions can often take place at lower temperatures.

3.2.2 Precursor Methods

At a simple level, **precursors** such as nitrates and carbonates can be used as starting materials instead of oxides: they decompose to the oxides upon heating at relatively low temperatures, losing gaseous species and leaving behind fine, more reactive powders. An even more intimate mixture of starting materials can be made by the **co-precipitation** of solids. A stoichiometric mixture of soluble salts of the metal ions is dissolved and then precipitated as hydroxides, citrates, oxalates, or formates. This mixture is filtered, dried, ground, and then heated to give the final product. It is necessary to be very careful not to introduce impurities during grinding.

The **precursor method** achieves mixing at the atomic level by forming a solid compound, the precursor, in which the metals of the desired compound are present in the correct stoichiometry. For example, if an oxide MM'_2O_4 is required, a mixed salt of an oxyacid, such as an acetate containing M and M′ in the ratio of 1:2, is prepared. The precursor is then heated to decompose it to the required product. Homogeneous products are formed at relatively low temperatures. A disadvantage is that it is not always possible to find a suitable precursor, but the preparation of barium titanate gives a good illustration of this method.

Barium titanate ($BaTiO_3$) is a ferroelectric material (Chapter 9) that is widely used in capacitors because of its high dielectric constant. It was initially prepared by heating barium carbonate and titanium dioxide at high temperature

$$BaCO_3(s) + TiO_2(s) = BaTiO_3(s) + CO_2(g).$$

However, for modern electronic circuits, it is important to have a product of a controlled grain size and the precursor method is one way to achieve this. (Another way is the sol–gel method, which has also been applied to this material.) The precursor used is an oxalate. The first step in the preparation is to prepare an oxo-oxalate of titanium. Excess oxalic acid solution is added to titanium butoxide, which initially hydrolyzes to give a precipitate and then redissolves in the excess oxalic acid

$$Ti(OBu)_4(aq) + 4H_2O(1) = Ti(OH)_4(s) + 4BuOH(aq)$$

$$Ti(OH)_4(s) + (COO)_2^{2-}(aq) = TiO(COO)_2(aq) + 2OH^-(aq) + H_2O(1).$$

Barium chloride solution is then added and barium titanyl oxalate precipitates

$$Ba^2 + (aq)(COO)_2^{2-}(aq) + TiO(COO)_2(aq) = Ba\left[TiO((COO)_2)_2\right](s).$$

This precipitate contains barium and titanium in the correct ratio and is easily decomposed by heat, to give the oxide. The temperature used for this final heating is 650°C

$$Ba\left[TiO((COO)_2)_2\right](s) = BaTiO_3(s) + 2CO_2(g) + 2CO(g).$$

The decomposition of oxalates is also involved in the preparation of ferrites MFe_2O_4, which are important as magnetic materials (Chapter 9).

A precursor can be used in the preparation of **promoted bimetallic catalysts**. The noble metal of a supported catalyst, for instance, Pt on carbon, is allowed to adsorb hydrogen and then to react with a solution of an organometallic complex of a second metal. The organometallic complex reacts with, and is reduced by, the hydrogen adsorbed in the Pt, to give a metal and a gaseous by-product. The reaction occurs only at the active catalyst sites where the hydrogen has been adsorbed and gives a homogeneous bimetallic product with a very close interface between the two metals.

Synthesis of $YBa_2Cu_3O_{7-x}$. This superconducting oxide can be prepared in undergraduate laboratories by the method of homogeneous co-precipitation from an aqueous urea/oxalic acid solution of the metal nitrates in 1:2:3 stoichiometry. At a temperature between 80 and 100°C hydrolysis of urea takes place with the simultaneous evolution of CO_2 and NH_3. As the urea hydrolyzes, the pH of the solution gradually rises and the metal ions precipitate out as their hydroxide, oxalate, or carbonate salts. The precipitate is then heated in air at 900°C for a least 16 hours in air to burn out all the residual carbon. The powder obtained at this stage is pressed into a pellet and sintered at 900°C for 4 hours, followed by annealing in oxygen at 500°C for another 16 hours. This results in a material with a composition of $YBa_2Cu_3O_{7-x}$ $(0 < x < 0.5)$.

The products of the precursor method are usually crystalline solids, often containing small particles of large surface area. For applications, such as catalysis and barium titanate capacitors, this is an advantage.

3.2.3 Sol–Gel Methods

Precipitation methods always have the disadvantage that the stoichiometry of the precipitate(s) may not be exact if one or more ions are left in the solution. The **sol–gel method** overcomes this, as the reactants never precipitate out. First, a concentrated solution or colloidal suspension of the reactants, the 'sol', is prepared, which is then concentrated or matured to form the 'gel'. This homogeneous gel is then heat-treated to form the product. The main steps in the sol–gel process are outlined in Figure 3.3.

The first investigation of a sol–gel process for synthesis was made in the mid-nineteenth century. This early investigation studied the preparation of silica glass from a sol made by hydrolysing an alkoxide of silicon. Unfortunately, in order to prevent the product cracking and forming a fine powder, an ageing period of a year or more was required. The sol–gel method was further developed in the 1950s and 1960s after it was realised that colloids, which contain small particles (1–1000 nm in diameter), can be highly chemically homogeneous and also that the particles have a very large surface area for reaction.

A sol is a colloidal suspension of particles in a liquid; for the materials being discussed here, these particles will typically be 1–100 nm in diameter. A gel is a semirigid solid in which the solvent is contained in a framework of material that is either colloidal (essentially a concentrated sol) or polymeric.

To prepare solids using the sol–gel method, a sol of reactants is first prepared in a suitable liquid. Sol preparation can be either simply the dispersal of an insoluble solid or the addition of a precursor that reacts with the solvent to form a colloidal product. A typical example of the first is the dispersal of oxides or hydroxides in water with the pH adjusted so that the solid particles remain in suspension rather than precipitate out. A typical example of the second method is the addition of metal alkoxides to water; the alkoxides are hydrolysed, giving the oxide as a colloidal product.

Subsequently, the sol is either treated or simply left to form a gel over time by dehydrating and/or polymerising. To obtain the final product, the gel is heated. This heating serves several purposes—it removes the solvent, it decomposes anions such as alkoxides or carbonates to give oxides, it allows rearrangement of the structure of the solid, and it allows crystallisation to occur. Both the time and the temperature needed for the reaction in sol–gel processes can be reduced from those of the direct ceramic method; in favourable cases, the time from days to hours and the temperature by several hundred degrees.

Cracking in the gel can be avoided if it is dried under supercritical conditions, that is, when the solvent is above its critical temperature. The solvent then stays as a gas throughout the drying process, leaving behind a so-called aerogel, which can be fired to give a porous monolith.

Several examples given below illustrate the sol–gel method; two of the examples have been chosen because they have interesting properties and uses, and are discussed later in this book. Many other materials have also been prepared by the

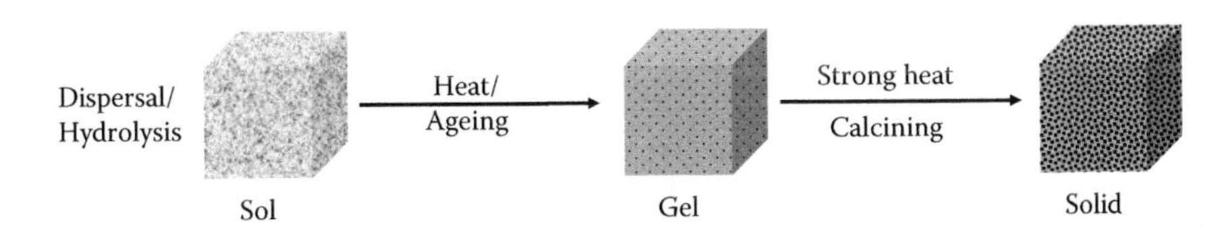

FIGURE 3.3 The steps in the sol–gel synthesis route.

sol–gel method, and other sol–gel preparations have also been employed for the materials chosen; therefore, these examples should be taken as illustrative and not as the main uses of the method.

Lithium niobate ($LiNbO_3$) is a ferroelectric material used as an optical switch (see Chapter 8). Preparation by the simple ceramic method leads to problems in obtaining the correct stoichiometry, and a mixture of phases often results. Several sol–gel preparations have been described, their advantage being the lower temperature required for the preparation and the greater homogeneity of the product. One such preparation starts with lithium ethoxide ($LiOC_2H_5$ (or LiOEt)) and niobium ethoxide ($Nb_2(OEt)_{10}$). Each metal ethoxide was dissolved in absolute ethanol and the two solutions were mixed. The addition of water leads to partial hydrolysis, giving hydroxy-alkoxides, for example:

$$Nb_2(OEt)_{10} + 2H_2O = 2Nb(OEt)_4(OH) + 2EtOH.$$

The hydroxy-alkoxides condense to form a polymeric gel with metal–oxygen–metal links. Lithium niobate is then formed when the gel is heated. Heating removes any remaining ethanol and also any water formed during the condensation. The remaining ethyl groups are pyrolysed (i.e., oxidised to carbon dioxide and water), leaving the oxides.

Zinc oxide, ZnO, is a common, widely used material. It is used for example in cosmetics, sensors, and solar cells. One sol–gel method for preparing ZnO for photocatalysis used zinc acetate and oxalic acid in ethanol to form the gel. Zinc acetate, $Zn(CH_3COO)_2$, was dissolved in ethanol and a mixture of oxalic acid and ethanol added to the solution. After refluxing, the mixture was allowed to cool to room temperature and then dried at 80°C for 20 hours. The dried mixture was then heated at 650°C in air.

Coatings for optical fibres are important because the core of an optical fibre carries the light signal, and the cladding on the outside of the core, with a lower refractive index, traps the light in the centre by total internal reflection (see Chapter 8). Optical fibres need to be extremely pure, completely free of impurities such as transition metal ions; the conventional methods of preparing silica glasses are inadequate. The fibres are made by using a modified form of **chemical vapour deposition** (CVD) (Section 3.7) and the coatings and core are drawn out together. However, coatings can also be added around the silica core as the fibre is drawn through a sol. As well as forming the outer lower refractive-index layer, coatings can be added to give protection and added strength and to provide sensor properties.

3.3 MECHANOCHEMICAL SYNTHESIS

Mechanochemical synthesis involves grinding solid starting materials together at high energy. One way to do this is to use a ball mill. In **ball milling**, the reaction mixture is rotated in a cylindrical device with a grinding medium such as ceramic or metal balls. No solvent is used, but care must be taken to avoid contamination of the product from the grinding medium. For example hydroxyapatite, $Ca_{10}(PO_4)_6(OH)_2$, the mineral in bone, can be synthesized by milling together calcium hydrogen phosphate and calcium hydroxide.

3.4 MICROWAVE SYNTHESIS

We are familiar with the use of microwave radiation in cooking food, where it increases the speed of reaction. This method has been used for the synthesis of solid-state materials such as mixed oxides. The first solid-state reaction experiments were performed in modified domestic ovens, which are still used, but more specialised ovens have also been developed to give more control over the conditions. We shall briefly consider how microwaves heat solids and liquids because this gives us an insight into which reactions will be good candidates for this method.

In a liquid or solid, the molecules or ions are not free to rotate, so the heating is not the result of the absorption of microwaves by molecules undergoing rotational transitions as they would in the gas phase. In a solid or liquid, the alternating electric field of the microwave radiation can act in two ways. If charged particles are present that can move freely through the solid or liquid, then these will move under the influence of the field, producing an oscillating electric current. Resistance to their movement causes energy to be transferred to the surroundings as heat. This is **conduction heating**. If there are no particles that can move freely but there are molecules or units with dipole moments, then the electric field acts to align the dipole moments. It is this effect that produces **dielectric heating**, and when it acts on water molecules in food, it is generally responsible for the heating/cooking in domestic microwave ovens. The electric field of microwave radiation, like that of all electromagnetic radiations, oscillates at the frequency of the radiation. The electric dipoles in the solid do not change their alignment instantaneously, but with a characteristic time (τ). If the oscillating electric field changes its direction slowly so that the time between changes is much greater than τ, then the dipoles can follow the changes. A small amount of energy is transferred to the surroundings as heat each time the dipole realigns, but this is only a small heating effect. If the electric field of the radiation oscillates very rapidly, then the dipoles cannot respond fast enough and do not realign. The frequency of microwave radiation is such that the electric field changes sign at a speed that is the same order of magnitude as τ. Under these conditions, the dipole realignment lags slightly behind the change of electric field and the solid absorbs microwave radiation. This absorbed energy is converted to heat. The quantities governing this process are the dielectric constant (see Chapter 9), which determines the extent of dipole alignment, and the dielectric loss, which governs how efficiently the absorbed radiation is converted to heat.

To use microwave heating in solid-state synthesis, at least one component of the reaction mixture must absorb microwave radiation. The speed of the reaction process is then increased by increasing both the rate of the solid-state reaction and the rate of diffusion, which, as we mentioned earlier, is often the rate-limiting step.

A number of oxides, mostly non-stoichiometric, including CuO, ZnO, V_2O_5, MnO_2, PbO_2, Co_2O_3, Fe_3O_4, NiO, and WO_3, are also efficiently heated by microwave radiation. Other strong absorbers are $SnCl_2$ and $ZnCl_2$. Oxides that do not absorb strongly include TiO_2, CeO_2, Fe_2O_3, Pb_3O_4, SnO, Al_2O_3, and La_2O_3. It is also possible to add a substance that does not take part in the reaction but absorbs microwave radiation, for example carbon.

High-temperature superconductors (see Chapter 10) were first prepared using a conventional ceramic method. For example $YBa_2Cu_3O_{7-x}$. (Figure 3.4) can be

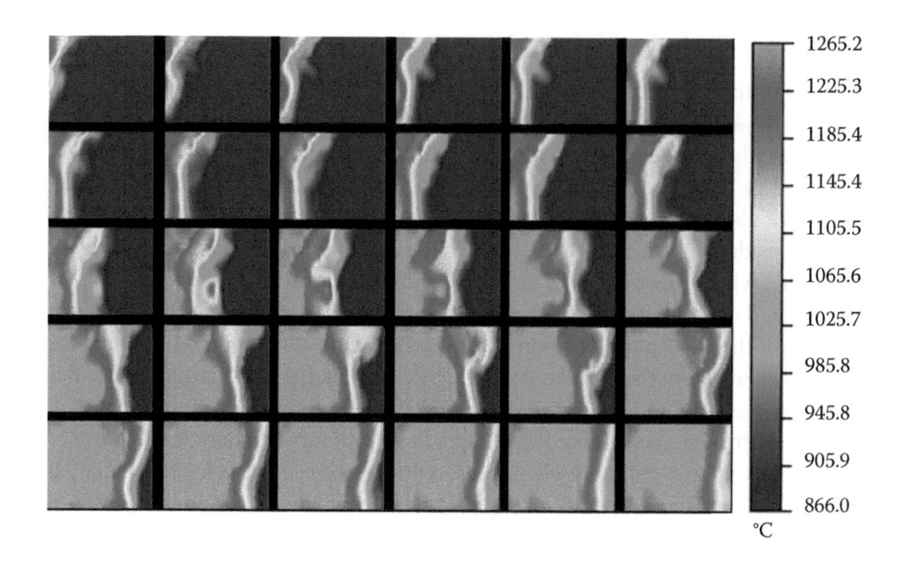

FIGURE 3.4 Thermal images of the synthesis wave moving through a pellet of MgO, Fe, Fe_2O_3, and $NaClO_4$. Each image is of dimension 3 × 2 mm. Images were captured at 0.06 s intervals. The first image is top left and the last is bottom right. (From Professor Ivan Parkin, University College, London. With permission.)

synthesised by baking together yttrium oxide, copper(II) oxide, and barium carbonate. The synthesis, however, takes 24 hours to complete. A microwave method that can prepare the superconductor in under 2 hours has been reported. A stoichiometric mixture of copper(II) oxide (CuO), barium nitrate ($Ba(NO_3)_2$), and yttrium oxide (Y_2O_3) was placed in a microwave oven that had been modified to allow safe removal of the nitrogen oxides formed during the reaction. The mixture was treated with 500 W of microwave radiation for 5 minutes, then reground and exposed to microwave radiation at 130–500 W for 15 minutes. Finally, the mixture was ground again and exposed to microwave radiation for a further 25 minutes. The microwaves in this example couple to the copper(II) oxide.

3.5 COMBUSTION SYNTHESIS

Combustion synthesis, also known as **self-propagating high-temperature synthesis**, has been developed as an alternative route to the ceramic method. The ceramic method depends on the diffusion of ions through the reactants, and thus for a uniform product, requires repeated heating and grinding for long periods. Combustion synthesis uses highly exothermic ($\Delta H \sim -170$ kJ mol^{-1}) and even explosive redox reactions to maintain a self-propagating, high reaction temperature and has been used to prepare many refractory materials, including borides, nitrides, oxides, silicides, intermetallics, and ceramics. In the solid state, the reactants are mixed together, formed into a cylindrical pellet, and then ignited at a single point on the surface (laser, electric arc, heating coil) at high temperature. Once ignited, the reaction propagates as a hot combustion wave, the **synthesis wave**, through the pellet, and the reaction must lose less heat than it generates or it will quench; high temperatures (1700°C–3700°C) are maintained during the fast reactions. Figure 3.5 shows the synthesis wave moving through a mixture of MgO, Fe, Fe_2O_3, and $NaClO_4$ (the images were captured by a thermal-imaging camera).

Ignition is sometimes initiated by heating the whole volume evenly, and self-ignition can sometimes be achieved by ball milling. The method is now used commercially in Russia, Spain, and Korea because it is fast, economical, and gives high-purity products. The ferrite $BaFe_{12}O_{19}$ has been used in magnetic stripes on credit cards (Chapter 9). Thermal ignition of a mixture of BaO_2, Fe_2O_3, Cr_2O_3, and Fe produces a Cr-substituted ferrite, $BaFe_{12-x}Cr_xO_{19}$. The exothermic reaction here is the oxidation of the metallic iron. Barium peroxide, BaO_2, decomposes during the reaction and provides a source of oxygen. For $x = 2$, the reaction is

$$BaO_2(s) + 2.5Fe_2O_3(s) + Cr_2O_3(s) + 5Fe(s) + 3.5O_2(g) = BaFe_{10}Cr_2O_{19}(s)$$

More unexpectedly perhaps, combustion synthesis can also take place in solutions, provided there is a sufficient quantity of a highly exothermic oxidiser present in the solution as a fuel. Various fuels have been used, the most commonly used being urea. The fuel not only provides the carbon and hydrogen necessary for heat production, but also forms a complex with the metal salts present, thus ensuring a very homogeneous mixture of reactants. The reactions are initiated by microwave heating and produce flaming or smouldering mixtures with temperatures from about 700°C to 1500°C.

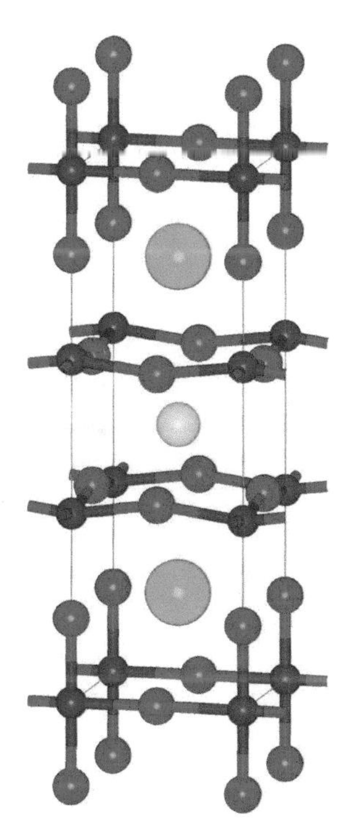

FIGURE 3.5 Idealised structure of $YBa_2Cu_3O_{7-x}$.

3.6 HIGH-PRESSURE METHODS

High pressures are employed directly and indirectly in various synthetic methods:

- High solvent pressures in an autoclave—the hydrothermal method
- Using a high pressure of a reactive gas
- Directly applied hydrostatic pressure on solids
- Ultrasound

3.6.1 Hydrothermal Methods

The hydrothermal method increases homogeneity and lowers the operating temperatures and is also used to grow single crystals, which are needed in some solid-state applications.

The original **hydrothermal method** involves heating the reactants in a closed vessel, an **autoclave**, with water (Figure 3.6). An autoclave is usually constructed from thick stainless steel to withstand the high pressures and is fitted with safety valves; it may be lined with nonreactive materials, such as the noble metals, quartz, or Teflon. The autoclave is heated, the pressure increases, and the water remains liquid above its normal boiling temperature of 100°C, so-called 'superheated water'. These conditions, in which the pressure is raised above atmospheric pressure and the temperature is raised above the boiling temperature of water but not to as high a temperature as used in the methods described previously, are known as hydrothermal conditions.

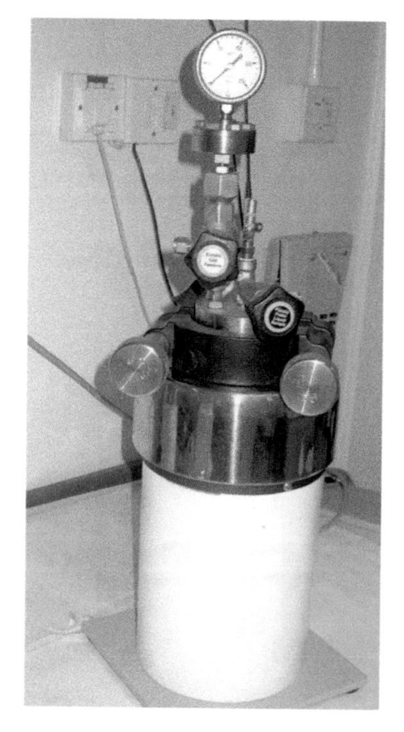

FIGURE 3.6 An autoclave.

Hydrothermal conditions exist in nature, and numerous minerals including naturally occurring zeolites and gemstones, are formed by this process. The term has been extended to other systems with moderately raised pressures and temperatures lower than those typically used in ceramic and sol–gel syntheses. The lower temperatures used are one of the advantages of this method. Others include the preparation of compounds in unusual oxidation states or phases, which are stabilised by the raised temperature and pressure. It is used industrially to prepare large crystals of quartz and synthetic gemstones. It is useful in metal oxide systems, where the oxides are not soluble in water at atmospheric pressure but dissolve in the superheated water of the hydrothermal setup. Where even these temperatures and pressures are insufficient to dissolve the starting materials, alkali or metal salts as mineralisers can be added, whose anions form complexes with the solid and render it soluble. Methods using solvents other than water are known as **solvothermal methods**.

The following examples have been chosen to illustrate these advantages and also the different variations of the method.

The first industrial process for using hydrothermal methods of synthesis was the production of quartz crystals for use as oscillators in radios. Quartz (SiO_2) can be used for generating a high-frequency alternating current via the piezoelectric effect (Chapter 9). The hydrothermal growth of quartz crystals employs a temperature gradient. In this variant of hydrothermal processing, the reactant dissolves at a higher temperature, is transported up the reaction vessel by convection, and crystallises out with the aid of seed crystals at a lower temperature in the second cooler chamber at the top of the autoclave. A schematic diagram of the apparatus is shown in Figure 3.7. One end of the autoclave is charged with silica and an alkaline solution. The autoclave is heated to typically between 300°C and 400°C, generating pressures of up to 1500×10^5 Pa; the silica dissolves in the solution and is transported along the autoclave to the cooler part (10°C–25°C cooler), where it crystallises out. The alkaline solution returns to the hotter region, where it can dissolve more silica. In this particular case, the autoclave, which is made of steel, can act as the reaction vessel. Because of the corrosive nature of the superheated solution and the possibility of contamination of the product with material from the autoclave walls, the autoclave is usually lined with an inert substance such as Teflon or the reaction is performed in a sealed ampoule in the autoclave.

The preparation of yttrium aluminium garnet ($Y_3Al_5O_{12}$; YAG) illustrates a variation of the hydrothermal method used if the starting materials all have solubilities that differ greatly from one another. In this case, yttrium oxide (Y_2O_3) is placed in a cooler section of the autoclave and aluminium oxide as sapphire in a hotter section to increase its solubility (Figure 3.8). YAG forms where the two zones meet

$$3Y_2O_3(s) + 5Al_2O_3(s) = 2Y_3Al_5O_{12}(s)$$

3.6.2 Using High-Pressure Gases

This method is usually used for preparing metal oxides and fluorides in less stable high-oxidation states. For instance, the perovskite $SrFeO_3$ containing Fe(IV) can be made from the reaction of $Sr_2Fe_2O_5$ and oxygen at 34 MPa. Fluorine at high pressures (>400 MPa) has also been used in the synthesis of fluorides containing high oxidation states, for example, Cs_2CuF_6 containing Cu(IV).

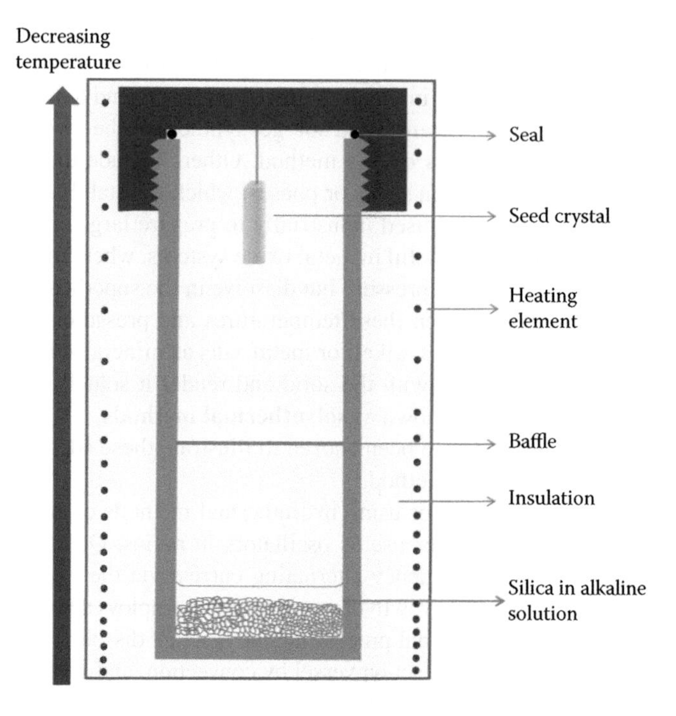

FIGURE 3.7 Growth of quartz crystals in an autoclave under hydrothermal conditions.

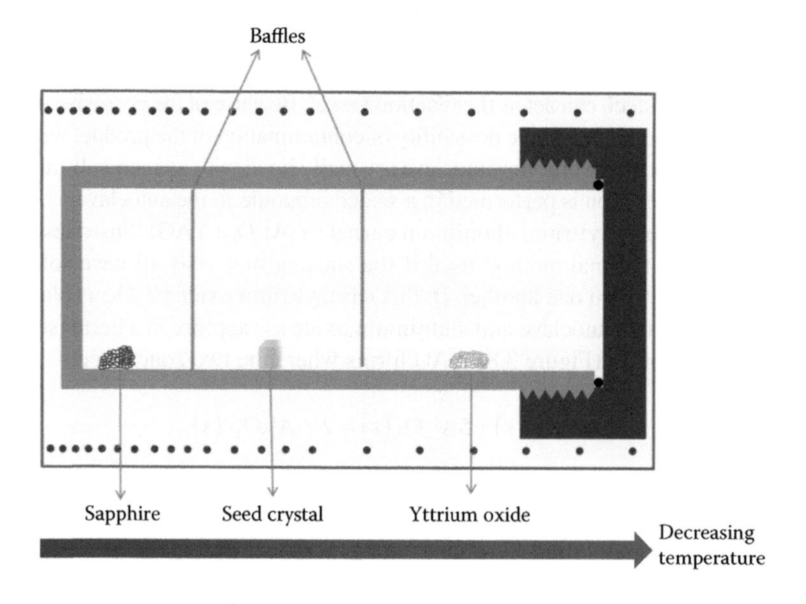

FIGURE 3.8 The hydrothermal arrangement for the synthesis of YAG starting from two reactants of different solubilities.

3.6.3 Using Hydrostatic Pressures

The application of high pressures and temperatures can induce reactions and phase changes that are not possible under ambient conditions. Applying very high pressures tends to decrease volume and thus improve the packing efficiency; as a consequence, coordination numbers tend to increase, so, for instance, Si can be transformed from the four-coordinate diamond structure to the six-coordinate white-tin structure, and NaCl (six-coordinate) can be changed to the CsCl (eight-coordinate) structure.

Various apparatus designs are used for achieving exceptionally high pressures. Originally, a **piston and cylinder** arrangement allowed synthesis at pressures up to 5 GPa (50,000 atm) and 1500°C, but the **belt apparatus** using two opposed tungsten carbide cylinders can reach 15 GPa and 2000°C and is used for making synthetic diamonds. Also used for manufacturing synthetic diamonds is the more efficient split-sphere or BARS apparatus, which consists of an inner capsule inside a cube. The capsule is pressed by both inner and outer anvils and heated by pressurised oil. Diamond anvils have been used to reach pressures of 20 GPa. Diamond anvils (Figure 3.9) can be arranged as two directly opposed, four tetrahedrally opposed, or six in an octahedral configuration. They achieve greater pressures than other methods, by applying a large force to small areas of sample between to diamond faces. The pressure is given by F/a where F is the force and a the area. Only milligram quantities of material can be processed, and so are more useful for investigating phase transitions and other properties than for synthesis.

The drive to produce **synthetic diamonds** arose during the Second World War; they were urgently needed for the diamond-tipped tools used for manufacturing military hardware, and it was feared that the South African supply of natural stones might dry up. GEC started a research programme to mimic the geothermal

FIGURE 3.9 A radial diamond-anvil-cell allows for in situ X-ray diffraction experiments at superbend beamline 12.2.2 of Berkeley Lab's Advanced Light Source. (Photo by Roy Kaltschmidt. Reproduced with the permission of The Regents of the University of California, Lawrence Berkeley National Laboratory.)

conditions that produce the natural stones. In 1955, they eventually succeeded in growing crystals up to 1 mm in length by dissolving graphite in a molten metal, such as nickel, cobalt or tantalum, in a pyrophyllite vessel and subjecting the graphite to pressures of 6 GPa and temperatures of ~2000°C until diamond crystals were formed. The molten metal acts as a catalyst, bringing down the working temperature and pressure. By refining the process, gemstone-quality stones were eventually made in 1970. Synthetic diamond can also be made by CVD methods (see Section 3.7), by detonation of carbon-containing explosives, and by high-energy ultrasound (see Section 3.6.4).

High-pressure synthesis has been used to make many high-temperature cuprate superconductors, as this facilitates the formation of the two-dimensional Cu–O sheets necessary for this type of superconductivity (see Chapter 10). For instance, high pressure increases the density of $SrCuO_2$ by 8%, and the Cu–O coordination changes from chains to two-dimensional CuO_2 sheets.

3.6.4 Using Ultrasound

When very high intensity sound waves—ultrasound—travel through a liquid, a phenomenon known as **cavitation** occurs. The travelling sound wave causes high-pressure areas (compression) of the liquid, which is followed by low pressure and sudden expansion (rarefaction) and the formation of tiny bubbles, which expand to an unsustainable size and then collapse. The expansion and collapse of the bubbles create very localised hot spots, which reach instantaneous pressures of more than 100 MPa (1000 atm) and temperatures of up to 5000°C. Figure 3.10 shows an experimental setup.

Since 2012, there has been a trend to use ultrasound in the preparation of nanoparticle (Chapter 11) catalysts such as coated TiO_2 and ZnO. Among its advantages are better control of particle size without agglomeration and improved size distribution. Ultrasonic methods have also been used to prepare zeolites (Chapter 7).

3.7 CHEMICAL VAPOUR DEPOSITION

So far, the preparative methods we have looked at have involved reactants that have mostly been in the solid state or in solution. In CVD, powders and microcrystalline compounds are prepared from reactants in the vapour phase and can, if required, be deposited on a substrate (such as a thin sheet of metal or ceramic, called a wafer) to form single-crystal films for devices (see Section 3.7.1). First, the volatile starting materials are heated to form vapours; these are then mixed at a suitable temperature and transported to the substrate for deposition using a carrier gas; finally, the solid product crystallises out. The whole vessel may be heated, which tends to deposit the product all over the walls of the vessel. More commonly, the energy to initiate the reaction is supplied to the substrate. Figure 3.11a shows a schematic setup for CVD. The conditions under which the reactions are carried out can be varied: pressures can be anything from atmospheric pressure to ultrahigh vacuum; ultrasound can be used to generate an aerosol to help transport; a plasma can be created to speed up reaction rates.

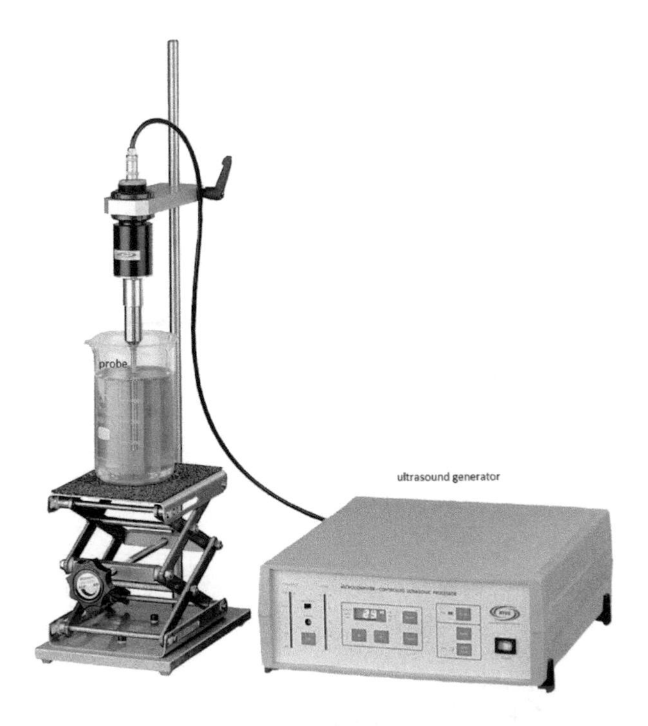

FIGURE 3.10 Set-up for an ultrasound synthesis. (Reproduced by permission of Roop Telsonic Ultrasonix Ltd.)

Typical starting materials are hydrides, halides, and organometallic compounds, as these compounds tend to be volatile; if an organometallic precursor is used, the method is often referred to as **metal organic CVD (MOCVD)**.

The advantages of this method are that reaction temperatures are relatively low, the stoichiometry is easily controlled, and dopants can also be incorporated.

3.7.1 Preparation of Semiconductors

The CVD method has been widely used for preparing the III–V semiconductors and silicon, for devices. Various reactions to prepare GaAs, involving chlorides and hydrides, have been tried. In one method for the preparation of gallium arsenide, arsenic(III) chloride ($AsCl_3$) (boiling temperature 103°C) is used to transport gallium vapour to the reaction site where gallium arsenide is deposited in layers. The reaction involved is

$$2Ga(g) + 2AsCl_3(g) = 2GaAs(s) + 3Cl_2(g)$$

An alternative to gallium vapour is an organometallic compound of gallium. One preparation reacts trimethyl gallium ($Ga(CH_3)_3$) with the highly volatile and toxic arsine (AsH_3)

$$Ga(CH_3)_3(g) + AsH_3(g) = GaAs(s) + 3CH_4(g).$$

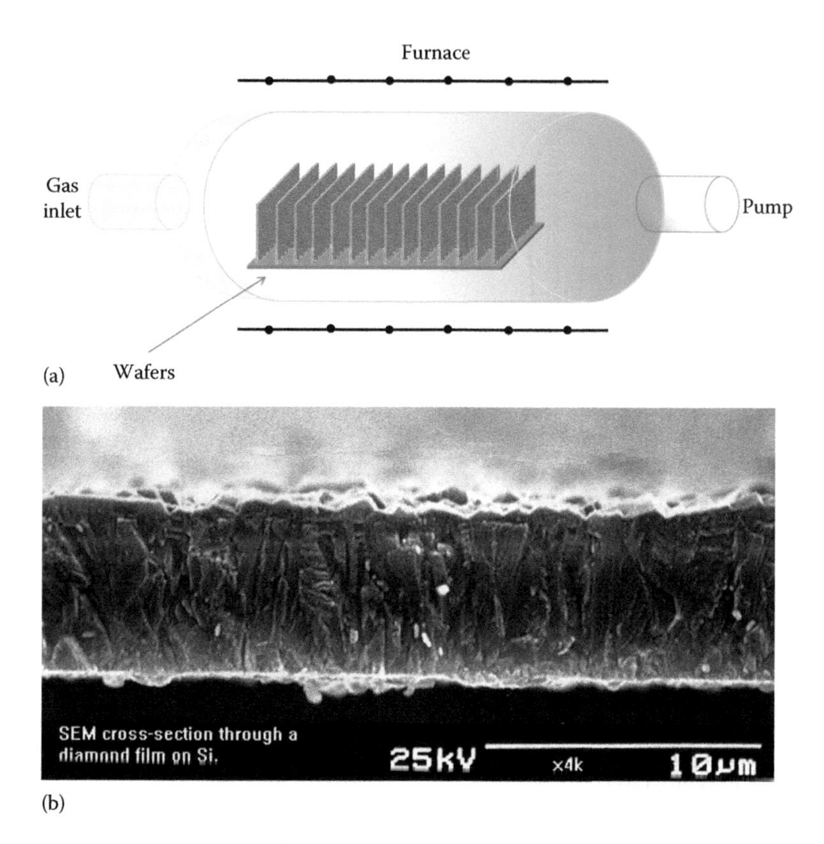

FIGURE 3.11 (a) A chemical vapour deposition reactor. (b) A cross-section of a 100-μm-thick CVD diamond film grown by DC arc jet. The columnar nature of the growth is clearly evident, as is the increase in film quality and grain size with growth time. (From Paul May and Mike Ashfold, Bristol University. With permission.)

Silicon for semiconductor devices can be prepared using this method from the decomposition of silane (SiH_4):

$$SiH_4(g) = Si(s) + 2H_2(g).$$

The silane can be either pure, or mixed with nitrogen or hydrogen at low pressures of about 100 Pa. Temperatures of around 600°C are used.

3.7.2 Diamond Films

Diamond has many useful properties. As well as being the hardest substance known, it also has the best thermal conductivity (at room temperature), is an electrical insulator, and has the highest transparency in the IR region of the spectrum of any known substance. Diamond films are made by a CVD process at lower pressures of ~27 kPa and temperatures of around 800°C. Carbon-containing gas mixtures, such as methane (or acetylene), and hydrogen, can be broken down in various ways using microwaves, laser or a hot filament; the carbon atoms and carbon-containing radical species are then deposited on a carefully cleaned and oriented substrate.

Early experiments managed to deposit monocrystalline layers onto a seed diamond and thus build up larger diamonds, layer by layer. Microcrystalline diamond layers have also been deposited onto silicon and metal surfaces. Such layers have been used for nonscratch optical coatings and for coating knives and scalpels so that they retain their sharpness. Additionally, such layers have the potential to be used as a hard, wear-resistant coating for moving parts or as an electrically insulating but heating-conducting layer.

3.7.3 Optical Fibres

Optical fibres now carry most of the world's communication traffic. In order to transmit light over distances of kilometres without loss of intensity, optical fibres have to be manufactured to very high specifications (see Chapter 8). Conventional methods of preparing silica glasses are inadequate and a modified form of CVD (MCVD) is usually employed.

The process uses volatile silicon tetrachloride, which can be easily purified, for example, by fractional distillation. Additives such as germanium tetrachloride are incorporated into the center of the fibre to give a higher refractive index. The process starts by making a 'preform'. A silica tube, which forms the outer low refractive layer, known as the cladding of the fibre, is heated to about 1600°C by a flame passing up and down the outside of the tube as it is rotated (Figure 3.12). $SiCl_4$ and O_2 pass down the tube and react together to form SiO_2, which deposits evenly along the tube wall as soot. Small amounts of $GeCl_4$ and other chemicals are gradually added into the process to deposit as oxides on the wall in the inner layers to increase the refractive index. The flame burner is then used to melt the soot deposit onto the glass preform. After cooling, the cylindrical preform is heated in a drawing tower to about 1900°C until it melts sufficiently to fall under gravity into a fine thread, which is slowly pulled and wound onto a spool. One preform can produce 2 km of fibre.

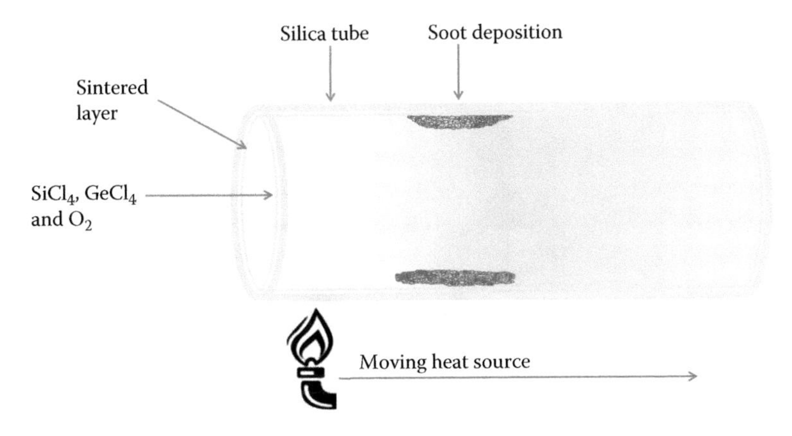

FIGURE 3.12 A diagram for the MCVD method of making optical fibres.

3.7.4 Lithium Niobate

The sol–gel method for preparing lithium niobate used lithium and niobium alkoxides. Alkoxides are often used in CVD methods, but unfortunately for the preparation of lithium niobate, lithium alkoxides are much less volatile than niobium alkoxides and to get the two metals deposited together, it is better to use compounds of similar volatility. One way around this problem is to use a more volatile compound of lithium. One reported synthesis uses a β-diketonate of lithium in which lithium is coordinated to 2,2,6,6-tetramethylheptan-3,5-dione ($Me_3CCOCH_2COCMe_3$) (Figure 3.13).

The lithium compound was heated at about 250°C in niobium pentamethoxide ($Nb(OMe)_5$) at 200°C in a stream of argon containing oxygen. Lithium niobate was deposited on the reaction vessel, which was heated to 450°C.

In this example, the volatile precursor compounds were heated to obtain the product. Other energy sources are also used, notably electromagnetic radiation. An example of vapour-phase deposition involving photodecomposition is given in the section on vapour-phase epitaxy.

3.8 PREPARING SINGLE CRYSTALS

Single crystals that are of high purity and are defect free may be needed for many applications, none more so than in the electronics and semiconductor industry. Various methods have been developed for preparing different crystalline forms, such as large crystals, films, etc. A few of the important methods are described below.

3.8.1 Epitaxy Methods

CVD methods are now used to make high-purity thin films in electronics where the deposited layers must have the correct crystallographic orientation. In epitaxial growth, a precursor is decomposed in the gas phase and a single crystal is built up layer by layer on a substrate, adopting the same crystal structure as the substrate. This is an important method for semiconductor applications where single crystals of a controlled composition are needed; semiconductors such as gallium nitride, gallium arsenide, and indium phosphide are made this way. Gallium arsenide (GaAs) has been prepared by several **vapour-phase epitaxial (VPE) methods** (Section 3.7.1).

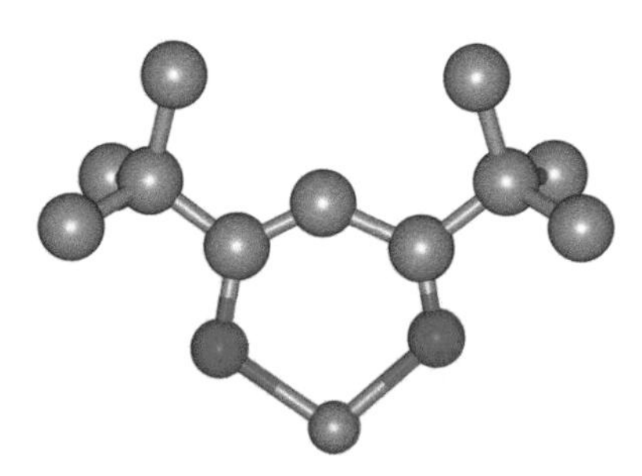

FIGURE 3.13 β-Diketonate of lithium from 2,2,6,6-tetramethylheptan-3,5-dione (lithium, purple; oxygen, red; carbon, grey).

A **molecular beam** is a narrow stream of molecules formed by heating a compound in an oven with a hole, which is small compared to the mean free path of the gaseous molecules produced. Very thin layers can be built up by directing a beam of precursor molecules onto the substrate. The system is kept under ultrahigh vacuum. Because of the very low pressure, the reactants need not be as volatile as in other vapour deposition methods. An application of this method is the growth of single crystals for **quantum cascade lasers**, semiconductor lasers that emit in the far infrared. The first quantum cascade laser crystal contained nanometer-thickness layers of $In_{0.52}Al_{0.48}As$ alternating with $In_{0.53}Ga_{0.47}As$, as shown in Figure 3.14. To make this crystal, beams of aluminium, gallium, arsenic, and indium are directed onto a substrate InP crystal. The substrate needs to be heated to allow the atoms deposited from the beams to migrate to their correct lattice position. The relative pressures of the component beams are adjusted for each layer to give the desired compositions.

3.8.2 Chemical Vapour Transport

In CVD, solids are formed from gaseous compounds. In **chemical vapour transport**, a solid or solids interact with a volatile compound and a solid product is deposited in a different part of the apparatus. It is used both for preparing compounds and for growing crystals from powders and for the purification of less pure crystalline material. The reactant solids are placed in a sealed quartz tube together with a transport/reactant gas, which is heated by a tube furnace; the tube is kept hotter at one end than the other (Figure 3.15); and the pure crystals deposit at the cooler end.

Magnetite crystals can be grown using the reaction of magnetite with hydrogen chloride gas:

$$Fe_3O_4(s) + 8HCl(g) = FeCl_2(g) + 2FeCl_3(g) + 4H_2O(g).$$

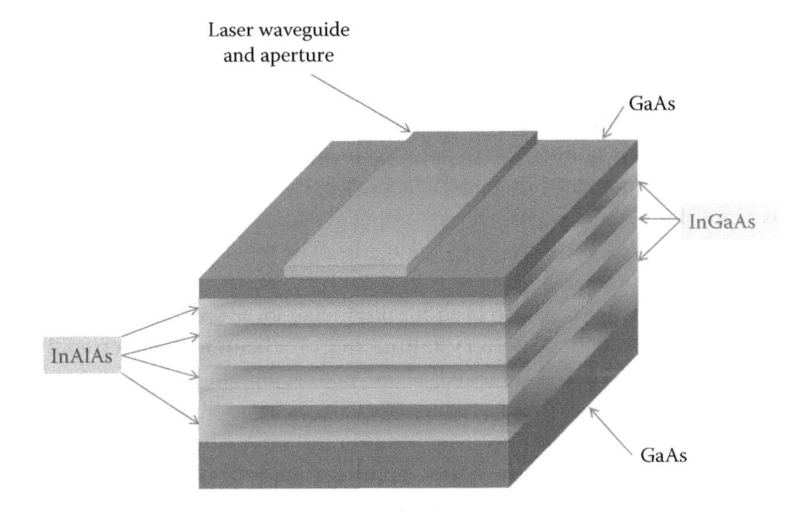

FIGURE 3.14 The layers of an active region of a quantum cascade laser.

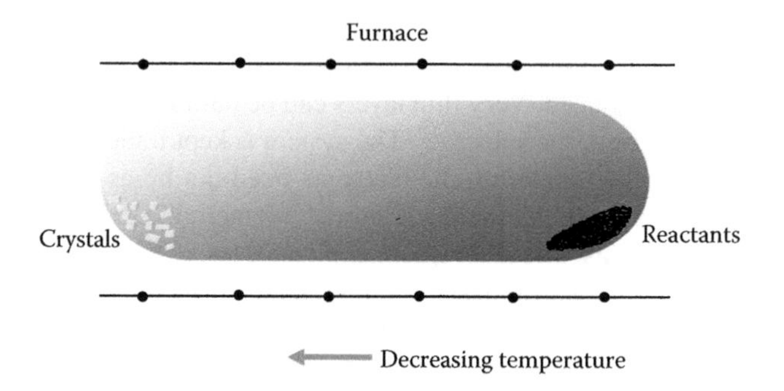

FIGURE 3.15 The growth of magnetite crystals using chemical vapour transport.

Powdered magnetite is placed at one end of the reaction vessel and the tube is evacuated and filled with HCl gas. The tube is sealed and then heated. The reaction is endothermic $\Delta H_m^\ominus$ positive, so the equilibrium moves to the right as the temperature is raised. Thus, at the hotter end of the tube, magnetite reacts with the hydrogen chloride gas and is transported down the tube as gaseous iron chlorides. At the cooler end of the tube, the equilibrium shifts to the left and magnetite is deposited.

3.8.3 Melt Methods

Melt methods depend on both the compound being stable in the liquid phase and the availability of a high-temperature, inert containing vessel.

The silicon required by the electronics industry for semiconductor devices has to have levels of key impurities, such as phosphorus and boron, of less than 1 atom in 10^{10} Si. Silicon is first converted to the highly volatile trichlorosilane ($SiHCl_3$), which is then distilled and decomposed on rods of high-purity silicon at 1000°C to give high-purity polycrystalline silicon. This is made into large single crystals by the **Czochralski process**. The silicon is melted in an atmosphere of argon, and then a single-crystal rod is used as a seed that is dipped into the melt and then slowly withdrawn, pulling with it an ever-lengthening single crystal in the same orientation as the original seed (Figure 3.16). In addition to silicon, this method is also used for preparing other semiconductor materials such as Ge and GaAs and for ceramics such as perovskites and garnets. Dopant impurity atoms can be added to the molten silicon to make *n*-type or *p*-type semiconductors. The ingot is sliced into thin wafers for electronics and photovoltaics.

The **float-zone process** is also used for producing large single crystals of silicon, which can then be sliced. A polycrystalline rod of silicon is held vertically and a single-crystal seed crystal is attached at one end. A moving heater then heats the area around the seed crystal, and a molten zone is caused to traverse the length of the rod, producing a single crystal as it moves (Figure 3.17). This process also has the advantage of refining the product—the so-called **zone refining**—as impurities tend to stay dissolved in the melt and are thus swept in front of the forming crystal to the end, where they can be cut off and rejected.

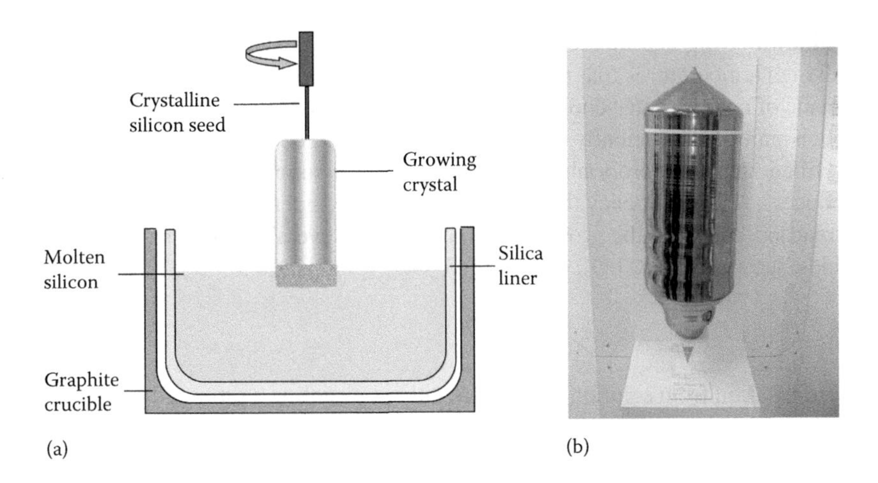

FIGURE 3.16 (a) The Czochralski process for producing a very pure single crystal of silicon, (b) Silicon Boule. (Courtesy of Massimiliano Lincetto [CC BY-SA 4.0]. From https://commons.wikimedia.org/wiki/File:Silicon_single_crystal.jpg.)

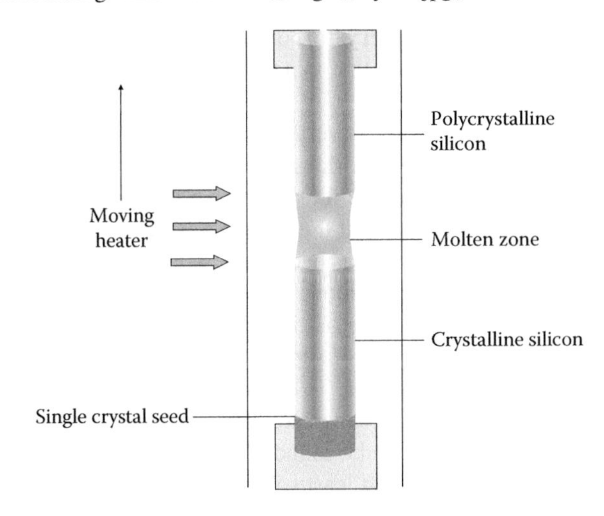

FIGURE 3.17 The float-zone method for producing very pure crystals of silicon.

Related methods include the **Bridgman** and **Stockbarger methods**, where a temperature gradient is maintained across a melt so that crystallisation starts at the cooler end on a seed crystal; this can be achieved either by using a furnace with a temperature gradient or by pulling the sample through a furnace. The crystallisation is in the same orientation as the seed. These methods are used when the Czochralski process proves difficult, for instance, for GaAs.

The old method used originally for producing artificial ruby and sapphire, the **Verneuil method**, is still widely used for making artificial gemstones. The powdered sample is melted in a high-temperature oxyhydrogen flame and droplets are allowed to fall onto a seed crystal where they crystallise. Using the same method but using a plasma torch to melt the powder can achieve even higher temperatures.

Skull melting is a method used for growing large oxide crystals, which melt at such high temperatures that normal containment vessels are not viable. The crucible is made of copper water-cooled fingers and base, which makes an outer skin on the molten material, sufficiently solid to hold it in place. The container can be evacuated and filled with an appropriate atmosphere. A power supply of up to 100 kW, 4 MHz produces a radio frequency that melts the sample. Temperatures of up to 3300°C can be reached, allowing the growth of large crystals (several centimetres) of refractory oxides, such as ZrO_2, ThO_2, CoO, and Fe_3O_4.

3.8.4 Solution Methods

Crystallisation from solution takes place either by cooling a saturated hot solution or by allowing evaporation to take place. Many crystalline substances have traditionally been grown from saturated solutions in a solvent; hot solutions are prepared and then cooled and crystals precipitate from the supersaturated solution.

Various other techniques can be used to induce crystallisation and these include: evaporating the solvent by heating, placing in a desiccator or using reduced pressure, freezing the solvent, and addition of other components (solvent or salt) to reduce solubility. Adding a crystal (**seed crystal**) of the desired product, altering the pH of the solution, and using a template that matches a surface of the desired crystal are also used.

For solids such as oxides that are very insoluble in water, it may be possible to dissolve them in melts of borates, fluorides, or even metals, in which case the solvent is generally known as a **flux** because it brings down the melting temperature of the solute. The melt is cooled slowly until the crystals form, and then the flux is poured off or dissolved away. This method has been successfully used for preparing crystalline silicates, quartz, and alumina, among many others.

Where a substance can crystallise out in different forms (polymorphs), the form produced can depend on the degree of supersaturation, the rate of cooling, and any other substance present in the solution. To obtain the thermodynamically most stable isomorph requires a low degree of supersaturation and slow cooling. A good example is calcium carbonate. This can be found as calcite, aragonite, and vaterite. The thermodynamically most stable and least soluble form is calcite, which would form as the final product if a saturated solution of calcium carbonate were allowed to cool slowly. However, shellfish such as pearl oysters also produce aragonite, which forms nacre or mother of pearl. The production of this polymorph is aided by a protein acting as a template.

3.9 INTERCALATION

Many layered solids form intercalation compounds, where a neutral molecule or ion is inserted between weakly bonded layers. These solids, produced by the reversible insertion of guest molecules into lattices, are known as **intercalation compounds**, and although originally applied to layered solids, it is now taken to include other solids with similar host–guest interactions. Interest in the lithium intercalation of graphite and of oxides has increased in recent years due to the use of the lightweight, rechargeable lithium–ion batteries in laptops and mobile devices and in batteries used for energy storage (see Chapter 6).

Halogens and alkali metals can easily be intercalated using **vapour phase** methods. In the two-zone technique, the host solid is held at a higher temperature than the substance to be intercalated. In the isothermal technique both host and intercalant are held at a temperature at which the guest is a vapour.

In **solution phase methods**, the host material is immersed in a solution of the guest. For example, graphite (see Chapter 10) can be intercalated with HSO_4^- by immersion in a solution of sulfuric acid. Another example is the intercalation of Li into titanium disulfide. Li_xTiS_2 is synthesised in the cathode material of a rechargeable battery described in Chapter 6 and can also be synthesised directly by the lithiation of TiS_2 with a solution of butyl lithium:

$$xC_4H_9Li + TiS_2 = Li_xTiS_2 + \frac{x}{2}C_8H_{18}.$$

While such methods are easily scalable and can give a high concentration of the guest species, it can be difficult to control and may result in damage to the host material.

For **electrochemical intercalation** the host must be an electrical conductor or placed on a conductive substrate. The host is then attached to a voltage source which supplies electrons to drive the intercalation process. For intercalation of metal atoms, the metal acts as the other electrode and between the electrodes is a solution of the metal ion acting as an electrolyte. For example K^+ can be intercalated into graphite using a metal electrode and a solution of KPF_6 in a mixture of ethylene carbonate and diethyl carbonate. This method allows more control over the intercalation than solution phase methods and the reaction is reversible. However the components of the electrochemical cell must be chosen carefully so that intercalation can occur at the applied voltage. Decomposition reactions can occur at too-high voltages. In addition the solvent can be intercalated as well as the desired species. Use of a solid electrolyte can overcome this.

3.10 GREEN CHEMISTRY

Sustainability is discussed in Chapter 12, but we briefly mention here how the concept can be applied to synthetic processes.

The sustainability of a process can be assessed using the principles of green chemistry developed by Paul Anastas and John Warner.

1. Prevention: It is better to prevent waste than to treat or clean up waste after it has been created.
2. Atom economy: Synthetic methods should be designed to maximize incorporation of all materials used in the process into the final product.
3. Less hazardous chemical synthesis: Wherever practicable, synthetic methodologies should be designed to use and generate substances that possess little or no toxicity to human health and the environment.

4. Designing safer chemicals: Chemical products should be designed to preserve efficacy of function while reducing toxicity.
5. Safer solvents and auxiliaries: The use of auxiliary substances (e.g. solvents, extraction agents, etc.) should be made unnecessary wherever possible and innocuous when used.
6. Design for energy efficiency: Energy requirements should be recognised for their environmental and economic impacts, and should be minimised. Synthetic methods should ideally be conducted at ambient temperature and pressure.
7. Use of renewable feedstocks: A raw material or feedstock should be renewable rather than depletable, wherever technically and economically practicable.
8. Avoid unnecessary derivatisation: Unnecessary derivatisation (e.g. blocking group, protection or deprotection, temporary modification of physical or chemical properties) should be avoided whenever possible.
9. Catalysis: The use of selective catalytic reagents is superior to stoichiometric reagents.
10. Design for degradation: Chemical products should be designed so that at the end of their function they do not persist in the environment, but break down into innocuous degradation products.
11. Real-time analysis for pollution prevention: Develop further analytical methodologies for real-time in-process monitoring and control prior to the formation of hazardous substances.
12. Minimise the potential for chemical accidents: Substances and the form of a substance used in a chemical process should be chosen so as to minimise the potential for chemical accidents, including releases, explosions, and fires.

These principles apply to any chemical reaction and some are more applicable to organic synthesis than to the solid-state syntheses in this chapter. We shall end by looking briefly at how these principles might apply to the preparation of barium titanate. In Section 3.2, we noted three different preparative methods for barium titanate.

The first was heating together barium carbonate and titanium dioxide at high temperature. This reaction produced carbon dioxide as well as the desired barium titanate. So this process produces a waste product, CO_2 (Principle 1). Not all the atoms in the starting material are in the product (Principle 2). Barium carbonate is toxic if swallowed and can cause irritation of the skin, eyes, and respiratory tract. Titanium dioxide as fine particles is a possible carcinogen (Principle 3). To satisfy Principle 4, a nontoxic material with the desired properties of barium titanate would have to be synthesised instead. This is a high-temperature process so does not conform with Principle 6.

The second method was precursor synthesis. The starting materials are barium chloride, titanium butoxide ($Ti(OC_4H_9)_4$), and oxalic acid. Overall, this process produces more carbon dioxide per molecule of barium salt and so more waste product than the first method (Principle 1). A lower percentage of the starting atoms are found in the product (Principle 2). Barium chloride is also toxic if swallowed and harmful if inhaled. Titanium butoxide is an irritant for eyes, skin, and respiratory and digestive tracts (Principle 3). This process requires a solvent, but the solvent is water (Principle 5). The temperature used is not as high as that for the first method (Principle 6).

The third method was sol–gel synthesis. In one reported method, barium acetate dissolved in acetic acid and titanium butoxide ($Ti(OC_4H_9)_4$) in glacial acetic acid and absolute ethanol were mixed to form a gel. The gel was heated at 120°C for 8 hours. The resulting dried gel was heated at 500°C for 2 hours. Overall, this process produces more carbon dioxide per molecule of barium salt and so more waste product than the first method (Principle 1). A lower percentage of the starting atoms is found in the product (Principle 2). Barium acetate is toxic if swallowed or inhaled. Titanium butoxide is an irritant for eyes, skin, and respiratory and digestive tracts. In addition glacial acetic acid causes severe burns to skin, eyes, and respiratory and digestive tracts. Absolute alcohol is toxic and irritating to skin and eyes (Principle 3). The solvents used are thus harmful (Principle 5). The temperature for the gel drying is relatively low and the temperature for the final process is lower than for the previous methods (Principle 6). Glacial acetic acid and absolute alcohol are flammable (Principle 12).

You might want to note that all the elements used in these syntheses are in plentiful supply.

3.11 CHOOSING A METHOD

Our choice of methods is not by any means exhaustive. Our aim here, therefore, is not to provide a way of choosing the method for a particular product. (Indeed, several of the examples given in this chapter show that several methods can be suitable for one substance.) Instead, we hope to give a few pointers to deciding whether a particular method is suitable for a particular material.

It is important that you consider the stability of the compounds under the reaction conditions and not at normal temperature and pressure. As we have illustrated in this chapter, a particular method may be chosen because the desired product is stable only at raised pressures or because it decomposes if the reaction temperature is too high.

What form do you want the product to be in? You might, for example, choose vapour phase epitaxy because an application requires a single crystal, or a method such as the precursor method or hydrothermal synthesis because you need a homogeneous product.

How pure must your product be? If you require high purity, you could choose a method that involves a volatile compound as a starting material because such compounds are generally easier to purify than solids.

You should consider the availability of the reactants required for a particular method. If you are considering a precursor method, is there a suitable precursor with the right stoichiometry? The CVD method needs reactants of similar volatility; do your proposed reactants meet this requirement? In microwave synthesis, does at least one of your starting materials absorb microwaves strongly and safely without sparking?

Should you be looking for a method that could be used on a large scale, you should consider in addition factors such as the cost, location of sources of the required elements, and sustainability

QUESTIONS

1. The Chevrel phase $CuMo_6S_8$ was prepared by a ceramic method. What would be suitable starting materials and what precautions would you have to take?
2. Which synthetic methods would be suitable for producing the following characteristics:
   a. A thin film of material?
   b. A single crystal?
   c. A single crystal containing layers of different material with the same crystal structure?
   d. A powder of homogeneous composition?
3. A compound $(NH_4)_2Cu(CrO_4)_2.2NH_3$ is known. How could this be used to prepare $CuCr_2O_4$? What would be the advantage of this method over a ceramic method? Suggest which solvent was used to prepare the ammonium compound.
4. β-TeI is a metastable phase formed at 465–470 K. Suggest an appropriate method of preparation.
5. In hydrothermal processes involving alumina (Al_2O_3), such as the synthesis of zeolites, alkali is added to the reaction mixture. Suggest a reason for this addition.
6. Which of the following oxides would be good candidates for microwave synthesis?

$$CaTiO_3, BaPbO_3, ZnFe_2O_4, Zr_{1-x}Ca_xO_{2-x}, KVO_3.$$

7. Crystals of silica can be grown using the chemical vapour transport method with hydrogen fluoride as a carrier gas. The reaction involved is $SiO_2(s) + 4HF(g) = SiF_4(g) + 2H_2O(g)$. This reaction is exothermic. In a temperature gradient method, will the silica crystals grow at the hotter or cooler end of the reaction vessel?
8. In the preparation of lithium niobate by CVD, argon-containing oxygen is used as a carrier gas. In the preparation of mercury telluride, HgTe, the carrier gas is hydrogen. Suggest reasons for these choices of carrier gas.
9. In Section 3.7.1, two methods of preparing gallium arsenide are given. Discuss the application of Principles 2, 3, and 4 of green chemistry to both methods.

4 Solids
Bonding and Electronic Properties

Neil Allan

4.1 INTRODUCTION

In Chapter 1, we introduced you to the physical structure of solids—the arrangement of their atoms and ions in space. We now turn to describe bonding in solids—their electronic structure.

Some solids consist of molecules bound together by very weak forces. We shall not be concerned with these because their properties are essentially those of the individual molecules. In the final section we shall be concerned with 'purely ionic' solids bound by electrostatic forces between ions, as discussed in Chapter 1. The solids considered here are those in which all the atoms can be regarded as bound together. We shall be looking at basic theories of bonding in these solids and how they account for the very different electrical conductivities of different groups of solids—metals, insulators, and semiconductors. In particular we explain *metallic* bonding, a type of bonding not found in small molecules and clusters, and the high conductivities of metals. We shall cover a theory based on the free-electron model, a view of a solid as an array of ions held together by a 'sea' of electrons; and another theory based on molecular orbital theory, a crystal viewed as a giant molecule. Some of the most important solids are semiconductors. Many solid-state devices—transistors, photocells, light-emitting diodes (LEDs), solid-state lasers, and solar cells—are based on semiconductors. We introduce examples of some devices and explain how the properties of semiconductors make them suitable for these applications. We start with the free-electron theory of solids and its application to metals and their properties.

4.2 BONDING IN SOLIDS: FREE-ELECTRON THEORY

Traditionally, bonding in metals has been approached through the idea of independent free electrons, a sort of electron gas.

The free-electron model regards a metal as a 'box' in which the electrons are free to roam unaffected by the atomic nuclei or by each other—electron–nuclear attractions and electron–electron repulsions are both neglected. The nearest approximation to this model is provided by the metals on the far left of the periodic table—Group 1 (Na, K etc.), Group 2 (Mg, Ca, etc.)—and aluminium. These metals are often referred to as simple metals.

The theory assumes that the nuclei stay fixed on their lattice sites surrounded by the inner or core electrons, while the outer or valence electrons travel freely through the solid. If we ignore the cores, then the quantum mechanical description of the outer electrons becomes very simple. Taking just one of these electrons, the problem becomes the well-known 'particle in a box'. We start by considering an electron in a one-dimensional solid.

The electron is confined to a line of length a (the length of the solid), which we shall call the x axis. The box starts at $x=0$ and so ends at $x=a$. Since we are ignoring the cores, there is nothing for the electron to interact with, so it experiences zero potential within the solid. The Schrödinger equation for the electron is

$$-\frac{\hbar^2}{2m_e}\frac{d^2\psi}{dx^2}=(E-V)\psi, \tag{4.1}$$

where $\hbar$ is Planck's constant divided by 2π, m_e is the mass of the electron, V the potential experienced by the electron, here equal to zero inside the box. ψ is the wave function of the electron and E is the corresponding energy of an electron with wave function ψ. When $V=0$, the solutions to this equation are simple sine or cosine functions, and we can verify this by substituting $\psi = \sin kx$ into Equation 4.1, as follows:

If $\psi = \sin kx$, where k is a constant then differentiating once gives

$$\frac{d\psi}{dx}=k\cos kx$$

Differentiating twice, we get

$$\frac{d^2\psi}{dx^2}=-k^2\sin kx$$

and multiplying $(-\hbar^2/2m_e)$ the left-hand side of Equation 4.1 becomes

$$-\frac{\hbar^2}{2m_e}\frac{d^2\psi}{dx^2}=\frac{\hbar^2k^2}{2m_e}\sin kx=\frac{\hbar^2k^2}{2m_e}\psi.$$

Comparison of this with Equation 4.1 shows that since $V=0$, $E=\hbar^2k^2/2m_e$.

The electron is not allowed outside the box so the potential is infinite outside. Since the electron cannot have infinite energy, the wave function must be zero outside the box and since it cannot be discontinuous, it must be zero at the boundaries of the box. If we take the sine wave solution, $\psi = \sin kx$ then this is zero at $x=0$ because $\sin 0=0$. To be zero at $x=a$ as well, $\sin ka = 0$ and there must be a whole number of half waves in the box. Sine functions have a value of zero at $n\pi$ radians ($\sin n\pi=0$ where n is an integer), and so $ka=n\pi$, or $k=n\pi/a$. Since $E=\hbar^2k^2/2m_e$, we substitute for k and obtain $E=n^2\pi^2\hbar^2/2m_ea^2$, or, equivalently since $\hbar=h/2\pi$, $E=n^2h^2/8m_ea^2$. Therefore, since the quantum number n can only take integer values, energy is quantized. As n can take *all* integral values, there are an infinite number of energy levels with larger gaps between each level as n increases. But when the length of the box a is very large, comparable to the length of any solid sample in an experiment, the levels are very close together and effectively continuous.

Most solids, of course, are three dimensional, although we shall meet some later where conductivity is confined to one or two dimensions, so we now need to extend the free-electron theory to three dimensions.

For three dimensions, the metal can be taken as a rectangular box, with sides a, b, and c. The appropriate wave function is now the product of three sine functions and the energy is given by

$$E = \frac{h^2}{8m_e}\left(\frac{n_a^2}{a^2} + \frac{n_b^2}{b^2} + \frac{n_c^2}{c^2}\right). \tag{4.2}$$

The energy difference between the two lowest energy levels if the volume of the solid is even only 10^{-6} m^3 is of the order of 10^{-34} J (10^{-15} eV), which is vanishingly small. Any set of three quantum numbers, n_a, n_b, and n_c, will give rise to an energy level. However, in three dimensions, there are many combinations of n_a, n_b, and n_c that will give the same energy. For example, the sets of numbers in Table 4.1 all give $\left(n_a^2/a^2 + n_b^2/b^2 + n_c^2/c^2\right) = 108$, and hence the same energy. The number of states with the same energy is known as the **degeneracy**. For small values of the quantum numbers, it is possible to write out all the combinations that will give rise to the same energy, but when we are dealing with a crystal of say 10^{20} atoms, it becomes difficult to work out all the combinations. However, we can estimate the degeneracy of any level by introducing a quantity called the **wave vector**. If we substitute k_x, k_y, and k_z for $n_a\pi/a$, $n_b\pi/b$, and $n_c\pi/c$, then the energy becomes

$$E = \left(k_x^2 + k_y^2 + k_z^2\right)\hbar^2 / 2m_e, \tag{4.3}$$

where k_x, k_y, and k_z can be considered as the components of a vector **k**; the energy is proportional to the square of the length of this vector, i.e., $|\mathbf{k}|^2$. The vector **k** is called the **wave vector**. **k-space** is the set of all possible **k** vectors. **k** is related to the momentum of the electron, as can be seen by comparing the classical

TABLE 4.1

Sets of Values of n_a/a, n_b/b, and n_c/c Such That $\left(n_a^2/a^2 + n_b^2/b^2 + n_c^2/c^2\right) = 108$

n_a/a	n_b/b	n_c/c
6	6	6
2	2	10
2	10	2
10	2	2

expression $E=p^2/2m$, where p is the momentum and m is the mass, with the expression above. This gives the electron momentum as $\pm\mathbf{k}\hbar$.

Since the number of energy levels is very large and depends on the volume of the solid, a very useful quantity that enables us to calculate the degeneracies is the **density of states** $N(E)$, defined as the number of energy levels with a given energy (strictly in an infinitesimal range of energies between E and $E+\mathrm{d}E$) per unit volume of the solid. To work towards this quantity, note that all the combinations of the three quantum numbers giving rise to one particular energy correspond to a wave vector of the same length $|\mathbf{k}|$. Thus, all possible combinations leading to the same given energy have wave vectors whose ends lie on the surface of a sphere of radius $|\mathbf{k}|$. The total volume of this sphere is is $4\pi k^3/3$, where $|\mathbf{k}|$ is now written as k. The states in k-space are quantized, because only integral values of n_a, n_b, and n_c are allowed, and if we increase n_a, n_b, or n_c by one, we obtain a new state. The volume of k-space *per state* is π^3/abc and the number of states per unit volume in k-space is the reciprocal of this, abc/π^3. The total number of states within the k-space volume of $4\pi k^3/3$ is $4\pi k^3/3 \times abc/\pi^3 = 4k^3abc/\pi^2$ and since the volume of the solid is abc the total number of states per unit volume (of the solid) is just $4k^3/\pi^2$. This is progress, as we now have the number of states with energies *from zero up to a given energy E*, but we still need to know the number of states *with a particular energy*, i.e., between E and $E+\mathrm{d}E$. The easiest way to do this is to assume k is continuous because the levels are so close together in a macroscopic crystal and use calculus to find the derivative of the total number of states with respect to k, and then substitute for k in terms of E. Each state in k-space can be occupied by two electrons with opposite spins (the Pauli Exclusion Principle) and the final result is that $N(E)$ is given by $\left(2m_e\right)^{3/2} E^{1/2} / 2\pi^2\hbar^3$.

A plot of $N(E)$ vs. E is given in Figure 4.1. Note that the density of states increases with increasing energy because it is proportional to $E^{1/2}$; the higher the energy, the greater the number of states. In metals, the valence electrons successively fill up the states with paired spins, starting from the lowest energy. For sodium, for example,

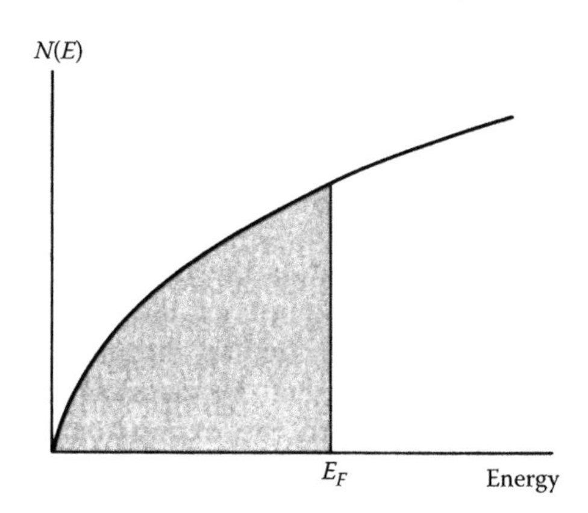

FIGURE 4.1 A density of states curve based on the free-electron model. The levels occupied at 0 K are shaded. Note that later in this book, energy is plotted on the vertical axis. In this figure, energy is plotted along the horizontal axis for comparison with the experiment. E_F is the Fermi level.

each atom contributes one 3s electron and the electrons from all the atoms in the crystal occupy the levels in Figure 4.1. This highest-occupied level in a metal is called the **Fermi level** and the corresponding energy, E_F, the **Fermi energy.**

How does this theoretical density of states compare with experiment? Experimentally, the density of states can be determined by X-ray emission spectroscopy, also called X-ray fluorescence (XRF) spectroscopy. When a beam of electrons or high-energy X-rays is incident on a metal, it can remove core electrons. In sodium, for example, the valence electron is 3s, so a 2s or a 2p electron might be removed. The core energy levels are essentially atomic levels, so the electron was removed from a discrete, well-defined energy level. Having a missing electron in a core energy level is an unstable situation, so an electron from an occupied valence level drops down to the vacated energy level, emitting photons as it does so. The photon energy depends on the difference in energies between the valence level from which the electron has come and that of the core level where it ends up. Due to the high energies involved, the emitted photon is usually in the X-ray region of the electromagnetic spectrum. The wavelengths (and hence energies) of all the emitted X-rays are then measured in a special spectrometer. A scan across the emitted X-ray energies corresponds to a scan across the filled levels. The intensity of the radiation emitted depends on the number of valence electrons with that particular energy, that is, the intensity depends on the density of states. Figure 4.2 shows X-ray emission spectra for sodium and aluminium, and we can see that the shape of these curves approximately resembles the occupied part of Figure 4.1, so that the free-electron model appears to describe these bands of energy levels (as we shall see later called conduction bands) reasonably well.

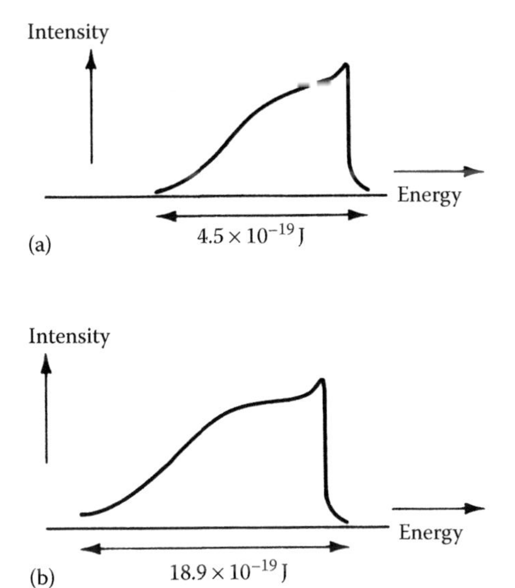

FIGURE 4.2 X-ray emission spectra obtained from (a) sodium metal and (b) aluminum metal when conduction electrons drop down into the 2p level. The slight tail at the high-energy end is due to thermal excitation of the electrons close to the Fermi level into higher k-states.

If we look at the density of states found experimentally for metals with more valence electrons per atom than the simple metals, however, the fit is not as good. We find that instead of continuing to increase with energy, as in Figure 4.1, the density reaches a maximum and then starts to decrease with energy. This is shown for the 3d band of Ni in Figure 4.3.

Extensions of this model, in which the atomic nuclei and core electrons are included by representing them by a potential function (V) in Equation 4.1 (plane wave methods), can account for the density of states in Figure 4.3 and can be used for semiconductors and insulators as well. To account for this, we shall use a different theoretical model, one based on the molecular orbital theory of molecules, described in the next section. We conclude this section by using our simple model to explain the electrical conductivity of metals.

4.2.1 Electronic Conductivity

The wave vector **k** is the key to understanding electrical conductivity in metals. For this purpose, it is important to note that a vector has direction as well as magnitude. It follows that there may be many different energy levels with the same value of $|k|$ and hence the same value of the energy, but with different components, k_x, k_y, and k_z, giving a different direction to the momentum. In the absence of an electric field, all directions of the wave vector **k** are equally likely, so there are equal numbers of electrons moving in all directions (Figure 4.4).

If we now connect our metal to the terminals of a battery producing an electric field, then an electron travelling in the direction of the field will be accelerated and the energies of those levels with a net momentum in this direction will be lowered. Electrons moving in the opposite direction will have their energies raised, so some of these electrons will drop down into levels of lower energy corresponding to the momentum in the opposite direction. Thus, more electrons will move in the direction of the field than in other directions. This net movement of the electrons in one direction is an electric current. The net velocity in an electric field is shown in Figure 4.5.

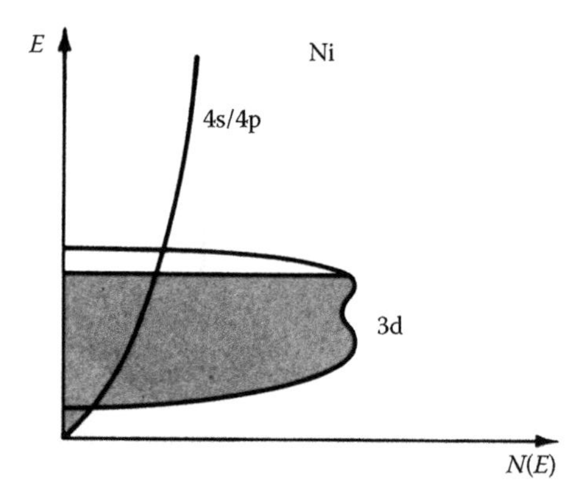

FIGURE 4.3 The band structure of nickel. Note the shape of the 3d band.

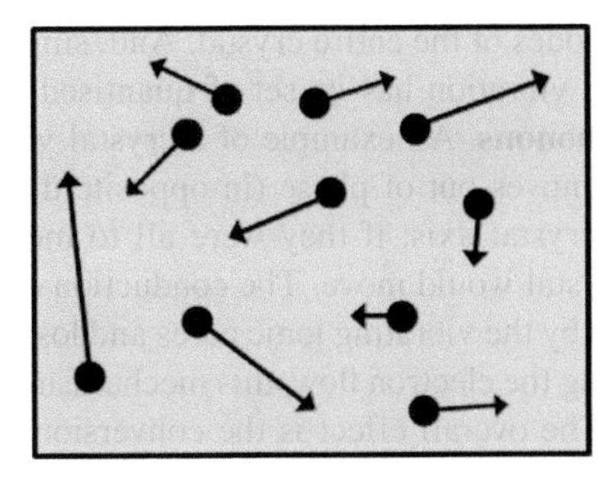

FIGURE 4.4 Electrons in a metal in the absence of an electric field. They move in all directions, but overall there is no net motion in any direction.

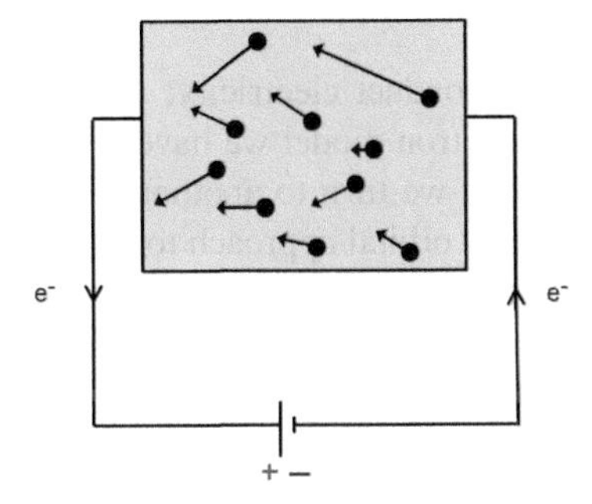

FIGURE 4.5 The metal in Figure 4.4 in a constant electric field, established by placing the sample between the terminals of a battery. The electrons can move in all directions, but now their velocities are modified so that each also has a net movement or drift velocity in the left-to-right direction.

An important point to note about this explanation is that it assumes that there are empty energy levels available that are close in energy to the Fermi level. You will see later that the existence of such levels is also crucial in explaining the conductivity of semiconductors.

The model as we have described it so far, explains why metals conduct electricity, but it does not account for the finite electrical resistance of metals. The current (I) flowing through a metal for a given applied voltage or electric potential difference (V) is given by Ohm's law, $V=IR$, where R is the resistance. It is characteristic of a metal that R increases with increasing temperature (or equivalently, the conductance ($\sigma=1/R$) decreases with increasing temperature). So for a given potential difference, the current decreases as the temperature is raised. So far, there is nothing in our theory that might impede the flow of electrons. To account for electrical resistance, we have to introduce the effect of the ionic cores. If these were arranged periodically on the lattice sites of a perfect crystal and did not move, then they would not interrupt the flow of electrons. However, most crystals contain some imperfections and these can scatter electrons. If the component of the electron's momentum in the field direction is reduced, then the current will drop. In addition, even in perfect crystals, the ionic cores will vibrate, just as in molecules the ionic cores vibrate, giving rise to normal vibrational modes of the entire crystal. And, similar to vibrations in small molecules, each crystal vibration has its set of quantised vibrational levels. These vibrations are called **phonons**. An example of a crystal vibrational mode is one in which each ionic core moves out of phase (in opposite directions) with its nearest neighbours along one crystal axis; if they were all to move in phase in the same direction, the whole crystal would move. The conduction electrons moving through the crystal are scattered by the vibrating ionic cores and lose some energy to the phonons. As well as reducing the electron flow, this mechanism increases the vibrational energy of the crystal. The overall effect is the conversion of electrical energy into heat energy. This ohmic heating effect is put to good use in, for example, the heating elements of kettles.

4.3 BONDING IN SOLIDS: MOLECULAR ORBITAL THEORY

We know that not all solids conduct electricity; no explanation for this has been offered by the simple free-electron model we have discussed above. To understand semiconductors and insulators, we turn to another description of solids, molecular orbital theory. In the molecular orbital approach to bonding in solids, we regard solids as a very large collection of atoms bonded together and, sometimes making large approximations, solve the Schrödinger equation for a periodically repeating system. For chemists, this has the advantage that solids are not treated differently from small molecules.

However, solving such an equation for a solid appears to be something of a tall order since it is far from straightforward to find accurate numerical solutions even for small molecules let alone even for a small crystal that could well contain of the order of 10^{16} atoms. An approximation often used for smaller molecules is that the molecular wave functions can be formed by combining (adding and subtracting) the wave functions of atomic orbitals. This **linear combination of atomic orbitals (LCAO)** approach can also be applied to solids and fortunately even at a qualitative level is capable of providing many powerful insights. In physics this approach is called the **tight-binding approximation** because in contrast to free-electron theory here the electrons are bound.

We start by reminding the reader how to combine atomic orbitals for a very simple molecule, H_2. For H_2, the molecular orbitals (Figure 4.6) are formed by combining 1s orbitals on the two hydrogen atoms H_A and H_B. If we add these together, there is constructive interference between the two, forming a bonding orbital with enhanced probability of the electron being found in the internuclear region. If we subtract the two orbitals, there is destructive interference between the two 1s orbitals, forming an antibonding orbital in which the electron density in the internuclear region is reduced. The bonding orbital is lower in energy than the atomic 1s orbitals while the antibonding orbital is higher. The amount by which the energy is lowered for a bonding orbital depends on the extent of overlap between the 1s orbitals on the two hydrogens. If the hydrogen nuclei are pulled further apart, the overlap decreases and the decrease in energy is less. If the nuclei are pushed together, the overlap increases but the electrostatic repulsion of the two nuclei becomes important and counteracts the effect of the increased overlap.

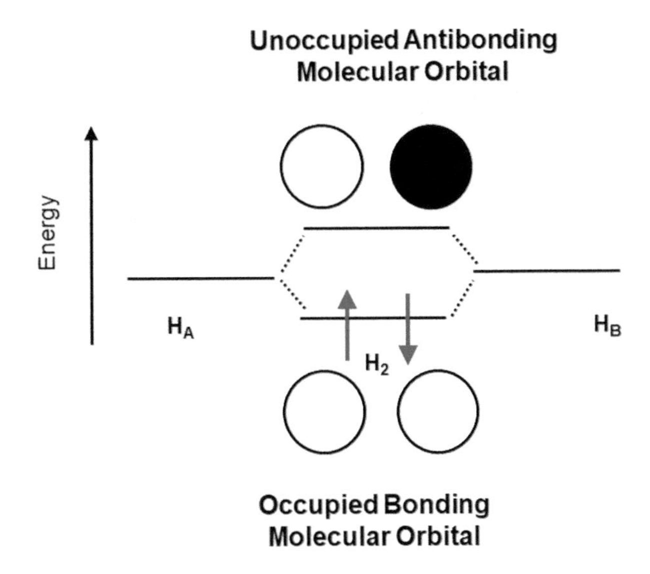

FIGURE 4.6 The bonding and antibonding molecular orbitals in H_2.

Suppose we form a linear chain of hydrogen atoms. For N hydrogen atoms, there are N molecular orbitals. In the lowest energy orbital all the 1s orbitals combine constructively—all the interactions between neighbouring 1s orbitals are bonding. In contrast, in the highest energy orbital all the orbitals combine destructively and all the interactions between neighbouring 1s orbitals are antibonding. These are shown in Figure 4.7 for $N=4$. Lying in between these two extremes are $(N - 2)$ molecular orbitals, in which there are some bonding and some antibonding interactions between neighbouring orbitals. The more bonding than antibonding interactions in a given molecular orbital, then the lower in energy is this orbital.

Figure 4.8 shows a plot of the energy levels as the length of the chain increases. Note in particular that as the number of atoms increases, the number of levels increases but the spread in energy between the highest and lowest increases only slowly and levels off for long chains. The addition of a single extra atom to an already long chain has a smaller effect the longer the chain. Extrapolating to chains with extremely large lengths similar to those found in macroscopic crystals, there are a very large number of levels within a comparatively small range of energies.

A chain of hydrogen atoms is a simple and somewhat artificial model; as an estimate of the energy separation of the levels, so let's instead now consider a chain of metal atoms. A metal crystal may contain 10^{16} atoms and the range of energies may be only 10^{-19} J. The average energy separation between the levels would thus be only 10^{-35} J. If we compare this figure with the separation of 10^{-18} J between the lowest energy levels in the hydrogen atom, we can see that the energy separation in a crystal is vanishingly small. The separation is in fact so small that, as in the free-electron model, we can think of the set of levels as a continuous range of energies. Such a continuous range of allowed energies is known as an **energy band**.

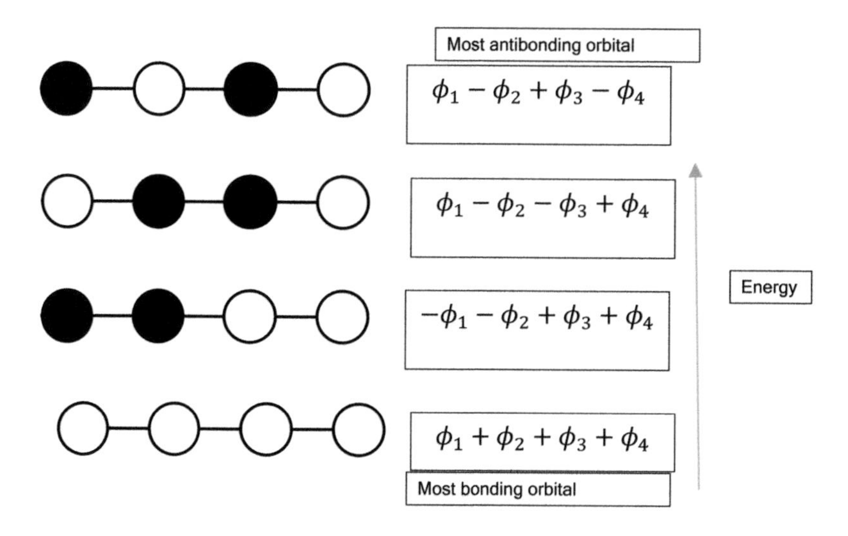

FIGURE 4.7 The N molecular orbitals of a chain of N atoms vary in energy from the most bonding to the most antibonding. Here we show the 1*s* orbitals in a chain of H atoms with $N=4$. In between there are many other orbitals with both bonding and antibonding interactions and when N is large as in a solid the separation between adjacent energy levels is vanishingly small.

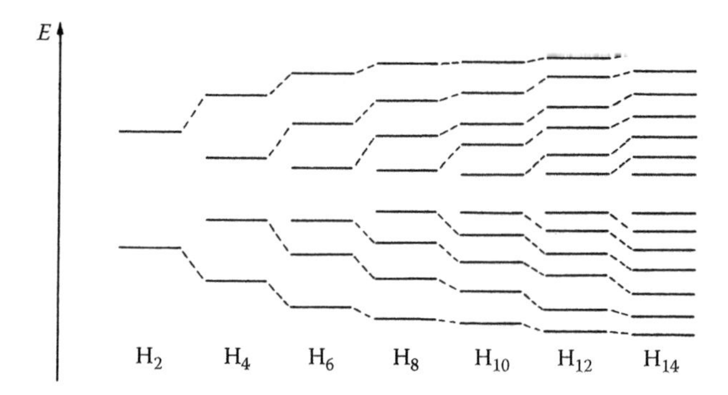

FIGURE 4.8 The orbital energies for a chain of N hydrogen atoms as N increases.

In actual calculations on crystals, it is impractical to include all 10^{16} atoms, so we use the periodicity of the crystal. The electron density in each unit cell must be identical, so we form linear combinations of orbitals for the unit cell that reflect the periodicity of the crystal. Such combinations have patterns like the sine waves that we obtained from the particle in the box calculation. For small molecules, the LCAO expression for molecular orbitals is

$$\psi(i) = \sum_n c_{ni}\phi_n,$$

where the sum is over all n atoms, $\psi(i)$ is the ith molecular orbital, and c_{ni} is the coefficient of the atomic orbital ϕ on the nth atom in the ith molecular orbital.

For solids, a detailed argument which we do not give here shows this is replaced by

$$\psi(\mathbf{k}) = \sum_n e^{i\mathbf{k}.\mathbf{r}_n} a_{nk}\phi_n,$$

where **k** is a wave vector which labels a given crystal (molecular) orbital. Also $i^2 = -1$ and $\mathbf{r}_n$ is the position vector of atom n in the lattice. a_{nk} is the coefficient of the orbital ϕ on the nth atom for wave vector **k**.

For further details see a solid-state physics text, but it is worth seeing that this equation is consistent with our qualitative model of the one-dimensional hydrogen linear chain, which also shows the significance of **k**. As we have seen, in one dimension the wave vector **k** is just one dimensional, and we denote it k. Let the distance between atoms in the chain be a, while the integer label n (=1,2,3,4... etc.) distinguishes the different atoms in the chain. The atoms are located at a, $2a$, $3a$..., na. The ϕ_n are just the 1s atomic orbitals on the different atoms n. The wave functions of the individual molecular (crystal) orbitals in our infinite chain of H atoms are

$$\psi(k) = \sum_n e^{inka}\phi(1s)_n,$$

where $\phi(1s)_n$ is the 1s atomic orbital on the nth atom in the chain.

Here k plays a role for solids similar to those of quantum numbers in atoms and molecules. When $k=0$, e^{inka} is one (since $e^0 = 1$) for all n and we obtain the wave function lowest energy orbital of the chain,

$$\psi(k=0) = \phi_1 + \phi_2 + \phi_3 + \phi_4 + \cdots$$

When, for example, $k = \pi/a$, e^{inka} equals $(-1)^n$ (since $e^{ix} = \cos x + i\sin x$) and we obtain the wave function of the highest energy orbital,

$$\psi(k = \pi/a) = -\phi_1 + \phi_2 - \phi_3 + \phi_4 + \cdots$$

These are both consistent with our qualitative argument and Figure 4.7. Other values of k correspond to the orbitals intermediate between these two. The **Brillouin zone** is the range of values of k (here $-\pi/a < k \leq -\pi/a$) which defines all the possible molecular orbitals. The concept of Brillouin zones is very important in the electronic theory of solids, and the books listed in the Further Reading section at the end of this book explain why in detail. Just as with quantum numbers in atoms and molecules, **k** is important for a detailed understanding of electronic transitions and spectra in solids, as transitions from one orbital to another with a different value of **k** are formally 'forbidden' and usually weak in intensity.

The hydrogen-chain orbitals we have considered are made up from only one type of atomic orbital (1s) and so only one energy band was formed. For most of the other atoms in the periodic table, it is necessary to consider other atomic orbitals in addition to the 1s orbitals, and we find that the allowed energy levels form a series of energy bands separated by ranges of forbidden energies. The ranges of forbidden energy between the energy bands are known as **band gaps**. Aluminium, for example, has the atomic configuration $1s^2 2s^2 2p^6 3s^2 3p^1$ and would be expected to form a 1s band, a 2s band, a 2p band, a 3s band, and a 3p band, all separated by band gaps. In fact, the lower energy bands, those formed from the aluminium core orbitals 1s, 2s, and 2p, are very narrow and for most purposes they can be regarded as a set of localised atomic orbitals. This arises because these orbitals are concentrated very close to the nuclei, so there is little overlap between the orbitals on the neighbouring nuclei. In small molecules, the greater the overlap, the larger the energy difference between the bonding and the antibonding orbitals. Similarly, in solids, the greater the overlap, the larger is the spread of energies or **bandwidth** of the resulting band. For aluminium, then, 1s, 2s, and 2p electrons can be taken to be core orbitals and only 3s and 3p bands are considered.

We can now start to consider some simple metals. Why is lithium, for example, a metal? The electron configuration of the Li atom is $1s^2\ 2s^1$. The 1s is a core orbital. The simplest picture, which we refine below in the next section, is that valence 2s orbitals overlap to form a band. Just as the 2s orbital in the Li atom is half-full, in the solid the 2s band is also half-full—thus we have a **metal** (Figure 4.9).

A metal has a partially filled band, called the **conduction** band, with no energy gap above the highest-occupied level. Electrons near this highest-occupied level are readily promoted into the empty orbitals in the band when an electric field is applied, and this gives rise to the high electrical conductivity of the metals. As in the free-electron model, the highest-occupied level in the metal at 0 K is the Fermi level.

In general, just as in small molecules, the available electrons are assigned to the levels in the energy bands starting with the lowest level. Each orbital can take two electrons of opposed spin. Thus, if N atomic orbitals are combined to make the band orbitals, then $2N$ electrons are needed to fill the band. For example, the 3s band in a crystal of aluminium containing N atoms can take up to $2N$ electrons, whereas the 3p band can accommodate up to $6N$ electrons. As aluminium has only one 3p electron per atom, however, there are only N electrons in the 3p band and only $N/2$ levels would be occupied.

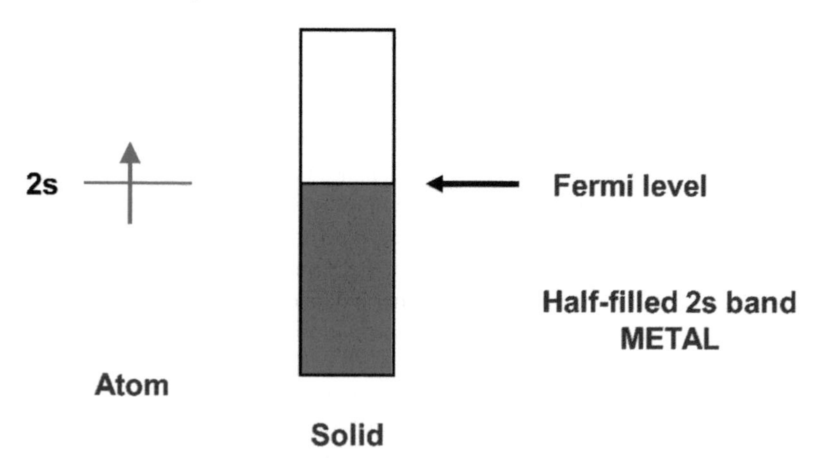

FIGURE 4.9 The 2s band in metallic lithium.

Magnesium is an interesting case as it has the atomic configuration $1s^22s^22p^63s^2$ and like aluminium the 1s, 2s, and 2p orbitals are core while the 3s is a valence orbital. The 3s orbital is filled, so we would expect the 3s band to be full and Mg to be an insulator. So why is magnesium a metal? The 3s and 3p orbitals are in fact close in energy so the 3s and 3p bands overlap and the result is a partially filled 3s/3p band.

Compounds of metals such as NaCl are classic insulators and we can readily qualitatively explain why this is so using our theory. In the simple ionic model an electron is transferred from a sodium atom forming Na^+ to a chlorine atom forming Cl^-. The highest occupied orbital is the chlorine 3p orbital, into which the electron has been transferred and the lowest unoccupied orbital is the sodium 3s, which has lost this electron. Similarly, in the molecular orbital theory, we have a full **valence band** made up of chlorine 3p orbitals and an empty **conduction band** higher in energy comprising sodium 3s orbitals. There is no partially full band, so NaCl is an insulator and does not conduct electricity. The band gap, the difference in energy between the full and empty bands, is very large (approximately 9 eV).

There are now several very large computer codes, the writing of which has taken decades, to calculate quantum mechanically the electronic structure of solids, calculating, among other properties, total energies, the corresponding wave function and electron density, and the detailed form of the bands and the corresponding densities of states There are many different methods, but at the time of writing by far the most popular technique, just as for studies of small molecules, is **density functional theory**. We give one example later. A longstanding problem has been to calculate successfully whether a given compound is a metal or an insulator.

Now let's return to the qualitiative model we've developed here and how it applies to some real solids—how we can apply the concept of energy bands to understanding some of their properties. First, we consider the simple metals in more detail.

4.3.1 Simple Metals

The crystal structures of the simple metals are such that the atoms have high coordination numbers. For example, the Group 1 elements have body-centered cubic structures with each atom surrounded by eight others (see Chapter 1). This high coordination number increases the number of ways in which the atomic orbitals can overlap. The *ns* and *np* bands of the simple metals are very wide because of the large amount of overlap, and, since the *ns* and *np* atomic orbitals are relatively close in energy in these elements, the two bands overlap and merge as we have already seen for magnesium above (Figure 4.10).

For the simple metals, then, we do not have an *ns* band or an *np* band, but rather one continuous band, which is labelled *ns*/*np*. For a crystal of N atoms, this *ns*/*np* band contains $4N$ energy levels and can hold up to $8N$ electrons.

However, the simple metals have far fewer than $8N$ electrons available; they have only N (Group 1), $2N$ (Group 2), or $3N$ (Al) electrons. Thus, the band is only partly full. There are empty energy levels just above the Fermi level and the metals are good electronic conductors.

If we now move to the right of the periodic table, we find elements that are semiconductors or insulators in the solid state; we now go on to consider these solids.

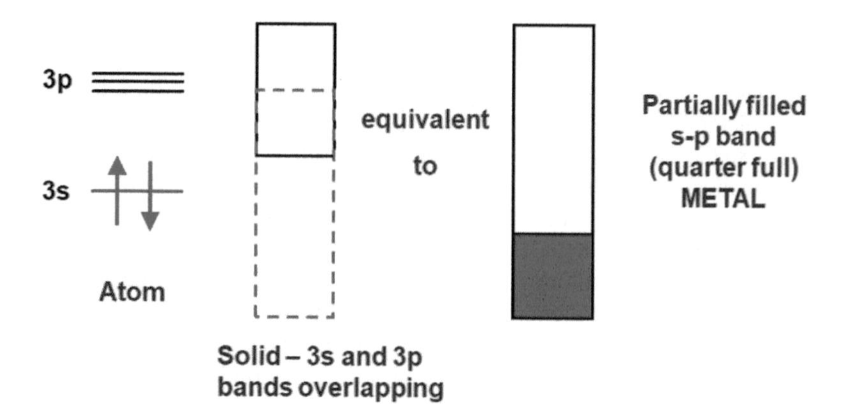

FIGURE 4.10 Mg is a metal because the 3p and 3s bands overlap and the resulting band is only a quarter-full.

4.4 DIAMOND, SI, AND GE: SEMICONDUCTORS

Carbon (in the form of diamond), silicon, and germanium, instead of forming structures of high coordination like the simple metals, form structures in which the atoms are tetrahedrally coordinated (see Chapter 1). With these structures, the *ns*/*n*p bands still overlap, but the *ns*/*n*p band splits into two. Each of the two bands contains 2*N* orbitals and can accommodate up to 4*N* electrons. We can think of the two bands as bonding and antibonding respectively; the tetrahedral symmetry does not give rise to any nonbonding orbitals. In chemical language, the lower energy band is made of bonding combinations of sp^3 hybrid orbitals, and the higher energy band the corresponding antibonding combinations. Carbon, silicon, and germanium have the electronic configurations ns^2np^2, thus they have 4*N* electrons available—exactly the correct number to fill the lower band. This lower occupied band is known as the **valence band**, the valence electrons in this band are essentially bonding the atoms in the solid together.

One question that may well have occurred to you is 'Why do these elements adopt the tetrahedral structure rather than one of the higher coordination structures?' If these elements adopted a structure like those of the simple metals, then the *ns*/*n*p band would be half-full. The electrons in the highest-occupied levels would be virtually nonbonding. In the tetrahedral structure, however, all 4*N* electrons would be in the bonding levels, which is energetically preferable. This is illustrated in Figure 4.11. Elements with few valence electrons will thus be expected to adopt high coordination structures and be metallic. Those with larger numbers (4 or more) will be expected to adopt lower coordination structures in which the *ns*/*n*p band is split and only the lower bonding band is occupied.

Tin (in its most stable room temperature form) and lead, although in the same group as silicon and germanium, are metals. For these elements, the atomic s–p energy separation is greater, and the overlap of s and p orbitals is much less than in silicon and germanium. For tin, the tetrahedral structure would have two s/p bands, but the band gap is almost zero. Below 291 K, tin undergoes a transition to the

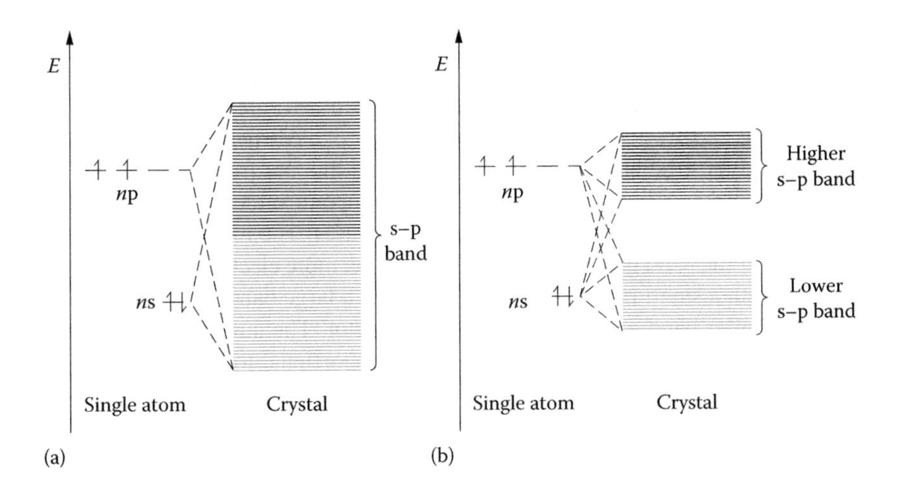

FIGURE 4.11 Energy bands formed from ns and np atomic orbitals for (a) a body-centered cubic crystal and (b) a crystal of diamond structure, showing filled levels for 4*N* electrons.

diamond structure, but above this temperature, it is more stable for tin to adopt a higher coordination structure. The advantage of having no nonbonding levels in the tetrahedral structure is reduced by the small band gap. In lead, the tetrahedral structure would give rise to an s band and a p band rather than *ns*/*np* bands, since the overlap of s and p orbitals is even smaller. It is more favourable for lead to adopt a cubic close-packed structure and, because lead has only 2*N* electrons to go in the p band, it is metallic. This is an example of the **inert pair effect** in the chemistry of lead, where the s electrons act as core electrons. Another example of the inert pair is the formation of divalent ionic compounds containing Pb^{2+}, rather than tetravalent covalent compounds like those of silicon and germanium.

Silicon and germanium thus have a completely full valence band and are expected to be insulators. However, they belong to a class of materials known as **semiconductors**.

The electronic conductivity, σ, is given by the expression:

$$\sigma = nZe\mu,$$

where n is the number of charge carriers per unit volume, Ze is their charge (for an electron, this is simply e, the electronic charge) and μ, the mobility, is a measure of the velocity of the electrons in the electric field.

As we have seen, the conductivity of metallic conductors *decreases* with temperature. As the temperature is raised, the phonons gain energy and the lattice vibrations have larger amplitudes. The displacement of the ionic cores from their lattice sites is larger, the electrons are scattered more, and the net current is reduced because the mobility μ of the electrons is reduced; in a metal n is effectively constant with temperature and changes in the conductivity is dominated by this change in μ.

In marked contrast, the conductivity of **intrinsic semiconductors** such as silicon or germanium *increases* with temperature. In these solids, conduction can only occur if electrons are promoted to the higher s/p band, the **conduction band**, because only then is there a partially full band. The current depends on n, in this case the number of electrons free to transport charge, and which is much smaller in semiconductors than in metals. In semiconductors, n is the number of electrons promoted to the conduction band plus the number in the valence band that are free to move into empty levels created due to this promotion. As the temperature increases, the number of electrons promoted increases, so the current increases. The increase depends on the magnitude of the band gap. At any one temperature, more electrons will be promoted in a solid with a small band gap, than in one with a large band gap, so that the solid with the smaller band gap will be a better conductor. The number of electrons promoted varies exponentially with temperature (proportional to the Boltzmann factor $\exp(-E_g/k_BT)$ where E_g is the band gap), so that a small change in band gap leads to a large difference in the number of electrons promoted and hence the number of current carriers, n. For example, tellurium has a band gap of about half that of germanium, but because of the exponential variation at a given temperature, the ratio of the number of electrons promoted in tellurium to that in germanium is of the order of 10^6. So germanium has an electrical resistivity of 0.46 Ω m at room temperature compared with that of tellurium of 0.0044 Ω m or with that of a typical insulator of around 10^{12} Ω m. The change in resistivity with temperature is used in thermistors, which are used as thermometers and also in fire alarm circuits.

Detailed quantum mechanical calculations have confirmed the qualitative picture presented here. Figure 4.12 shows a recent calculation of the band structure of diamond using the CRYSTAL17 code and one type of density functional theory called HSE06. Energies in the figure are all relative to the energy of the top of the valence band, which is arbitrarily set to zero. The full valence band on the left and the empty conduction band in Figure 4.12 are both made up from the carbon 2s and

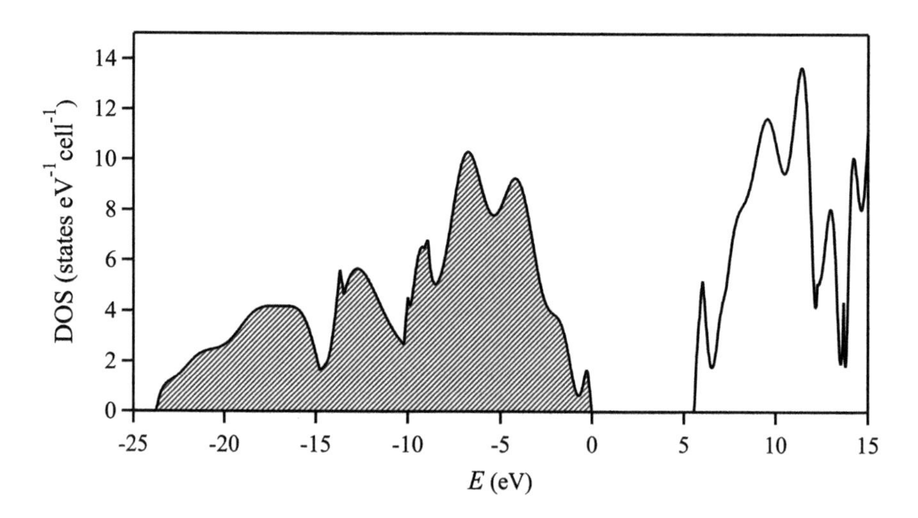

FIGURE 4.12 Calculated density of states for diamond using density functional theory and the HSE06 functional. The filled valence band is on the left, the empty conduction band on the right of the figure. (Courtesy of Dr. Sergio Conejeros.)

2p orbitals, in agreement with Figure 4.11. The calculated band gap, the difference in energy between the valence and the conduction band, is the band gap calculated with HSE06 is 5.19 eV, in good agreement with the experimental value (5.45 eV).

4.4.1 Photoconductivity

Electrons can also be promoted across a semiconductor band gap by forms of energy other than heat, for example, by light. If the photon energy (*hv*) of light shining on a semiconductor is greater than the energy of the band gap, then valence band electrons will be promoted to the conduction band and conductivity will increase. The promotion of electrons by light is illustrated in Figure 4.13.

Semiconductors with band gap energies corresponding to photons of visible light are **photoconductors** because they are, essentially, nonconducting in the dark, but conduct electricity in the light. One use of such photoconductors is in **electrophotography**. In the xerographic (photocopying) process, there is a positively charged plate covered with a film of semiconductor material. The semiconductor used is not silicon, but a solid with a more suitable band-gap energy, such as selenium, the compound As_2Se_3, or a conducting polymer. During copying, the light reflected from the white parts of the page to be copied shines onto the semiconductor film; the illuminated areas of the film become conducting, and an electron is promoted to the conduction band. This electron cancels the positive charge on the film and the positive hole in the valence band is removed by an electron from the metal backing plate entering the valence band. Now the previously illuminated areas of the film are no longer charged, but the areas underneath the black lines are still positively charged. In the next stage of the process, tiny, negatively-charged, plastic capsules of ink (toner) are spread onto the semiconductor film, but are only attracted to the

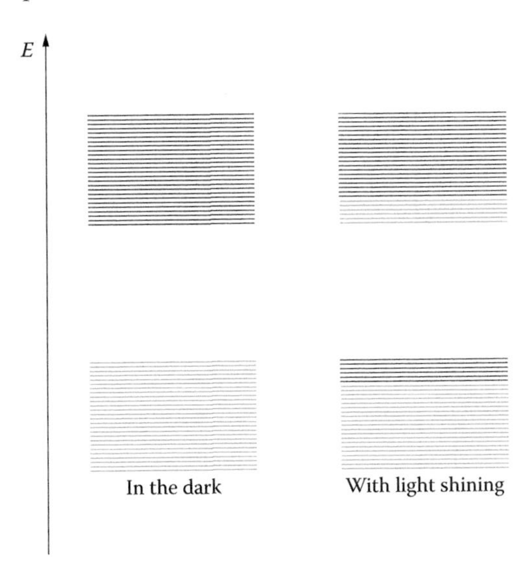

FIGURE 4.13 The promotion of electrons from the valence band to the conduction band by light.

charged areas of the film, to which they stick. A piece of positively charged white paper removes the toner from the semiconductor film and hence acquires an image of the black parts of the original. Finally, the paper is heated to melt the plastic coating and fix the ink.

4.4.2 Doped Semiconductors

The properties of semiconductors are extremely sensitive to the presence of impurities at concentrations as low as 1 part in 10^{10}. For this reason, silicon manufactured for transistors and other devices must be very pure. The deliberate introduction of a very low concentration of certain impurities into the very pure semiconductor, however, alters the properties in a way that has proved invaluable in constructing semiconductor devices. Such semiconductors are known as **doped** or **extrinsic semiconductors** as opposed to natural semiconductors such as Si and Ge, which are called **intrinsic semiconductors**.

Consider a crystal of silicon containing a few substitutional phosphorus atoms, i.e., the P atoms sit on lattice sites that would normally house Si atoms. Phosphorus has five valence electrons—one more than silicon. Four of these electrons form bonds with the four neighbouring Si atoms, but the fifth is only loosely bound to the atom. These extra electrons occupy energy levels between the valence and conduction bands and these levels are very close to the conduction band (Figure 4.14). Electrons in these levels are readily thermally promoted into the conduction band itself. The conductivity can increase by several orders of magnitude just by adding one part per million of phosphorus because of the extra electrons entering the conduction band. Such semiconductors are called ***n*-type**; ***n*** for negative charge carriers (electrons) and the new energy levels are called **donor levels** because they donate extra electrons.

Now suppose that, instead of phosphorus, the silicon is doped with an element with less valence electrons than silicon: for example, boron. Boron has three valence electrons, one less than silicon, so now there is an incomplete covalent bond. This

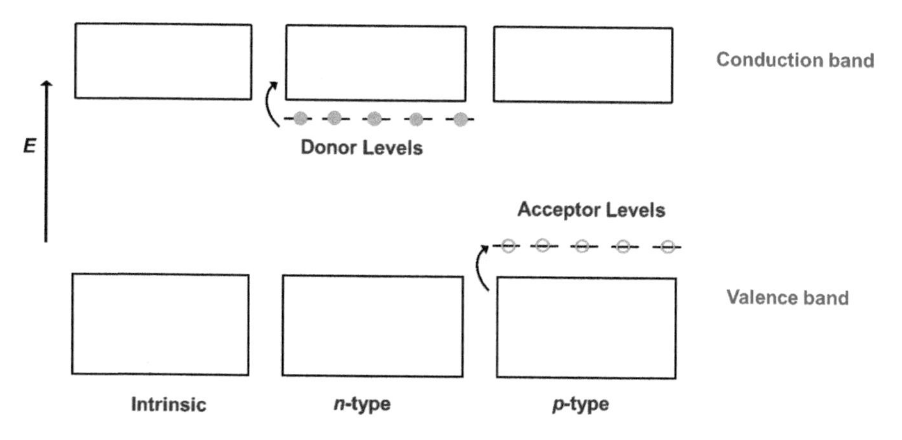

FIGURE 4.14 Intrinsic, n-type, and p-type semiconductors showing negative charge carriers (electrons in the conduction band) and positive holes (where electrons are missing).

produces vacant orbitals just above the valence band, and these new energy levels are called **acceptor levels** because electrons are thermally excited from the valence band into these levels. The dopants are thus withdrawing electrons from the valence band, leaving behind **holes** (empty orbitals) in the valence band. Because the valence band is no longer full, the electrons near the top of the valence band can move and so the doped silicon will be a better conductor than the pure material. This gives rise to ***p*-type** semiconductivity, the ***p***- indicating that the holes in the valence band are positively charged relative to the filled orbitals in the undoped material—an electron is missing from a place where there would normally be one.

Figure 4.14 schematically shows the energy bands in intrinsic, *p*-type, and *n*-type semiconductors.

The *n*-type and *p*-type semiconductors in various combinations make up many electronic devices, such as rectifiers, transistors, photovoltaic cells, and LEDs.

4.4.3 *P–N* Junction and Field Effect Transistors

Semiconductor *p–n* junctions are prepared either by doping different regions of a single crystal with different atoms or by depositing one type of material on top of another, using techniques such as chemical vapour deposition (Chapter 3). The use of these junctions stems from what happens where the two differently doped materials, *n*-type and *p*-type, meet. At this interface, conduction electrons in the *n*-type material diffuse across into the *p*-type material, dropping down into the holes (empty orbitals) in the latter. This drift produces an overall positive electric charge on the *n*-type side of the junction and an overall negative electric charge on the *p*-type. The electric field thus set up hinders further movement of electrons from the *n*-type to the *p* type. As electron transfer proceeds, the energies of the bands in the *p*-type rise and those in the *n*-type fall so that eventually an equilibrium is reached in which the bands 'bend' across the junction as shown in Figure 4.15.

The regions immediately on either side of the junction are known as **depletion regions** because there are fewer current carriers (electrons or holes) here and so these regions are insulators with a high resistance. Applying an external electric field across such a junction disturbs the equilibrium and the consequences of this are exploited in LEDs and transistors. If the positive terminal of a battery is connected to the *n*-type side of the junction while a negative terminal is connected to the *p*-type side, in a process called **reverse bias**, the electric field at the junction is reinforced. The positively charged holes in the *p*-type semiconductor are attracted to the negative terminal, and the negatively charged electrons in the *n*-type are attracted to the positive terminal. Charge does not flow across the junction, the depletion layer become wider, and no current flows (Figure 4.16). Conversely, under **forward bias** the polarity of the battery is reversed so the positive terminal of the battery is connected to the *p*-type side and the negative to the *n*-type. This reduces the electric field at the junction allowing the electrons in the *n*-type semiconductor to move towards the positive terminal and holes to move in the opposite direction. Current flows freely. This is so in LEDs, which we discuss in Chapter 8. Thus the *p–n* junction permits considerable control of the direction (and magnitude) of the current flowing through the material

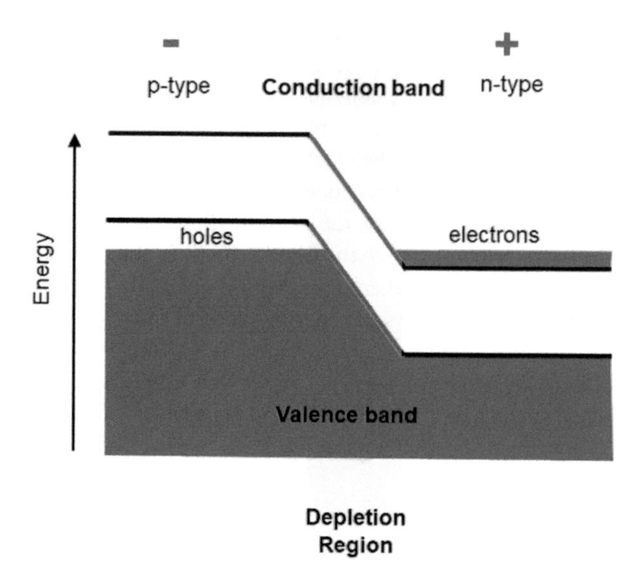

FIGURE 4.15 The bending of energy levels across a p–n junction, as electrons move across from the n-type into the p-type region, generating a positive charge on the former and a negative charge on the latter.

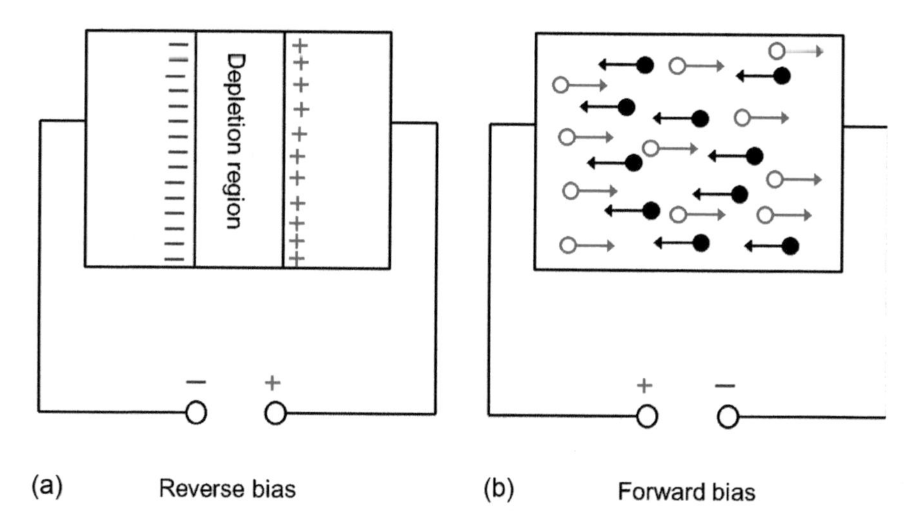

FIGURE 4.16 (a) left and (b) right. A p–n junction under (a) reverse bias (no flow of current) (b) forward bias (free flow of current carriers). In (b) black circles denote electrons, empty red circles holes.

depending on the direction of the applied voltage. This control is essential for the operation of diodes and transistors, the basis of silicon chips and all our modern electronic devices.

The 1956 Nobel Prize in Physics was awarded to Bardeen, Brattain, and Shockley for their work on developing transistors. Here, we briefly consider a very important class of transistors, the MOSFET transistor (metal-oxide-semiconductor).

The most common type of field effect transistor (FET) used in microelectronics is based on the metal–oxide–semiconductor (MOS) structure shown in Figure 4.17. In their most basic form, they consist of three electrodes, a Source (S), a Gate (G), and a Drain (D), although to be strictly accurate there is a fourth electrode attached to the back of the device called the Back gate (B) which carries a reference voltage (*e.g.* GND), but this is often omitted in some descriptions for simplicity. 'MOS' refers to the materials that were used to make up the gate electrode when these devices were first invented: a conducting metal contact sat on top of a very thin (10s of nm) insulating silicon dioxide layer, which sat on the silicon semiconductor substrate. Nowadays, 'MOS' is a bit of a misnomer because materials have improved such that the 'metal' is now highly doped polycrystalline silicon which is easier to deposit and pattern, the silicon dioxide is being replaced by better-performing 'high-*k* dielectric' oxides such as hafnium oxide, and the semiconductor can be Si, Ge, GaAs, or even doped diamond. Nevertheless, the term MOS is still used for modern devices, probably more for its historical legacy than its accuracy.

The source and drain contacts connect to isolated, highly doped (and therefore highly conducting) regions of the substrate, shown as either n^+ or p^+, where the + sign indicates highly doped.

The two common versions of MOS device, PMOS and NMOS (see Figure 4.17), get their names from their mode of operation, because they conduct current using either positive holes or negative electrons, respectively, but their principles of operation are the same. For the NMOS device, the source is held at a higher negative voltage than the drain, such that electrons attempt to travel from the S to the D—however, their route is blocked by a poorly conducting region underneath the gate electrode, and so no current can flow. Applying a positive voltage to the G changes the situation, because this bias attracts electrons from the bulk of the semiconductor

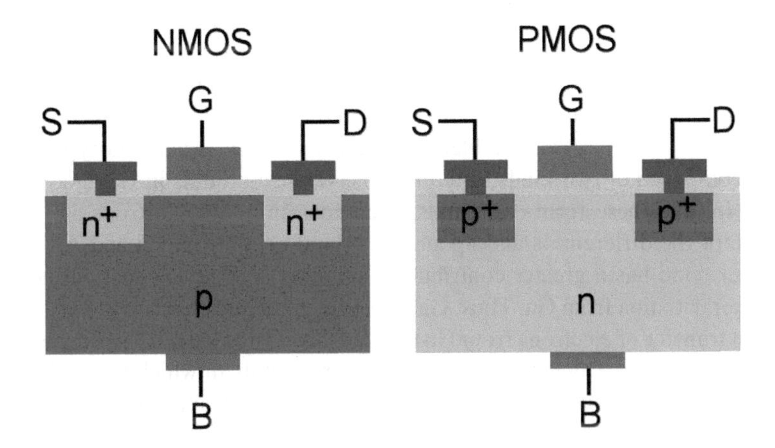

FIGURE 4.17 Schematic diagram of the most common type of field effect transistor (FET) used in microelectronics, which is based on the metal–oxide–semiconductor (MOS) structure shown. The two common versions of MOS device, PMOS and NMOS, are so called because they conduct current using either positive holes or negative electrons, respectively, but their principles of operation are the same. (Courtesy Professor Paul May).

toward the G, but they cannot complete their journey because they are blocked by the insulating oxide layer under the G. Thus, these electrons accumulate under the gate oxide, and the region become more conducting. With a sufficiently high gate voltage, the region becomes sufficiently conductive that the blockage preventing the S electrons from travelling to the D is effectively removed and a conductive '*n*-type channel' is opened connecting S to D. So long as there is a voltage on the G, current can flow freely from S to D; but as soon as the gate voltage is turned off, the S-D current ceases also.

For the PMOS device, everything is reversed, the S and D regions are doped p^+ so they conduct via holes. A negative gate voltage now repels electrons away from the gate oxide, creating a *p*-type conducting channel which now connects the S and D regions allowing a (hole) current to flow. Depending on the semiconductor, the mobilities of holes and electrons can be very different, and so the performance of PMOS and NMOS devices also vary. In most modern microprocessors, PMOS and NMOS transistors are connected together in pairs to create Complementary MOS (or CMOS) devices, a term which may be familiar to readers from computer usage. The reason for this is that CMOS devices only draw current (and hence use power) for the few microseconds when the gate voltage is switched on or off—at all other times they consume no power.

MOSFET devices can be used in two modes, as a switch or as an amplifier. Because a S–D current will flow only when a gate voltage is present, it can be used as a binary **switch**. Voltage on = current on = '1' in binary logic. Voltage off = current off = '0'. Connecting the individual FETs together with clever circuitry allows logic gates (AND, NAND, NOR, etc.), to be constructed, and these are the building blocks for modern computers. Low-power CMOS circuits are the linchpin of the computer industry.

4.5 BANDS IN COMPOUNDS: GALLIUM ARSENIDE

Gallium arsenide (GaAs) is a rival to silicon in some semiconductor applications, including solar cells, and it is also used for LEDs and in solid-state lasers, as you will see in Chapter 8. It has the zinc-blende structure which is a diamond-type structure, except composed of two kinds of atoms. The valence orbitals in Ga and As are the 4s and 4p, and these form two bands, each containing $4N$ electrons as in silicon. Because of the different 4s and 4p atomic orbital energies in Ga and As, however, the lower band has a greater contribution from As and the conduction band has a higher contribution from Ga. Thus, GaAs has partial ionic character because there is a partial transfer of electrons from Ga to As. The valence band has more As than Ga character, so all the valence electrons end up in orbitals in which the probability of being close to an As nucleus is greater than that of being near a Ga nucleus. The band energy diagram for GaAs is shown in Figure 4.18. GaAs is an example of a class known as III/V semiconductors, in which an element with one more valence electron than the silicon group is combined with an element with one less valence electron. Many of these compounds are semiconductors, for example, GaSb, InP, InAs, and InSb. Moving farther along the periodic table, there are II/VI semiconductors such as CdTe and ZnS. Towards the top of the periodic table and further out towards the

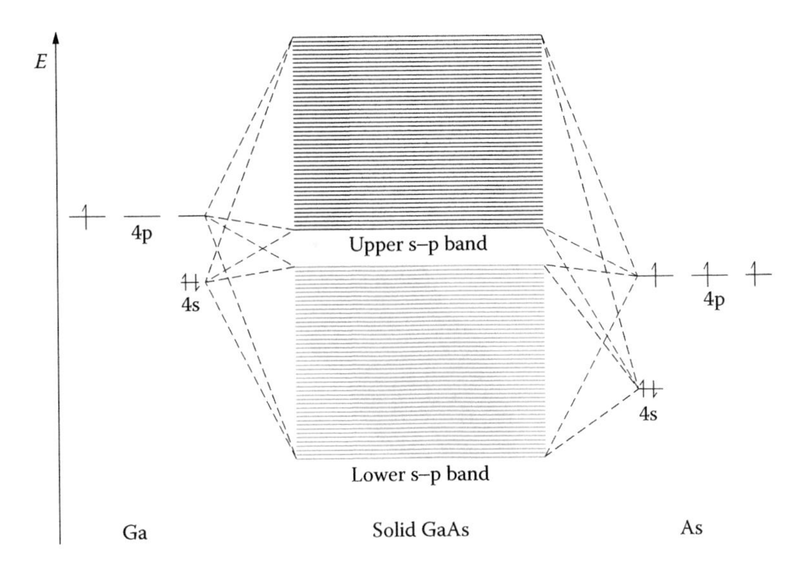

FIGURE 4.18 An orbital energy level diagram for gallium arsenide.

edges, for example, AlN and AgCl, the solids tend to adopt different structures and become more ionic. The band gaps of the semiconductor solids decrease down a group, for example, GaP > GaAs > GaSb and AlAs > GaAs > InAs.

4.6 BANDS IN D-BLOCK COMPOUNDS: TRANSITION METAL MONOXIDES

Monoxides (MO) with structures based on that of sodium chloride* are formed by the first- row transition elements Ti, V, Mn, Fe, Co, and Ni. TiO and VO are metallic conductors, and the other monoxides are semiconductors. In all these compounds the O 2p orbitals form a filled valence band while the 4s orbitals on the metal form an unoccupied band high in energy. So why are TiO and VO metallic and what is the role of the 3d orbitals?

In the sodium chloride structure, the octahedral coordination enables three of the five d orbitals (d_{xy}, d_{xz}, d_{yz}, i.e., the t_{2g} orbitals) on neighbouring metal ions to overlap since they oriented towards each other (Figure 4.19). Because the atoms are not *nearest* neighbours, the overlap is not as large as in pure metals and so the bands are narrow. The other two d orbitals ($d_{x^2-y^2}$, d_{z^2}, i.e. the e_g orbitals) point towards the adjacent oxygens and so the overlap between the e_g orbitals on different metal ions is much smaller than that between the t_{2g} orbitals. Thus, there are two 3d bands (Figure 4.20). The lower one, the t_{2g}, can contain up to 6*N* electrons and the upper one, labelled e_g, can take up to 4*N* electrons. In octahedral coordination titanium(II) has two d electrons in the t_{2g} orbitals (t_{2g}^2), so there are only 2*N* electrons to fill the 3*N* available levels of the lower band. Similarly, vanadium(II) has three d electrons in the t_{2g} orbitals (t_{2g}^3), so the lower band is half-full. As we have seen, a partly filled band leads to metallic conductivity.

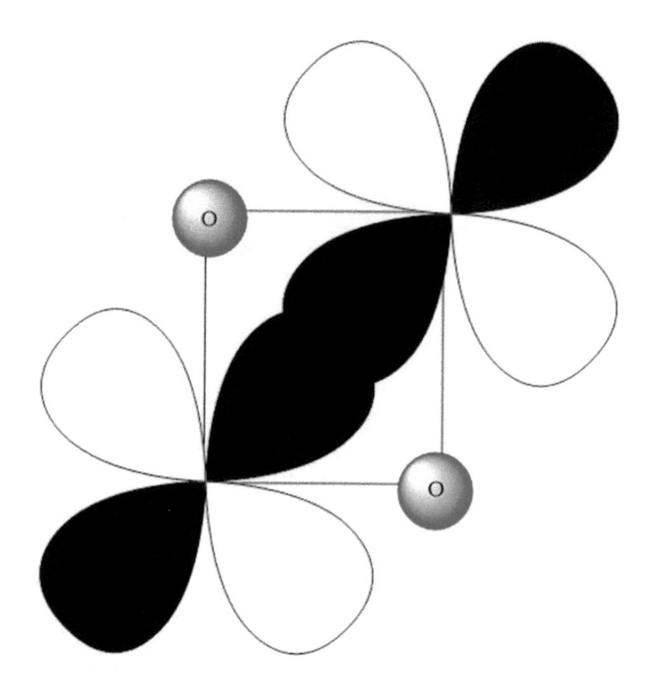

FIGURE 4.19 In a transition metal in octahedral coordination such as in the oxides MO, the d-orbitals are split by the crystal field into two sets, t_{2g} and e_g, and each set of orbitals forms its own band.

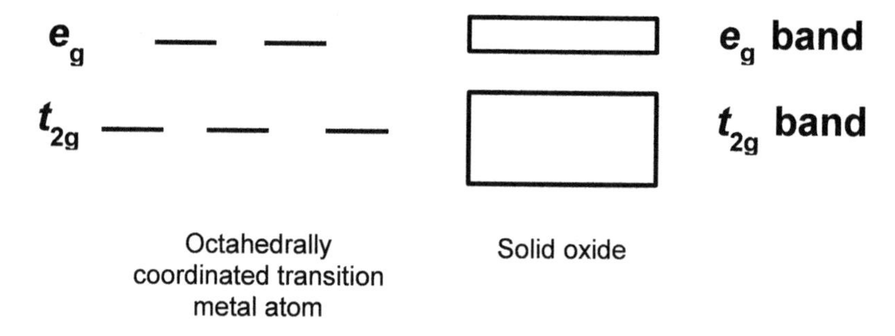

FIGURE 4.20 Overlap of metal 3d t_{2g} orbitals in transition metal oxides MO such as TiO and VO.

In marked contrast, MnO with only five electrons per manganese (Mn^{2+}, d^5, t_{2g}^3 e_g^2) is an insulator even though it has partially full t_{2g} and e_g orbitals. FeO (Fe^{2+}, d^6, t_{2g}^5 e_g^2) and CoO (Co^{2+}, d^7, t_{2g}^5 e_g^2) similarly are insulators. NiO (Ni^{2+}, d^8, t_{2g}^6 e_g^2), which has partially full e_g orbitals, is also an insulator, and understanding this proved a challenge to theoreticians for many years. It is easier to understand these oxides using a localised d-electron model.

Across the first transition series, there is a contraction in the size of the 3d orbitals due to the increasing effective nuclear charge experienced by these orbitals; the 3d orbital overlap therefore decreases and the width of the 3d band narrows. In a wide band such as the s/p bands of the alkali metals, the electrons are essentially free to move through the crystal, keeping away from the nuclei and from each other. In a narrow band, by contrast, the electrons are more tightly bound to the nuclei. Electron–electron repulsion in these circumstances becomes important; in particular, the repulsion between electrons on the same atom. Consider an electron in a partly filled band, moving from one atom to another. In an alkali metal, the electron would already be in the sphere of influence of the surrounding nuclei and would not be greatly repelled by the electrons on these nuclei. The 3d electron moving from one atom to another adds an extra electron near to the nucleus on the second atom, which already has 3d electrons close to it. Thus, electron repulsion near the nucleus is increased. For narrow bands, therefore, we must balance the gain in energy from band formation against the electron repulsion. For MnO, FeO, CoO, and NiO, electron repulsion dominates and it becomes more favourable for the 3d electrons to remain in localised orbitals than to be delocalised. The pure oxides are insulators. However, they are often found to be nonstoichiometric, that is, their formulae are not exactly MO, and the presence of more than one oxidation state leads to semiconducting properties, which will be discussed in Chapter 5.

The MOs are not unique in displaying a variation in their properties across the transition series. The behaviour can be complex. The dioxides form another series, and CrO_2 will be discussed later because of its magnetic properties. Some compounds are metallic or non-metallic depending on the temperature, such as VO_2 which becomes metallic at a phase transition from a monoclinic to the rutile structure at 340 K. Ti_2O_3 is semiconducting below 390 K with a small band gap (0.2 eV) but is metallic at higher temperatures. Several classes of mixed oxides also exhibit a range of electronic properties. Some ternary series, the perovskites for instance, show a transition along the series from metal to insulator early on, and then a transition back to metallic behavior. The compounds $LaMO_3$ (M = V, Cr, Mn, Fe) are all insulators whereas $LaNiO_3$ and $LaCuO_3$ are metallic. In general, compounds with broader d bands are metallic. Metallic compounds are commoner in the 4d and 5d series than for 3d oxides, reflecting the greater ability of 4d and 5d orbitals on one metal atom to overlap with those on neighboring metal atoms. Metallic behavior is found more often in compounds in which the metals are in lower oxidation states and which contain less electronegative anions. Metallic sulfides, for instance, are more common than metallic oxides.

* As will be explained in Chapter 5, non-stoichiometric TiO and VO have structures based on sodium chloride but with some of each element missing in an ordered manner. The other oxides, when stoichiometric, adopt the sodium chloride structure.

4.7 SUMMARY

We covered

1. The free-electron theory of metals and electronic conductivity.
2. Bands, densities of states, and Fermi energies of metals.
3. The molecular-orbital (tight-binding) theory of solids and how this can also explain metallic and insulating behavior. Metals have partly filled bands; insulators and intrinsic semiconductors have a filled valence band and an empty conduction band higher in energy.

4. Photoconductivity.
5. n- and p-type doped semiconductors and their applications in p–n junctions and MOSFET transistors.
6. Bands in compounds—why many metal transition metal oxides and other compounds show metallic conductivity while others do not.

QUESTIONS

1. In the free-electron model, the electron energy is entirely kinetic. Using the formula $E = 1/2\ mv^2$, calculate the velocity of electrons at the Fermi level in sodium metal (take $E_F = 3.2$ eV)
2. The density of magnesium metal is 1740 kg m^{-3}. A typical crystal has a volume of 10^{-12} m^3 (corresponding to a cube of side 0.1 mm). How many atoms would such a crystal contain?
3. An estimate of the total number of occupied states can be obtained by integrating the density of states from 0 to the Fermi level.

$$N = \int_0^{E_F} N(E)\,dE = \int_0^{E_F} E^{1/2}\left(2m_e\right)^{3/2} /2\pi^2\hbar^3\,dE = \left(2m_e E_F\right)^{3/2} V/3\pi^2\hbar^3$$

   Calculate the total number of occupied states for a magnesium crystal of volume (a) 10^{-12} m^3, (b) 10^{-6} m^3 and (c) 10^{-29} m^3 (approximately atomic size). Compare your results with the number of electrons available and comment on the different answers to (a), (b), and (c). Take $E_F = 7.1$ eV.
4. The elements calcium, zinc, cadmium and mercury all have outer electron configuration ns^2 which is just the right number of electrons to fill an s-band, so might be expected to be insulating. How does band theory explain why these elements are metals and not insulators?
5. The energy associated with one photon of visible light ranges from 2.4 to 5.0×10^{-19} J. The band gap in selenium is 2.9×10^{-19} J. Explain why selenium is a good material to use as a photoconductor in applications such as photocopiers.
6. The band gaps of several semiconductors and insulators are given below. Which substances would be photoconductors over the entire range of visible wavelengths?

Substance	Si	Ge	CdS
Band gap (10^{-19} J)	1.9	1.3	3.8

7. Which of the following doped semiconductors will be p-type and which will be n-type? (a) arsenic in germanium, (b) germanium in silicon, (c) indium in germanium, (d) silicon on antimony sites in indium antimonide (InSb) and (e) Mg on gallium sites in gallium nitride (GaN).

8. There is much interest in the use of doped diamond as a potential semiconductor in electrical devices. Which of the following dopants might be expected to be *p*-type and which *n*-type (a) nitrogen in diamond where the nitrogen occupies a carbon site (b) boron in diamond where the boron occupies a carbon site (b) lithium in diamond where the lithium occupies a carbon site (c) lithium in diamond where the lithium sits in a interstitial site, i.e., a site normally unoccupied in the crystal structure?
9. Would you expect carborundum (SiC) to adopt a diamond structure or one of a higher coordination? Explain why.
10. Comment on the electrical properties of the perovskites $LaTiO_3$ (metallic), $LaVO_3$ (semimetal) and $LaMnO_3$ (insulator) in terms of trends in the width of the d bands across the transition series. (A semimetal is a special case where the top of the full valence band is at the same energy as the bottom of an empty band but there is zero density of states at the energy where they meet. Another important example is graphite.)

5 Defects and Nonstoichiometry

5.1 INTRODUCTION

In a perfect crystal, all atoms would be in their correct lattice positions in the structure. This situation only exists at the absolute zero of temperature, 0 K. Above 0 K, **defects** occur in the structure. These defects may be **extended defects**, such as **dislocations**. The strength of a material depends very much on the presence (or absence) of extended defects, such as dislocations and **grain boundaries**; however, discussion of these types of phenomena lie more in the realm of materials science and will not be discussed further. Defects can also occur at isolated atomic positions and are known as **point defects**, which can be due to the presence of a foreign atom at a particular site, or to a vacancy where normally one would expect an atom to be present. Point defects can have significant effects on the chemical and physical properties of a solid, for instance, the beautiful colours of many gemstones are due to defect impurity atoms in the crystal structure. Tiny amounts of impurity atoms are crucial for silicon-based electronics. Another example is ionic solids, which are able to conduct electricity by a mechanism that is due to the movement of ions through vacant ion sites within the lattice (this is in contrast to the electronic conductivity explored in the previous chapter, which depends on the movement of electrons). Such solids are discussed in Chapter 6.

5.2 POINT DEFECTS AND THEIR CONCENTRATION

Point defects fall into two main categories: **intrinsic defects**, which are integral to the crystal in question—they do not change the overall composition and because of this, they are also known as **stoichiometric defects**; and **extrinsic defects**, which are created when a foreign atom is inserted into the lattice.

5.2.1 INTRINSIC DEFECTS

Intrinsic defects fall into two categories: **Schottky defects**, which consist of vacancies in the lattice; and **Frenkel defects**, where a vacancy is created by an atom or an ion moving into an interstitial position.

For a 1: 1 solid, MX, a Schottky defect, consists of a pair of vacant sites, a cation vacancy, and an anion vacancy. This is shown in Figure 5.1 for an alkali halide–type structure: the number of cation vacancies and that of anion vacancies have to be equal in order to preserve electrical neutrality. A Schottky defect for an MX_2-type structure will consist of the vacancy caused by the M^{2+} ion together with two X^- anion vacancies, thereby balancing the electrical charges. Schottky defects are more

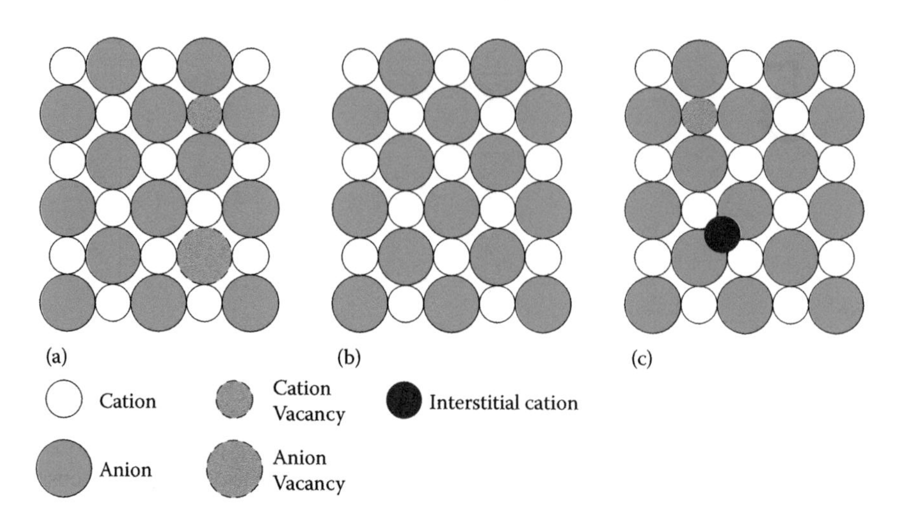

FIGURE 5.1 Schematic illustration of intrinsic point defects in a crystal of composition MX: (a) Schottky pair; (b) perfect crystal; (c) Frenkel pair.

common in 1: 1 stoichiometry, and examples of crystals that contain them include rock salt (NaCl), wurtzite (ZnS), and caesium chloride (CsCl).

A Frenkel defect usually occurs only on one sublattice of a crystal and consists of an atom or an ion moving into an interstitial position, thereby creating a vacancy. This is illustrated in Figure 5.1c for an alkali halide–type structure such as NaCl, where one cation is shown as having moved out of the lattice and into an interstitial site. This type of behaviour is seen, for instance, in AgCl, where we observe such a cation Frenkel defect when Ag^+ ions move from their octahedral coordination sites into tetrahedral coordination, and this is illustrated in Figure 5.2.

It is less common to observe an anion Frenkel defect when an anion moves into an interstitial site. This is because the anions are commonly larger than the cations in the structure, therefore it is more difficult for them to enter a crowded low-coordination interstitial site.

An important exception to this generalisation lies in the formation of anion Frenkel defects in compounds with the fluorite structure, such as calcium fluoride (CaF_2; see Chapter 1). (Other compounds adopting this structure are strontium

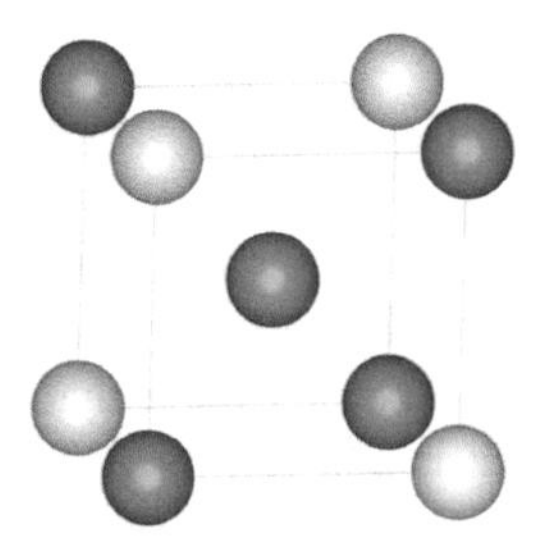

FIGURE 5.2 The tetrahedral coordination of an interstitial Ag^+ ion in AgCl.

and lead fluorides [SrF_2, PbF_2] and also thorium uranium, and zirconium oxides [ThO_2, UO_2, ZrO_2], which we will discuss again later in this chapter.) One reason for this is that the anions have a lower electrical charge than the cations and therefore can move nearer to each other. The other reason lies in the nature of the fluorite structure, which is shown again in Figure 5.3. You may recall that we can think of this structure as based on a cubic close-packed (*ccp*) array of Ca^{2+} ions with all the tetrahedral holes occupied by the F^- ions. This leaves all of the larger octahedral holes unoccupied, giving a very open structure. This can be seen clearly if we redraw the structure as in Figure 5.3c based on a simple cubic array of F^- ions. The unit cell now consists of eight small octants with the Ca^{2+} ions now occupying every other octant. The two different views are completely equivalent, but the cell shown in Figure 5.3c shows the possible interstitial sites more clearly.

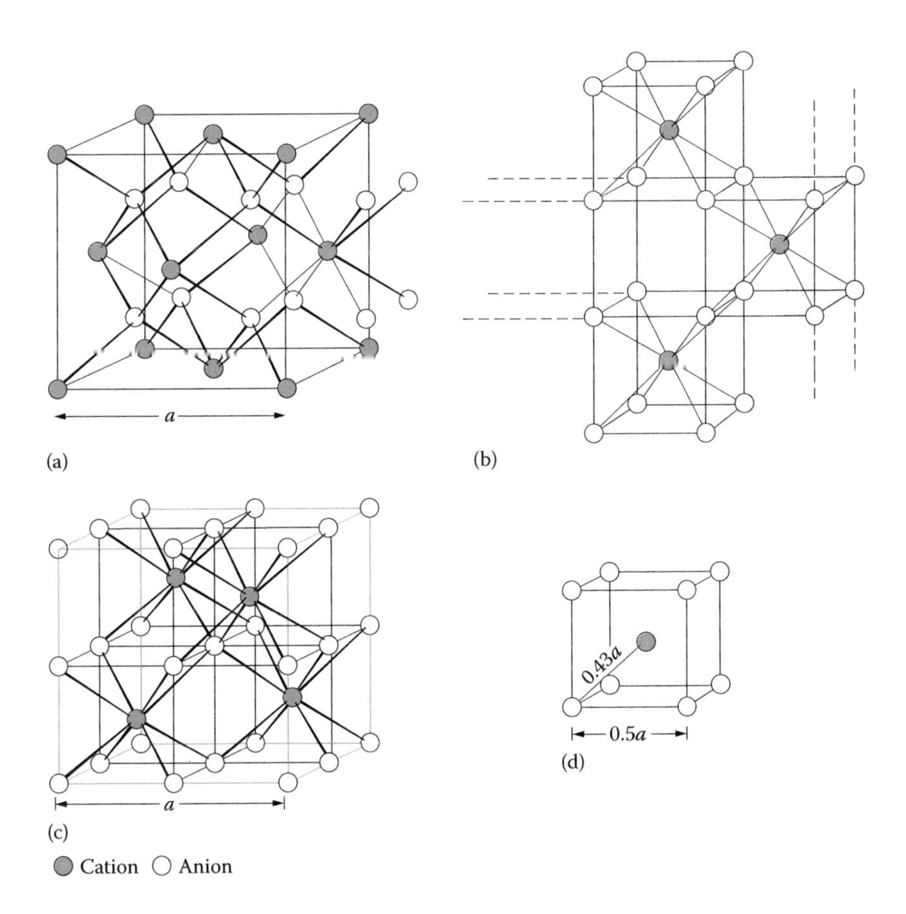

FIGURE 5.3 The crystal structure of fluorite MX_2. (a) A unit cell as a *ccp* array of cations. (b, c) The same structure redrawn as a simple cubic array of anions: the unit cell is marked by a coloured outline in (c). (d) Cell dimensions.

5.2.2 Concentration of Defects

Energy is required to form a defect; this means that the formation of defects is always an **endothermic process**. It may therefore seem surprising that defects exist in crystals at all. The reason for this is that although it costs energy to form defects, there is a gain in entropy. The enthalpy of formation of the defects is thus balanced by the gain in entropy such that, at equilibrium, the overall change in the free energy of the crystal due to the defect formation is zero according to the equation:

$$\Delta G = \Delta H - T\Delta S$$

The interesting point is that, thermodynamically, we do not expect a crystalline solid to be perfect, contrary, perhaps, to our 'commonsense' expectation of symmetry and order. At any particular temperature, there will be an equilibrium population of defects in the crystal.

The number of Schottky defects in a 1: 1 crystal of composition MX is given by

$$n_S \approx N \exp\left(\frac{-\Delta H_S}{2kT}\right) \tag{5.1}$$

where n_S is the number of Schottky defects per unit volume, at T/K, in a crystal with N cation and N anion sites per unit volume; ΔH_S is the enthalpy required to form one defect. It is relatively straightforward to derive these equations for the equilibrium concentration of Schottky defects by considering the change in entropy of a perfect crystal due to the introduction of defects. The change in entropy will be due to the change in the frequency of the normal mode vibrations of atoms around the defects (the phonons) and also to the arrangement of the defects, the **configurational entropy**. It is possible to estimate this latter quantity, using the methods of statistical mechanics.

If the number of Schottky defects is n_S per unit volume at T/K, then there will be n_S cation vacancies and n_S anion vacancies in a crystal containing N possible cation sites and N possible anion sites per unit volume. The **Boltzmann formula** tells us that the entropy of such a system is given by

$$S = k \ln W \tag{5.2}$$

where W is the number of ways of distributing n_S defects over N possible sites at random and k is the Boltzmann constant (1.380662×10^{-23} J K^{-1}). The number of cation vacancies (W_c) equals the number of anion vacancies (W_a), and both are given by probability theory, which shows that

$$W_c = W_a = \frac{N!}{(N - n_S)!\,n_S!} \tag{5.3}$$

where $N!$ is called 'factorial N' and is mathematical shorthand for $N \times (N-1) \times (N-2) \times (N-3) \ldots \times 1$.

The total number of ways of distributing these defects (W) is given by the product of W_c and W_a:

$$W = W_c W_a$$

It follows that when a perfect crystal has no defects, that is, $n_S = 0$, then $W = 1$ and $S = 0$; a perfect crystal has zero configurational entropy. The change in configurational entropy due to introducing defects into a perfect crystal is thus

$$\Delta S = k \ln W = k \ln \left(\frac{N!}{(N - n_S)! n_S!} \right)^2$$

$$= 2k \ln \left(\frac{N!}{(N - n_S)! n_S!} \right)$$

We can simplify this expression using **Stirling's approximation** that, for values of $N \gg 1$,

$$\ln N! \approx N \ln N - N$$

and the expression becomes (after manipulation)

$$\Delta S = 2k \left\{ N \ln N - (N - n_S) \ln (N - n_S) - n_S \ln n_S \right\}$$

If the enthalpy change for the formation of a single defect is ΔH_S and we assume that the enthalpy change for the formation of n_s defects is $n_s \Delta H_S$, then the Gibbs free energy change is given by

$$\Delta G = n_S \Delta H_S - 2kT \left\{ N \ln N - (N - n_S) \ln (N - n_S) - n_S \ln n_S \right\}$$

At equilibrium, at constant T, the Gibbs free energy of the system must be a minimum with respect to changes in the number of defects n_S; thus,

$$\left(\frac{d\Delta G}{dn_S} \right) = 0$$

Therefore,

$$\Delta H_S - 2kT \frac{d}{dn_S} \left[N \ln N - (N - n_S) \ln (N - n_S) - n_S \ln n_S \right] = 0$$

$N \ln N$ is a constant and hence its differential is zero; the differential of $\ln x$ is $1/x$ and of $(x \ln x)$ is $(1 + \ln x)$. On differentiating, we get

$$\Delta H_S - 2kT \left[\ln (N - n_S) + 1 - \ln n_S - 1 \right] = 0$$

hence, $\Delta H_S - 2kT\ln\left[(N-n_S)/n_S\right]$ and $n_S = (N-n_S)\exp(-\Delta H_S/2kT)$. As $N \gg n_S$, we can approximate $(N-n_S)$ by N, finally giving

$$n_S \approx N\exp\left(\frac{-\Delta H_S}{2kT}\right) \tag{5.1}$$

If we express this equation in molar quantities, it becomes

$$n_S \approx N\exp\left(\frac{-\Delta H_S}{2RT}\right) \tag{5.4}$$

where now ΔH_s is the enthalpy required to form one mole of Schottky defects and R is the universal gas constant, 8.314 J K mol^{-1}; the units of ΔH_s are in joules per mole.

By a similar analysis, we find that the number of Frenkel defects present in a crystal MX is given by the expression:

$$n_F \approx (NN_i)^{1/2}\exp\left(\frac{-\Delta H_F}{2kT}\right) \tag{5.5}$$

where n_F is the number of Frenkel defects per unit volume, N is the number of lattice sites, and N_i is the number of interstitial sites available. ΔH_F is the enthalpy of formation of one Frenkel defect. If ΔH_F is the enthalpy of formation of 1 mol of Frenkel defects, the expression becomes

$$n_F \approx (NN_i)^{1/2}\exp\left(\frac{-\Delta H_F}{2RT}\right) \tag{5.6}$$

Note that in these derivations, we have ignored the phonon correction due to the lattice vibrations. This correction tends to increase the value of n_s and n_F a little, because the presence of vacancies tends to lower the frequency and thus the energy required by the lattice vibrations.

The exponential form of these equations means that n_S and n_F are very temperature-dependent. The energy of formation of a defect can be calculated. In Chapter 1, we saw that using Coulomb's law, Buckingham potentials to represent the short-range interactions, and a core–shell model, the lattice energies of the solids can be reproduced to a high degree of accuracy. For defects, the lattice energy calculation is then repeated with either one ion missing or one ion moved from its lattice site to an interstitial position. Ions in a spherical region around the defect are allowed to alter their positions in response to the defect, but beyond a certain distance, the lattice is assumed to be unaffected. The difference between the lattice energies with and without the defect is the **defect energy**.

Table 5.1 lists some enthalpy of formation values for Schottky and Frenkel defects in various crystals. Using the information in Table 5.1 and Equation 5.1, we can now get an idea of how many defects are present in a crystal. If we assume that ΔH_S has a middle-of-the-range value of 5×10^{-19} J and substitute in Equation 5.1, we find that the proportion of vacant sites n_S/N at 300 K is 6.12×10^{-27}. This illustrates how

TABLE 5.1
Formation Enthalpy of Schottky and Frenkel Defects in Some Selected Compounds

Compound	ΔH (10^{-19} J)	ΔH (eV)[a]
	Schottky Defects	
MgO	10.57	6.60
CaO	9.77	6.10
LiF	3.75	2.34
LiCl	3.40	2.12
LiBr	2.88	1.80
LiI	2.08	1.30
NaCl	3.69	2.30
KCl	3.62	2.26
	Frenkel Defects	
UO2	5.45	3.40
ZrO2	6.57	4.10
CaF2	4.49	2.80
SrF2	1.12	0.70
AgCl	2.56	1.60
AgBr	1.92	1.20
β-AgI	1.12	0.70

[a] The literature often quotes values in electron volts (eV), so these are included for comparison; 1 eV= 1.60219×10^{-19} J.

few Schottky defects are present at room temperature. At $T = 1000$ K, this rises to 1.37×10^{-8}, a huge increase of a factor of ~10^{-19}, but still only one or two vacancies per hundred million sites.

Whether Schottky or Frenkel defects are found in a crystal depends mainly on the value of ΔH, the defect with the lower ΔH value predominating. In some crystals, it is possible for both types of defects to be present.

In order to change the properties of crystals, particularly their ionic conductivity, we may wish to introduce more defects into the crystal. It is important, therefore, at this stage, to consider how this might be done.

First, we have seen from the calculation above that raising the temperature introduces more defects. We would have expected this to happen because defect formation is an endothermic process and **Le Chatelier's principle** tells us that increasing the temperature of an endothermic reaction will favour the products—in this case, defects. Second, if it were possible to decrease the enthalpy of formation of a defect, ΔH_S or ΔH_F, this would also increase the proportion of defects present. A simple calculation as we did before, again using Equation 5.1, but now with a lower value for ΔH_S say 1×10^{-19} J, allows us to see this; Table 5.2 compares the results. The effect on the numbers of defects is dramatic—at 1000 K, there are now approximately

TABLE 5.2
Values of n_S/N

T/K	$\Delta H_S = 5 \times 10^{-19}$ J	$\Delta H_S = 1 \times 10^{-19}$ J
300	6.12×10^{-27}	5.72×10^{-6}
1000	1.37×10^{-8}	2.67×10^{-2}

three defects for every hundred sites. It is difficult to see how the value of ΔH could be manipulated within a crystal, but we do find crystals where the value of ΔH is lower than usual due to the nature of the structure, and this can be exploited. This is true for one of the systems that we shall look at in detail in Chapter 6, that of α-AgI. Third, if we introduce impurities selectively into a crystal, we can increase the defect population.

5.2.3 Extrinsic Defects

We can introduce vacancies into a crystal by **doping** it with a selected impurity: for instance, low concentrations of P, As, and/or B are added to very pure silicon to manufacture semiconductor devices such as transistors and solar cells. For ionic solids, the doped solid must be electrically neutral. For example, if we add $CaCl_2$ to an NaCl crystal, in order to preserve electrical neutrality, each Ca^{2+} ion replaces two Na^+ ions, so one cation vacancy is created. Such created vacancies are known as **extrinsic**. An important example is that of **zirconia (ZrO_2)**: this structure can be stabilised by doping with CaO, where the Ca^{2+} ions replace the Zr(IV) atoms in the lattice. The charge compensation here is achieved by the production of anion vacancies on the oxide sublattice.

5.2.4 Defect Nomenclature

We have seen that there is often more than one way to compensate for a desired defect. The possible ways of doing this for a particular compound can be expressed as balanced equations using the **Kröger–Vink notation**. This labels the point defect by the defect species, the species initially on the site, and the change in charge on the site. Charges are indicated by ′ for positive charge and • for negative charge. Table 5.3 gives the notation for common point defects.

Such equations can be used to predict whether a dopant ion will substitute for an ion in the undoped solid or occupy an interstitial site. Like chemical equations, both charge and number of atoms must balance. A simple example is Al-doped α-Fe_2O_3. In Chapter 1, we noted that in the α-Fe_2O_3 structure, Fe occupied two-thirds of the octahedral holes formed by close-packing of the oxygen. Al could in principle replace Fe or occupy a vacant octahedral site to give an interstitial Al. To calculate which is more favourable, we assume Al is added as Al_2O_3. Simple defect equations could then be

$$Al_2O_3(s) + 2Fe_{Fe} = 2Al_{Fe} + \alpha\text{-}Fe_2O_3(s)$$

TABLE 5.3
Notation for Common Point Defects

Defect type	Notation	Defect type	Notation
non-metal vacancy at non-metal (X) site	V_X	non metal (L) at non-metal (X) site	L_X
metal vacancy at metal (M) site	V_M	metal (A) substituting at metal (M) site	A_M
metal ion (charge +1) vacancy	V_M'	non-metal ion (charge -1) vacancy	$V_X^{\bullet}$
metal (A^{4+}) substituting at metal (M^{2+}) site	$A_M^{\bullet\bullet}$	non metal (L^-) at non-metal (X^{2-}) site	$L_X^{\bullet}$
interstitial atom, X	X_I	interstitial anion X charge −1	X_I'
free electron	e'	free positive hole	$H^{\bullet}$

and

$$Al_2O_3(s) + 2Fe_{Fe} = 2Al_i^{\circ\circ\circ} + 2V_{Fe}''' + \alpha\text{-}Fe_2O_3(s).$$

Note that the interstitial atom introduces extra charge. We have balanced this by removing an Fe^{3+} from its lattice site, but there are other possibilities such as reducing some Fe^{3+} to Fe^{2+}.

The case of Ti-doped α-Fe_2O_3 where Ti is introduced as Ti^{4+} is more complex. It has been found that Ti occupies both Fe lattice sites and interstitial sites. The defect equation is

$$3TiO_2(s) + 4Fe_{Fe} = Ti_i^{\circ\circ\circ\circ} + 2V_{Fe}''' + 2Ti_{Fe}^{\bullet} + 2\alpha\text{-}Fe_2O_3(s)$$

5.3 NONSTOICHIOMETRIC COMPOUNDS

We saw in the previous sections in this chapter that it is possible to introduce defects into a perfect crystal by adding an impurity. Such an addition causes point defects of one sort or another to form, but they no longer occur in complementary pairs. Impurity-induced defects are said to be **extrinsic**. We have also noted that when assessing which defects have been created in a crystal, it is important to remember that the overall charge on the crystal must always be zero.

A careful analysis of many substances, particularly inorganic solids, shows that it is common for the atomic ratios to be nonintegral: uranium dioxide, for instance, can range in composition from $UO_{1.65}$ to $UO_{2.25}$, certainly not the perfect UO_2 that we might expect. There are many other examples, some of which we discuss in some detail. What kind of compounds are likely to be nonstoichiometric? 'Normal' covalent compounds are assumed to have a fixed composition where the atoms are usually held together by strong covalent bonds formed by the pairing of two electrons. Breaking these bonds usually takes quite a lot of energy, and therefore under normal circumstances, a particular compound does not show a wide range of composition; this is true for most molecular organic compounds, for instance. Additionally, ionic compounds are usually stoichiometric, because to remove or add ions requires a considerable amount of energy. It is possible to make ionic crystals nonstoichiometric by doping them with an impurity. There is also another mechanism whereby ionic crystals can become nonstoichiometric: if the crystal contains an element with a variable valency, then a change in the number of ions of that element can be compensated for by changes in ion charge; this maintains the charge balance but alters the stoichiometry. Elements with a variable valency mostly occur in the transition elements, the lanthanoids and actinoids, and the heavier main group elements.

In summary, nonstoichiometric compounds can have formulae that are not simple integer ratios of atoms; they also usually exhibit a range of composition. They can be made by introducing impurities into a system, but are frequently a consequence of the ability of the metal to exhibit variable valency. Table 5.4 lists a few nonstoichiometric compounds together with their composition ranges.

Until more sophisticated structure determination became possible, defects in both stoichiometric and nonstoichiometric crystals were treated entirely from the point of view that point defects are randomly distributed. However, isolated point defects are not scattered at random in nonstoichiometric compounds, but are often dispersed throughout the structure in some kind of regular pattern. X-ray diffraction

TABLE 5.4
Approximate Composition Ranges for Some Nonstoichiometric Compounds

Compound		Composition range[a]
TiO_x	[$\approx TiO$]	$0.65 < x < 1.25$
	[$\approx TiO_2$]	$1.998 < x < 2.000$
VO_x	[$\approx VO$]	$0.79 < x < 1.29$
Mn_xO	[$\approx MnO$]	$0.848 < x < 1.000$
Fe_xO	[FeO]	$0.833 < x < 0.957$
Co_xO	[$\approx CoO$]	$0.988 < x < 1.000$
Ni_xO	[$\approx NiO$]	$0.999 < x < 1.000$
CeO_x	[$\approx Ce_2O_3$]	$1.50 < x < 1.52$
ZrO_x	[$\approx ZrO_2$]	$1.700 < x < 2.004$
UO_x	[$\approx UO_2$]	$1.65 < x < 2.25$
$Li_xV_2O_5$		$0.2 < x < 0.33$
Li_xWO_3		$0 < x < 0.50$
TiS_x	[$\approx TiS$]	$0.971 < x < 1.064$
Nb_xS	[$\approx NbS$]	$0.92 < x < 1.00$
Y_xSe	[$\approx YSe$]	$1.00 < x < 1.33$
V_xTe_2	[$\approx VTe_2$]	$1.03 < x < 1.14$

[a] Note that all composition ranges are temperature-dependent and that the figures here are intended only as a guide.

is the usual method for the determination of the structure of a crystal; however, this method yields an average structure for a crystal. For pure, relatively defect-free structures, this is a good representation, but for nonstoichiometric and defect structures, it avoids precisely the information that we want to know. For this kind of structure determination, a technique that is sensitive to local structure is needed, and such techniques are scarce. Structures are often elucidated from a variety of sources of evidence: X-ray and neutron diffraction, density measurements, spectroscopy (when applicable), and high-resolution electron microscopy (HREM); magnetic measurements have also proved useful in the case of FeO. The electron microscopy techniques have done most to clarify our understanding of defect structures, as under favourable circumstances information on an atomic scale is now possible. Such methods were described in Chapter 2.

Nonstoichiometric compounds are of use to industry because their electronic, optical, magnetic, and mechanical properties can be modified by changing the proportions of the atomic constituents.

5.3.1 Nonstoichiometry in Wüstite (FeO) and MO-Type Oxides

Ferrous oxide is known as **wüstite (FeO)**, and it has the NaCl (rock-salt) crystal structure. Accurate chemical analysis shows that it is nonstoichiometric: it is always deficient in iron. The FeO phase diagram (Figure 5.4) shows that the compositional range of wustite increases with temperature and that stoichiometric FeO is not included in the range of stability. Below 570°C, wustite disproportionates to α-iron and Fe_3O_4.

There are two ways in which an iron deficiency could be accommodated by a defect structure: there could either be iron vacancies giving a formula $Fe_{1-x}O$, or, alternatively, there could be an excess of oxygen in interstitial positions, with the formula FeO_{1+x}. A comparison of the theoretical and measured densities of the crystal distinguishes between the alternatives. The easiest method of measuring the density of a crystal is the flotation method. Liquids of differing densities (that dissolve in one another) are mixed together until a mixture is found that will just suspend the crystal so that it neither floats nor sinks. The density of that liquid mixture must then be the same as that of the crystal, and it can be found by weighing an accurately measured volume.

The theoretical density of a crystal can be obtained from the volume of the unit cell and the mass of the unit cell contents. The results of an X-ray diffraction structure determination gives both of these data, as the unit cell dimensions are accurately measured and the type and number of formula units in the unit cell are also determined. An example of this type of calculation for FeO follows:

> A particular crystal of FeO was found to have a unit cell dimension of 430.1 pm, a measured density of 5.728×10^3 kg m^{-3}, and an iron-to-oxygen ratio of 0.945. The unit cell volume (which is a cube) is thus $(430.1\text{ pm})^3 = 7.956 \times 10^7\ (\text{pm})^3 = 7.956 \times 10^{-29}\ \text{m}^3$.

There are four formula units of FeO in a perfect unit cell with the rock-salt structure. The mass of these four units can be calculated from their relative atomic masses:

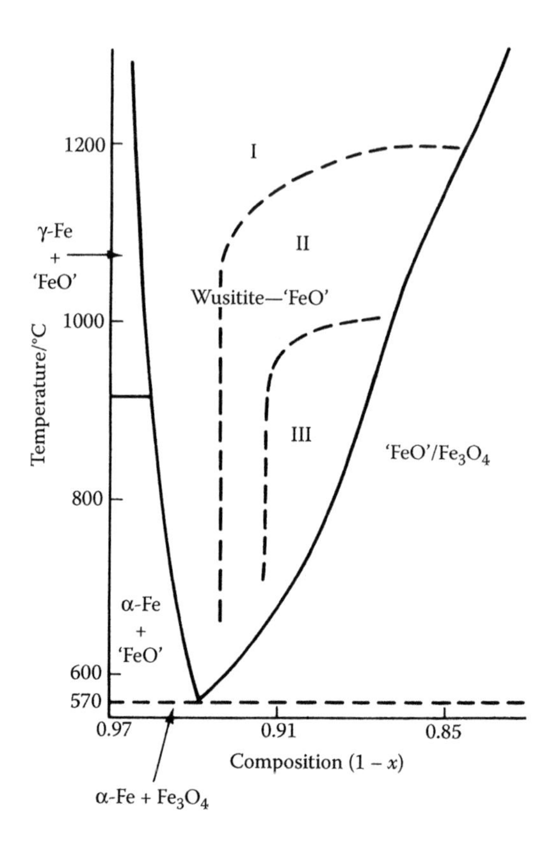

FIGURE 5.4 A phase diagram of the FeO system. I, II, and III together comprise the wüstite region.

Fe, 55.85 and O, 16.00. One mole of FeO weighs (55.85 + 16.00) g=0.07185 kg, 4 moles weigh (4 × 0.07185) kg, and four formula units weigh $(4\times0.07185)/N_A$ kg $=4.773\times10^{-25}$ kg; where Avogadro's number, $N_A=6.022\times10^{23}$ mol^{-1}. The perfect cell would have a density of 4.773×10^{-25} kg/7.956×10^{-29} m^3 $=6.000\times10^3$ kg m^{-3}.

The sample under consideration has an Fe/O ratio of 0.945. Assume, in the first instance, that it has iron vacancies: the unit cell contents in this case will be (4 × 0.945) = 3.78 Fe and 4 O. The mass of the contents will be $[(3.78\times55.85)+(4\times16.00)]/(N_A\times10^3)$ kg. Dividing by the volume of the unit cell, we get a value of 5.742×10^3 kg m^{-3} for the density. If, instead, the sample possesses interstitial oxygens, the ratio of oxygens to iron in the unit cell will be given by 1/0.945 = 1.058. The unit cell in this case will contain 4 Fe and (4 × 1.058) = 4.232 O. The mass of this unit cell is given by: $[(4\times55.85)+(4.232\times16.00)]/(N_A\times10^3)$ kg, giving a density of 6.076×10^3 kg m^{-3}. Comparing the two sets of calculations with the experimentally measured density of 5.728×10^3 kg m^{-3}, it is clear that this sample contains iron vacancies and that the formula should be written as $Fe_{0.945}O$. A table of densities is drawn up for FeO in Table 5.5.

It is found to be characteristic of most nonstoichiometric compounds that their unit cell size varies smoothly with composition but the symmetry is unchanged. This is known as **Vegard's law**.

TABLE 5.5
Experimental and Theoretical Densities (10^3 kg m^{-3}) for FeO

O:Fe ratio	Fe:O ratio	Lattice parameter (pm)	Observed density	Theoretical density	
				Interstitial O	Fe vacancies
1.058	0.945	430.1	5.728	6.076	5.742
1.075	0.930	429.2	5.658	6.136	5.706
1.087	0.920	428.5	5.624	6.181	5.687
1.099	0.910	428.2	5.613	6.210	5.652

In summary, nonstoichiometric compounds are found to exist over a range of composition, and throughout that range the unit cell length varies smoothly with no change in symmetry. It is possible to determine whether the nonstoichiometry is accommodated by vacancy or interstitial defects using density measurements.

So far, the discussion of the defects in FeO has been structural. Now we turn our attention to the balancing of the charges within the crystal. In principle, the compensation for the iron deficiency can be made either by the oxidation of some Fe(II) ions *or* by the reduction of some oxide anions. It is energetically more favourable to oxidise Fe(II). For each Fe^{2+} vacancy, two Fe^{2+} cations must be oxidised to Fe^{3+}. In the overwhelming majority of cases, defect creation involves changes in the cation oxidation state. In the case of metal excess in simple compounds, we would usually expect to find that neighbouring cation(s) would be reduced.

In a later section, we will look at some general cases of nonstoichiometry in simple oxides, but before we do that, we will complete the FeO story with a look at its detailed structure.

FeO has an NaCl structure with Fe^{2+} ions in the octahedral sites. The iron deficiency manifests itself as cation vacancies, and the electronic compensation made for this is that for every Fe^{2+} ion vacancy, there are two neighbouring Fe^{3+} ions, and this is confirmed by Mössbauer spectroscopy. If the rock-salt structure were preserved, the Fe^{2+}, Fe^{3+}, and the cation vacancies would be distributed over the octahedral sites in the *ccp* O^{2-} array. However, Fe^{3+} is a high-spin d^5 ion and has no crystal field stabilisation in either the octahedral or the tetrahedral sites and therefore no preference. Structural studies (X-ray, neutron, and magnetic) have shown that some of the Fe^{3+} ions are in the tetrahedral sites. If a tetrahedral site is occupied by an Fe^{3+} ion, the immediately surrounding four Fe^{2+} octahedral sites must be vacant as they are too close to be occupied at the same time, and this type of defect is found for low values of x. At higher values of x, the structure contains various types of **defect cluster**, which are distributed throughout the crystal. A defect cluster is a region of the crystal where the defects form an ordered structure. One possibility is known as the **Koch–Cohen cluster** (Figure 5.5). This shows a standard NaCl-type unit cell at the centre, but with four interstitial, tetrahedrally coordinated Fe^{3+} ions in the tetrahedral holes; the thirteen immediately surrounding octahedral Fe^{2+} sites must be vacant. Surrounding this central unit cell, the other octahedral iron sites in the cluster are occupied, but they may contain either Fe^{2+} or Fe^{3+} ions. We therefore

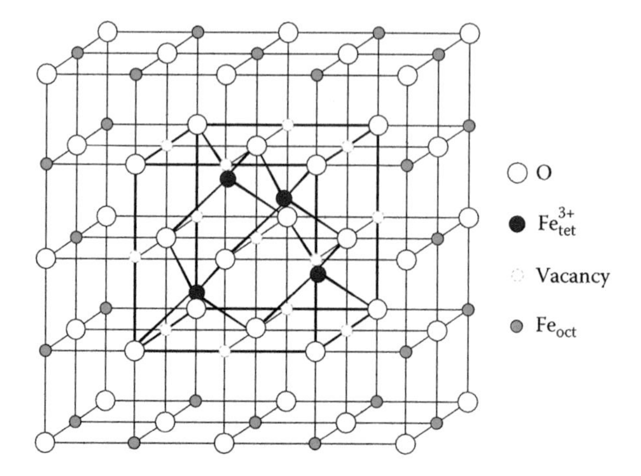

FIGURE 5.5 The Koch–Cohen cluster shown with the back and front planes cut away for clarity. The central section with four tetrahedrally coordinated Fe^{3+} ions is highlighted in bold.

designate them simply as Fe_{oct}. The front and back planes have been cut away from the diagram in Figure 5.5 to make the central section more visible.

It is instructive to consider the composition of a cluster such as the one shown in Figure 5.5 in some detail. Clusters such as this are often referred to by the ratio of cation vacancies to interstitial Fe^{3+} in tetrahedral holes, in this case 13: 4. The complete cluster, allowing for the back and front faces, which are not illustrated here, contains 8 NaCl-type unit cells, and thus 32 oxide ions. A stoichiometric array with no defects would also contain 32 Fe^{2+} cations on the octahedral sites: taking into account the 13 octahedral vacancies, there must be 19 Fe_{oct} ions and the 4 interstitial Fe^{3+}_{tet} ions, making 23 iron cations in all. The overall formula for the cluster is thus $Fe_{23}O_{32}$, almost Fe_3O_4. It bears a strong resemblance to the structure of Fe_3O_4, the next highest oxide of iron.

Having determined the atomic contents of the cluster, we now turn our attention to the charges. There are 32 oxide ions, so to balance them, the Fe cations overall must have 64 positive charges. Twelve are accounted for by the four Fe^{3+} ions in the tetrahedral positions, leaving 52 to find from the remaining 19 Fe_{oct} cations. This accounting is difficult to do by inspection. Suppose that there are $x$$Fe^{2+}$ ions and $y$$Fe^{3+}$ ions in the octahedral sites, we know that

$$x + y = 19$$

We also know that their total charges must equal 52, therefore,

$$2x + 3y = 52$$

giving two simultaneous equations. Solving gives $y = 14$ and $x = 5$. The octahedral sites surrounding the central (bold) unit cell are thus occupied by 5 Fe^{2+} ions and by 14 Fe^{3+} ions. By injecting such clusters throughout the FeO structure, the nonstoichiometric structure is built up. The exact formula of the compound (the value of x in $Fe_{1-x}O$) will depend on the average separation of the randomly injected clusters. Neutron scattering experiments indicate that as the concentration of the defects increases, the clusters order into a regular repeating pattern with its own unit cell of lower symmetry; the new structure is referred to as a **superstructure** or **superlattice** of the parent. In the oxygen-rich limit when the whole structure is composed of these clusters, there would be a new ordered structure of the formula $Fe_{23}O_{32}$ based on the structure of the parent defect cluster.

Manganese, cobalt, and nickel oxides (MnO, CoO, and NiO) have similar structures to FeO. **ZnO** has the würtzite crystal structure where half of the tetrahedral holes of the hexagonally packed oxygen anions are filled with Zn^{2+} ions. ZnO absorbs in the ultraviolet (340–400 nm in the UVA range) and is widely used in sunscreen and cosmetics. When it is heated to 800°C; oxygen is lost, leaving nonstoichiometric $Zn_{1+x}O$ with oxide vacancies. To compensate, Zn^{2+} ions migrate to interstitial positions and are reduced to Zn^{+} ions or Zn atoms. The formation of oxide vacancies leads to the appearance of energy levels in the band gap and hence effectively reduces the band gap. Now the oxide can absorb visible light and thus at high temperatures is yellow.

5.3.2 Uranium Dioxide

Above 1127°C, a single oxygen-rich nonstoichiometric phase of UO_2 is found with the formula UO_{2+x}, ranging from UO_2 to $UO_{2.25}$. Unlike FeO, where a metal-deficient oxide was achieved through cation vacancies, in this example the metal deficiency arises from interstitial anions.

$UO_{2.25}$ corresponds to U_4O_9, which is a well-characterised oxide of uranium known at low temperature. UO_2 has the fluorite structure. The unit cell is shown in Figure 5.6a and contains four formula units of UO_2. (There are four uranium ions contained within the cell boundaries; the eight oxide ions come from: $\left[8\times\frac{1}{8}\right]=1$ at the corners, $\left[6\times\frac{1}{2}\right]=3$ at the face centres, $\left[12\times\frac{1}{4}\right]=3$ at the cell edges, and 1 at the cell body-centre.)

As more oxygen is taken into UO_2, the extra oxide ions go into interstitial positions. The most obvious site available is in the middle of one of the octants (the vacant octahedral holes), where there is no metal atom. This site, however, is not ideal for an extra oxide ion, as not only is it crowded, but it is also surrounded by eight ions of the same charge. Neutron diffraction shows that an interstitial oxide anion does not sit exactly at the centre of an octant, but it is displaced sideways; this has the effect of moving two other oxide ions from their lattice positions by a very small amount, leaving two vacant lattice positions. This is illustrated in Figure 5.6b, where three vacant octants are picked out and the positions of one additional interstitial oxide and the two displaced oxides with their vacancies are shown. This defect cluster can be thought of as two vacancies, one interstitial of one kind (O′) and two of another (O″), and this is called the 2: 1: 2 Willis cluster. The movement of the ions from the 'ideal' positions is shown by small arrows: the movement of the interstitial oxide O′ from

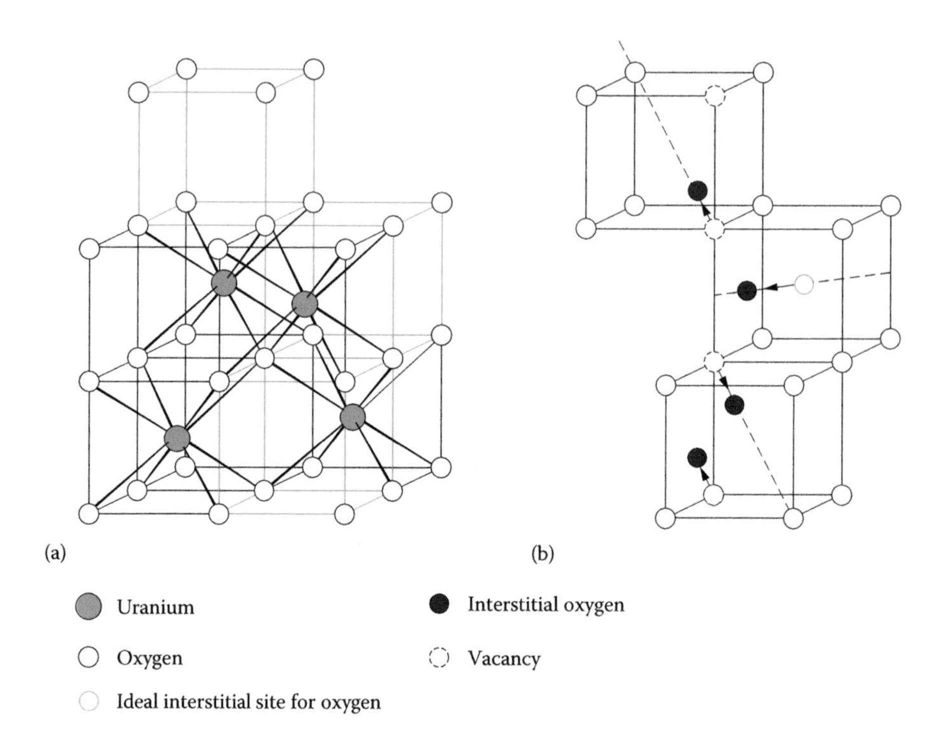

FIGURE 5.6 (a) The fluorite structure of UO_2. (b) Interstitial defect cluster in UO_{2+x}. The uranium positions (not shown) are in the centre of every other cube.

the centre of an octant is along the direction of a diagonal of one of the cube faces (the *110* direction), whereas the movement of the oxide ions O″ on lattice positions is along cube diagonals (the *111* direction). The atomic composition of a fluorite unit cell (Figure 5.6a) when modified by this defect cluster is U_4O_9; the net oxygen gain is just the single new interstitial in the central octant of Figure 5.6b. A more commonly found defect cluster contains two vacancies, two interstitial oxides (O′) and two interstitial oxides (O″), leading to a 2: 2: 2 Willis cluster (not shown here). In the oxygen-rich limit for the UO_{2+x} nonstoichiometric structure, $UO_{2.25}$, these are ordered throughout the structure, and the structure has a very large unit cell based on 4 × 4 × 4 fluorite unit cells (with a volume of 64 times that of the fluorite unit cell). We can think of UO_{2+x} as containing **microdomains** of the U_4O_9 structure within that of UO_2. The electronic compensation for the extra interstitial oxide ions will most likely be the oxidation of the neighbouring U(IV) atoms to either U(V) or U(VI).

5.3.3 Titanium Monoxide Structure

Titanium and oxygen form nonstoichiometric phases that exist over a range of composition centered about the stoichiometric 1: 1 value, from $TiO_{0.65}$ to $TiO_{1.25}$. We shall look at what happens in the upper range from $TiO_{1.00}$ to $TiO_{1.25}$.

At the stoichiometric composition of $TiO_{1.00}$, the crystal structure can be thought of as a NaCl-type structure with vacancies in both the metal and the oxygen sublattices: one-sixth of the titaniums and one-sixth of the oxygens are missing. Above 900°C, these vacancies are randomly distributed, but below this temperature they are ordered, as shown in Figure 5.7.

In Figure 5.7a, we show a layer through an NaCl-type structure. Every third vertical diagonal plane has been picked out with a dashed line. In the $TiO_{1.00}$ structure, every other atom along those dashed lines is missing. This is illustrated in Figure 5.7b. If we consider that in these first two diagrams we are looking at the structure along the y axis, and that this layer is the top of the unit cells, $y=0$ or 1, then the layer below this and parallel to it will be the central horizontal plane of the unit cell at $y=\frac{1}{2}$. This is drawn in Figure 5.7c, and again we notice that every other atom

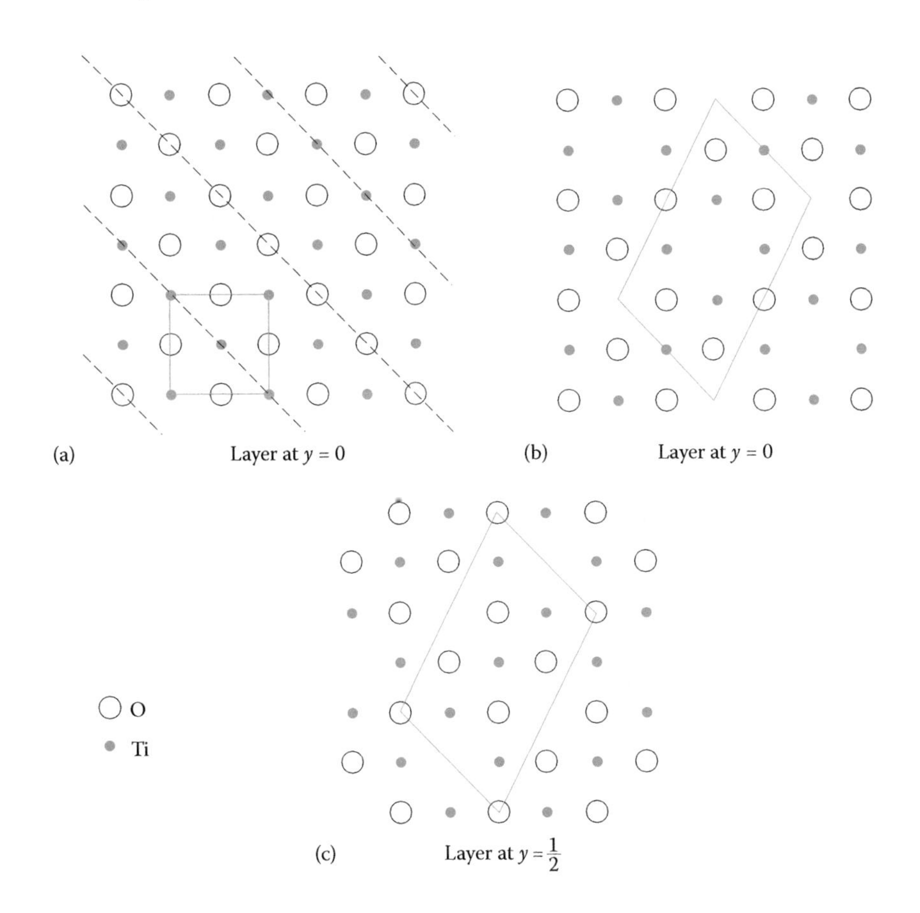

FIGURE 5.7 Layers parallel to the horizontal planes of Figure 1.31. (a) The hypothetical TiO structure of the NaCl type shown in Figure 1.31; the line of intersection of every third vertical diagonal plane is marked by a dashed line. (b) The same plane in the observed structure of TiO; every alternate atom is removed along the diagonal lines in (a). (c) The plane directly beneath the layer in (b); again, every alternate atom is removed along the cuts made by the planes whose intersection lines are shown in (a). In (b) and (c), the cross-section of a monoclinic unit cell is indicated.

along every third diagonal plane is missing. This is true throughout the structure. In Figure 5.7, the unit cell of a perfect NaCl-type structure is marked on (a), whereas the boundaries of the new unit cell, taking the ordered defects into account, are marked on (b) and (c). The new unit cell of the superlattice is monoclinic (see Chapter 1) because the angle in the *xz* plane (β) is not equal to 90°. This structure is unusual in that it appears to be stoichiometric, but in fact it contains defect vacancies on both the anion and the cation sublattices.

As we noted in Chapter 4, unusually for a transition metal monoxide, $TiO_{1.00}$ shows metallic conductivity. The existence of the vacant sites within the TiO structure is thought to permit sufficient contraction of the lattice that the 3d orbitals on titanium overlap, thus broadening the conduction band and allowing electronic conduction.

When titanium monoxide has the limiting formula $TiO_{1.25}$, it has a different defect structure, still based on the NaCl structure, but with all the oxygens present, and one in every five titaniums missing (Figure 5.8). The pattern of the titanium vacancies is shown in Figure 5.9, which shows a layer of the type in Figure 5.8 but with the oxygens omitted; only the titaniums are marked. If we draw lines through the titaniums, every fifth one is missing. The ordering of the defects has again produced a superlattice. Where samples of titanium oxide have formulae that lie between the two limits discussed here, $TiO_{1.00}$ and $TiO_{1.25}$, the structure seems to consist of portions of the $TiO_{1.00}$ and $TiO_{1.25}$ structures intergrown. Note that although most texts refer to the structure as $TiO_{1.25}$, as we have here, on the definitions that we have used previously when discussing 'FeO', it is more correctly written as $Ti_{0.8}O$ ($Ti_{1-x}O$), as this indicates that the structure contains titanium vacancies rather than interstitial oxygens.

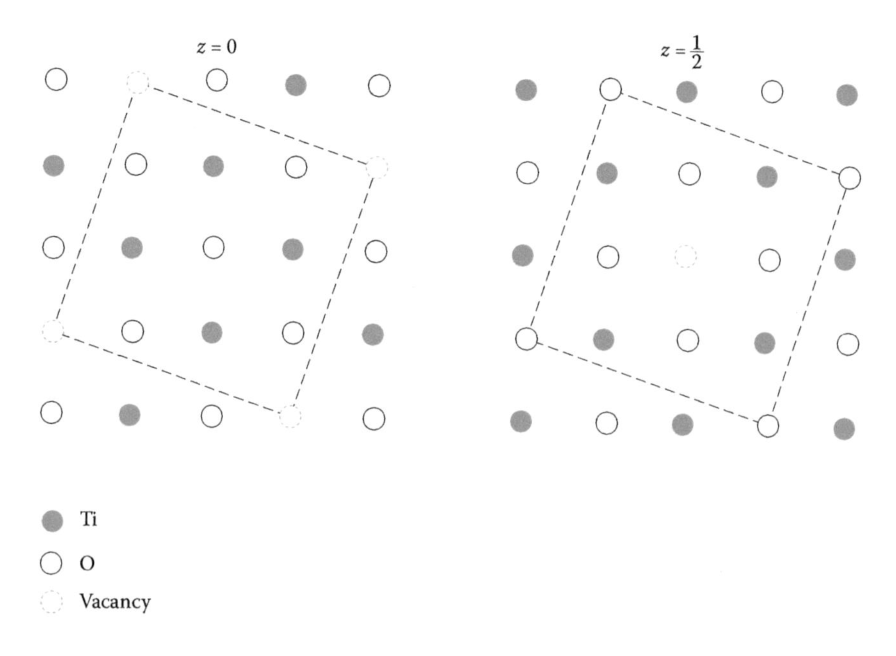

FIGURE 5.8 The structure of $TiO_{1.25}$ showing both O and Ti positions.

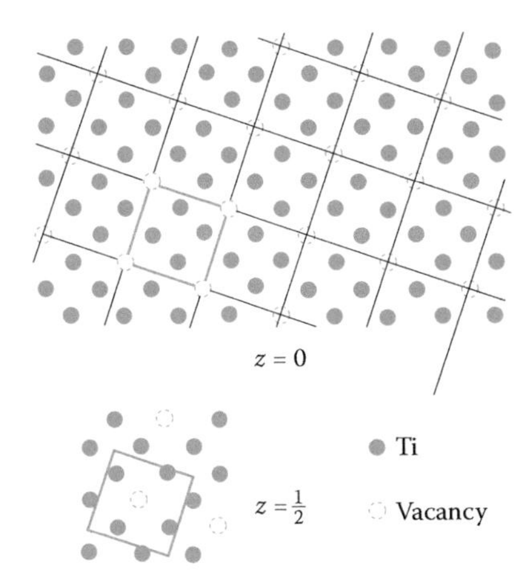

FIGURE 5.9 Successive Ti layers in the structure of $TiO_{1.25}$.

5.4 EXTENDED DEFECTS

We mentioned in the introduction to this chapter that crystals often contain **extended defects** as well as point defects. The simplest line or **linear defect** is an **edge dislocation** where there is a fault in the arrangement of the atoms in a line through the crystal lattice where an extra plane of atoms extends part way through the structure, thus distorting it (Figure 5.10a). Another linear defect is the **screw dislocation**, which occurs when a stress is applied to the crystal; the dislocation of the line of atoms is now perpendicular to the stress (Figure 5.10b). New material arriving at the surface of the crystal during crystallisation preferentially anchors itself to the edge formed and the planes of atoms now form a single helical surface (Figure 5.11). It has been suggested that such defects are important in the formation of crystals from weakly supersaturated solutions. Linear dislocations greatly reduce the strength of a crystal, as the planes can glide over one another more easily, the dislocation occurring along the densest planes of atoms, but they are the site of enhanced chemical activity and have been shown to be very important for the enhancement of the activity of catalysts.

There are different types of **planar defects**, and examples include **grain boundaries**, **stacking faults**, and **chemical twinning** which can contain unit cells mirrored about the twin plane through the crystal.

Polycrystalline substances are made of small crystallites orientated randomly to one another; the surface that separates each crystallite is known as the grain boundary (Figure 5.10c). Grain boundaries tend to hinder the movement of dislocations so that a polycrystalline metal is usually much stronger than a single crystal. To make metals stronger, interstitial atoms and grain boundaries may be deliberately introduced to 'pin' dislocations in place. Grain sizes vary hugely from the order of micrometres to millimetres in diameter. Metal alloys are often stronger than the host metal—take steel for example—because the atoms 'foreign' to the structure have a tendency to congregate on the grain boundaries, thus hindering further the movement of dislocations. An **antiphase domain** can also occur

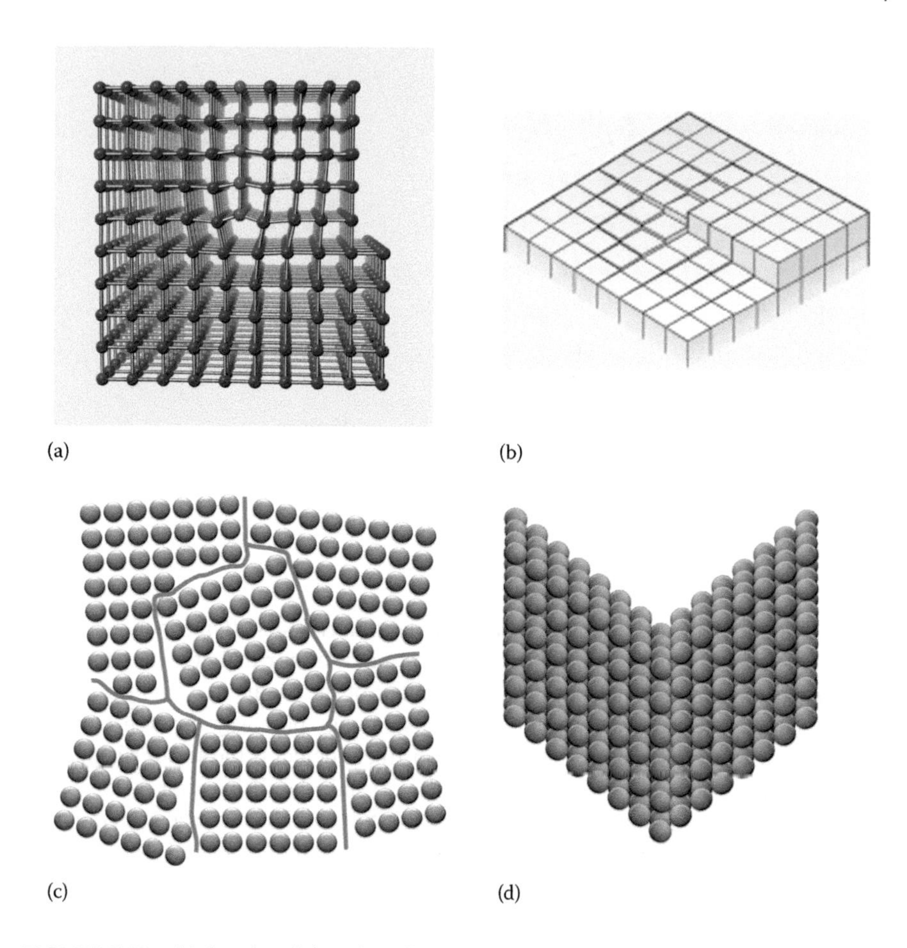

FIGURE 5.10 (a) An edge dislocation. (Image generated using CrystalMaker X, http://crystalmaker.com. With permission.) (b) A screw dislocation. (Copyright The Open University. With permission.) (c) Grain boundary. (d) Crystal twinning.

with a crystal, where this grain has the reverse structure from the surrounding structure. At the antiphase boundary, each side of the boundary has the opposite phase: for instance, in a *ccp* (see Chapter 1), the sequence at the boundary would be ABCAB**CC**BACBA….

A stacking fault occurs in a crystal when there are either one or two extra planes or missing planes of atoms. This can occur quite easily in close-packed structures. Think of a *ccp* structure ABCABCABC…; if one of the planes is missing, an intrinsic fault, we get a faulty stacking sequence such as ABC**ACA**BCABC…. Similarly, in a *ccp* structure if an extra layer interposes itself, we could get the sequence ABCA**BAC**ABCABCA…, now called an extrinsic fault.

The occurrence of stacking faults over many layers in a crystal can result in twinning. In a twinned crystal, two crystals meet at a common plane of atoms and are often mirror images of one another (Figure 5.10d).

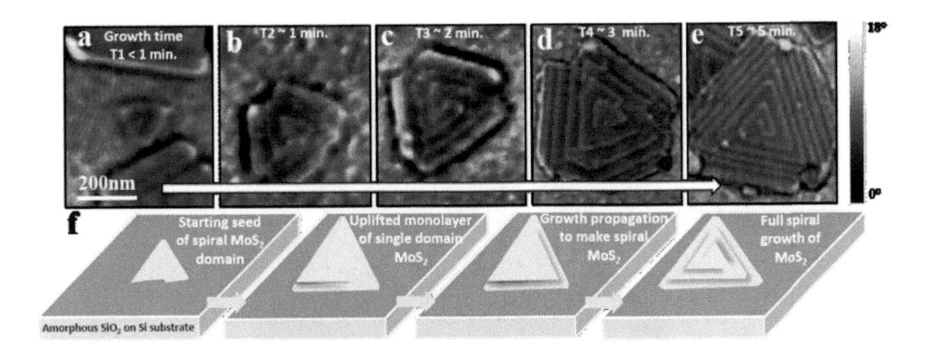

FIGURE 5.11 Growth evolution of screw dislocation driven spiral MoS_2 flakes. (a–e) AFM phase images of a different growth stage (micron marker is same for all images) and (f) corresponding schematics explaining the growth phenomena using screw dislocation. Approximate growth durations for each stage are as (a–b) ≤ 1 min, (c) ~2 min, (d–e) ~5 min. (Reprinted with permission from Kumar, P., and Viswanath, B. (2016). *Effect of Sulfur Evaporation Rate on Screw Dislocation Driven Growth of MoS2 with High Atomic Step Density*. *Cryst. Growth Des.*, 2016, 7145–7154. Copyright 2016 American Chemical Society.)

Three-dimensional faults occur as small voids in a structure, where there are no atoms, or as small regions of a different phase, known as **precipitates**.

The above-mentioned defects are very important in the study of materials as they affect their physical properties, such as tensile strength and ductility. However, there are other planar defects in particular that affect the chemical and crystallographic structure of a compound:

- Crystallographic shear (CS) planes, where the oxygen vacancies effectively collect together in a plane that runs through the crystal
- Intergrowth structures, where two different but related structures alternate throughout the crystal

5.4.1 CS Planes

Nonstoichiometric compounds are found for the higher oxides of tungsten, molybdenum, and titanium, WO_{3-x}, MoO_{3-x}, and TiO_{2-x}, respectively. The reaction of these systems to the presence of point defects is entirely different from what has been discussed previously; in fact, the point defects are eliminated by a process known as **CS**.

In these systems, a series of closely related compounds with very similar formulae and structures exist. The formulae of these compounds all obey a general formula, which for the molybdenum and tungsten oxides can be Mo_nO_{3n-1}, Mo_nO_{3n-2}, W_nO_{3n-1}, and W_nO_{3n-2} and for titanium dioxide it is Ti_nO_{2n-1}; n can vary taking values of 4 and above. The resulting series of oxides is known as a **homologous series** (as for the alkanes in organic chemistry). The first seven members of the molybdenum trioxide series are Mo_4O_{11}, Mo_5O_{14}, Mo_6O_{17}, Mo_7O_{20}, Mo_8O_{23}, Mo_9O_{26}, $Mo_{10}O_{29}$, and $Mo_{11}O_{32}$.

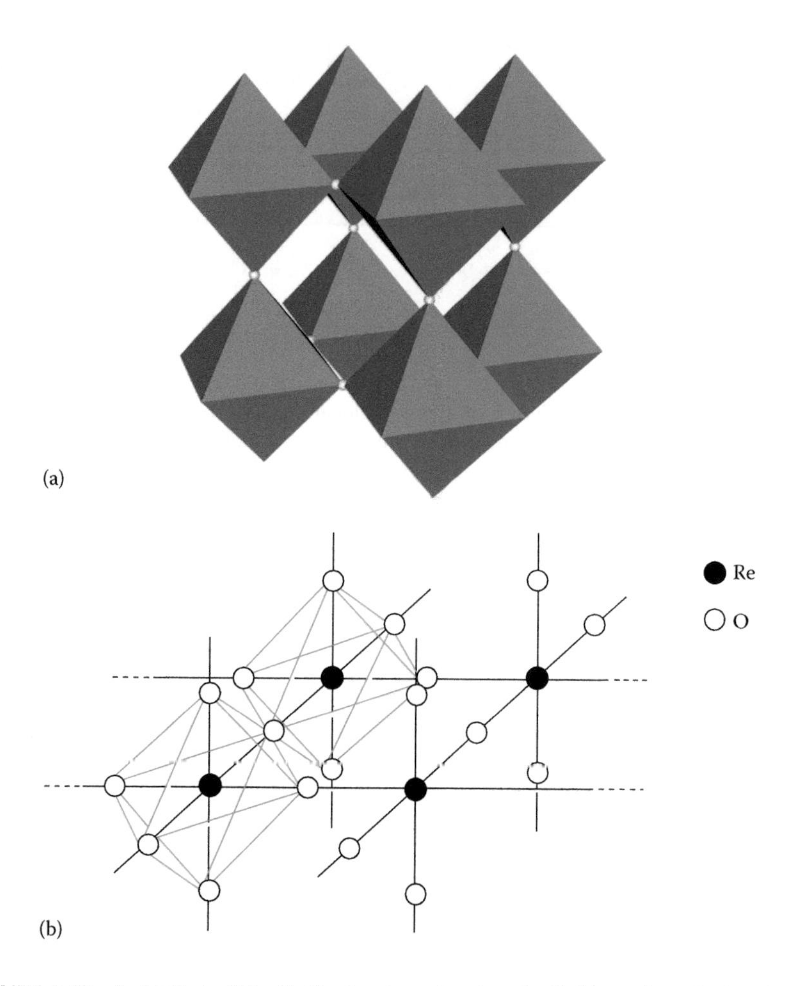

FIGURE 5.12 (a, b) Part of the ReO_3 structure showing the linking of octahedra through the corners.

In these compounds, we find regions of corner-linked octahedra separated from each other by thin regions of a different structure known as the CS planes. The different members of a homologous series are determined by the fixed spacing between the CS planes. The structure of a shear plane is quite difficult to understand, and these structures are usually depicted by the linking of octahedra, as described in Chapter 1.

WO_3 has several polymorphs, but above 900°C the WO_3 structure is that of ReO_3, which is shown in Figure 5.12. (The structures of the other polymorphs are distortions of the ReO_3 structure.) ReO_3 is made up of $[ReO_6]$ octahedra that are linked together through their corners; each corner of an octahedron is shared with another. Part of the ReO_3 structure is drawn in Figure 5.12b, and we can see that every oxygen atom is shared between two metal atoms: as six oxygens surround each Re, the overall formula is ReO_3. Figure 5.12a shows part of one layer of linked octahedra in

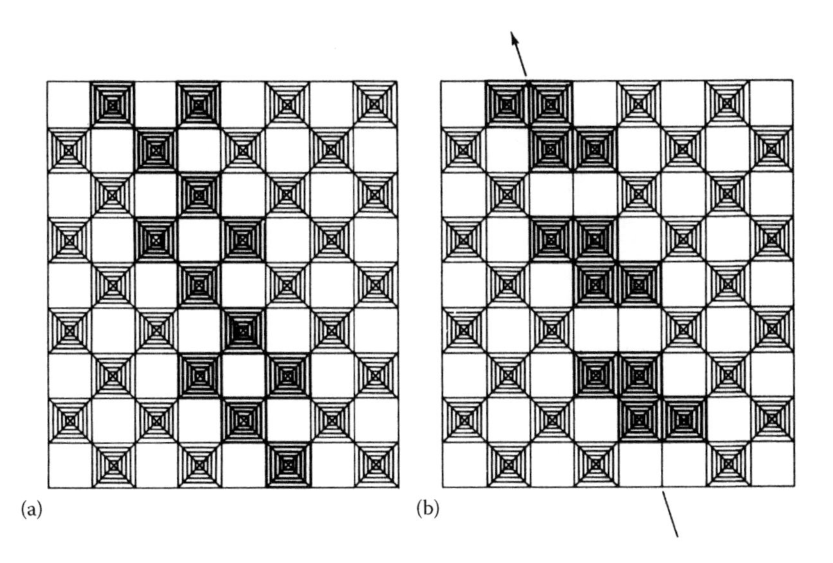

FIGURE 5.13 (a, b) The formation of a shear structure.

the structure. Notice that within the layer, any octahedron is linked to four others; it is also linked, through its upper and lower corners, to octahedra in the layers above and below.

The nonstoichiometry in WO_{3-x} is achieved by some of the octahedra in this structure changing from corner-sharing to edge-sharing. Look now to the octahedra marked in bold in Figure 5.13a. The edge-sharing corresponds to shearing the structure so that the chains of bold octahedra are displaced to the positions in Figure 5.13b. This shearing occurs at regular intervals in the structure and is interspersed with slabs of the 'ReO_3' structure (corner-linked [WO_6] octahedra). It creates groups of four octahedra that share edges. The direction of maximum density of the edge-sharing groups is called the CS plane and is indicated by an arrow in Figure 5.13b.

In order to see how CS alters the stoichiometry of WO_3, we need to find the stoichiometry of one of the groups of four octahedra that are linked together by sharing edges; part of the structure depicting one of these groups is shown in Figure 5.14a. The four octahedra consist of 4 W atoms and 18 O atoms (Figure 5.14b). Fourteen of the oxygen atoms are linked out to other octahedra (these bonds are indicated) so each are shared by two tungsten atoms, while the remaining four oxygens are only involved in the edge-sharing within the group. The overall stoichiometry is given by $\left[4W+\left(14\times\frac{1}{2}\right)O+4O\right]$, giving W_4O_{11}.

Clearly, if groups of four octahedra with stoichiometry W_4O_{11} are interspersed throughout a perfect WO_3 structure, then the amount of oxygen in the structure is reduced and we can write the formula as WO_{3-x}. The effect of introducing the groups of four in an ordered way can be quantified. If the structure sheared in such a way that the entire structure was composed of these groups, the formula would become W_4O_{11}. If there is one [WO_6] octahedron for each group of four, then the overall

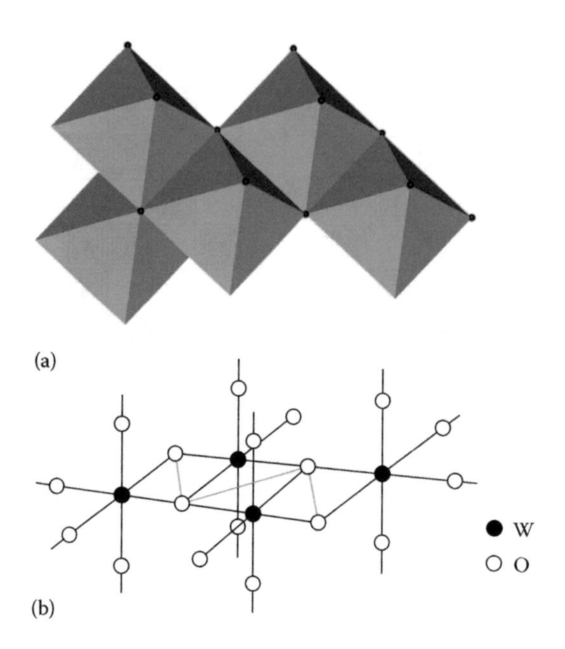

FIGURE 5.14 (a, b) A group of $[WO_6]$ octahedra-sharing edges formed by the creation of shear planes in W_nO_{3n-1}.

formula becomes $[W_4O_{11} + WO_3] = W_5O_{14}$. Clearly, we can extend this process to any number of $[WO_6]$ octahedra regularly interspersed between the groups:

$$W_4O_{11} + 2WO_3 = W_6O_{17};$$
$$W_4O_{11} + 4WO_3 = W_8O_{23};$$
$$W_4O_{11} + 6WO_3 = W_{10}O_{29};$$

$$W_4O_{11} + 3WO_3 = W_7O_{20}$$
$$W_4O_{11} + 5WO_3 = W_9O_{26}$$
$$W_4O_{11} + 7WO_3 = W_{11}O_{32}$$

The basic formula of the group of four, W_4O_{11}, can be written as W_nO_{3n-1} where $n=4$. This formula also holds for all the other formulae that are listed above, and we have produced the general formula for the homologous series simply by introducing set ratios of the edge-sharing groups in among the $[WO_6]$ octahedra.

The shear planes are found to repeat throughout a particular structure in a regular and ordered fashion, so any particular sample of WO_{3-x} will have a specific formula corresponding to one of those listed above. The different members of the homologous series are determined by the fixed spacing between the CS planes. An example of one of the structures is shown in Figure 5.15. A unit cell has been marked so that the ratio of $[WO_6]$ octahedra to the groups of four is clear. Within the marked

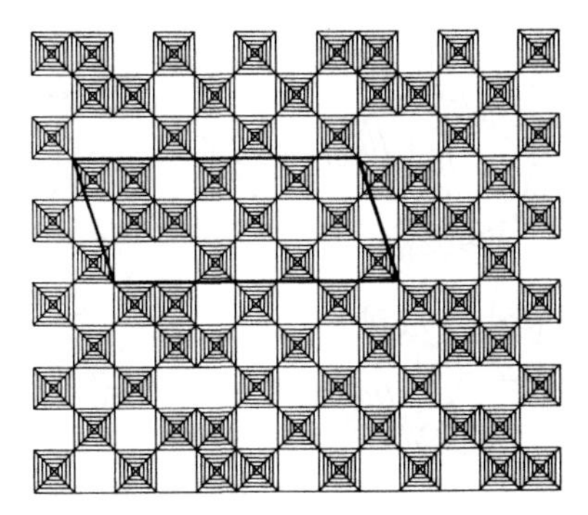

FIGURE 5.15 A member of the W_nO_{3n-1} homologous series with the projection of a unit cell marked.

unit cell there is one group of four and seven octahedra, giving the overall formula $W_4O_{11}+7WO_3=W_{11}O_{32}$.

Members of the Mo_nO_{3n-1} series have the same structure as their W_nO_{3n-1} analog, even though unreduced MoO_3 does not have the ReO_3 structure, but a layer structure.

If the structures shear in such a way that groups of six octahedra share edges regularly throughout the structure, then homologous series with the general formula M_nO_{3n-2} are formed.

The homologous series for oxygen-deficient TiO_2 is given by the formula Ti_nO_{2n-1}. In this case, the octahedra along the CS planes are joined to each other by sharing faces, whereas in the unreduced parts of the TiO_2 structure, the octahedra share edges as in rutile.

5.4.2 Planar Intergrowths

Many systems show examples of intergrowth where a solid contains regions of more than one structure with clear solid–solid interfaces between the regions. Epitaxy (Chapter 3) is an example of such a phenomenon with great technological interest in the production of circuits and 'smart' devices. The zeolites ZSM-5 and ZSM-11 can form intergrowths (Chapter 7), as can the barium ferrites (Chapter 9). We will only look at one example here, that of intergrowths in the **tungsten bronzes**. The term bronze is applied to metallic oxides that have a deep colour, a metallic lustre, and are either metallic conductors or semiconductors. The sodium–tungsten bronzes (Na_xWO_3) have colours that range from yellow to red and deep purple, depending on the value of x.

We have already seen that WO_3 has the rhenium oxide (ReO_3) structure, with $[WO_6]$ octahedra joined through the corners. This is illustrated in Figures 5.13a and 5.14. The structure contains a three-dimensional network of channels throughout the structure and it has been found that alkali metals can be incorporated into the structure in these channels. The resultant crystal structure depends on the proportion of alkali metal in the particular compound. The structures are based on three main types: cubic phase where the alkali metal occupies the centre of the unit cell (similar to perovskite—see Chapters 1 and 10) and the tetragonal and hexagonal phases.

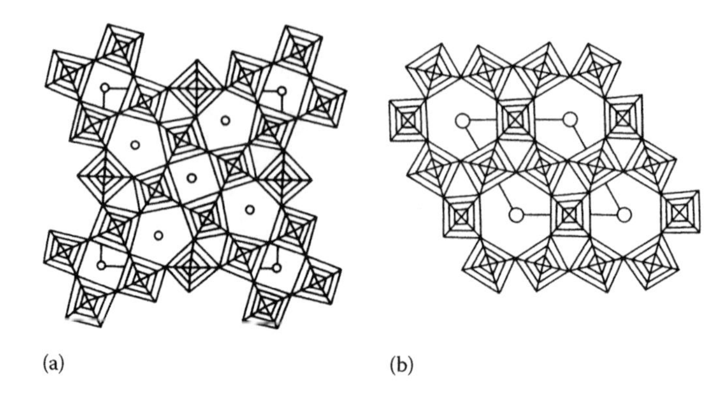

FIGURE 5.16 (a) The tetragonal tungsten bronze structure and (b) the hexagonal tungsten bronze structure. The shaded squares represent WO_6 octahedra, which are linked to form pentagonal, square, and hexagonal tunnels. These are able to contain a variable population of metal atoms, shown as open circles.

The basic structures of two of these are illustrated in Figure 5.16. The electronic conductivity properties of the bronzes arise because charge compensation has to be made for the presence of M^+ ions in the structure: this is achieved by the change in the oxidation state of some of the tungsten atoms from VI to V (such processes are discussed in more detail in the final section of this chapter).

The hexagonal bronze structure illustrated (Figure 5.16b) is formed when potassium reacts with WO_3 (K needs a bigger site) and the composition lies in the range $K_{0.19}WO_3$ to $K_{0.33}WO_3$. If the proportion of potassium in the compound is less than this, the structure is found to consist of WO_3 intergrown with the hexagonal structure in a regular fashion. The layers of the hexagonal structure can be either one or two tunnels wide, as shown in Figure 5.17. Similar structures are observed for tungsten bronzes containing metals other than potassium, such as Rb, Cs, Ba, Sn, and Pb. A high-resolution electron micrograph of the barium tungsten bronze clearly shows its preferred single tunnel structure (Figure 5.18). In other samples, the separation of the tunnels increases as the concentration of barium decreases.

5.4.3 Block Structures

In oxygen-deficient Nb_2O_5 and in mixed oxides of Nb and Ti, and Nb and W, the CS planes occur in two sets at right angles to each other. The intervening regions of perfect structure now change from infinite sheets to infinite columns or blocks. These structures are known as **double shear** or **block structures** and are characterised by the cross-sectional size of the blocks: the block size is expressed as the number of octahedra sharing vertices. In addition to having phases built of blocks of one size, the complexity can be increased by having blocks of two or even three different sizes arranged in an ordered fashion. The block size(s) determines the overall stoichiometry of the solid. An example of a crystal showing two different block sizes is shown in Figure 5.19.

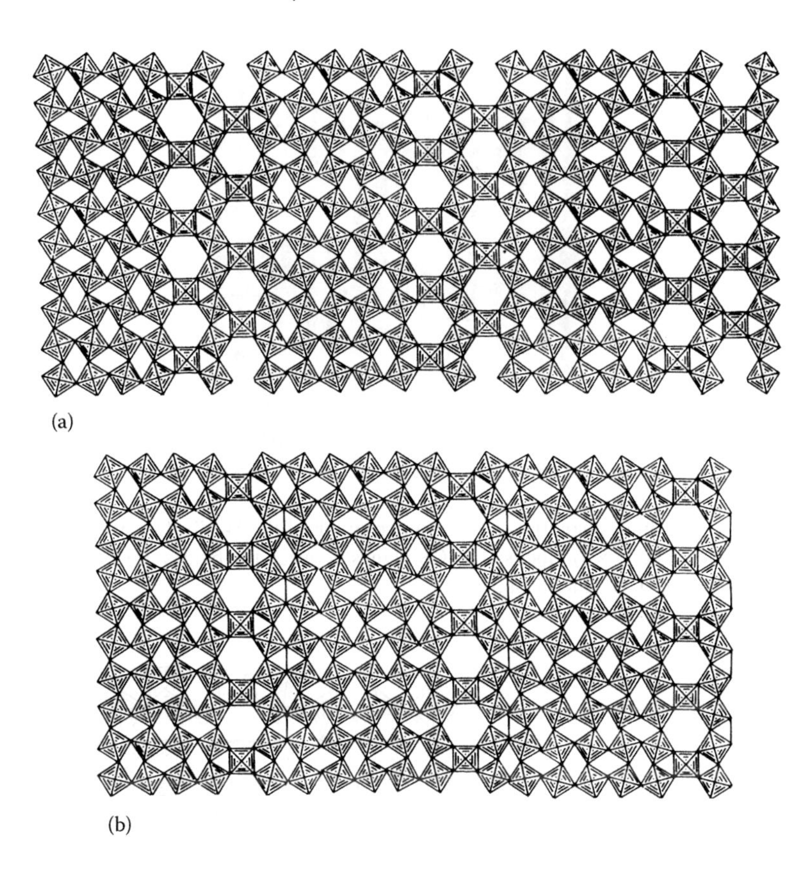

FIGURE 5.17 The idealised structures of two intergrowth tungsten bronze phases, (a) containing double rows of hexagonal tunnels and (b) containing single rows of tunnels. The tungsten trioxide matrix is shown as shaded squares and the hexagonal tunnels are shown empty, although in the known intergrowth tungsten bronzes the tunnels contain variable amounts of metal atoms.

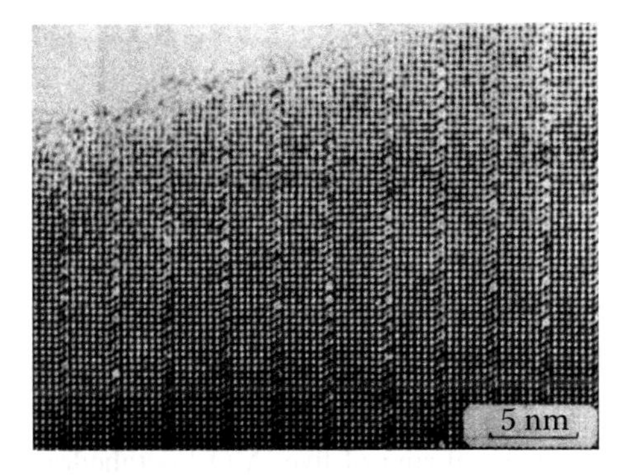

FIGURE 5.18 Electron micrograph of the intergrowth tungsten bronze phase Ba_xWO_3, clearly showing the single rows of tunnels. Each black spot on the image represents a tungsten atom, and many of the hexagonal tunnels seem to be empty or only partly filled with barium.

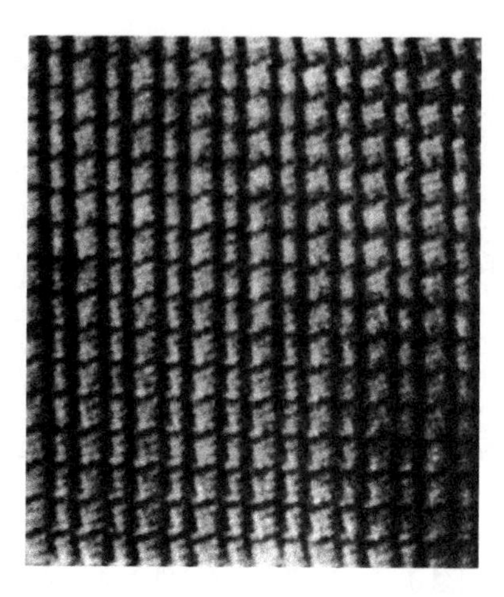

FIGURE 5.19 A high-resolution electron micrograph of the $W_4Nb_{26}O_{77}$ structure showing strings of (4×4) and (3×4) blocks. The CS planes between the blocks are shown as darker contrast. (From Dr. J. L. Hutchison.)

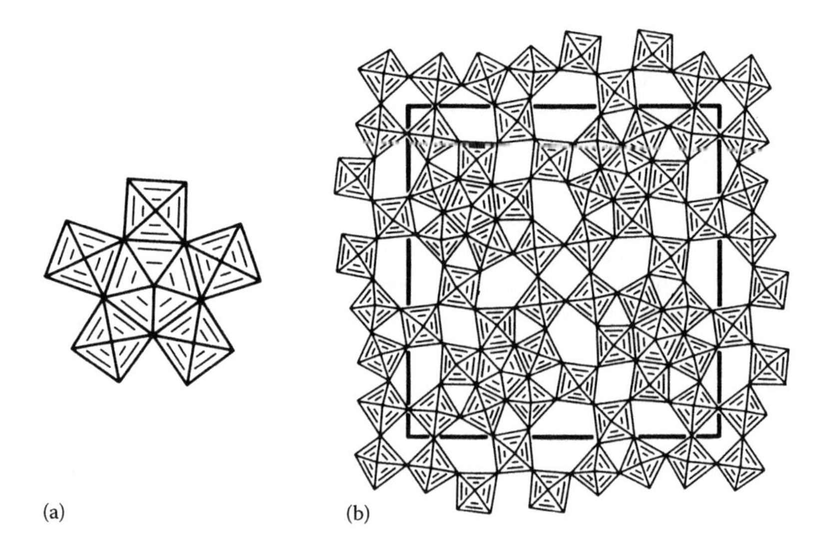

FIGURE 5.20 (a) A PC and (b) the structure of Mo_5O_{14}.

5.4.4 Pentagonal Columns

Three-dimensional faults also occur in the so-called **pentagonal column (PC) structures**. These structures contain the basic repeating unit shown in Figure 5.20a, which consists of a pentagonal ring of five $[MO_6]$ octahedra. When these stack on top of one another, a PC is formed that contains chains of alternating M and O atoms. These PCs can fit inside an ReO_3-type structure in an ordered way, and depending on the spacing, a homologous series is formed.

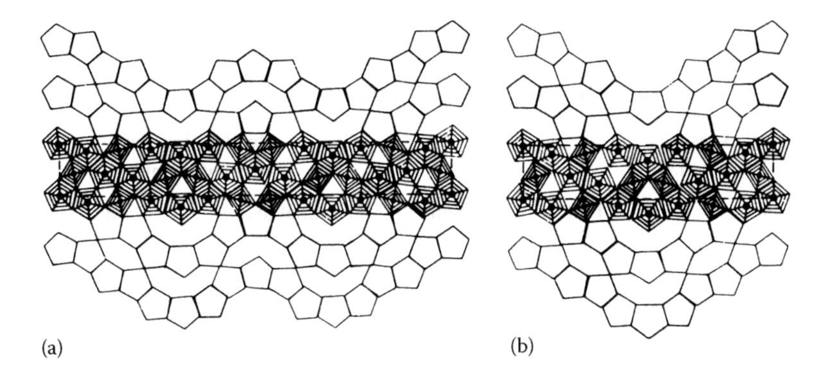

FIGURE 5.21 The idealised structures of (a) $Ta_{22}W_4O_{67}$ and (b) $Ta_{30}W_2O_{81}$. These phases are built from PCs, shown as shaded pentagons, and octahedra, shown as shaded squares. The wavelength of the chains of the PCs varies with composition in such a way that any given anion-to-cation ratio can be accommodated by an ordered structure.

An example is shown in Figure 5.20b for the compound Mo_5O_{14}. This type of structure is also found in the tetragonal tungsten bronzes.

5.4.5 Infinitely Adaptive Structures

In Section 5.4.3, we saw that the mixed oxides of niobium and tungsten could have a range of different compositions made by fitting together rectangular columns or blocks. The closely related Ta_2O_5–WO_3 system does something even more unusual. A large number of compounds form, but they are built up by fitting together PCs. The idealised structures of two of these compounds are shown in Figure 5.21; the structures have a wavelike skeleton of PCs. As the composition varies, so the wavelength of the backbone changes, giving rise to a huge number of possible ordered structures, known as **infinitely adaptive compounds**.

5.5 ELECTRONIC PROPERTIES OF NONSTOICHIOMETRIC OXIDES

Earlier, we considered the structure of nonstoichiometric FeO in some detail. If we apply the same principle to other binary oxides, we can define four types of compounds:

Metal excess (reduced metal)

Type A: anion vacancies present; formula MO_{1-x}
Type B: interstitial cations; formula $M_{1+x}O$

Metal deficiency (oxidised metal)

Type C: interstitial anions; formula MO_{1+x}
Type D: cation vacancies; formula $M_{1-x}O$

The four types are summarised in Table 5.6, and Figure 5.22 illustrates some of the structural possibilities for simple oxides with A-, B-, C-, and D-type nonstoichiometry, assuming an NaCl-type structure.

TABLE 5.6
Types of Nonstoichiometric Oxides (MO)

Metal excess Reduced metal M		Metal deficiency Oxidised metal M	
A anion vacancies	**B interstitial cations**	**C interstitial anions**	**D cation vacancies**
MO_{1-x}	$M_{1+x}O$	MO_{1+x}	$M_{1-x}O$
TiO, VO, (ZrS)	CdO, ZnO	—	TiO, VO, MnO, FeO, CoO, NiO

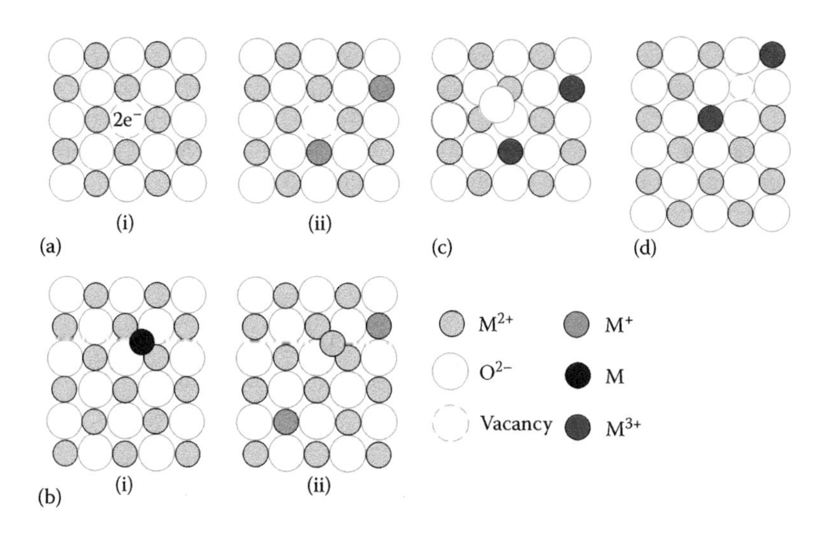

FIGURE 5.22 Structural possibilities for binary oxides. (a) Type A oxides: metal excess/anion vacancies. (i) This shows the two electrons that maintain charge neutrality, localised at the vacancy. (ii) The electrons are associated with the normal cations making them into M^+. (b) Type B oxides: metal excess/interstitials. (i) This shows an interstitial atom, whereas in (ii) the atom has ionised to M^{2+}, and the two liberated electrons are now associated with two normal cations, reducing them to M^+. (c) Type C oxides: metal deficiency/interstitial anions. The charge compensation for an interstitial anion is by way of two M^{3+} ions. (d) Type D oxides: metal deficiency/cation vacancies. The cation vacancy is compensated for by two M^{3+} cations.

FeO falls into the **type D** category, with metal deficiency and cation vacancies resulting in oxidised metal. (Other compounds falling into this category are MnO, CoO, and NiO.)

Type A oxides compensate for metal excess with anion vacancies. In order to maintain the overall neutrality of the crystal, two electrons have to be introduced for each anion vacancy. These can be trapped at a vacant anion site, as is shown in Figure 5.22a(i). However, it is an extremely energetic process to introduce electrons into the crystal, so we are more likely to find them associated with the metal cations, as is shown in Figure 5.22a(ii), which we can describe as reducing those cations from M^{2+} to M^+. VO_{1-x} is an example of this type of system.

Type B oxides have a metal excess that is incorporated into the lattice in **interstitial** positions. This is shown in Figure 5.36b(i) as an interstitial atom, but it is more likely that the situation in Figure 5.36b(ii) will hold, where the interstitial atom has ionised and the two electrons so released are now associated with two neighbouring ions, reducing them from M^{2+} to M^+. Cadmium oxide (CdO) has this type of structure. Oxygen is lost when zinc(II) oxide is heated, forming $Zn_{1+x}O$, oxide vacancies form and to compensate, Zn^{2+} ions migrate to interstitial positions and are reduced to Zn^+ ions or Zn atoms.

Type C oxides compensate for the lack of metal with **interstitial anions**. The charge balance is maintained by the creation of two M^{3+} ions for each interstitial anion (Figure 5.36c), each of which we can think of as M^{2+} associated with a positive hole.

Before considering the conductivity of these nonstoichiometric oxides, it is probably helpful to recap what we know about the structure and the properties of the stoichiometric binary oxides of the first-row transition elements. A summary of the properties of binary oxides is given in Table 5.7.

We discussed the conductivity of the stoichiometric oxides in Chapter 4, and we saw that their conductivity is dependent on two competing effects: on the one hand, the d orbitals overlap to give a band—the bigger the overlap the greater is the bandwidth—and electrons in the band are delocalised over the whole structure; on the other hand, interelectronic repulsion tends to keep the electrons localised on the individual atoms. TiO and VO behave as metallic conductors and must therefore have good overlap of the d orbitals producing a d electron band. This overlap arises partly because Ti and V are early in the transition series (the d orbitals have not suffered the contraction due to increased nuclear charge, seen later in the series) and partly because of the crystal structures (TiO has one-sixth of the titaniums and oxygens missing from an NaCl-type structure), which allow contraction of the structures and thus better d orbital overlap.

Further along the series, we saw that stoichiometric MnO, FeO, CoO, and NiO are insulators. This situation is not easily described by band theory because the d orbitals are now too contracted to overlap much, typical bandwidths are 1 eV, and the overlap is not sufficient to overcome the localising influence of interelectronic repulsions. (It is this localisation of the d electrons on the atoms that gives rise to the magnetic properties of these compounds, which are discussed in Chapter 9.)

Going back now to the nonstoichiometric oxides, in the excess metal monoxides of type A and type B, we saw that extra electrons have to compensate for the excess metal in the structure. Figure 5.22 shows that these could be associated with an anion vacancy or, alternatively, they could be associated with the metal cations within the structure. Although we have described this association as reducing neighbouring cations, this association can be quite weak, and these electrons can be free to move through the lattice; they are not necessarily strongly bound to particular atoms. Thermal energy is often sufficient to make these electrons move, so conductivity increases with temperature. We associate semiconductivity with such behaviour (metallic conductivity decreases with temperature).

TABLE 5.7

Properties of the First-Row Transition Element Monoxides

Element	Ca	Sc	Ti	V	Cr	Mn	Fe	Co	Ni	Cu	Zn
Structure of stoichiometric oxide MO	NaCl structure	Does not exist	Defect NaCl 1/6 vacancies	Defect NaCl	Does not exist	NaCl structure	NaCl structure[a]	NaCl structure	NaCl structure	PtS structure	Wurtzite structure (NaCl at high pressure)
Defect structure			$Ti_{1-\delta}O$: Ti vacancies (intergrowths of $TiO_{1.00}$ and $TiO_{1.25}$ structures)	$V_{1-\delta}0$:V vacancies and tetrahedral V interstitials in defect clusters		$Mn_{1-\delta}O$: Mn vacancies	$Fe_{1-\delta}O$: Fe vacancies and tetrahedral Fe interstitials in defect clusters	$Coj_{1-\delta}O$: Co vacancies	$Nii_{1-\delta}O$: Ni vacancies		$Zn_{1+\delta}0$: interstitial Zn
Conductivity of stoichiometric compound			Metallic	Metallic <120 K		Insulator	Insulator	Insulator	Insulator		Insulator
Conductivity of nonstoichiometric compound			Metallic	Metallic		*p*-Type hopping semiconductor	*p*-Type	*p*-Type	*p*-Type		*n*-Type
Magnetism (Chapter 8)			Diamagnetic	Diamagnetic		Paramagnetic $\mu=5.5\ \mu_B$ (antiferromagnetic when cooled, $T_N=122$ K)	Paramagnetic (antiferromagnetic when cooled, $T_N=198$ K)	Anti ferromagnetic ($T_N=292$ K)	Antiferromagnetic (paramagnetic when cooled, $T_N=530$ K)		

[a] Exactly stoichiometric FeO is never found.

In Chapter 4, we discussed semiconductors in terms of band theory. An intrinsic semiconductor has an empty conduction band lying close above the filled valence band. Electrons can be promoted into this conduction band by heating, leaving positive holes in the valence band; the current is carried by both the electrons in the conduction band and by the positive holes in the valence band. Semiconductors such as silicon can also be doped with impurities to enhance their conductivity. For instance, if a small amount of phosphorus is incorporated into the lattice, the extra electrons form impurity levels near the empty conduction band and are easily excited into it. The current is now carried by the electrons in the conduction band and the semiconductor is known as ***n*-type** (*n* for negative). Correspondingly, doping with Ga increases the conductivity by creating positive holes in the valence band and such semiconductors are called ***p*-type** (*p* for positive).

Compounds of type A and B would produce *n*-type semiconductors because the conduction is produced by electrons. Conduction in these nonstoichiometric oxides is not easily described by band theory, for the reasons given earlier for their stoichiometric counterparts—the interelectronic repulsions have localised the electrons on the atoms. Therefore, it is easiest to think of the conduction electrons (or holes) as localised or trapped at atoms or at defects in the crystal rather than delocalised in bands throughout the solid. Conduction then occurs by jumping or **hopping** from one site to another under the influence of an electric field. In a perfect ionic crystal where all the cations are in the same valence state, this would be an extremely energetic process. However, when two valence states are available, as in these transition metal nonstoichiometric compounds, the electron jump between them does not take much energy. Although we cannot develop this theory here, we can note that the conduction in these so-called **hopping semiconductors** can be described by the equations of diffusion theory in much the same way as we did earlier for ionic conduction: we find that the mobility of a charge carrier (either an electron or a positive hole), μ, is an activated process and we can write

$$\mu \propto \exp\left(\frac{-E_a}{kT}\right) \tag{5.7}$$

where E_a is the activation energy of the hop and is of the order of 0.1–0.5 eV. The hopping conductivity is given by the expression:

$$\sigma = ne\mu \tag{5.8}$$

where *n* is the number of mobile charge carriers per unit volume and *e* is the electronic charge. The density of mobile carriers, *n*, depends only on the composition of the crystal and does not vary with temperature. From Equation 5.7, we can see that the hopping electronic conductivity increases with temperature.

In the type C and D monoxides, we have shown the lack of metal as being compensated for by the oxidation of the neighbouring cations to M^{3+}. The M^{3+} ions can be regarded as M^{2+} ions associated with a positive hole. Accordingly, if sufficient energy is available, conduction can be thought to occur via the positive hole hopping to another M^{2+} ion and the electronic conductivity in these compounds will be *p*-type. MnO, CoO, NiO, and FeO are materials that behave in this way. Regarding the charge carriers as positive holes is simply a matter of convenience and the description of a positive hole moving from Ni^{3+} to Ni^{2+} is the same as saying that an electron moves from Ni^{2+} to Ni^{3+}.

Nonstoichiometric materials that cover the whole range of electrical activity from metal to insulator can be listed. Here, we have considered some metallic examples that can be described by band theory (TiO, VO) and others (such as MnO) that are better described as hopping semiconductors. Other cases, such as WO_3 and TiO_2, fall in between these extremes and a different description again is needed. It is thus difficult to make generalisations about this complex behaviour and each case is best treated individually.

The semiconductor properties are extremely important to the modern electronics industry, which is constantly searching for new and improved materials. Much of their research is directed at extending the composition range and therefore the properties of these materials. The composition range of a nonstoichiometric compound is often quite narrow (the so-called line phase), so to extend it (and thus extend the range of its properties also) the compound is doped with an impurity. To take one example: if we add Li_2O to NiO and then heat to high temperatures in the presence of oxygen, Li^+ ions become incorporated in the lattice and the resulting black material has the formula $Li_xNi_{1-x}O$, where x lies in the range 0–0.1. The equation for the reaction (using stoichiometric NiO for simplicity) is given by

$$\frac{1}{2}x\mathrm{Li_2O}+(1-x)\mathrm{NiO}+\frac{1}{4}x\mathrm{O_2}=\mathrm{Li}_x\mathrm{Ni}_{1-x}\mathrm{O}$$

To compensate for the presence of the Li^+ ions, the Ni^{2+} ions will be oxidised to Ni^{3+} or the equivalent of a high concentration of positive holes located at Ni cations.

This process of creating electronic defects is called **valence induction** and it increases the composition range of 'NiO' tremendously. Indeed, at high Li concentrations, the conductivity approaches that of a metal (although it still exhibits semiconductor behaviour in that its conductivity increases with temperature).

5.6 SUMMARY

1. The crystals of most ionic solids are not perfect but contain defects. These defects can be isolated point defects, clusters of point defects, or extended defects.
2. Point defects include missing ions giving a vacant site, substitution of an ion by a different one, occupation of an interstitial site, and a change in the valency of an ion. Such defects must be balanced by compensating defects so that the crystal has no net charge. The presence of these defects can lead to materials that are nonstoichiometric. That is, they have formulas such as MO_{2+x} rather than MO_2.
3. On a larger scale, crystals can contain dislocations due to faults in the arrangements of the atoms. These can weaken the crystal but can also act as active sites for catalysis and crystal growth.

TABLE 5.8
Defect Concentration Data for CsI

T/K	$\frac{n_s}{N}$
300	1.08×10^{-16}
400	1.06×10^{-12}
500	2.63×10^{-10}
600	1.04×10^{-8}
700	1.43×10^{-7}
900	4.76×10^{-6}

4. Some nonstoichiometric transition metal oxides can act as semiconductors using a hopping mechanism. Localised electrons 'hop' from a metal ion in one oxidation state to a higher oxidation state ion.

QUESTIONS

1. If $\Delta H_m^\ominus$ for the formation of Schottky defects in a certain MX crystal is 200 kJ mol^{-1}, calculate n_S/N and n_S per mole for the temperatures 300, 500, 700, and 900 K.
2. Table 5.8 gives the variation of defect concentration with temperature for CsI. Determine the enthalpy of formation for one Schottky defect in this crystal.
3. Make a simple estimate of the energy of the defect formation in the fluorite structure:
   a. Describe the coordination by nearest neighbours and next-nearest neighbours of an anion both for a normal lattice site and for an interstitial site at the centre of the unit cell shown in Figure 5.3a.
   b. Use the potential energy of two ions, given by

$$E = -\frac{e^2 Z}{4\pi\varepsilon_0 r}$$

$$= -\left(2.31 \times 10^{-28}\ \text{Jm}\right)\frac{Z}{r}$$

   where Z is the charge on the other ion and r is the distance between them, to estimate the energy of formation of a Frenkel defect in fluorite; a = 537 pm.
4. Using Kröger–Vink notation, write a balanced equation for the substitution of Mg^{2+} for Fe^{3+} in Fe_2O_3. Assume the Mg^{2+} ions are added as MgO and that Fe^{3+} is not reduced.
5. Confirm the presence of iron vacancies for a sample of wüstite that has a unit cell dimension of 428.2 pm, an Fe:O ratio of 0.910, and an experimental density of 5.613×10^3 kg m^{-3}.

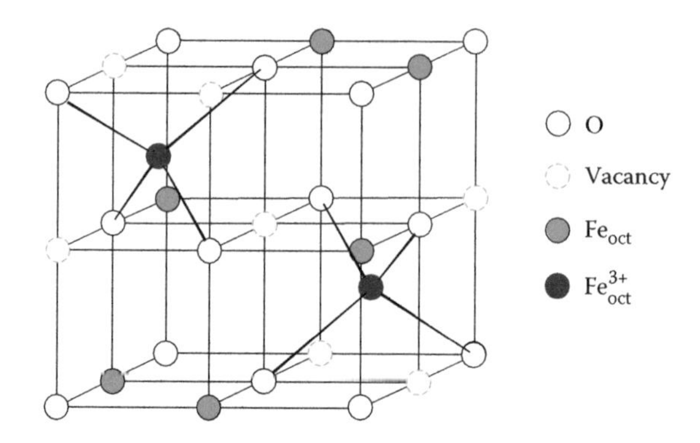

FIGURE 5.23 A possible cluster in $Fe_{1-x}O$.

6. A compound AgX has the same crystal structure as NaCl, a density of 6477 kg m^{-3}, and a unit cell dimension of 577.5 pm. Identify X.
7. How does the change in the lattice parameter of 'FeO' with iron content corroborate the iron vacancy model and refute an oxide interstitial model?
8. Figure 5.23 shows the central section of a possible defect cluster for FeO.
   a. Determine the vacancy:interstitial ratio for this cluster.
   b. Assuming that this section is surrounded by Fe ions and oxide ions in octahedral sites as in the Koch–Cohen cluster, determine the formula of a sample made totally of such clusters.
   c. Determine the numbers of Fe^{2+} and Fe^{3+} ions in octahedral sites.
9. Use Figure 5.7 to confirm that TiO is a one-sixth defective NaCl structure, by counting up the atoms in the monoclinic cell.
10. Figure 5.8 shows layers in the $TiO_{1.25}$ structure with both the Ti and O sites marked. Use this and Figure 5.9 to demonstrate that the unit cell shown has the correct stoichiometry for the crystal.
11. How would you expect charge neutrality to be maintained in $TiO_{1.25}$?

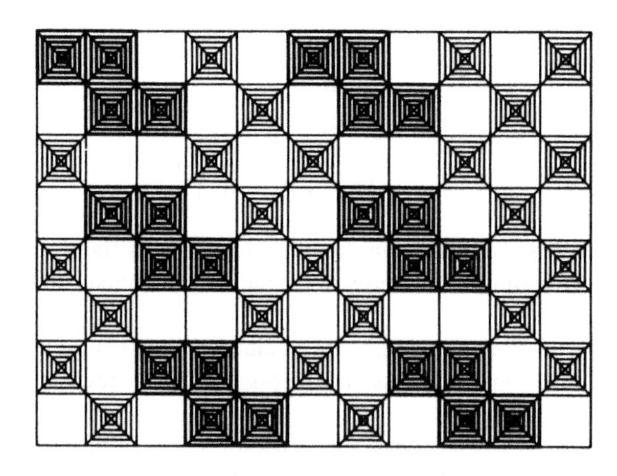

FIGURE 5.24 A member of the W_nO_{3n-1} homologous series.

12. Take a simple case where two metal oxide octahedra wish to eliminate oxygen by sharing. How does the formula change as they (i) share a corner, (ii) share an edge, and (iii) share a face?
13. Figure 5.24 shows a member of the homologous series, W_nO_{3n-1}. To what formula does it correspond?
14. ZnO is a type B (excess metal) material. What do you expect to happen to its electronic properties if it is doped with Al_2O_3 under vacuum?

6 Solid-State Materials for Batteries

6.1 INTRODUCTION

A **battery** or **electrochemical cell** produces an electric current at a constant voltage as a result of a chemical reaction and consists of two **electrodes** separated by an electrolyte (Figure 6.1). The ions taking part in the reaction pass through the electrolyte and are then either oxidised or reduced at one of the electrodes. The electrode at which oxidation takes place, usually a metal, is called the **anode**, and reduction takes place at the **cathode**; this usually consists of a metal current collector together with an active component which is in a high oxidation state and can be reduced, such as a metal oxide (Figure 6.1a).

Rechargeable batteries are reversible, and the reactant concentrations can be restored by reversing the cell reaction using an external source of electricity (Figure 6.1b). In the recharging mode, note that the electrons in the external circuit are now being driven *towards* the negatively charged electrode where reduction is now taking place; the negatively charged electrode becomes the cathode.

When the cell is in the discharge mode, electrons are released at the negatively charged electrode and they travel around the external circuit where they can be harnessed for useful work. The electromotive force (emf) or voltage produced by the cell under standard open-circuit conditions is related to the standard Gibbs free energy change for the reaction by the following equation:

$$\Delta G^{\ominus} = -nE^{\ominus}F \tag{6.1}$$

where n is the number of electrons transferred in the reaction, $E^{\ominus}$ is the standard emf of the cell (the voltage delivered under standard, zero-current conditions), and F is the Faraday constant (96,485 C mol^{-1} or 96,485 J V^{-1}). The energy stored in a battery is a factor of the energy generated by the cell reaction and the amount of materials used; it is usually expressed in watt-hours (Wh; current × voltage × discharge time).

Solid-state batteries are useful in that they can perform over a wide temperature range, they have a long shelf life, and it is possible to manufacture them so that they are extremely small.

In this chapter, we also look at some important types of battery such as lithium-ion used to power electric vehicles, including an electric plane being developed by NASA (Figure 6.2).

As well as a source of power, batteries can also be used to store energy. Figure 6.3 shows a demonstration project of a bank of batteries which can store excess energy produced by solar cells and wind farms, and release it to the grid when needed.

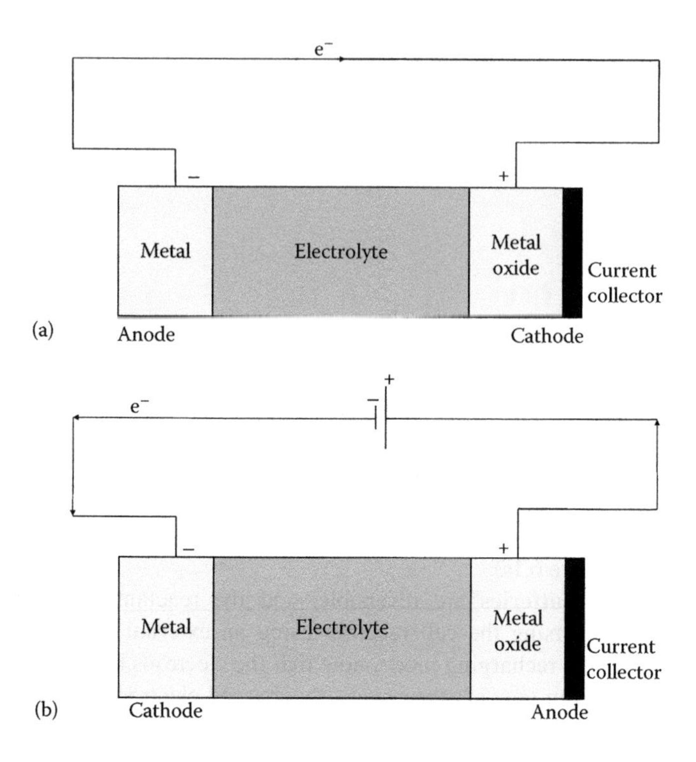

FIGURE 6.1 (a) A general electrochemical cell based on a metal and its oxide in the discharge mode. (b) The same cell under recharging.

The system consists of two types of storage batteries, lithium ion and NaS batteries (total capacity: 11.5MW/22.5MWh) to take advantage of their different features. Lithium-ion batteries have a high power charge and discharge output, and NaS batteries have high energy density, high efficiency of charge and discharge, and a long lifetime.

While many of the batteries we look at have solid electrodes, the electrolyte is a liquid solution. The electrolyte in **solid-state batteries** is a solid. To be useful as a solid electrolyte, the conducting solid must have a high ionic conductivity, but it must be an electronic insulator, so that it can separate the two reactants of the device allowing only ions, and not electrons, to travel through the solid—or the device would short-circuit. Before we consider individual batteries, we shall first look at ionic conductivity in solids and the types of solid that are good ionic conductors. We then describe some important types of batteries.

FIGURE 6.2 NASA's X-57 Maxwell will be powered by a battery system that consists of 16 battery modules. The system will comprise 800 pounds of the aircraft's total weight. X-57 will demonstrate that electric propulsion can make planes quieter and more efficient, with fewer carbon emissions in flight. (Courtesy of NASA.)

FIGURE 6.3 Large-scale hybrid power storage system in the City of Varel in Niedersachsen, Germany. The project is developed by NEDO (New Energy and Industrial Technology Development Organization). (With permission from NGK Insulators, Ltd. https://www.ngk.co.jp/nas/)

6.2 IONIC CONDUCTIVITY IN SOLIDS

Ionic conductivity, that is, ion transport under the influence of an external electric field, is made possible by the presence of point defects (Chapter 5). Two possible mechanisms for the movement of ions through a lattice are sketched in Figure 6.4. In Figure 6.4a, an ion hops or jumps from its normal position on the lattice to a neighbouring equivalent but vacant site. This is called the **vacancy mechanism** (it can equally well be described as the movement of a vacancy rather than the movement of the ion). Figure 6.4b shows the **interstitial mechanism**, where an interstitial ion jumps or hops to an adjacent equivalent site. These simple pictures of movement in an ionic lattice are known as the **hopping model**, and they ignore the more complicated cooperative motions.

Ionic conductivity (σ) is defined in the same way as electronic conductivity:

$$\sigma = nZe\mu \tag{6.2}$$

where n is the number of charge carriers per unit volume, Ze is their charge (expressed as a multiple of the charge on an electron, $e = 1.602189 \times 10^{-19}$ C), and μ is their **mobility**, a measure of the drift velocity in a constant electric field. Table 6.1

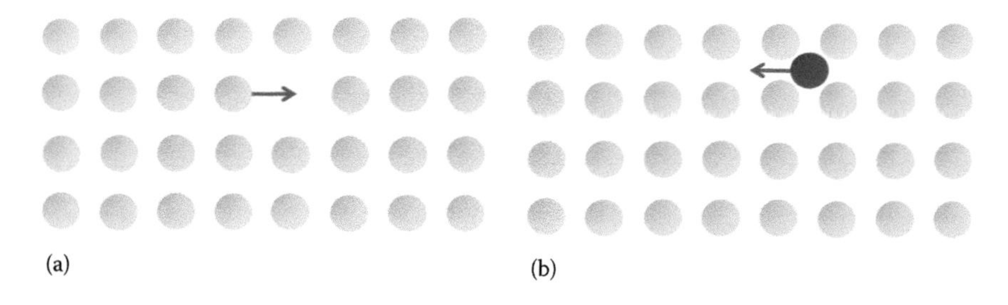

FIGURE 6.4 A schematic representation of ionic motion by (a) a vacancy mechanism and (b) an interstitial mechanism.

TABLE 6.1
Typical Values of Electrical Conductivity

Material	Conductivity (S m^{-1})
Ionic conductors	
Ionic crystals	$<10^{-16}$–10^{-2}
Solid electrolytes	10^{-1}–10^{3}
Strong (liquid) electrolytes	10^{-1}–10^{3}
Electronic conductors	
Metals	10^{3}–10^{7}
Semiconductors	10^{-3}–10^{4}
Insulators	$<10^{-10}$

shows the sort of conductivity values one might expect to find for different materials. As we might expect, ionic crystals, although they do conduct, are poor conductors compared with metals. This is a direct reflection of the difficulty that the charge carrier (in this case an ion, although sometimes an electron) has in moving through the crystal lattice.

Equation 6.2 is a general equation defining conductivity in all conducting materials. In order to understand why some ionic solids conduct better than others, it is useful to look at the definition more closely in terms of the hopping model that we have set up. First, we have said that an electric current is carried in an ionic solid by the defects. In the cases of crystals, where the ionic conductivity is carried by the vacancy or interstitial mechanism, n, the concentration of charge carriers, will be closely related to the concentration of defects in the crystal, n_S or n_F; thus, μ will refer to the mobility of these defects in such cases.

Let us look more closely at the mobility of the defects. Take the case of NaCl, which contains Schottky defects. The Na^+ ions are the smallest and therefore move most easily; however, they also meet quite a lot of resistance, as shown in Figure 6.5.

We have used dotted lines to illustrate two possible routes that the Na^+ could take from the centre of the unit cell to an adjacent vacant site. The direct route (labelled **4**) is clearly going to be very unlikely as it leads directly between two Cl^- ions, which, in a close-packed structure such as this, are going to be very close together. The other pathway first passes through one of the triangular faces of the octahedron (point **1**), then through one of the tetrahedral holes (point **2**), and finally through another triangular face (point **3**), before arriving at the vacant octahedral site. The coordination of the Na^+ ion taking this path changes from $6 \rightarrow 3 \rightarrow 4 \rightarrow 3 \rightarrow 6$ as it jumps from one site

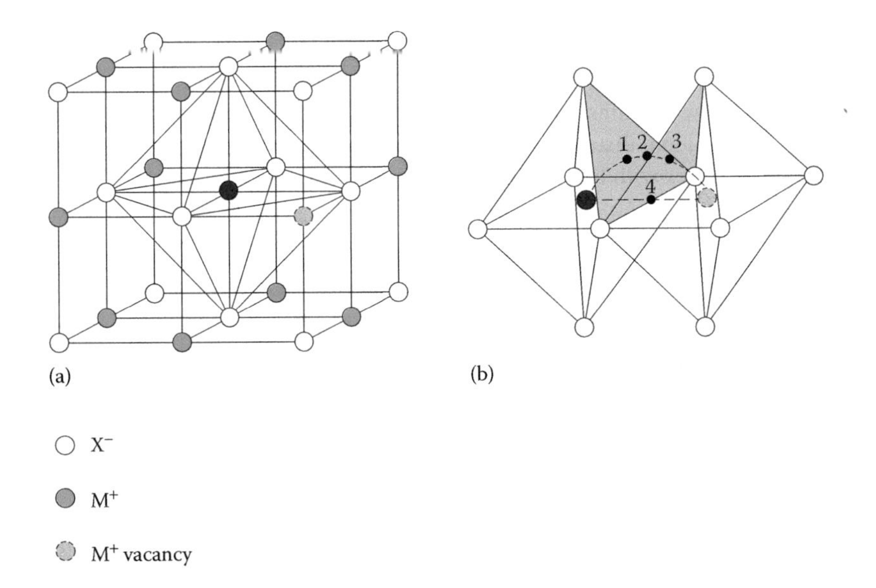

FIGURE 6.5 A sodium chloride-type structure, showing (a) a coordination octahedron of the central cation and (b) a coordination octahedra of the central cation and an adjacent vacancy.

to the other. While there is clearly going to be an energy barrier to this happening, it will not be as large as for the direct path where the Na^+ ion becomes two-coordinate. In general, we would expect the ion to follow the lowest energy path available. A schematic diagram of the energy changes involved in such a pathway is shown in Figure 6.6; notice that the energy of the ion is the same at the beginning and the end of the jump. The energy required to make the jump (E_a) is known as the **activation energy**. The temperature dependence of the mobility of the ions can be expressed by an **Arrhenius equation**:

$$\mu \propto \exp\left(\frac{-E_a}{kT}\right) \tag{6.3}$$

or

$$\mu = \mu_0 \exp\left(\frac{-E_a}{kT}\right) \tag{6.4}$$

where μ_0 is a proportionality constant, known as a pre-exponential factor, which depends on several factors: the number of times per second that the ion attempts the move (ν), called the **attempt frequency** (this is the frequency of the lattice vibration, of the order of 10^{12}–10^{13} Hz); the distance moved by the ion; and the size of the external field. If the external field is small (up to about 300 V cm^{-1}), a temperature dependence of 1/T is introduced into the pre-exponential factor.

If we combine all this information in Equation 6.4, we arrive at an expression for the variation of the ionic conductivity with temperature that has the form:

$$\sigma = \frac{\sigma_0}{T} \exp\left(\frac{-E_a}{T}\right) \tag{6.5}$$

The term σ_0 now contains n and Ze, as well as the information on the attempt frequency and the jump distance. This expression accounts for the fact that ionic conductivity *increases* with temperature. If we now take logs of Equation 6.5, we get

$$\ln \sigma T = \ln \sigma_0 - \left(\frac{-E_a}{T}\right) \tag{6.6}$$

Plotting ln σT against $1/T$ should produce a straight line with a slope of $-E_a$. The expression in Equation 6.6. is sometimes plotted empirically as ln σ against $1/T$

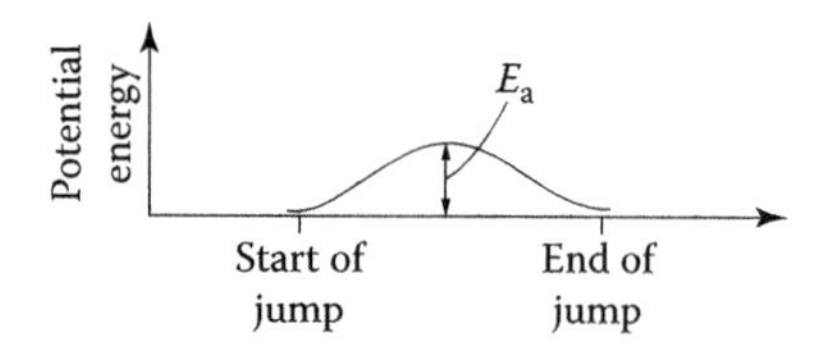

FIGURE 6.6 A schematic representation of the change in energy during the motion of an ion along the lowest-energy path.

because plotting either ln σT or ln σ makes little difference to the slope; both types of plot are found in the literature. The results of doing this for several compounds are shown in Figure 6.7.

Ignore the AgI line for the moment, which we will discuss in the next section. The other lines on the plot are straight lines, apart from the one for LiI, where we can clearly see two lines of differing slope. So, it looks as though the model we have set up does describe the behaviour of many systems. But how about LiI? In fact, other crystals also show this kink in the plot (some experimental data is shown for NaCl in Figure 6.8 where it can also be seen). Is it possible to explain this using the equations that we have just set up?

The explanation for the two slopes in the plot lies in the fact that even a very pure crystal of LiI contains some impurities and the line corresponding to low temperatures (on the right of the plot) is due to the extrinsic vacancies: at low temperatures, the concentration of intrinsic vacancies is so small that it can be ignored because it is dominated by the defects created by the impurity. For a particular amount of impurity, the number of vacancies present will be essentially constant. In this **extrinsic region**, μ thus depends only on the cation mobility due to these extrinsic defects, whose temperature dependence is given by

$$\mu = \mu_0 \exp\left(\frac{-E_a}{kT}\right) \tag{6.4}$$

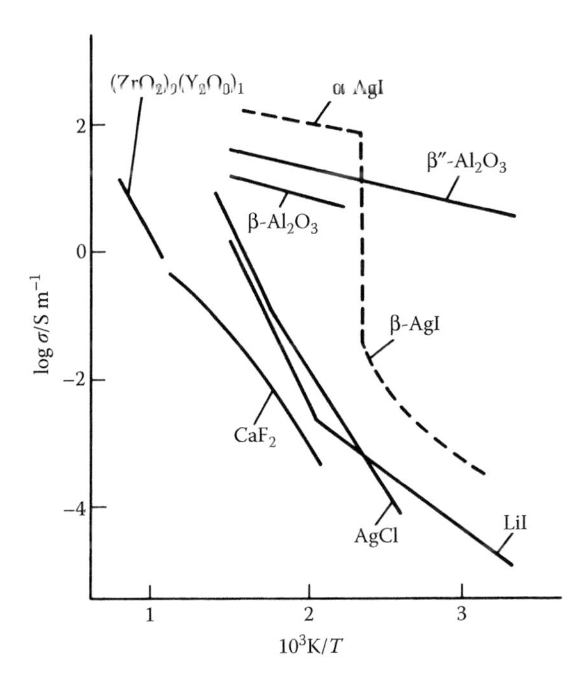

FIGURE 6.7 The conductivities of selected solid electrolytes over a range of temperatures.

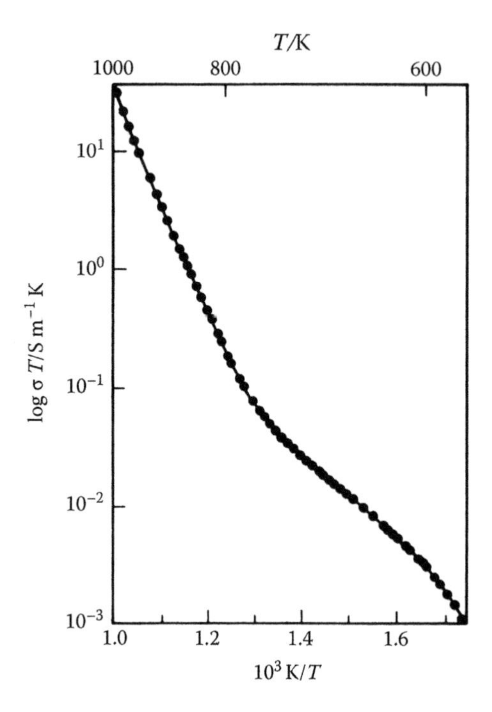

FIGURE 6.8 The ionic conductivity of NaCl plotted against reciprocal temperature.

However, at the higher temperatures on the left-hand side of the graph, the concentration of the intrinsic defect increases to such an extent that it is now similar to or greater than the concentration of the extrinsic defects. The concentration of the intrinsic defects, unlike that of the extrinsic defects, will *not* be constant; indeed, it will vary according to Equation 6.7, as we showed in Chapter 5:

$$n_s \approx N \exp\left(\frac{-\Delta H_S}{2kT}\right) \tag{6.7}$$

so the conductivity in this **intrinsic region** on the left-hand side of the plot is given by

$$\sigma = \frac{\sigma'}{T} \exp\left(\frac{-E_a}{kT}\right) \exp\left(\frac{-\Delta H_S}{2kT}\right) \tag{6.7}$$

A plot of ln σT versus $1/T$ in this case will give a greater value for the activation energy, which we label E_S because it will actually depend on two terms, the activation energy for the cation jump (E_a) and the enthalpy of the formation of a Schottky defect:

$$E_S = E_a + \frac{1}{2} \Delta H_S \tag{6.8}$$

Similarly, for a system with Frenkel defects:

$$E_F = E_a + \frac{1}{2}\Delta H_F \tag{6.9}$$

From plots such as these, we find that the activation energies lie in the range 0.05–1.1 eV, rather lower than the enthalpies of the defect formation. As we have seen, raising the temperature increases the number of defects, thereby increasing the conductivity of a solid. But better than increasing the temperature to increase the conductivity is to find materials that have low activation energies, less than about 0.2 eV. We find such materials in the top right-hand corner of Figure 6.7.

6.3 SOLID ELECTROLYTES

Ionic solids that have a much higher conductivity than is typical for such compounds are variously known as **fast-ion conductors**, **superionic conductors**, and **solid electrolytes**. These are useful in applications such as solid-state batteries and thus are currently a major subject for research.

6.3.1 Silver Ion Conductors

One of the earliest fast-ion conductors to be noticed, in 1913, by Tubandt and Lorenz, was a high-temperature phase of **silver iodide**.

Below 147°C, there are two phases of AgI: γ-AgI, which has the zinc blende structure, and β-AgI with the wurtzite structure; both are based on a close-packed array of iodide ions with half of the tetrahedral holes filled. However, above 147°C, a new phase, α-AgI, is observed where the iodide ions now have a body-centred cubic (*bcc*) lattice. If we look at Figure 6.6, we can see that a dramatic increase in conductivity is observed for this phase: the conductivity of α-AgI is very high, 131 S m^{-1}, a factor of 10^4 higher than that of β- or γ-AgI and comparable with the conductivity of the best conducting liquid electrolytes. How can we explain this startling phenomenon?

The explanation lies in the crystal structure of α-AgI. The structure is based on a *bcc* array of I^- ions, as shown in Figure 6.9a. Each I^- ion in the array is surrounded by eight equidistant I^- ions. In order to see where the Ag^+ ions fit into the structure, we need to look at the *bcc* structure in a little more detail. Figure 6.9b shows the same unit cell but with the next-nearest neighbours added in; these are six at the body-centres of the surrounding unit cells and are only 15% further away than the eight immediate neighbours. This means that the atom marked **A** is effectively surrounded by *14* other identical atoms, although not in a completely regular way: these 14 atoms lie at the vertices of a **rhombic dodecahedron** (Figure 6.9c). It can also be convenient to describe the structure in terms of the space-filling **truncated octahedron**, shown in Figure 6.9d, which has six square faces and eight hexagonal faces corresponding to the two sets of neighbours. This is called the **domain** of an atom. Each vertex of the domain lies at the centre of an interstice (like the tetrahedral and octahedral holes found in the close-packed structures), which in this structure is a distorted tetrahedron. Figure 6.9e shows two such adjacent distorted tetrahedral holes, and we

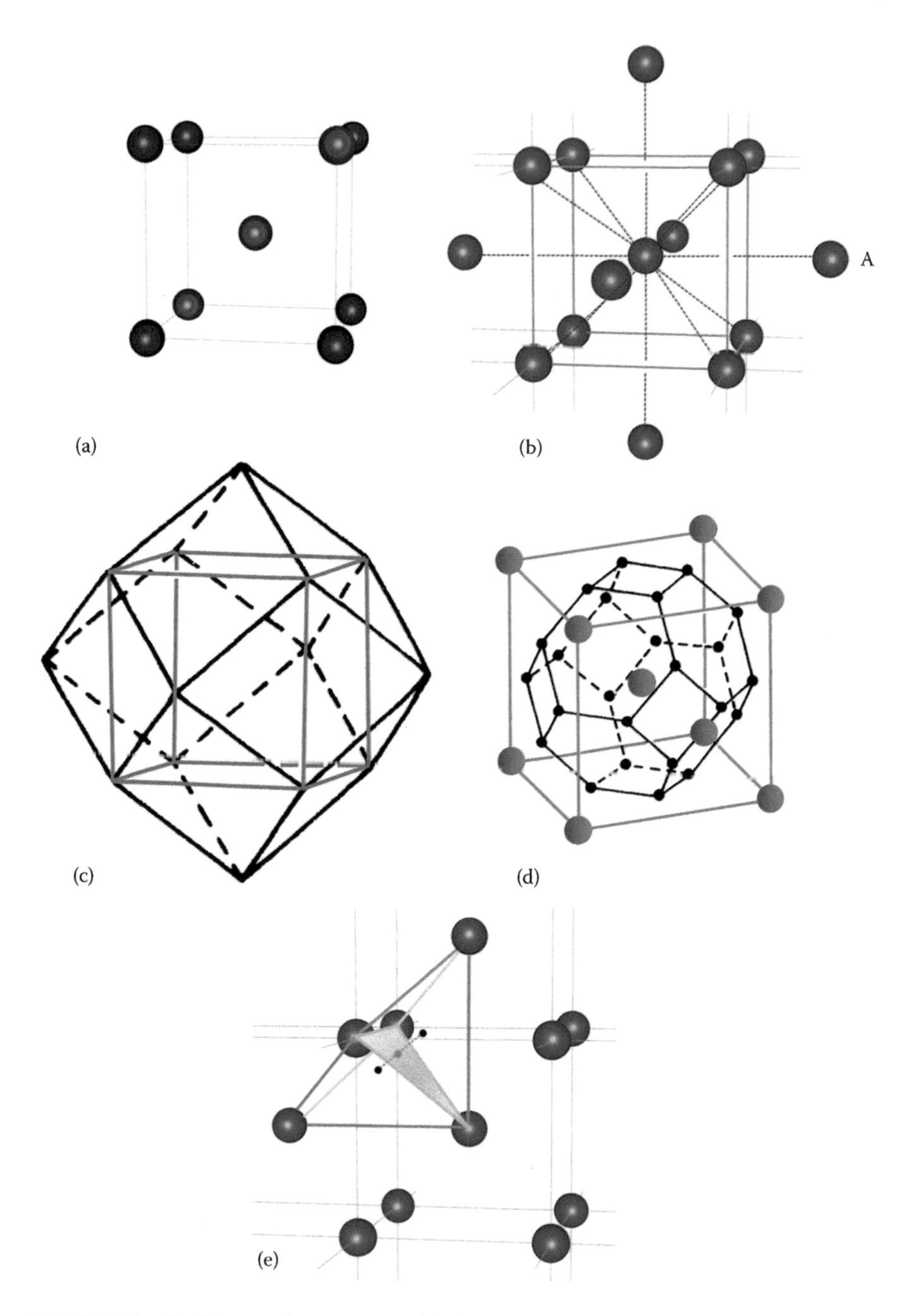

FIGURE 6.9 Building up the structure of AgI. (a) The body-centered cubic array of I^- ions. (b) The *bcc* array extended to the next-nearest neighbours. (c) A rhombic dodecahedron. (d) The *bcc* array with enclosed truncated octahedron—the 24 vertices (black dots) lie at the centres of the distorted tetrahedra. (Because these 24 sites lie on the unit cell faces, they are each shared between two unit cells, making 12 such sites per unit cell on average.) (e) The positions of two tetrahedral sites and a trigonal site between them.

can see that each of these has a 'tetrahedral' site lying on the face of the unit cell. Where the two tetrahedra join, they share a common triangular face—and there is a trigonal site at the centre of this face, which is also marked in Figure 6.9e. A third type of site at the centre of each face and at each edge of the unit cell can also be defined: these sites have distorted octahedral coordination. The structure thus possesses the unusual feature that there are a great variety of positions that can be adopted by the Ag^+ ions, and Figure 6.10 sketches many of them. Each AgI unit cell possesses two I^- ions $\left(8 \times \frac{1}{8} \text{ at the corners and one at the body centre}\right)$, so only two positions for the counterbalancing Ag^+ ions need to be available. From the possible sites that we have described, there are 6 distorted octahedral sites, 12 'tetrahedral' sites, and 24 trigonal sites—42 possible sites—so the Ag^+ ions have a huge choice of positions open to them. The structure determinations indicate that the Ag^+ ions are statistically distributed among the 12 tetrahedral sites, all of which have the same energy; therefore, counting these sites only, we find that there are five spare sites per Ag^+ ion. We can visualise the Ag^+ ions moving from tetrahedral site to tetrahedral site by jumping through the vacant trigonal site, following the paths marked with solid lines in Figure 6.10, continually creating and destroying the Frenkel defects and able to move easily through the lattice. The paths marked with thin dotted lines in Figure 6.10 require higher energy, as they pass through the vacant 'octahedral' sites that are more crowded. The jump that we have described from tetrahedral → trigonal → tetrahedral only changes the coordination number from $4 \rightarrow 3 \rightarrow 4$ and, experimentally, the activation energy is found to be very low, 0.05 eV. This easy movement of the Ag^+ ions through the lattice has often been described as a **molten sublattice** of Ag^+ ions.

The very high conductivity of α-AgI seems to arise because of a conjunction of favourable factors, and we can list the features that have contributed to this:

- The charge on the ions is low, the mobile Ag^+ ions are monovalent.
- There are a large number of vacant sites for the cations to move into.
- The structure has an open framework with pathways that the ions can move through.
- The coordination around the ions is also low, so that when they jump from one site to another, the coordination only changes a little, affording a route through the lattice with a low activation energy.
- The anions are rather **polarisable**; this means that the electron cloud surrounding an anion is easily distorted, making the passage of a cation past an anion rather easier.

These properties are important when looking for other fast-ion conductors.

The special electrical properties of α-AgI inevitably led to a search for other solids exhibiting high ionic conductivity, preferably at temperatures lower than 147°C. The partial replacement of Ag by Rb forms the compound **$RbAg_4I_5$**. This compound has an ionic conductivity at room temperature of 25 S m^{-1}, with an activation energy of only 0.07 eV. The crystal structure is different from that of α-AgI, but similarly the

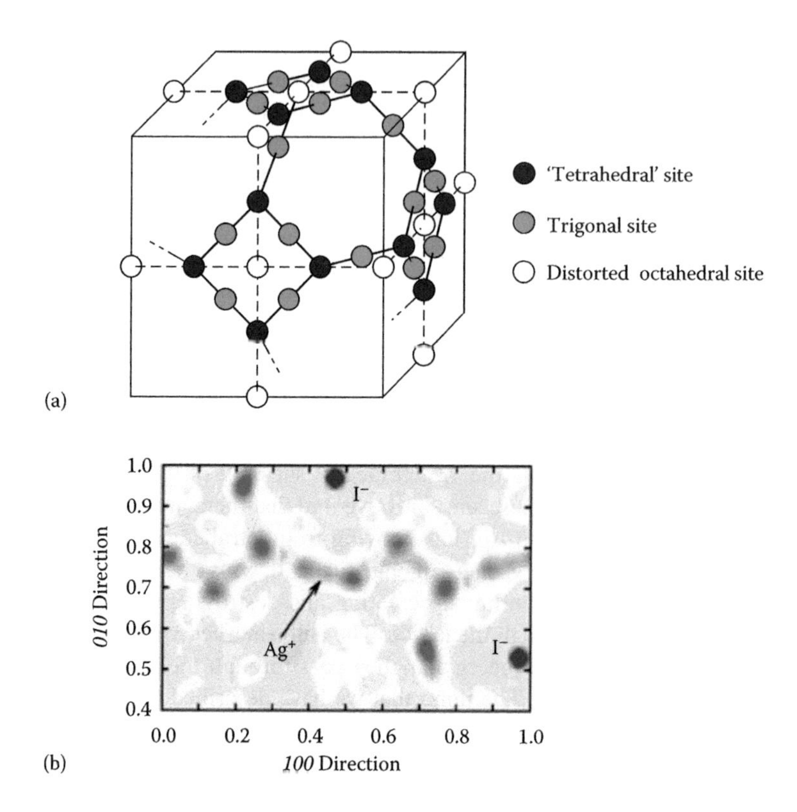

FIGURE 6.10 (a) Possible cation sites in the *bcc* structure of α-AgI. The thick solid and dashed lines mark possible diffusion paths. (b) The crystal structure of $RbAg_4I_5$ highlighting the time-averaged distribution of Ag^+ density in the ionic conductivity channels. (Reprinted from *J. Solid State Chem.*, **165**, Hull, S., et al., Crystal structures and ionic conductivities of ternary derivatives of the silver and copper monohalides, 363–371, Copyright 2002, with permission from Elsevier.)

Rb^+ and I^- ions form a rigid array while the Ag^+ ions are randomly distributed over a network of tetrahedral sites through which they can move (Figure 6.10).

If a conducting ionic solid is to be useful as a solid electrolyte in a battery, not only must it possess a high conductivity, but it also must have negligible electronic conductivity. This is to stop the battery short-circuiting: the electrons must only pass through the external circuit, where they can be harnessed for work. $RbAg_4I_5$ has been used as the solid electrolyte in batteries with electrodes made of Ag and rubidium triiodide ($Rb[I_3]$). Such cells operate over a wide temperature range (from −55°C to +200°C), have a long shelf-life, and can withstand mechanical shock.

A table of ionic conductors that behave in a similar way to α-AgI is given in Table 6.2. Some of these structures are based on a close-packed array of anions and this is noted in the table; the conducting mechanism in these compounds is similar to that in α-AgI. The chalcogenide structures, such as silver sulfide and selenide, tend to show electronic conductivity as well as ionic conductivity, although this can be quite useful in an electrode material as opposed to an electrolyte.

TABLE 6.2
α-AgI-Related Ionic Conductors

Anion structure			
bcc	ccp	hcp	Other
α-AgI	α-CuI	β-CuBr	$RbAg_4I_5$
α-CuBr	α-Ag_2Te		
α-Ag_2S	α-Cu_2Se		
α-Ag_2Se	α-Ag_2HgI_4		

6.3.2 Lithium Ion Conductors

Lithium iodide has a fairly low ionic conductivity (see Figure 6.6). Nevertheless, it is used in the LiI battery used in heart pacemakers. These batteries only need to provide a low current, but can be made very small, last a long time, produce no gas so they can be hermetically sealed and, above all, they are very reliable. In these batteries, LiI is the solid electrolyte that separates a lithium electrode from an electrode of iodine embedded in a conducting polymer (poly-2-vinyl-pyridine), which, based on Figure 6.1a, we can depict as

A		B		C
Li	//	LiI	//	I_2 and polymer

The electrode reactions are:

$$\textit{\textbf{Anode A}}: 2\text{Li(s)} = 2\text{Li}^+(\text{s}) + 2\text{e}^-$$

$$\textit{\textbf{Cathode C}}: \text{I}_2(\text{s}) + 2\text{e}^- = 2\text{I}^-(\text{s})$$

Because LiI contains intrinsic Schottky defects, the small Li^+ cations are able to pass through the solid electrolyte, while the released electrons go around an external circuit.

An implantable pacemaker powered by such batteries was first proposed by Wilson Greatbatch, after the earlier use of mercury cells, and they went into production in the early 1970s.

An area of much current research activity is lithium ion conducting solids which can act as electrolytes in the absence of a solvent.

LISICON-like conductors have a structure based on γ-Li_3PO_4. This structure can be thought of as based on a distorted hcp arrangement of oxide ions with the other ions occupying two distinct types of tetrahedral site. The structure is shown in Figure 6.11.

In the lithium ion conducting solids, phosphorus is partially or wholly substituted for by one or more metals. The substituted ions often have lower oxidation states than

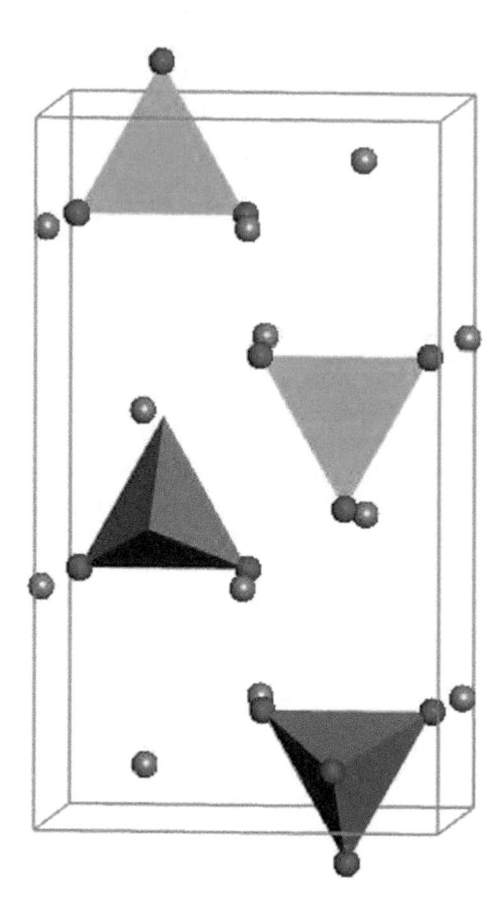

FIGURE 6.11 Structure of γ-Li_3PO_4—orange, tetrahedral phosphorus; red spheres, oxygen; mauve spheres, lithium. (Based on data from Ayu, N. I. P., Kartini, E., Prayogi, L. D., Faisal, M., Supardi, Z. (2016) Crystal structure analysis of Li_3PO_4 powder prepared by wet chemical reaction and solid-state reaction by using X-ray diffraction (XRD). *Ionics* 22, 1051–1057.)

phosphorus and charge is balanced by the addition of excess Li^+ ions. The excess Li^+ ions occupy interstitial sites. One example is $Li_{3.6}Ge_{0.6}V_{0.4}O_4$, which was reported to have a conductivity at room temperature of 4×10^{-3} S m^{-1}. Even higher conductivities have been found when oxygen is substituted for by sulfur. An example is $Li_{10}GeP_2S_{12}$ (LGPS), which has a conductivity at 300 K of 1.2 S m^{-1}.

Lithium ion stuffed garnets are another group currently being studied. Garnets are an example of minerals containing discrete SiO_4^{4-} units (see Chapter 7). They have the formula $M_3^{II}M_2^{III}\left(SiO_4\right)_3$ (where M^{II} can be Ca^{2+}, Mg^{2+}, or Fe^{2+} and M^{III} can be Al^{3+}, Cr^{3+}, or Fe^{3+}), in a framework of $M^{III}O_6$ octahedra, which are joined to six others through vertex-sharing SiO_4 tetrahedra—the M^{II} ions are coordinated by eight oxygens in dodecahedral interstices. The definition can be extended to include M^{II} = Y, La or a lanthanoid, M^{III} = Ga, Ge, Mn, Ni, or V and Si replaced by Ga or Al.

Li-stuffed garnets contain excess lithium ions, typically 5 to 7 Li per unit formula. The first reported Li ion conducting garnets had the formula $Li_5La_3M_2O_{12}$ with M = niobium or tantalum. Neutron diffraction has shown that the La^{3+} ions occupy the 8-coordinate sites and the Nb or Ta the octahedral sites. There are too many Li^+ ions for the tetrahedral sites and these ions were found to occupy both tetrahedral sites and distorted octahedral sites. The ionic conductivity of these solids was about 10^{-4} S m^{-1} at room temperature. Li-stuffed garnets have been synthesized with higher conductivities. One such example is $Li_7La_3Zr_2O_{12}$ with a room temperature conductivity of 10^{-1} to 10^{-2} S m^{-1}.

The conductivity in the stuffed garnets is due to movement of the lithium ions. Solid-state NMR and computational studies suggest that the ions hop between the distorted octahedral sites.

Other systems being studied are perovskites, nitrides, and NASICON-like structures. The NASICON structure is described in the next section. Like the LISICON structure, this is based on a phosphate. However in NASICON, the metal ions occupy octahedral sites.

6.3.3 Sodium Ion Conductors

β-Alumina is the name given to a series of compounds that show fast-ion conducting properties. The parent compound is sodium β-alumina ($Na_2O{\cdot}11Al_2O_3$ [$NaAl_nO_{17}$]), and it is found as a by-product of the glass industry. (The compound was originally thought to be a polymorph of Al_2O_3 and was named as such; it was only later found to contain sodium ions, but the original name has stuck.) The general formula for the series is $M_2O{\cdot}nX_2O_3$, where n can range from 5 to 11; M is a monovalent cation, such as (alkali metal)$^+$, Cu^+, Ag^+, or NH_4^+ and X is a trivalent cation Al^{3+}, Ga^{3+}, or Fe^{3+}. The real composition of β-alumina actually varies quite considerably from the ideal formula and the materials are always found to be rich in Na^+ and O^{2-} ions to a greater or lesser extent. The interest in these compounds started in 1966 when research at the Ford Motor Co. showed that the Na^+ ions were very mobile both at room temperature and above.

The high conductivity of the Na^+ ions in β-alumina is due to the crystal structure. This can be thought of as close-packed layers of oxide ions, but in every fifth layer, three-quarters of the oxygens are missing (Figure 6.12). The four close-packed layers contain the Al^{3+} ions in both octahedral and tetrahedral holes (they are known as the 'spinel blocks' because of their similarity to the crystal structure of the mineral spinel $MgAl_2O_4$—discussed in Chapter 1 and illustrated in Figure 1.44). The groups of four close-packed oxide layers are held apart by a rigid Al–O–Al linkage; this O atom constituting the fifth oxide layer contains only a quarter of the number of oxygens of each of the other layers. The Na^+ ions are found in these fifth oxide layers, which are mirror planes in the structure. The overall stoichiometry of the structure is also shown layer by layer in Figure 6.13. The sequence of layers is

$$\mathrm{B(ABCA)C(ACBA)B\ldots}$$

where the brackets enclose the four close-packed layers and the intermediate symbols refer to the fifth oxide layer.

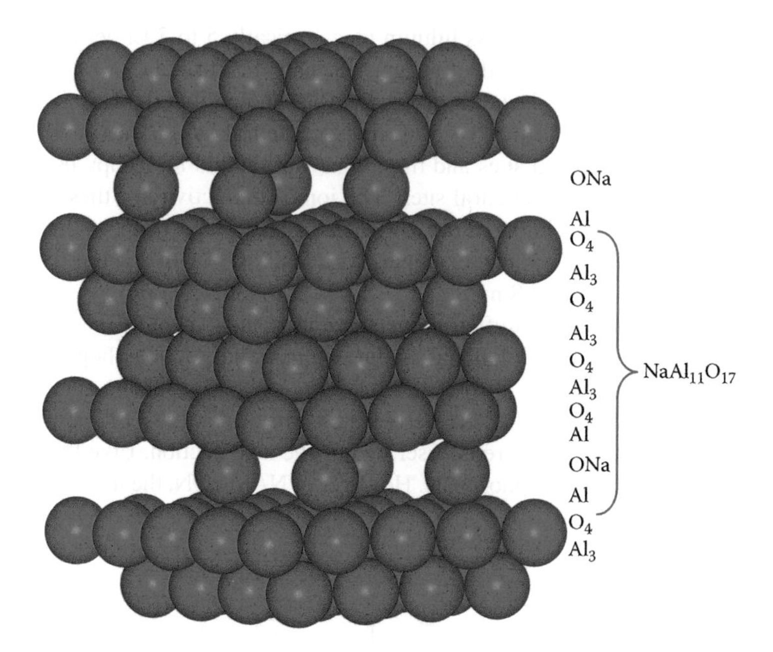

FIGURE 6.12 The oxide layers in β-alumina with the ratio of atoms in each layer of the structure.

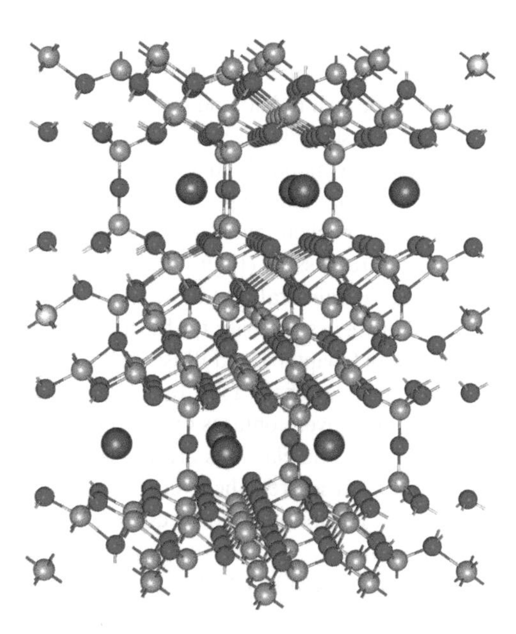

FIGURE 6.13 The structure of stoichiometric β-alumina (Na purple spheres Al brown spheres O red spheres).

The crystal structure of β-alumina is shown in Figure 6.13. The Na^+ ions can move around easily as there are plenty of vacancies, so there is a choice of sites. Conduction in the β-aluminas only occurs within the planes containing the oxygen vacancies; these are known as the **conduction planes**. The alkali-metal cations cannot penetrate the dense 'spinel blocks', but can move easily from one site to another within the plane. β-Alumina is not found in the stoichiometric form—it is usually Na_2O rich, and the sodium-rich compounds have a much higher conductivity than stoichiometric β-alumina. The extra sodium ions have to be compensated by a counter defect in order to keep the overall charge on the compound at zero. There is more than one possibility for this, but in practice it is found that extra oxide anions provide the compensation and the overall formula can be written as $(Na_2O)_{1+x}{\cdot}11Al_2O_3$. The extra sodium and oxide ions both occupy the fifth oxide layer; the O^{2-} ions are locked into position by an Al^{3+} moving out from the spinel block, and the Na^+ ions become part of the mobile pool of ions. The Na^+ ions are so fluid that the ionic conductivity in β-alumina at 300°C is close to that of typical liquid electrolytes at ambient temperature.

β-Aluminas are used as electrolytes in the manufacture of molten salt electrochemical cells, particularly for power supplies.

In 1968, Hagman and Kierkegaard found an exciting new material, sodium zirconium phosphate ($NaZr_2(PO_4)_3$), known as **NZP** (Figure 6.14). The structure consists of corner-linked ZrO_6 octahedra joined by PO_4 tetrahedra, each of which corner-shares to four of the octahedra. This creates a three-dimensional system of channels through the structure containing two types of vacant site: Type I, a single distorted octahedral site (occupied by Na^+ ions in NZP), and three larger Type II sites (vacant in NZP). This has proved to be an incredibly versatile and stable structure with the general formula $A_xM_2((Si,P)O_4)_3$, which is adopted by hundreds of compounds. The balancing cations (A) are usually alkali or alkaline earth metals, and the structural metal(s) (M) a transition metal such as Ti, Zr, Nb, Cr, or Fe. The phosphorus can be substituted by silicon. The most famous member of this family is known as **NASICON** (from Na SuperIonic CONductor). This has proved to be a very good Na^+ fast-ion conductor with a conductivity of 20 S m^{-1} at 300°C. It has the formula $Na_3Zr_2(PO_4)(SiO_4)_2$ and three out of the four vacant sites are occupied by Na^+, allowing a correlated motion as the ions diffuse through the channels.

6.4 LITHIUM-BASED BATTERIES

The 2019 Nobel Prize in Chemistry was awarded to M. Stanley Whittingham, John Goodenough, and Akira Yashino for their contributions to the development of **lithium-ion batteries**. In the 1970s, Whittingham developed a battery using lithium metal as the anode and titanium disulfide, TiS_2 as the cathode. **Titanium disulfide** has a CdI_2 structure (Chapter 1). The solid is golden-yellow and has a high electrical conductivity along the titanium layers. The conductivity of titanium disulfide can be increased by forming intercalation compounds with electron donors, the best-known example being with lithium, Li_xTiS_2. This compound is synthesised in the cathode material of the rechargeable battery and can also be synthesised directly by the lithiation of TiS_2 with a solution of butyl lithium:

FIGURE 6.14 The structure of NZP (Na purple Si blue P brown/yellow O red).

$$x C_4H_9Li + TiS_2 = Li_xTiS_2 + \frac{x}{2} C_8H_{18}.$$

Goodenough found that replacing TiS_2 by a transition metal oxide, CoO_2 produced a higher-voltage battery.

However, making reliable rechargeable lithium batteries has proved to be a very difficult problem because the lithium re-deposits in a finely divided state or forms dendritic (branch-like) growths on the electrode. These are very reactive and can even catch fire. Lithium reacts vigorously with water to form lithium hydroxide and liberating hydrogen, so batteries must be sealed absolutely from leakage. Yoshino produced the first commercially viable lithium-ion battery. This replaced the lithium anode by lithium intercalated into heated petroleum coke, a mixture of graphite and non-crystalline carbon. The electrolyte was a solution of lithium perchlorate ($LiClO_4$) in a non-aqueous solvent propylene carbonate.

Electric cars are currently (2019) using lithium-ion batteries to power them. They have several advantages over internal combustion engine cars in that they are more efficient (of the order of 90% compared with 30%) and have zero emissions of CO_2 and particulate matter.

In current batteries, the anode is made of lithium embedded in graphite, forming an intercalation compound, typically C_6Li; in discharge mode, this easily releases Li^+. The lithium ions travel through a Li^+-containing electrolyte to a cathode, where it intercalates. The cathode can be made of mixed metal oxides, e.g. $Ni_xMn_yCo_zO_2$ or other materials that intercalate lithium, notably $FePO_4$, NiO_2, and TiS_2. The battery used in the second-generation Nissan Leaf electric car, for example, has a layered cathode consisting of layers of oxide ions, lithium ions, and mixed Mn/Co/Ni ions. The structure of a layered cathode, $LiMO_2$, is shown in Figure 6.15. The electrolyte is a nonaqueous solvent, such as ethylene carbonate, mixed with a lithium complex salt; lithium hexafluorophosphate ($LiPF_6$), lithium tetrafluoroborate ($LiBF_4$), and lithium triflate ($LiCF_3SO_3$), being commonly used. The Li^+ ions 'rock' between the two intercalation compounds and no lithium metal is ever present, eliminating many of the hazards associated with lithium batteries.

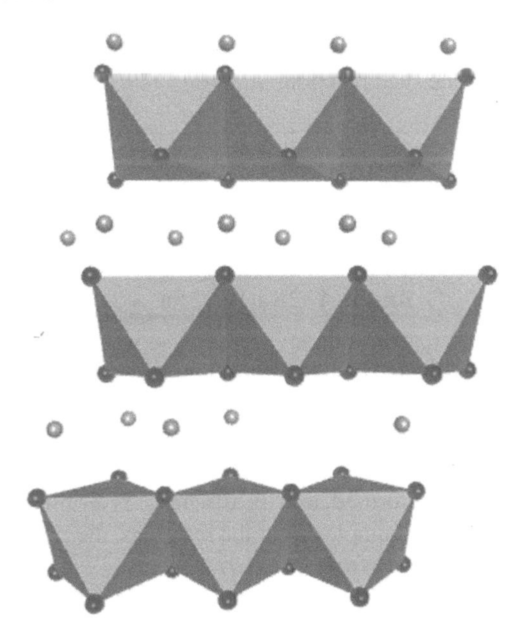

FIGURE 6.15 The structure of $LiMO_2$. Li, small purple spheres. O, small red spheres attached to blue octahedra (M).

The safety of these batteries is ensured by the use of a nonaqueous electrolyte and a carefully sealed container that completely excludes water. These containers are usually metal and can take a variety of forms such as cylinders, pouches, and rectangular prisms.

Lithium ion polymer batteries have evolved from the lithium ion battery. The cell reaction is the same, but the electrolyte is no longer suspended in a nonaqueous solvent; instead, the Li salt is suspended in a conducting polymer gel such as polyethylene oxide (PEO) or polyacrylonitrile. These batteries can be enclosed in a flexible foil casing that makes them very light, easily shaped, and very thin. Great care has to be taken with not overcharging them as they can overheat and expand.

6.5 SODIUM-BASED BATTERIES

Sodium-containing batteries are used in electrical storage (see Figure 6.3). Na^+ conduction has been put to good use in a secondary battery that operates at high temperatures—the **sodium sulfur battery**. This system uses either NASICON or β-alumina as the electrolyte. From Equation 6.1, we know that in order to obtain a large voltage from a cell, we must have a cell reaction with a large negative Gibbs free energy change, such as a reaction between an alkali metal and a halogen might give. In terms of specific energy, such a reaction can yield up to 150 Wh kg^{-1} of material because it incorporates a highly energetic reaction between light substances. Such reactive materials have to be separated by an electrolyte that is impermeable to electrons, but can be crossed by ions. The electrolyte separates molten sodium from molten sulfur, and at the sulfur/electrolyte interface a complex reaction forming polysulfides of sodium takes place. The heat required to maintain the cell at the operating temperature of 300°C is supplied by the cell reaction itself. The cell reaction is given here for one such reaction:

A		B		C
Na(1)	//	Β-Alumina	//	S(I)and graphite

$$\textbf{\textit{Anode A}}: 2\text{Na}(1) = 2\text{Na}^+ + 2\text{e}^-$$

$$\textbf{\textit{Cathode C}}: 2\text{Na}^+ + 5\text{S}(1) = 2\text{e}^- = \text{Na}_2\text{S}_5(1)$$

$$\text{Overall reaction}: 2\text{Na}(1) + 5\text{S}(1) = 2\text{e}^- = \text{Na}_2\text{S}_5(1)$$

Later, low polysulfides are formed, and the discharge is terminated at a composition of about Na_2S_3. Despite the complexity of the reactions, the electrode process can be reversed by applying a current from an external source.

These cells have the advantages of high power and low weight, and in many countries the components of these cells are abundantly available and relatively cheap. However, they require stringent safety features as they contain highly reactive and corrosive chemicals at 300°C, and this has meant that they are not well suited for their original development purpose, that of use in the electric car. However, they are used for energy storage.

Interest has also been directed towards a similar high-temperature system, the **ZEBRA battery** or the Na–$NiCl_2$ battery, which also uses β-alumina as an Na^+ ion conductor. The sulfur electrode is replaced by nickel chloride. The contact between the $NiCl_2$ electrode and the solid electrolyte is poor as they are both solids, and current flow is improved by adding a second liquid electrolyte (molten $NaAlCl_4$) between this electrode and the β-alumina. The overall cell reaction is now:

$$2Na + NiCl_2 = Ni + 2NaCl$$

The high specific energy for this cell of >100 Wh kg^{-1} gives electric vehicles powered by these batteries a range of up to 250 km, sufficient for daily use in a city. These batteries are fully rechargeable, robust, and safe, and they have been found to need no maintenance over 100,000 km.

The **sodium-ion battery** is analogous to the lithium ion battery that uses the interaction with a graphite anode, in this case, to form Na_xC_6. This battery is currently (2019) at a research stage as it does not maintain its capacity after many cycles of charging and recharging. If it can be developed to perform consistently, it would have many advantages over ZEBRA and Na–S batteries as it operates at low temperatures.

6.6 SUMMARY

1. In some solids, defects allow ions to move through the structure. When an electric field is applied, the ions move in response causing a current to flow. These solids are known as ionic conductors.
2. Simple (hopping) models for the movement of ions through the solid are the vacancy mechanism and the interstitial mechanism.
3. Ionic conductors can act as electrolytes in solid-state batteries. For this purpose, the solids should not be electronic conductors.
4. Uses of solid-state batteries include heart pacemakers and energy storage.
5. Lithium-ion batteries are used in electronic devices and electric vehicles. The electrolyte here is a lithium salt in a non-aqueous solvent.
6. In lithium ion polymer batteries, the lithium salt is suspended in a conducting polymer gel.
7. Sodium-based batteries include the sodium-sulfur battery, the ZEBRA battery, and sodium-ion batteries.

QUESTIONS

1. Figure 6.16 shows the fluorite structure with the tetrahedral environment of one of the anions shaded. (The anion behind this tetrahedron has been omitted for clarity.) Suppose this anion jumps to the octahedral hole at the body centre. Describe, and sketch, the pathway it takes in terms of the changing coordination by cations.

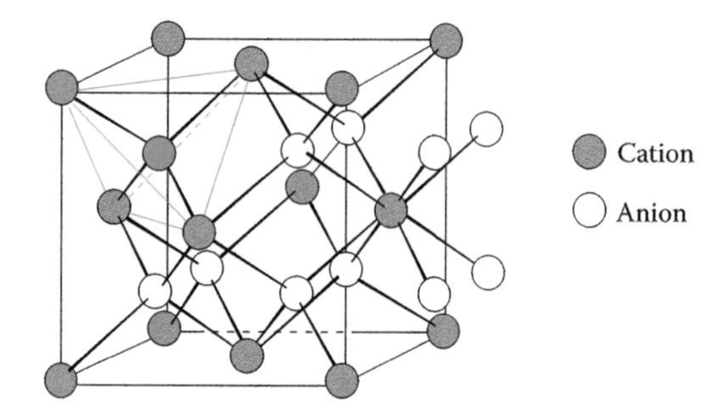

FIGURE 6.16 The fluorite structure, showing the coordination tetrahedron around one of the anions.

2. The zinc blende structure of γ-AgI, a low-temperature polymorph of AgI, is shown in Figure 6.17. Discuss the similarities and differences between this structure and that of α-AgI. Why do you think that the conductivity of the Ag^+ ions is lower in γ-AgI?
3. The compounds in Table 6.2 mostly contain either I^- ions or ions from the heavier end of Group 6. Explain.
4. The positions of lithium ions in lithium-stuffed garnets were determined using neutron diffraction. Suggest why this technique was preferred to X-ray diffraction.
5. Undoped β-alumina shows a maximum conductivity and minimum activation energy when the sodium excess is around 20–30 mol%. Thereafter, a further increase in the sodium content causes the conductivity to decrease. By contrast, β-alumina crystals doped with Mg^{2+} have a much higher conductivity than do undoped crystals. Explain these observations.
6. Compare the advantages and drawbacks of lithium- and sodium-based batteries.

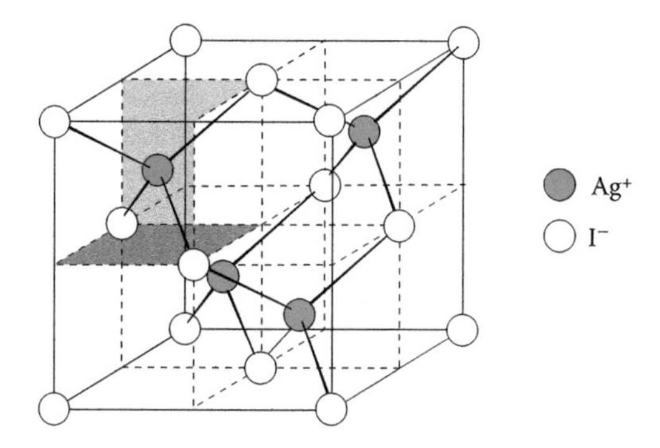

FIGURE 6.17 The zinc blende structure of γ-AgI.

7 Microporous and Mesoporous Solids

7.1 INTRODUCTION

Microporous and mesoporous solids have cavities (cages) and channels (pores) within their structure. **Microporous** structures have pores of less than 2 nm in diameter and **mesoporous** structures pores of between 2 and 50 nm. Solids with pores >50 nm are **macroporous.** These pores can contain ions and/or molecules and it is this that makes these solids valuable in industrial applications, for example as absorbents and catalysts, and even in medical applications.

Several of the porous solids we discuss are aluminosilicates, in which aluminium atoms replace some silica in a silicate network. We start therefore by looking at the structures of silicates. The first aluminosilicate microporous solids discussed are zeolites. Some of these occur naturally and have been known since 1756, but there also many synthetic zeolites. We look at the structure and composition of these solids, their synthesis, and their uses.

Metal-organic frameworks (MOFs) are a class of porous solids that have attracted much research and come to prominence in the 21st century. These form the second main area of study in this chapter.

Even more recently, Covalent Organic Frameworks (COFs) have aroused much interest. These porous solids are covalently bonded and composed of light elements.

Finally we will look at some other porous solids, mesoporous aluminosilicates, clays, and mesoporous organosilicas that have found uses or are currently an active area of research.

7.2 ZEOLITES

In 1756, the Swedish mineralogist Baron Axel Fredrik Cronstedt first described zeolites as a mineral group. They are a class of crystalline **aluminosilicates** based on rigid anionic frameworks with well-defined **pores (channels)** running through them, which intersect at **cavities (cages)**. These cavities contain exchangeable metal cations (Na^+, K^+, etc.) and can also hold removable and replaceable guest molecules (water in naturally occurring zeolites). It is their ability to lose water on heating that has earned them their name: Cronstedt observed that on heating with a blowtorch, they hissed and bubbled as though they were boiling and thus named them zeolites from the Greek words *zeo*, to boil, and *lithos*, stone. This loss of water is completely reversible.

Over 60 naturally occurring zeolites have been characterised but only seven (analcime, chabazite, clinoptilolite, erionite, ferrierite, mordenite, and phillipsite) zeolites occur in large deposits, mostly in China and Cuba. However over 200 artificial zeolites have been synthesized. Before considering the structure of zeolites themselves, we look at silicates.

7.2.1 Silicates

The **silicates** form a large group of crystalline compounds with rather complex but interesting structures. A great part of the Earth's crust is formed from these complex oxides of silicon.

Silicon itself crystallises with the same structure as diamond. There are two crystalline forms of silica at atmospheric pressure—**quartz** and **cristobalite**. Each of these also exists in low- and high-temperature forms, α and β, respectively. We have already discussed the structure of β-cristobalite in terms of close packing in Section 1.8.2. Quartz is commonly encountered in nature. The structure of β-quartz is illustrated in Figure 7.1, and it consists of SiO_4 tetrahedra linked so that each oxygen atom is shared by two tetrahedra, thereby giving the overall stoichiometry of SiO_2. Notice how, once again, the covalency of each atom dictates the coordination around itself, silicon having four bonds and oxygen two, rather than the larger coordination numbers that are found for metallic and some ionic structures. Quartz is unusual in that the linked tetrahedra form spirals or **helices** throughout the crystal, which are all either left- or right-handed, producing laevorotatory or dextrorotatory crystals, respectively; these are known as **enantiomorphs**.

Quartz is one of the most common minerals on the Earth, occurring as sand on the seashore, as a constituent in granite and flint, and, in less pure form, as agate and opal. The silicon atom in all these structures is tetrahedrally coordinated.

The silicate structures are most conveniently discussed in terms of the SiO_4^{4-} unit. The SiO_4^{4-} unit has tetrahedral coordination of silicon by oxygen and is represented in these structures by a small tetrahedron, as shown in Figure 7.2a. The silicon–oxygen bonds possess considerable covalent character.

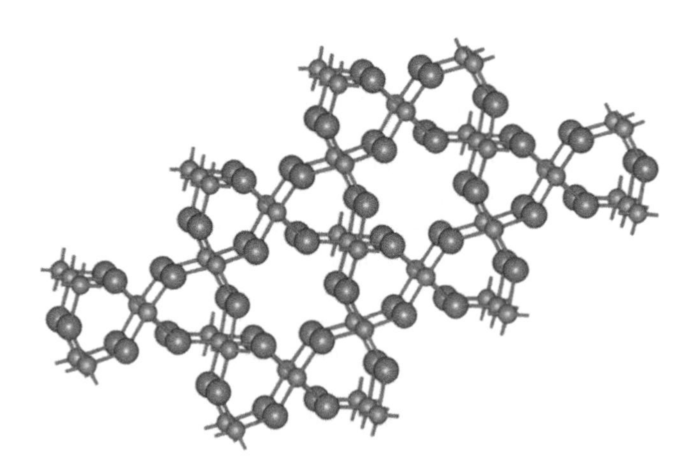

FIGURE 7.1 β-Quartz structure. Key: Si, blue; O, grey.

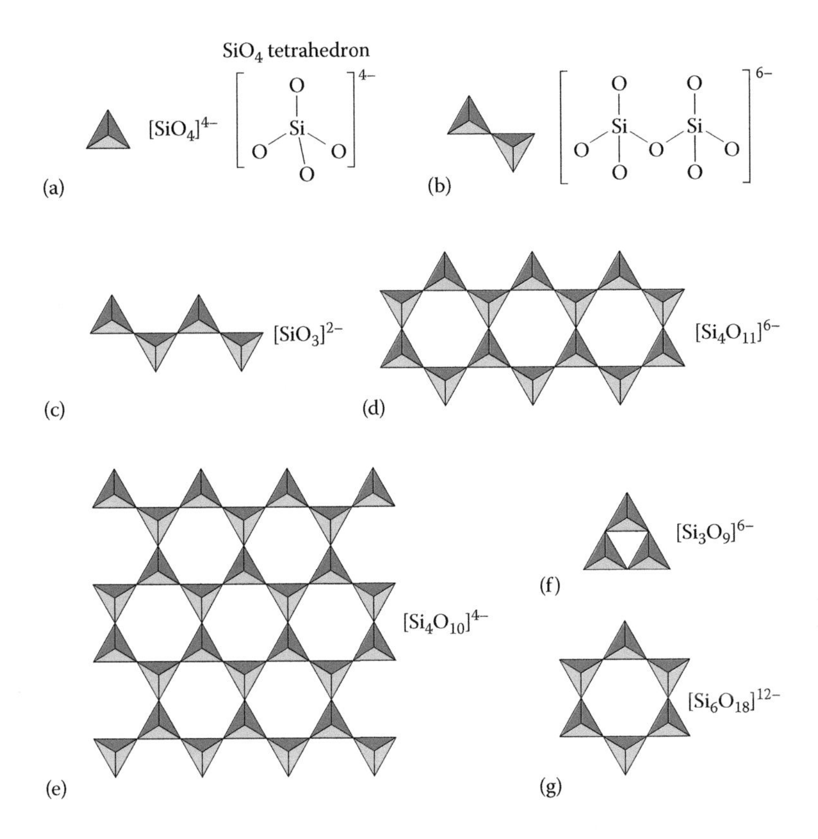

FIGURE 7.2 (a–g) Structural classification of mineral silicates.

Some minerals—**olivines** for instance—contain discrete SiO_4^{4-} tetrahedra (Figure 7.3a): these compounds do not contain Si–O–Si–O–Si–... chains, but there is considerable covalent character in the metal–silicate bonds. These are often described as **orthosilicates** or salts of orthosilicic acid ($Si(OH)_4$) or (H_4SiO_4), which is a very weak acid. The structure of olivine itself ($(Mg,Fe)_2SiO_4$), which can be described as an assembly of SiO_4^{4-} ions and Mg^{2+} (or Fe^{2+}) ions, appeared in Table 1.8 because it can be alternatively described as an *hcp* array of oxygens with silicons occupying one-eighth of the tetrahedral holes and magnesium ions occupying one-half of the octahedral holes.

In most silicates, however, the SiO_4^{4-} tetrahedra are linked by oxygen sharing through a vertex, as illustrated in Figure 7.2b for two linked tetrahedra to give $Si_2O_7^{6-}$. Notice that each terminal oxygen confers a negative charge on the anion and the shared oxygen is neutral. The diagrams of silicate structures showing the silicate frameworks, such as are shown in Figure 7.2, omit these charges because they can be readily calculated. By sharing one or more oxygen atoms through the vertices, the tetrahedra are able to link up to form chains, rings, layers, etc. The negative charges on the silicate framework are balanced by metal cations in the lattice. Some examples are discussed below.

FIGURE 7.3 (a) Unit cell of olivine. Key: Mg and Fe, green; Si, grey; O, red. (b) Structure of an amphibole. Key: Mg, green; Si, grey; O, red; Na, purple. (c) Structure of biotite. Key: Mg, green; Si, grey; O, red; K, purple; Al, pink; Fe, blue.

Examples of **discrete SiO_4^{4-} units** are found in: **olivine** (see Table 1.8 and Figure 7.3a), an important constituent of basalt; **Ca_2SiO_4**, found in mortars and Portland cement; **zircon** ($ZrSiO_4$), which has eight-coordinate Zr and **garnets**, $M_3^{II}M_2^{III}(SiO_4)_3$ (where M^{II} can be Ca^{2+}, Mg^{2+}, or Fe^{2+} and M^{III} can be Al^{3+}, Cr^{3+}, or Fe^{3+}), the framework of which is composed of $M^{III}O_6$ octahedra, which are joined to six others through vertex-sharing SiO_4 tetrahedra—the M^{II} ions are coordinated by eight oxygens in dodecahedral interstices.

Structures containing **disilicate units**, $Si_2O_7^{6-}$ (Figure 7.3b), are not common but occur in **thortveitite**, the primary source of scandium ($(Sc,Y)_2Si_2O_7$), and **hemimorphite**, a rare zinc silicate ($Zn_4(OH)_2Si_2O_7$).

SiO_4^{4-} units share two corners to form infinite **chains** (see Figure 7.3c). The repeat unit is SiO_3^{2-}. Minerals with this structure are called **pyroxenes**, which together with olivine, form the primary constituents of the Earth's upper mantle, for example, **diopside** ($CaMg(SiO_3)_2$) and **enstatite** ($Mg_2(SiO_3)_2$). The silicate chains lie parallel to one another and are linked together by the cations that lie between them.

In **double chains** alternate tetrahedra share two and three oxygen atoms, respectively, as in Figure 7.2d. This class of minerals are known as the **amphiboles**, usually containing magnesium and/or iron (Figure 7.3b), an example of which is **tremolite** ($Ca_2Mg_5(OH)_2(Si_4O_{11})_2$). Most of the asbestos minerals fall into this class. The repeat unit is $Si_4O_{11}^{6-}$.

In **infinite layers** the tetrahedra all share three oxygen atoms (see Figure 7.2e). The repeat unit is $Si_4O_{10}^{4-}$. Examples are the **mica** group, named for their glittering appearance and used as insulators in high-voltage appliances, of which **biotite** ($K(Mg,Fe)_3(OH)_2Si_3AlO_{10}$) (Figure 7.3c) is a member containing a sandwich of two layers with octahedrally coordinated cations between the layers; and clay minerals such as **kaolinite**, found in china clay ($Al_4(OH)_8Si_4O_{10}$) and **talc** ($Mg_3(OH)_2Si_4O_{10}$).

In **rings,** each SiO_4^{4-} unit shares two corners as in the chains. Figure 7.2f and 7.2g shows three and six tetrahedra linked together; these have the general formula $(SiO_3)_n^{-2n}$; rings may also be made from four tetrahedra. An example of a six-tetrahedra ring is **beryl** ($Be_3Al_2Si_6O_{18}$); here, the rings lie parallel to the metal ions between them. Other examples include the uncommon desert mineral **dioptase** ($Cu_6Si_6O_{18}.6H_2O$) and **benitoite**, California's very rare state gemstone ($BaTiSi_3O_9$).

If SiO_4^{4-} tetrahedra share all four oxygens, then the **three-dimensional structure** of silica (SiO_2) is produced. However, if some of the silicon atoms are replaced by the similarly sized atoms of the Group III element aluminium (i.e., if SiO_4^{4-} is replaced by AlO_4^{5-}, then other cations must be introduced to balance the charges). Such minerals include the **feldspars** (general formula $M(Al,Si)_4O_8$), the most abundant of the rock-forming minerals; the **zeolites**; and the **ultramarines**, which are coloured silicates manufactured for use as pigments, **lapis lazuli** being a naturally occurring mineral of this type.

As one might expect, there is an approximate correlation between the solid-state structure and the physical properties of a particular silicate. For instance, cement contains discrete SiO_4^{4-} units and is soft and crumbly; asbestos minerals contain double chains of SiO_4^{4-} units and are characteristically fibrous; mica contains infinite layers of SiO_4^{4-} units and the weak bonding between the layers is easily broken giving cleavage parallel to the layers; granite contains feldspars, which are based on three-dimensional SiO_4^{4-} frameworks and are very hard.

7.2.2 Composition and Structure of Zeolites

The general formula for the composition of a zeolite is $M_{x/n}\left[\left(AlO_2\right)_x\left(SiO_2\right)_y\right]\cdot mH_2O$, where cations M of valence n neutralise the negative charges on the aluminosilicate framework.

The primary building units of zeolites are $[SiO_4]^{4-}$ and $[AlO_4]^{5-}$ tetrahedra linked together by **corner sharing**, forming oxygen bridges. The oxygen bridge is not usually linear, but the Si/Al-O-Si/Al linkage is very flexible and the angle can vary between 120° and 180°. The silicon–oxygen tetrahedra are electrically neutral when connected together in a three-dimensional network as in quartz (SiO_2) (Figure 7.1). The substitution of Si(IV) by Al(III) in such a structure, however, creates an electrical imbalance, and to preserve overall electrical neutrality, each $[AlO_4]$ tetrahedron needs a balancing positive charge. This is provided by exchangeable cations, such as Na^+, held electrostatically within the zeolite.

It is possible for the tetrahedra to link by sharing two, three, or all four corners, thus forming a variety of different structures. The linked tetrahedra are usually illustrated by drawing a straight line to represent the oxygen bridge connecting two tetrahedral units. In this way, the six linked tetrahedra in Figure 7.4a and 7.4b are simply represented by a hexagon (Figure 7.4c). This is known as a **6-ring**, and a tetrahedrally coordinated Si or Al atom occurs at each intersection between two straight lines. As we will see later, many different ring sizes are found in the various zeolite structures.

Many zeolite structures are based on a secondary building unit that consists of 24 silica or alumina tetrahedra linked together; here we find 4-rings and 6-rings linked together to form a basket-like structure called the **sodalite unit** (also known as the **β-cage**) shown in Figure 7.5, which has the shape of a **truncated octahedron**

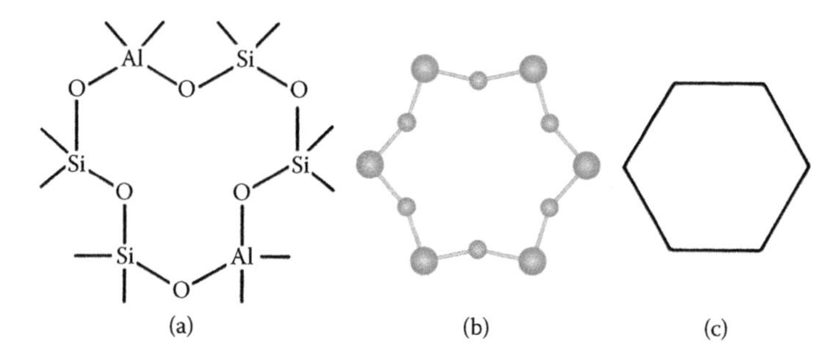

FIGURE 7.4 (a) A 6-ring containing two Al and four Si atoms, (b) a computer model of the 6-ring and (c) a shorthand version of the same 6-ring.

(Figure 7.6). Several of the most important zeolite structures are based on the sodalite unit (Figure 7.7).

The mineral sodalite is composed of these units, with each 4-ring shared directly by two β-cages in a primitive array. Note that the cavity or cage enclosed by the eight sodalite units shown in Figure 7.7a is itself a sodalite unit, that is, sodalite units are space-filling. In this three-dimensional structure, a tetrahedral Si or Al atom is located at the intersection of four lines, as oxygen bridges are made by corner sharing from all four vertices of the tetrahedron. Sodalite is a highly symmetrical structure and the cavities link together to form the channels or pores, which run parallel to all three cubic crystal axes; the entrance to these pores is governed by the 4-ring window.

A synthetic zeolite, **zeolite A** (also called Linde A), is shown in Figure 7.7b. Here, the sodalite units are again stacked in a primitive array, but now they are linked by oxygen bridges between the 4-rings. A three-dimensional network of

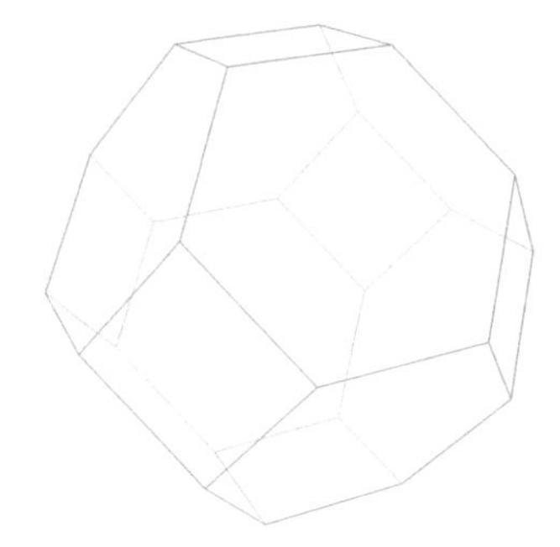

FIGURE 7.5 The sodalite unit.

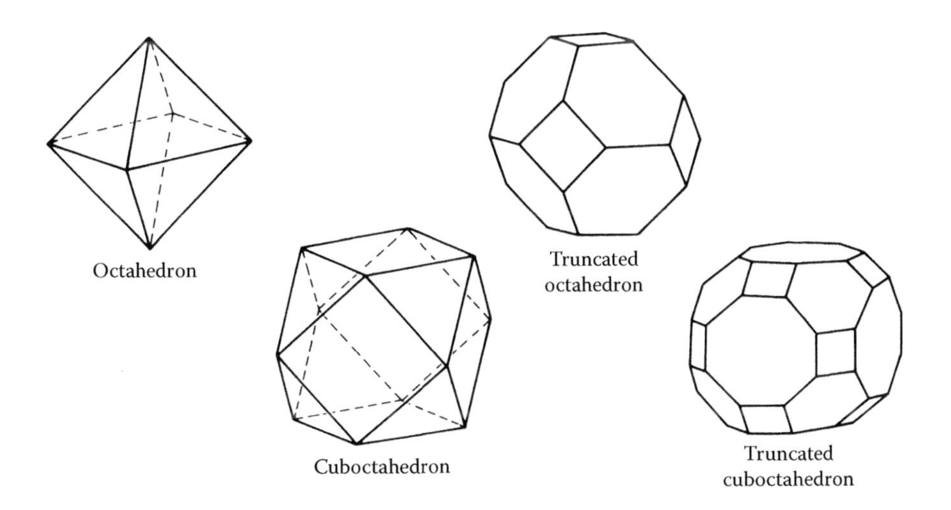

FIGURE 7.6 The relationship between an octahedron, a truncated octahedron, a cuboctahedron, and a truncated cuboctahedron.

linked cavities, each with a **truncated cuboctahedron** shape (Figure 7.6) through the structure, is thus formed; the truncated cuboctahedra are also space-filling (each shares its octagonal face with six others), forming channels that run parallel to the three cubic axial directions through these large cavities. The computer-drawn models in Figure 7.8 are of the zeolite A framework, showing the cavity and its 8-ring window more clearly and how it links to a sodalite cage. The formula of zeolite A is given by $Na_{12}\left[(SiO_2)_{12}(AlO_2)_{12}\right]\cdot 27H_2O$. In this fairly typical example, the Si/Al ratio is unity, and we find that in the crystal structure, the Si and Al atoms strictly alternate.

The structure of **faujasite**, a naturally occurring mineral, is shown in Figure 7.7c. The sodalite units are linked by oxygen bridges between four of the eight 6-rings in a tetrahedral array. The tetrahedral array encloses a large cavity (sometimes known as the α-cage) entered through a 12-ring window. The synthetic zeolites **zeolite X** and **zeolite Y** (Linde X and Linde Y) also have this basic underlying structure. The zeolite X structures have an Si/Al ratio between 1 and 1.5, whereas the zeolite Y structures have Si/Al ratios between 1.5 and 3.

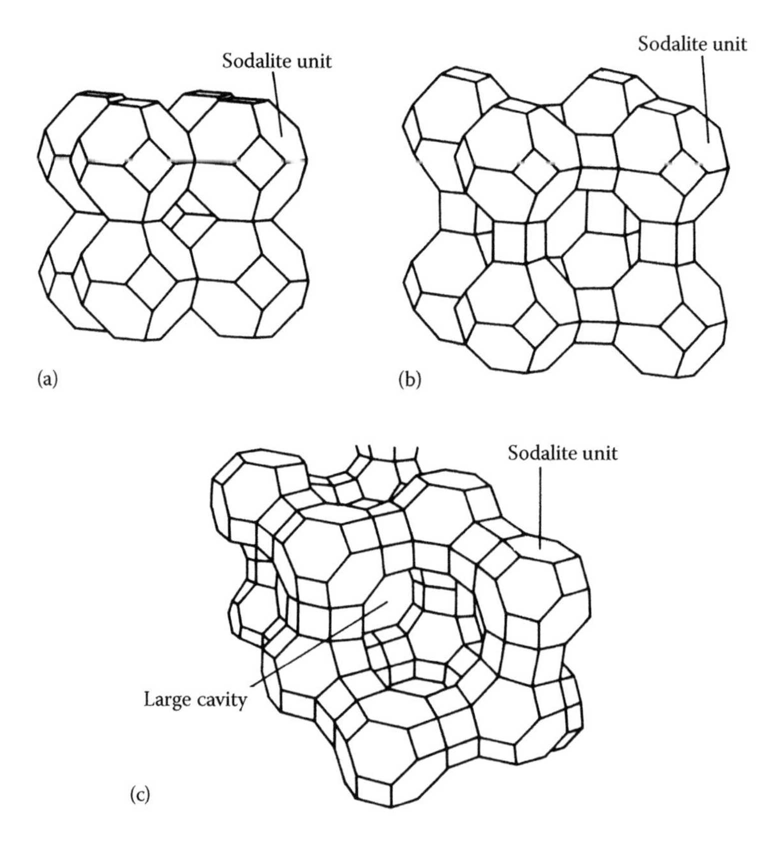

FIGURE 7.7 The zeolite frameworks built up from sodalite units: (a) sodalite (SOD), (b) zeolite A (LTA), and (c) faujasite (FAU).

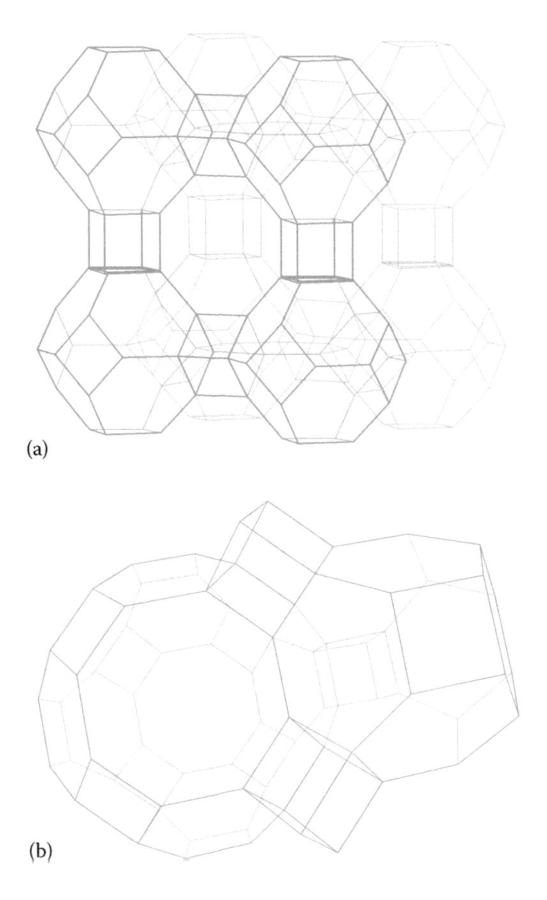

FIGURE 7.8 (a) The zeolite A framework and (b) a sodalite unit in zeolite A, showing the linkage to the truncated cuboctahedral cavity.

7.2.3 Zeolite Nomenclature

You may have noticed that the naming of zeolites has been somewhat unsystematic. Some structures are named after the parent minerals, for example, sodalite and faujasite, while others were named by researchers, or after the projects that synthesised them, for example, ZSM (Zeolite Socony Mobil). Unfortunately, this led to the same zeolites being synthesised by different routes, bearing different names—in some cases, up to 20 different trade names!

The International Zeolite Association (IZA), with the permission of the International Union of Pure and Applied Chemistry (IUPAC), introduced a three-letter structure code for unique framework topologies to try and simplify matters; zeolite A and the more silicon-rich zeolite ZK-4 have the same framework structure and are designated LTA. Similarly, ZSM-5 and its silicon-rich relation, silicalite, have the same framework and are both designated MFI.

The zeolites are also often written as M-[zeolite], where M refers to the particular cation in the structure, for example, Ca-zeolite A or even simply CaA or NaA.

7.2.4 Si/Al Ratios in Zeolites

The structures of the zeolite frameworks have been determined by X-ray and neutron diffraction techniques. Unfortunately, it is extremely difficult for diffraction techniques to determine a structure unequivocally because Al and Si are next to each other in the periodic table and therefore have very similar atomic scattering factors (Chapter 2). One of the techniques used for successfully elucidating zeolite structures is magic angle spinning NMR spectroscopy (MAS-NMR) (see Chapter 2). Five peaks can be observed for the ^{29}Si spectra of various zeolites, which correspond to the five possible, different Si environments.

Zeolites A and X are 'aluminium-rich' zeolites with Si/Al ratios of close to 1. These zeolites have very useful absorbent properties and are still used industrially, but are unstable and lose aluminium when attacked by acid, steam, or water. The natural zeolite mordenite, which is the most siliceous of the naturally occurring zeolite minerals, was found to be more stable and because it has an Si/Al ratio of >5, research turned to synthesising zeolites with higher ratios in the hope of greater stability. Many synthetic zeolites that have been developed for catalysis are thus highly siliceous: zeolite Y has an improved stability with a ratio of 2.25; zeolite ZK-4 (LTA), with the same framework structure as zeolite A, has a ratio of 2.5; and **ZSM-5** (MFI) can have an Si/Al ratio that lies between 20 and ∞ (the latter, called silicalite [see above], being virtually pure SiO_2), which far outstrips the ratio of 5.5 found in mordenite.

Clearly, changing the Si/Al ratio of a zeolite also changes its cation content; the fewer the aluminium atoms, the fewer cations will be present to balance the charges. The highly siliceous zeolites are inherently hydrophobic in character and their affinity is for hydrocarbons.

7.2.5 Exchangeable Cations

The zeolite Si/Al-O framework is rigid, but the cations are not an integral part of this framework and are often called **exchangeable cations**: they are fairly mobile and readily replaced by other cations (hence their use as cation-exchange materials).

The presence and position of the cations in zeolites are important for several reasons. The cross-section of the rings and the channels in the structures can be altered by changing the size or the charge (and thus the number) of the cations, and this significantly affects the size of the molecules that can be adsorbed. A change in the cationic occupation also changes the charge distribution within the cavities and therefore the adsorptive behaviour and the catalytic activity.

The balancing cations in a zeolite can have more than one possible location in the structure. Figure 7.9 shows the available sites in the Na^+ form of zeolite A. Favourable sites can be found using a computer package such as that discussed in Section 1.9.2. Some sites occupy most of the centres of the 6-rings, while other sites are in the 8-ring entrances to the β-cages. The presence of cations in these positions effectively reduces the size of the rings and the cages to any guest molecules that are trying to enter. In order to alter a zeolite to allow organic molecules, for instance, to diffuse into or through the zeolite, a divalent cation such as Ca^{2+} can be exchanged

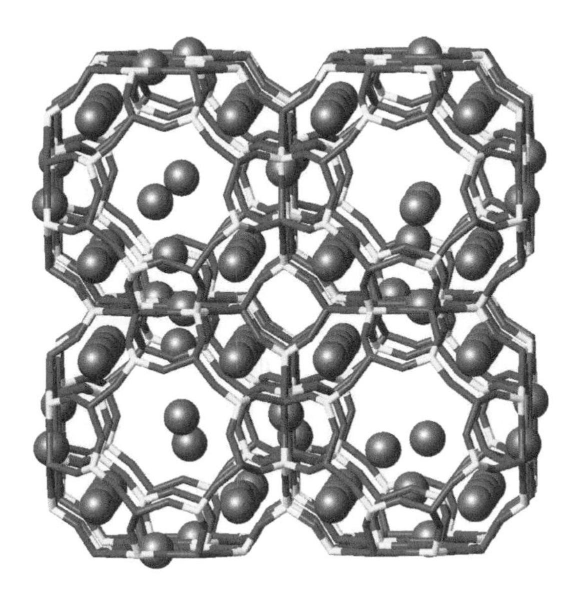

FIGURE 7.9 Available cation sites in the sodium form of zeolite A (LTA). (From Robert G. Bell, UCL, with permission.)

for the univalent Na^+ or K^+ ions, not only halving the number of cations present but also replacing them with a smaller ion. As the divalent cations tend to occupy the sites in the 6-rings, this opens up the 8-ring windows, thus leaving the channels free for diffusion.

The normal crystalline zeolites contain water molecules that are coordinated to the exchangeable cations. These structures can be dehydrated by heating under vacuum, and in these circumstances, the cations move position at the same time, frequently settling on the sites with a lower coordination number. The dehydrated zeolites are extremely good drying agents, absorbing water to get back to their preferred hydrated condition.

The exchange of the metal cations for protons, usually by exchanging the metal ion with ammonium $\left(NH_4^+\right)$, and then heating to drive off ammonia (NH_3), produces strongly acidic materials that have proved to be highly active acid catalysts.

7.2.6 Channels and Cavities

The important structural feature of zeolites, which can be exploited for various uses, is the network of linked cavities forming a system of channels throughout the structure. These cavities are of molecular dimensions and can adsorb species small enough to gain access to them. A controlling factor in whether the molecules can be adsorbed in the cavities is the size of the **window** or the **port** into the channel; hence the importance of the number of tetrahedra forming the window, that is, the ring size. Figure 7.10 shows how the window sizes can vary.

The windows to the channels thus form a three-dimensional sieve with mesh widths of between about 300 and 1000 pm; hence the well-known name **molecular**

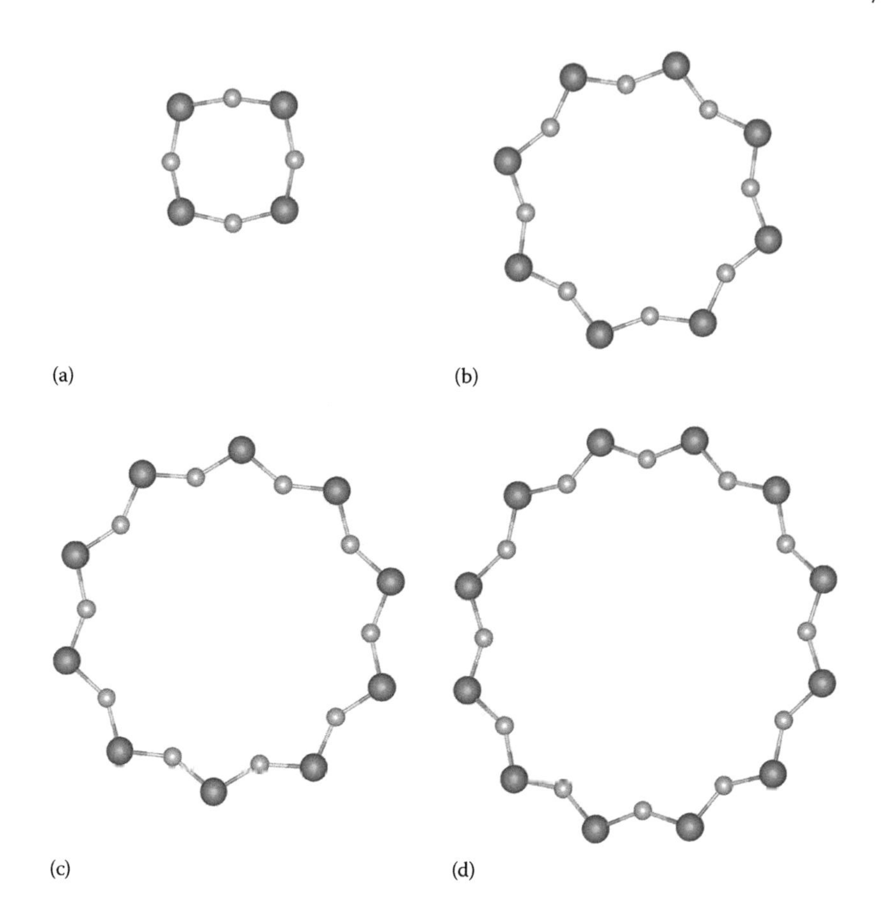

FIGURE 7.10 Computer models of various window sizes in zeolites (a) 4-ring, (b) 8-ring, (c) 10-ring, and (d) 12-ring.

sieve for these crystalline aluminosilicates. The zeolites thus have large internal surface areas and high sorption capacities for molecules small enough to pass through the window into the cavities. They can be used to separate mixtures such as straight-chain and branched-chain hydrocarbons.

The zeolites fall into three main categories. The channels may be parallel to (a) a single direction, so that the crystals are fibrous; (b) two directions arranged in planes, so that the crystals are lamellar; or (c) three directions, such as cubic axes, in which there is strong bonding in three directions. The most symmetrical structures have cubic symmetry. By no means do all zeolites fall neatly into this classification; some, ZSM-11, for instance, have a dominant two-dimensional structure interlinked by smaller channels. A typical fibrous zeolite is edingtonite (EDI, $Ba[(AlO_2)_2(SiO_2)_3]4H_2O$), which has a characteristic chain formed by the regular repetition of five tetrahedra. The lamellar zeolites occur frequently in sedimentary rocks, for example, phillipsite $\left(PHI, (K/Na)_5\left[(SiO_2)_{11}(AlO_2)_5\right]\cdot 10H_2O\right)$, which is a well-known example. In terms of their useful properties, zeolites are conveniently discussed based on their pore size.

Sodalite is a good example of a **small-pore zeolite**. A cavity in sodalite (the β-cage) is bounded by a 4-ring with a diameter of 260 pm (Figure 7.5); although this is a very small opening, it can admit water molecules and can be used to dry gas streams.

The channels in zeolite A run parallel to the three cubic axial directions and are entered by a port of diameter 410 pm, determined by an 8-ring window; this is still considerably smaller than the diameter of the internal cavity, which measures 1140 pm across. The computer model of zeolite A in Figure 7.11 clearly shows the 8-ring windows, the channels running through the structure, and the cavities created by their intersection.

Small-pore zeolites can accommodate linear-chain molecules, such as straight chain hydrocarbons and primary alcohols and amines, but not branched chain molecules. The port size can be enlarged to about 500 pm in diameter by replacing sodium ions with calcium ions. The values for channel and cavity sizes for various zeolites and zeotypes are shown in Table 7.1.

In the mid-1970s, some completely novel **medium-pore zeolite** structures were synthesised, which led to significant new developments. This family of framework structures, comprising the zeolites synthesised by the oil company Mobil, ZSM-5 and ZSM-11 (MEL), silicalite (MFI), and some closely related natural zeolites, have been given the generic name **pentasil**.

ZSM-5 is a catalyst that is now widely used in the industrial world. Its structure is generated from the pentasil unit shown in Figure 7.12 (as are the others of this group). These units link into chains, which join to make layers. The appropriate stacking of these layers gives the various pentasil structures. Both ZSM-5 and ZSM 11 are characterised by channels controlled by 10-ring windows with diameters of about 550 pm. The pore systems in these zeolites do not link big cavities, but they do contain intersections where larger amounts of free space are available for

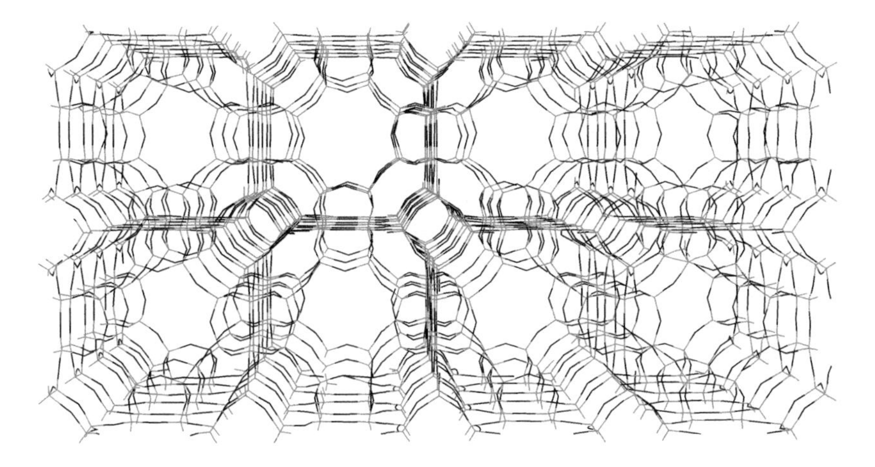

FIGURE 7.11 A computer model of the zeolite A (LTA) structure showing the channels and 8-ring windows.

TABLE 7.1
Window and Cavity Diameters in Zeolites

Zeolite/zeotype	Structure type code	No. of tetrahedra in ring	Window diameter/pm	Cavity diameter/pm
Sodalite	SOD	4	260	600
Zeolite A	LTA	8	410	1140
Erionite-A	ERI	8	360×520	
ZSM-5	MFI	10	510×550	
			540×560	
Faujasite	FAU	12	740	1180
Mordenite	MOR	12	670×700	
			290×570	
Zeolite-L	LTL	12	710	

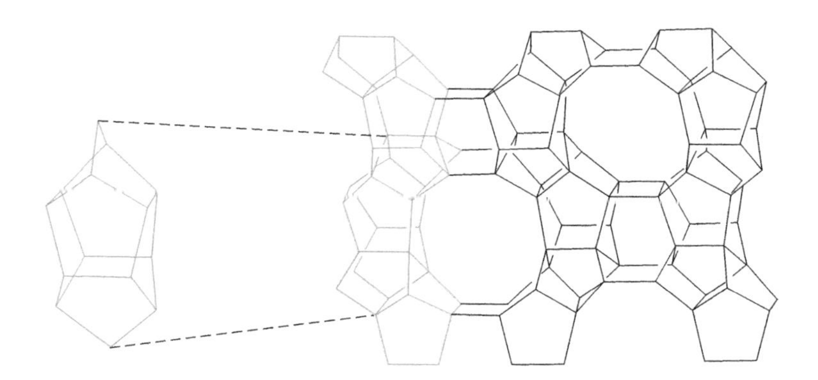

FIGURE 7.12 A pentasil unit together with a slice of the structure of ZSM-5 showing a linked chain of pentasil units (highlighted).

molecular interactions to take place. Figure 7.13a shows the pore system of ZSM-5 with nearly circular zigzag channels intersecting with straight channels of elliptical cross section; this contrasts with the ZSM-11 structure, which just has intersecting straight channels of almost circular cross-section (Figure 7.13b); in both cases, the two-dimensional system of pores is linked by much smaller channels.

Faujasite is a **large-pore zeolite** based on four sodalite cages in a tetrahedral configuration, linked by the oxygen bridges through the 6-ring windows. This leads to a structure with large cavities of diameter 1180 pm entered by 12-ring windows of diameter 740 pm, resulting in a three-dimensional network of channels, as shown in Figure 7.14.

The channel system for **mordenite** (MOR) is shown in Figure 7.15. Mordenite has an orthorhombic structure and there are two types of channels, governed by 8-ring and 12-ring windows, respectively, all running parallel to each other and

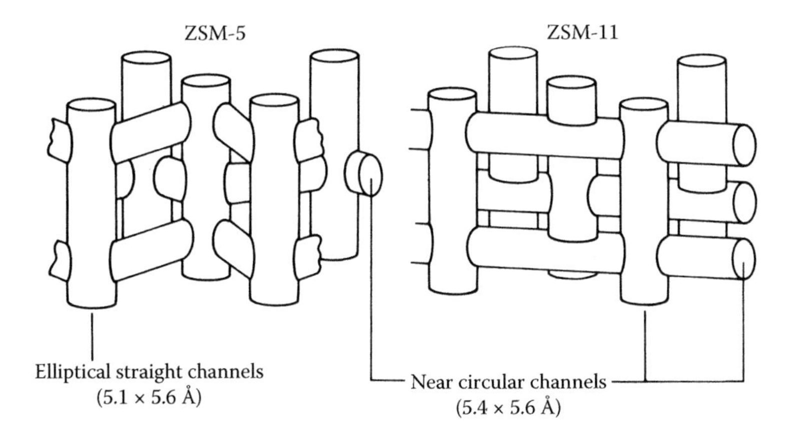

FIGURE 7.13 The interconnecting channel systems in ZSM-5 and ZSM-11.

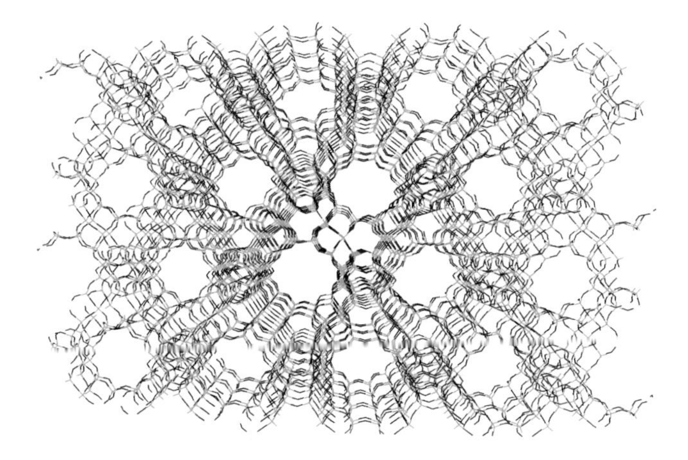

FIGURE 7.14 The cubic faujasite structure showing the channels that lie parallel to each of the face diagonals (oxygen bridges are included in this model).

interconnected only by the smaller 5-ring and 6-ring systems. Interestingly, mordenite synthesised at low temperatures (<260°C) is found to have larger ports than natural mordenite or mordenite synthesised at higher temperatures, and this is due to crystal defects blocking some of the large pores.

7.2.7 Synthesis of Zeolites

It was the pioneering work of Richard Barrer at the University of Aberdeen and Bob Milton at Union Carbide that led the way for the synthesis of zeolites. They were able to prepare zeolites using reactive silica and alumina reagents, such as sodium silicates and aluminates ($[Al(OH)_4]^-$), under hydrothermal conditions, at a high pH obtained by using an alkali metal hydroxide and/or an organic base. A gel forms by a process of copolymerisation of the silicate and aluminate ions. The gel is then heated

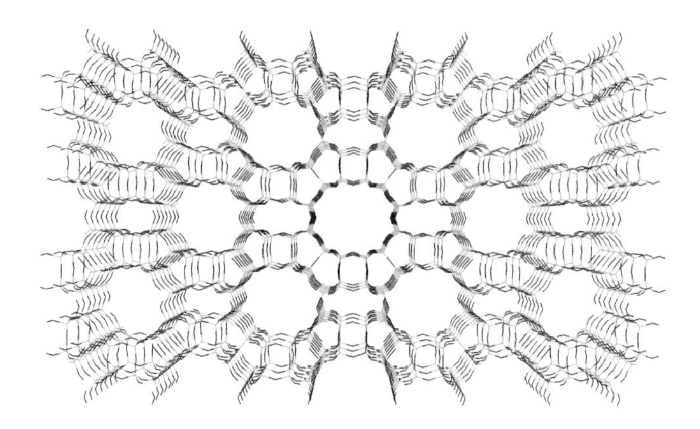

FIGURE 7.15 The channel structure of mordenite showing the larger channels running along the z direction.

gently (60°C–100°C) in an autoclave for several days, producing a condensed zeolite. The product obtained is determined by the synthesis conditions; temperature, time, pH, and mechanical movement are all possible variables. The presence of organic bases is useful for synthesising silicon-rich zeolites. The organic bases ae removed by calcining or washing with water.

The formation of novel, silicon-rich synthetic zeolites has been facilitated by the use of **templates**, such as large quaternary ammonium cations instead of Na^+. For instance, the tetramethylammonium cation ($[(CH_3)_4N]^+$) is used in the synthesis of ZK-4. The aluminosilicate framework condenses around this large cation, which can subsequently be removed by chemical or thermal decomposition.

However calcining at high temperature requires considerable energy input and the emission of carbon dioxide and nitrogen oxides. The organic bases flushed out with water may be toxic. Recently more sustainable syntheses (green chemistry, Section 3.10) have been developed such as template-free or solvent-free processes or the use of a nontoxic template. Use of seed crystals of the desired zeolite avoids the need for a template. This has been demonstrated for a number of zeolites including MFI.

7.2.8 Uses of Zeolites

7.2.8.1 Adsorbents

Zeolite A is a commonly used drying agent, and it can be regenerated by heating after use. Its commercial uses are in drying natural gas, refrigerant gases, and hydrocarbons. Normal crystalline zeolites contain water molecules that are coordinated to the exchangeable cations. As we noted previously, these structures can be dehydrated by heating under vacuum; in these circumstances, the cations move position, frequently settling on sites with a much lower coordination number. The dehydrated zeolites are very good drying agents, absorbing water to get back to the preferred high coordination condition.

The cations M^{n+} in a zeolite will exchange with others in a surrounding solution. In this way, the Na^+ form of zeolite A can be used as a water softener: the

Na^+ ions exchange with the Ca^{2+} ions from the hard water. When used as a water softener, the zeolite is reusable because it can be regenerated by running through a very pure saline solution; this is a familiar procedure to anyone who has used a dishwasher.

Some zeolites have a strong affinity for particular cations. Clinoptilolite (HEU) is a naturally occurring zeolite that sequesters caesium and is used by British Nuclear Fuels (BNFL) to remove ^{137}Cs from radioactive waste, exchanging its own Na^+ ions for the radioactive Cs^+ cations. Similarly, zeolite A can be used for recovering radioactive strontium. Zeolites were heavily used in the clean-up operations after the Chernobyl and Three Mile Island incidents.

As dehydrated zeolites have very open, porous structures, they have large internal surface areas and are capable of adsorbing large amounts of substances other than water. The ring sizes of the windows leading into the cavities determine the sizes of the molecules that can be adsorbed. An individual zeolite has a highly specific sieving ability, which can be exploited for purification or separation. This was first noted for chabazite (CHA) as long ago as 1932, when it was observed that it would adsorb and retain small molecules such as formic acid and methanol, but would not adsorb benzene and larger molecules. CHA has been used commercially to adsorb polluting SO_2 emissions from chimneys. Similarly, the 410-pm pore opening in zeolite A (determined by an 8-ring and much smaller than the 1140-pm diameter of the cavity) can admit a methane molecule, but excludes a larger benzene molecule.

Zeolites have been extensively studied for carbon capture (that is, the removal of carbon dioxide produced as waste by industrial processes or from the air). Al-rich zeolites have a higher adsorption capacity for CO_2 due to their basicity.

7.2.8.2 Catalysts

Zeolites are very useful catalysts, displaying several important properties that are not found in traditional amorphous catalysts. The amorphous catalysts have always been prepared in a highly divided state in order to give a high surface area and thus a large number of catalytic sites. The presence of the cavities in zeolites provides a very large internal surface area that can accommodate as many as 100 times more molecules than the equivalent amount of amorphous catalysts. Zeolites are also crystalline and can be prepared with improved reproducibility, so they tend not to show the varying catalytic activity of the amorphous catalysts. Furthermore, their molecular sieve action can be exploited to control which molecules have access to (or which molecules can depart from) the active sites. This is generally known as **shape-selective catalysis**. The introduction of zeolite catalysts have also made a significant difference in protecting the environment, as the liquid catalysts they replaced presented industrial pollution problems.

The catalytic activity of decationised zeolites is attributed to the presence of acidic sites arising from the $[AlO_4]$ tetrahedral units in the framework. These acid sites may be Brønsted or Lewis in character. Zeolites that are normally synthesised usually have Na^+ ions balancing the framework charges, but these can be readily exchanged for protons by a direct reaction with an acid, giving surface Si–OH and Al–OH hydroxyl groups—the **Brønsted sites**. Alternatively, if the zeolite is not stable in an acid solution, it is common to form the ammonium (NH_4^+) salt, and then heat it so that the ammonia is driven off, leaving a proton. The oxygen of the hydroxyl was thought to be three-coordinate, forming a bridge between Si and Al. Further heating removes water from the Brønsted site, exposing a tricoordinated aluminium ion which has electron-pair acceptor properties to hydroxyl groups on adjacent silicon atoms; this is identified as a **Lewis acid site**. A scheme for the formation of these sites is shown in Figure 7.16. The surfaces of the zeolites can thus display either Brønsted or Lewis acid sites, or both, depending on how the zeolite is prepared. Brønsted sites are converted into Lewis sites as the temperature is increased above 600°C and water is driven off.

Not all zeolite catalysts are used in the decationised or acid form; it is also quite common to replace the Na^+ ions with lanthanoid ions such as La^{3+} or Ce^{3+}. These ions now place themselves so that they can best neutralise the three separated negative charges on the tetrahedral Al in the framework. The separation of the charges causes high electrostatic field gradients in the cavities which are sufficiently large to polarise the C–H bonds or even to ionise them, enabling a reaction to take place. This effect can be strengthened by a reduction in the aluminium content of the zeolite so that the [AlO_4] tetrahedra are farther apart. If one thinks of a zeolite as a solid ionising solvent, the difference in the catalytic performances of various zeolites can be likened to the behaviour of different solvents in solution chemistry. A third way of using zeolites as catalysts is to replace the Na^+ ions with other metal ions, such as Ni^{2+}, Pd^{2+}, or Pt^{2+}, and then reduce them in situ so that the metal atoms are deposited within the framework. The resultant material displays the properties associated with a supported metal catalyst and extremely high dispersions of the metal can be achieved.

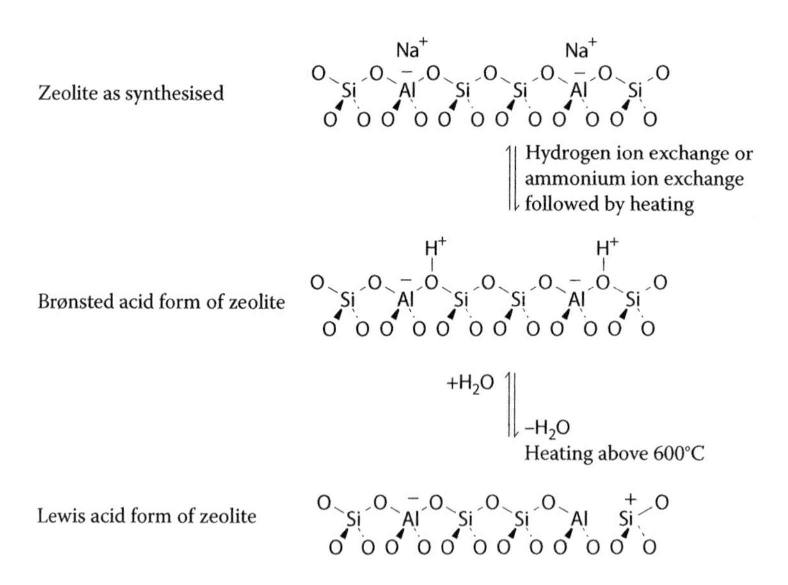

FIGURE 7.16 A scheme for the generation of Brønsted and Lewis acid sites in zeolites.

There are several types of shape-selective catalysis:

- **Reactant shape-selective catalysis**: Only molecules with dimensions less than a critical size can enter the pores and reach the catalytic sites, so they react there. This is illustrated diagrammatically in Figure 7.17a, in which a straight-chain hydrocarbon is able to enter the pore and react, but the branched-chain hydrocarbon is not able to do so.
- **Product shape-selective catalysis**: Only products less than a certain dimension can leave the active sites and diffuse out through the channels, as illustrated in Figure 7.17b for the preparation of xylene. A mixture of all three isomers is formed in the cavities, but only the para form is able to escape.
- **Transition-state shape-selective catalysis**: Certain reactions are prevented because the transition state requires more space than is available in the cavities, as shown in Figure 7.17c, for the transalkylation of dialkylbenzenes.

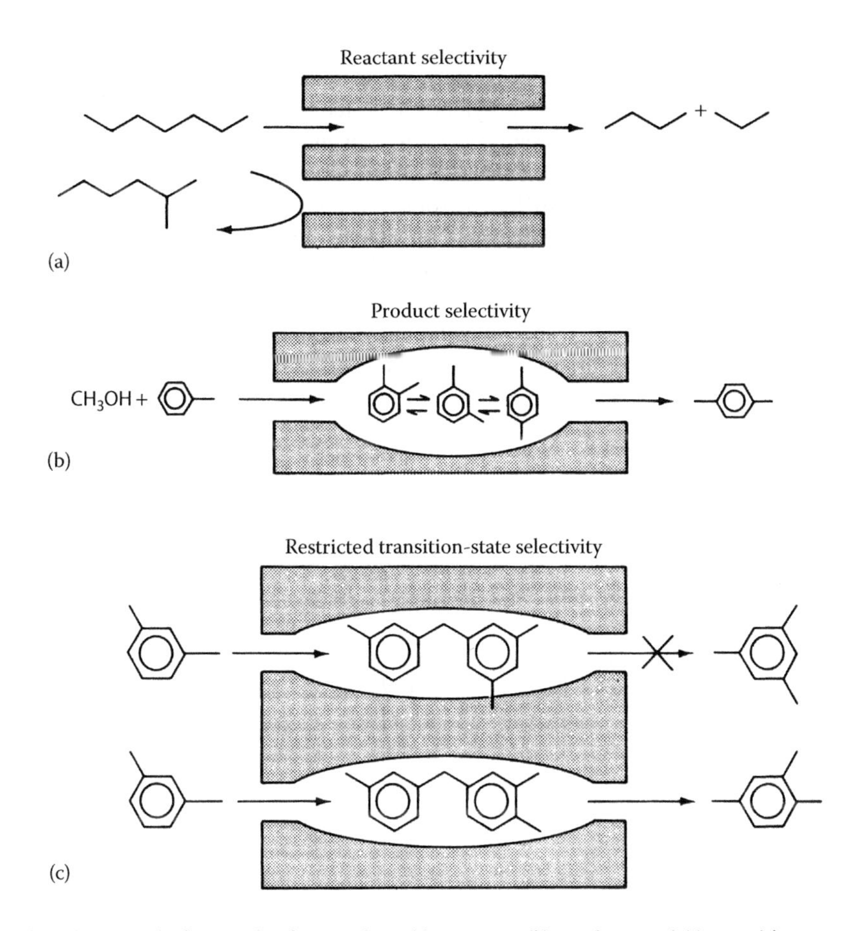

FIGURE 7.17 A shape-selective catalyst: (a) reactant, (b) product, and (c) transition state.

- **Reactant shape-selective catalysis** is demonstrated in the dehydration of butanols. If butanol (*n*-butanol) and butan-2-ol (*iso*-butanol) are dehydrated either over Ca-zeolite A or over Ca-zeolite X, we see a difference in the products formed.

Zeolite X has windows large enough to admit both of the alcohols easily, and both undergo conversion to the corresponding alkene. Over zeolite A, however, the dehydration of the straight chain alcohol is straightforward, but virtually none of the branched-chain alcohols are converted, as these are too large to pass through the smaller windows of zeolite A. These results are summarised in Figure 7.18. Notice that, at higher temperatures, curve d begins to rise. This is because the lattice vibrations increase with temperature, making the pore opening slightly larger and thus beginning to admit butan-2-ol. The very slight conversion at lower temperatures is thought to take place on external sites.

One of the industrial processes using ZSM-5 provides us with an example of **product shape-selective catalysis**: the production of 1,4-(para-)xylene. Para-xylene is used in the manufacture of terephthalic acid, the starting material for the production of polyester fibres such as 'Terylene'.

Xylenes are produced in the alkylation of toluene by methanol (Figure 7.19a). When ZSM-5 catalyst is impregnated with phosphoric acid (to produce the acid catalyst), the reaction is selective because of the difference in the rates of diffusion through the channels, of the different isomers produced. This is confirmed by the observation that selectivity increases with increasing temperature, indicating the increasing importance of diffusion limitation. The diffusion rate of para-xylene is approximately 1000 times faster than that of the other two isomers. The computer models in Figure 7.16b and c show why meta-xylene cannot enter a 10-ring window easily and hence cannot diffuse along the channels. The xylenes isomerise within

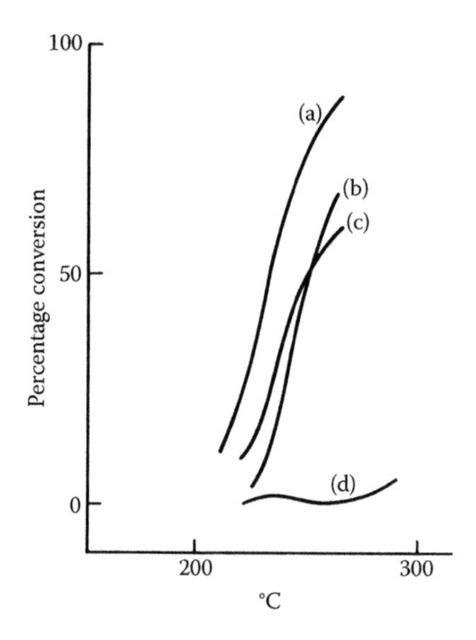

FIGURE 7.18 Dehydration of (a) 2-methylpropanol on Ca-X, (b) butanol on Ca-X, (c) butanol on Ca-A, and (d) 2-methylpropanol on Ca-A.

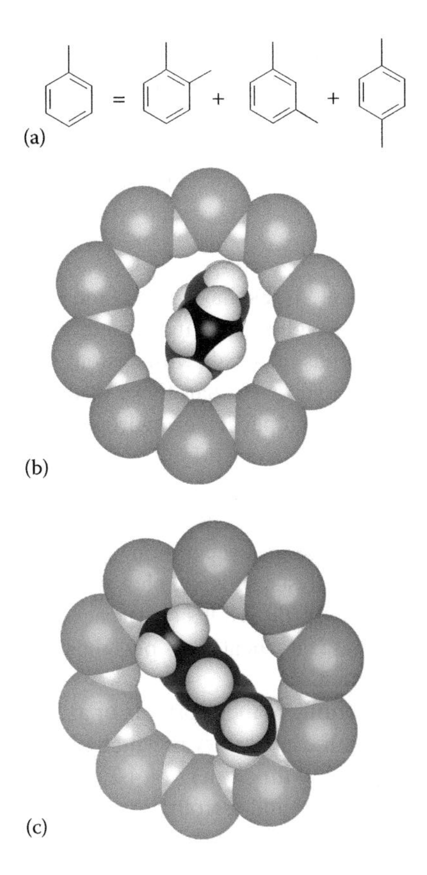

FIGURE 7.19 (a) The structures of ortho-xylene, meta-xylene, and para-xylene. Computer models showing how (b) the para-xylene fits neatly into the pores of ZSM-5 whereas (c) meta-xylene is too large to diffuse through.

the pores, so para-xylene diffuses out while the ortho-isomers and meta-isomers are trapped and take more time to convert to the para form before escaping. Up to 97% of selective conversion to para-xylene has been achieved by suitable treatment of this catalyst. X-ray studies have demonstrated that the para-xylene molecules are located both in the sinusoidal channels and at the junction of the straight and sinusoidal channels. Using a zeolite catalyst makes this a much greener process than using the conventional alkylation reaction using CH_3I over an $AlCl_3$ or $FeCl_3$ catalyst, as a pure product is produced in good yield, and the zeolite catalyst can be regenerated.

Transition-state shape-selective catalysis occurs in the acid-catalysed transalkylation of dialkylbenzenes, where one of the alkyl groups is transferred from one molecule to another. This bimolecular reaction involves a diphenylbenzene transition state. When the transition state collapses, it can split to give either the 1,2,4-isomer or the 1,3,5-isomer, along with the monoalkylbenzene. When the catalyst used for this reaction is mordenite (Figure 7.17c), the transition state for the formation of the symmetrical 1,3,5-isomer is too large for the pores, and the 1,2,4-isomer is formed in almost 100% yield. (This compares with the equilibrium mixtures, in which the symmetrically substituted isomers tend to dominate.)

The synthesis of the open-framework zeolites dramatically improved the number of accessible active sites for catalysis. It is estimated that ZSM-5 has a turnover up to 10^7 molecules per active site per minute for xylene isomerisation.

7.3 METAL ORGANIC FRAMEWORKS

Metal Organic Frameworks (MOFs) have only come to prominence in the twenty-first century but by 2020 over eighty thousand structures of MOFs had been deposited. They form a class of porous materials with customisable adaptable frameworks with surface areas that are larger even than those of the zeolites. They can absorb large volumes of various gases, in particular hydrogen, and can selectively capture carbon dioxide. With the need for a clean fuel, such as hydrogen, to be used in fuel cells and the rapid increase of carbon dioxide in the atmosphere, this makes the storage and capture properties of MOFs of great interest. Most of the earlier research concentrated on their gas storage, gas separation, and gas purification abilities, but they also show promise as catalysts, sensors, photocatalysts for water splitting, and even for drug delivery. However, the MOFs synthesized so far are usually unstable with respect to water, acid/base, and heat. This plus the current cost has slowed progress in commercial exploitation.

MOFs can be named according to their formula or as MOF followed by a number. They can also be named after the laboratory in which they were first synthesized. For example, $Zn_4O(bdc)_3$ is a MOF with clusters of 4 Zn ions linked by 3 bdc^{2-} (where bdc^{2-} = 1,4-benzenedicarboxylate) and surrounding an O. It is also known as MOF-5. $Cu_3(btc)_2$ (btc^{3-} = 1,3,5-benzenetricarboxylate) is also known as HKUST-1 after the Hong Kong University of Science and Technology.

7.3.1 Composition and Structure of MOFs

MOFs basically consist of inorganic metal ions or metal atom clusters, joined together by (usually rigid) organic linker groups. A few examples of the types of organic ligands that are being used as linkers are given in Table 7.2; most commonly, they are derivatives of carboxylic acids and azoles.

Figure 7.20 shows the structure of six MOFs and their linker molecules.

The structures are diverse and can be 'tuned' to produce different pore sizes by choosing different metal cluster precursors and/or organic linkers. While most have pores in the microporous range, mesoporous and macroporous pores have been achieved. For example, a set of one type of MOF (MOF-74) was generated by increasing the length of the organic ligand. A microporous MOF-74 could accommodate small molecules such as CO_2, whereas a macroporous MOF-74 could accommodate small proteins.

7.3.2 Synthesis of MOFs

Many diverse syntheses of MOFs have been developed. Initially there were two main approaches, one of which was to assemble 'building blocks' of metal clusters and

TABLE 7.2
Some Ligands Used in MOF Synthesis

Common name of parent	IUPAC name	Formula anion	Ligand abbreviation	MOF example and common name	Structure anion
Succinic acid	Butanedioate	$^{-}OOC{-}(CH_2)_2{-}COO^{-}$ $C_4H_4O_4$			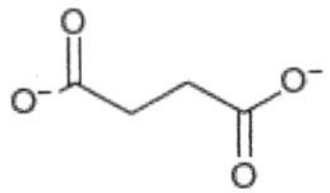
Isophthalic acid	Benzene-1,3-dicarboxylate	$C_6H_4(COO)_2^{2-}$ $C_8H_4O_4$			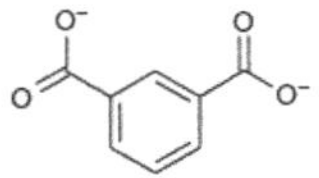
Terephthalic acid	Benzene-1,4-dicarboxylate	$C_6H_4(COO)_2^{2-}$ $C_8H_4O_4$	BDC	$Zn_4O(BDC)_3$ MOF-5	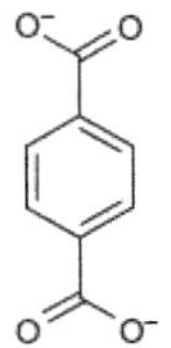
1,3,5-Tris(4-carboxyphenyl) benzene	Benzene-1,3,5-tribenzoate	$C_6H_3(C_6H_4COO)_3^{3-}$ $C_{27}H_{18}O_6$	BTB	$Zn_4O(BTB)_2$ MOF-177	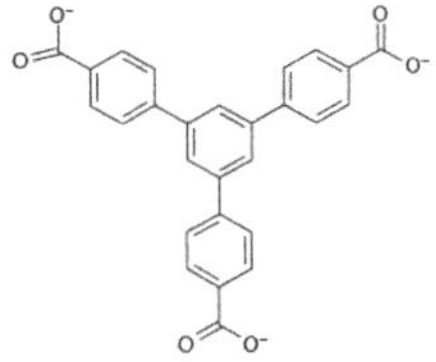

(Continued)

TABLE 7.2 (CONTINUED)
Some Ligands Used in MOF Synthesis

Common name of parent	IUPAC name	Formula anion	Ligand abbreviation	MOF example and common name	Structure anion
Trimesic acid	Benzene-1,3,5-tricarboxylate	$C_9H_3O_6$	BTC	$Cu_3(BTC)_2(H_2O)_3$ HKUST-1	
Imidazole 1,3-diazole	1*H*-1,3-Imidazole	$C_3H_4N_2$			
1,2,3-Triazole	1*H*-1,2,3-Triazole	$C_2H_3N_3$			
1,2,3,4-Tetrazole	1*H*-1,2,3,4-Tetrazole	CH_2N_4			
	Benzene-1,3,5-tris(1*H*-tetrazole)	$C_9H_6N_{12}$	BTT	$Mn_3[(Mn_4Cl)_3(BTT)_8]_2$ Mn-BTT	

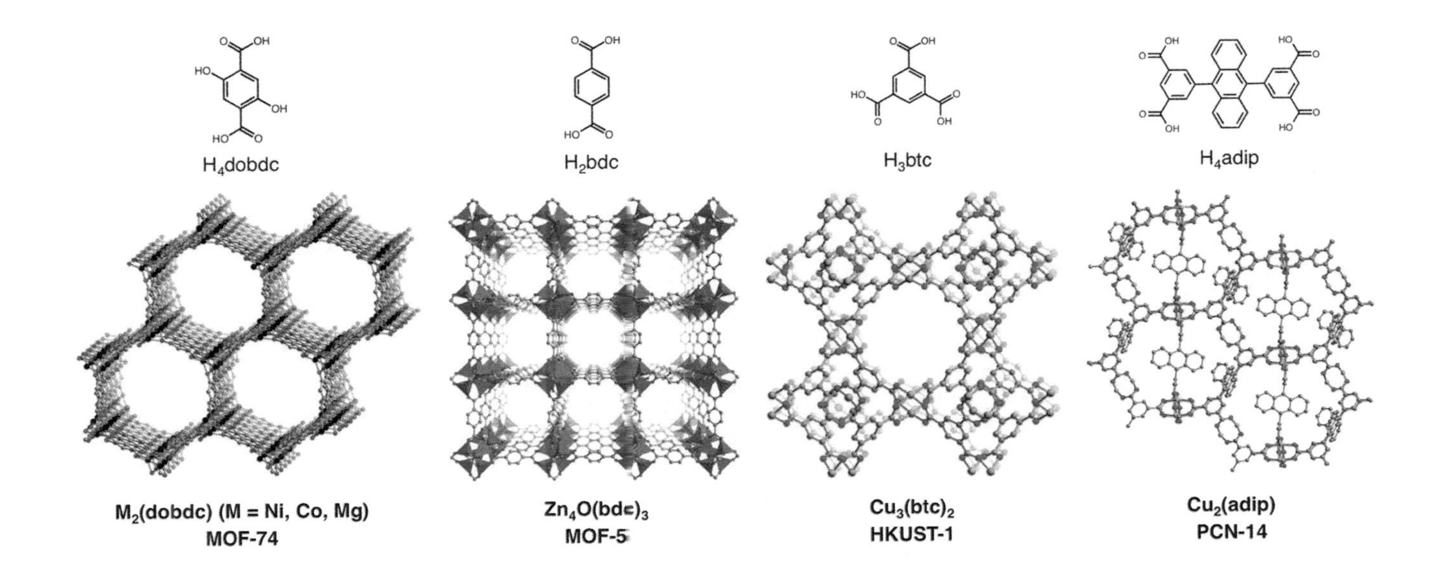

FIGURE 7.20 Crystal structures and organic bridging ligands for the six metal–organic frameworks evaluated in this work: M_2(dobdc) (M = Ni, Co, Mg; dobdc^{4-} = 2,5-dioxido-1,4-benzenedicarboxylate; M-MOF-74, CPO-27-M), $Zn_4O(bdc)_3$ (bdc^{2-} = 1,4-benzenedicarboxylate; MOF-5, IRMOF-1), $Cu_3(btc)_2$ (btc^{3-} = 1,3,5-benzenetricarboxylate; HKUST-1), and Cu_2(adip) (adip^{4-} = 5,5′-(9,10-anthracenediyl)di-isophthalate; PCN-14). Green, grey, and red spheres represent Cu, C, and O atoms, respectively; H atoms have been omitted for clarity. Black spheres represent Ni, Co, or Mg atoms, and blue tetrahedra represent Zn atoms. (Figure 2 in J.A. Mason, M. Veenstra and Jeffrey R. Long, Evaluating Metal–Organic Frameworks for Natural Gas Storage, *Chem. Sci.*, 2014, 5, 32. DOI: 10.1039/c3sc52633j. Licensed under a Creative Commons Attribution 3.0 unported licence.)

organic linkers. Table 7.2 gave examples of organic linkers. Metal cluster building blocks, known as **secondary building units** (SBU), contain two or more metal ions linked by several ligands. A common type of SBU is a 'paddle wheel'. Figure 7.21 shows an SBU with this form. This one has four 'paddles' consisting of acetate ligands which coordinate to both Cu atoms. The linking ligands can be more complex and there may be other ligands present attached to a single metal centre.

Alternatively, similar methods to those used for zeolites were employed, the crystals being grown hydrothermally or solvothermally from a hot solution of the organic ligand and a metal precursor such as nitrate. These latter generally involved high temperatures.

Different methods and conditions can produce different MOFs from the same reaction mixture. For example a study of the reaction mixture $Al(NO_3)_3/9H_2O/H_4BTEC/H_2O/NaOH$ at different starting pH values showed that low pH favoured corner-sharing AlO_6 polyhedra, whereas higher pH favoured edge-sharing. Increasing the pH led to a higher degree of deprotonation of H_4BTEC (H_4BTEC = 1,2,4,5-benzene-tetracarboxylic acid). The ligand after deprotonation at pH 1.4 and pH 9 is shown in Figure 7.22. At pH 1.4 there are 4 sites to which Al ions can attach, and this leads to the MOF MIL (Matérial Institut Lavoisier) -121 with corner-sharing AlO_6. At pH 9 there are 8 Al coordination sites, giving MIL-120 with edge-sharing AlO_6.

Different synthesis types are also needed to produce crystals of particular sizes and/or morphology. Templating techniques can also be used to control the pore size and to give access to metal sites that are exposed when a templating solvent such as *N,N*-dimethylformamide or *N,N* diethylformamide is removed; it is these metal sites that then provide sites for gases, particularly hydrogen, to adsorb. Ultrasound- and microwave-assisted syntheses have been used to produce MOFs, for example, for

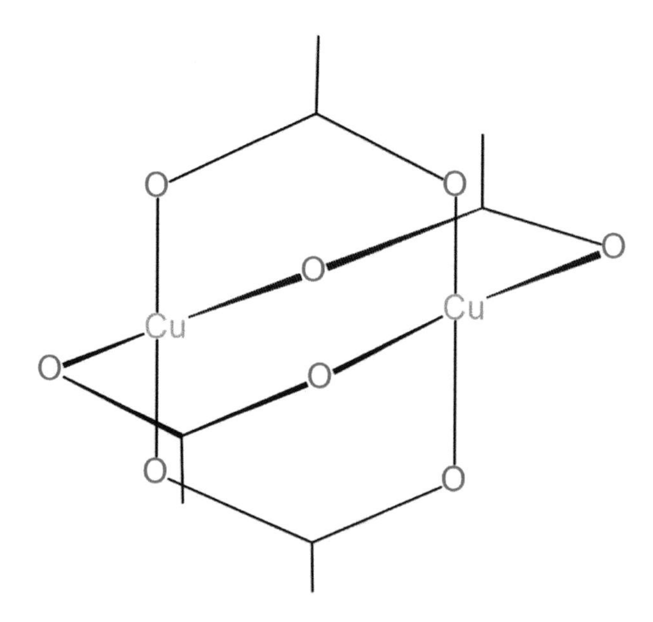

FIGURE 7.21 A simple secondary building unit (SBU) of paddle wheel form.

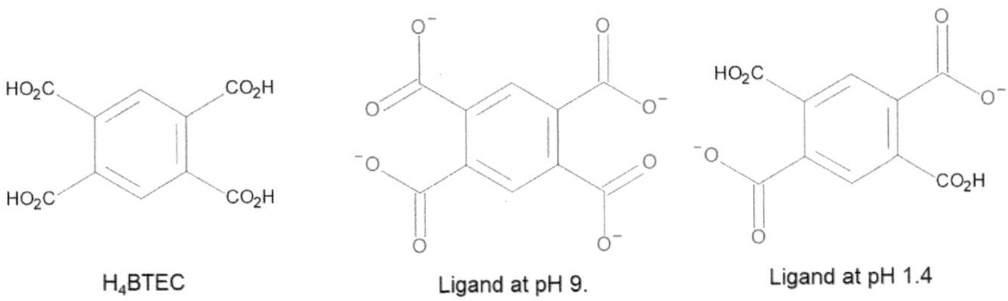

FIGURE 7.22 The structure of (a) H_4BTEC and its deprotonated forms at (b) pH 9 and at (c) pH 1.4.

potential biomedical use. Such MOFs need to be nontoxic and nanocrystalline and have high loading capacities.

7.3.3 Uses of MOFs

7.3.3.1 Storage and Separation

The production of carbon dioxide by fossil fuel burning has produced an increased concentration of this greenhouse gas in the atmosphere and is of worldwide concern. MOFs have been investigated for **carbon capture**. The most-studied MOFs for CO_2 absorption have the metals Mg, Zn, or Co in their structure. Performance can be enhanced by tuning the MOF by altering functional groups, pore size, and the presence of open metal sites. A particularly high-performing MOF with open Mg^{2+} sites is Mg-MOF-74 or Mg_2DOT (DOT=2,5-dioxodoterephthalate). This has a CO_2 storage capacity of 32.5 wt% at room temperature and atmospheric pressure. Performance of this MOF can be increased by incorporating hydrazine into its structure. At the partial pressure of CO_2 in the atmosphere the capacity was increased ~40-fold.

For carbon capture, it is also important that CO_2 is absorbed in preference to other small molecules such as N_2 and CH_4. NJU-Bai35 ($[Cu_4(\mu_4\text{-}O)Cl_2(IN)_{4]}][CuCl_2]$) has an outstanding specificity of 275.8 CO_2:N_2 (NJU-Bai stands for Nanjing University Bai group).

Currently, hydrogen is usually stored and transported as a liquid or compressed gas. It is not only the extra cost of hydrogen-fuelled vehicles that prevents more sales, but also the public perception of safety with general concern about using hydrogen as a fuel. A solid containing a very high volume of adsorbed hydrogen that can be easily and safely released becomes a very attractive option. Hydrogen physisorbs to the surface of an MOF with weak van der Waals forces, and thus the uptake increases with the surface area; it can also form stronger bonds by chemisorption. Usually, adsorption of hydrogen increases with decreasing temperature and increasing pressure up to about 20–40 bar (up to 100 bar is deemed safe for automotive uses). This means that MOFs are most efficient at hydrogen storage at low temperatures, mostly operating at liquid nitrogen temperatures (77 K), and finding MOFs that are efficient at room temperature is a very active area of research. For onboard storage of hydrogen for vehicles, the US Department of Energy set a goal of 4.5 wt%

hydrogen by 2020. However by 2019, the highest room temperature storage system was $Be_{12}(OH)_{12}$(1,3,5-benzenetribenzoate)$_4$ with 2.3 wt% under a pressure of 95 bar.

Methane produces lower CO_2 emissions than liquid hydrocarbons such as petrol. MOFs are better than zeolites at methane storage, and Table 7.3 shows some of the best-performing (2019) MOFs for methane storage. The formulas are complicated but you should note that these MOFs contain clusters of metal ions linked by aromatic ligands.

The pore sizes in an MOF dictate which gaseous molecules are able to diffuse through them, making them useful candidates for **gas separation**. Zn-based MOFs that retain CO_2 but allow through O_2 and N_2 have been made, whereas others will retain water but keep out larger molecules such as N_2, O_2, CO_2, and CH_4.

The metal centres in MOFs, in particular open metal sites, can act as Lewis acids to bases such as amines, sulfur-containing compounds, phosphines, etc. This makes them potentially useful for the clean-up of trace contaminants of these compounds.

7.3.3.2 Heterogeneous Catalysis

Heterogeneous catalysis is very important to industry, with about nine out of ten industrial processes using such catalysts, overall worth tens of millions of pounds to global economies. Catalysis can take place at three types of sites—metal centres, organic linkers, and pores. The pore structure enables size/shape-selective catalysis as in zeolites. In addition catalytically active molecules can be inserted into the pores. However, an industrial catalyst has to be robust enough to withstand the temperature and pressure conditions of the reaction, and to be a viable economic concern, it must also cycle reliably hundreds of times.

TABLE 7.3

Adsorption of Methane by MOFs at Standard Temperature and Pressure

MOF	Formula	Total CH_4 adsorption m^3/kg
DUT-49	$Cu_{2(}$bbcdc)[a]	0.363
NU-111	Cu_2(tcepbb)[b]	0.333
PCN-68	Cu_2(ptei)[c]	0.332
MOF-210	$Zn_4O(bbc)_2$[d]	0.331
DUT-23(Co)	$Co_2(bipy)_3)(btb)_4$[e]	0.331

[a] bbcdc=9,90-([1,10-biphenyl]-4,40-diyl)bis(9Hcarbazole-3,6-dicarboxylate); [b] tcepbb=1,3,5-tris[(1,3-carboxylate-5-(4-(ethynyl) phenyl))butadiynyl]-benzene; [c] ptei=5,50-((50-(4-((3,5-dicarboxy phenyl)ethynyl)phenyl)-[1,10,30,100-terphenyl]-4,40 0-diyl)-bis(ethyne-2,1-diyl))diisophthalic acid; [d] bbc=4,40,400-[benzene-1,3,5-triyl-tris(benzene-4,1-diyl)]tribenzoate; [e] btb=benzene-1,3, 5-tribenzoate.

7.3.3.3 Other Applications

MOFs have potential as chemical sensors for a variety of molecules including biomolecules, metal ions, explosives, and toxins. Guest molecules can change the luminosity of MOFs and this has been used as a sensing method for a variety of molecules. One example was distinguishing between normal and mismatched DNA with the MOF UiO-66-NH_2 ($Zr_6O_4(OH)_4(BDCNH_2)_6$).

Use of MOFs as electrodes in Li-ion and Na-ion batteries (Chapter 6) has been investigated. Li ions can be incorporated in MOFs either by a chemical reaction forming a lithium alloy or by intercalation. The formation of lithium alloys can cause the framework to collapse and thus limit the number of charge/discharge cycles that can be achieved. A greater number of cycles can be achieved when the lithium ions intercalate. MOFs with N-rich functional groups are particularly effective.

Drug delivery is another possible use of the storage ability of MOFs, with the possibility of the slow release of drugs.

7.3.4 Zeolite-like MOFs

Zeolite-like MOFs (**ZMOF**s) are a class of MOFs with zeolite topology. They have 4-connected nodes, for example tetrahedrally connected metal ions and 2-connected linkers, the nodes acting as Si atoms in zeolites and the linkers as linking oxygen. A famous set of ZMOFs are the **Zeolitic Imididazolate Frameworks** (**ZIF**s). One of the major problems facing research into MOF catalysts has been their lack of stability at high temperatures compared with zeolites. Stability up to >500°C and for a pH range of 2–14 has been found for some benzene-1,3,5-tris(1*H*-imidazole) MOFs. (Zn, Cu, Co, Ni) ZIF-7 ($Zn(bim)_2$), where bim = benximidazole is chemically stable in benzene, water, and methanol and thermally stable up to about 400°C depending on the morphology of the crystal. ZIF-8 ($Zn(2\text{-mim})_2$ where 2-mim is 2-methylimidazole) is of high thermal and chemical stability and has large pore sizes. It can be prepared simply by mixing solutions of a Zn salt (e.g. $Zn(NO_3)_2$) and 2-methylimidazole. Varying the reaction conditions produces particles of different sizes. The structures of ZIF-7 and ZIF-8 are shown in Figure 7.23.

ZIFs also show potential as anodes in Li-ion batteries (Chapter 6) due to the Li ions binding to the imidazole groups.

A subset of ZMOFs use metal clusters such as Cu_4X_4, where X is a halogen, as the 4-connected nodes to form porous **Zeolite-like Cluster Organic Frameworks** (**ZCOF**s).

7.4 COVALENT ORGANIC FRAMEWORKS

Covalent Organic Frameworks (COFs) were first reported only in 2005, but the number of papers has increased rapidly year by year. Potential uses include heterogeneous catalysis, environmental remediation, gas storage and separation, and sensors. They are porous solids composed of covalently bonded light atoms (C, H, N, O, B) formed by linking organic building blocks. The chemical and thermal stability of these materials due to their covalent bonding and the ready availability (see Chapter 12) of these elements (apart from boron) make them particularly attractive.

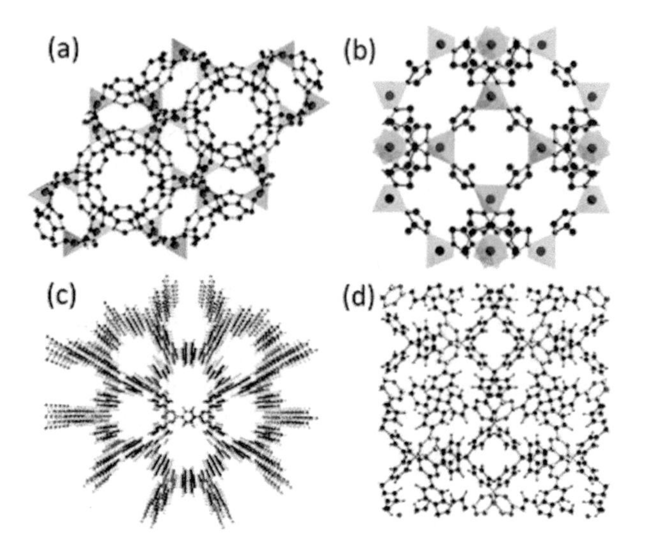

FIGURE 7.23 Crystal structures of investigated ultramicroporous frameworks for quantum sieving: (a) ZIF-7, (b) ZIF-8, (c) COF-1, and (d) COF-102. In the case of ZIF-7 and ZIF-8, hydrogen atoms are omitted for clarity. Carbon (black), zinc (blue tetrahedral), nitrogen (green), oxygen (red), boron (yellow), and hydrogen (grey). (Republished with permission of Royal Society of Chemistry from Quantum cryo-sieving for hydrogen isotope separation in microporous frameworks: an experimental study on the correlation between effective quantum sieving and pore size, Michael Hirscher et al., *J. Mater. Chem. A*, 2013, 1, 3244; permission conveyed through Copyright Clearance Center, Inc.)

7.4.1 Structure of COFs

COFs can be divided into 2-D and 3-D structures. In 2-D structures, a two-dimensional covalently bonded layer is stacked so that periodic columns are formed. The bonding between layers is due to π-π bonding. 3-D structures are covalently bonded throughout. Two COFs are shown in Figure 7.23(c) COF-1 (2D) and (d) COF-102 (3D). A number of linkages forming covalent bonds in COFs have been reported, particularly boron linkages, imine linkages, and triazine linkages. Simple examples of these are shown in Figure 7.24.

COFs are mostly microporous but some with pore sizes in the mesoporous range have been synthesized.

7.4.2 Synthesis of COFs

The most common synthetic method is solvothermal synthesis. Microwave-assisted synthesis reduces the reaction time. For example COF-5 has been synthesized by this method in 20 minutes compared to 3 days for a solvothermal synthesis using a mixture of 1,3,5-trimethylbenzene and dioxane as solvent. Some COFs have been prepared under ambient conditions. These generally involve making a solution of the reactants, mixing, and allowing to crystallise, but even grinding the reacting solids in a mortar and pestle has been reported.

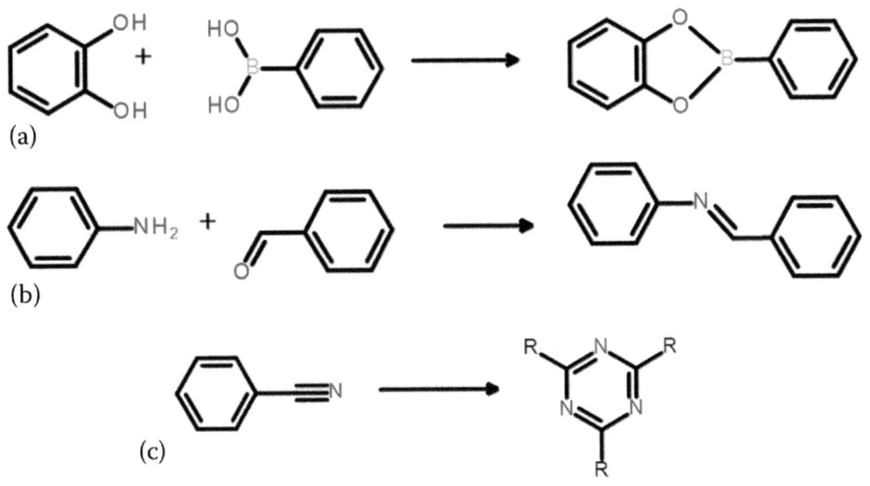

FIGURE 7.24 (a) boron linkage, (b) imine linkage, and (c) triazene linkage.

7.4.3 Uses of COFs

COFs have potential use in sensors. Many COFs are fluorescent and the fluorescence can be modified or quenched by molecules and ions absorbed by the material. This property can be used to detect explosives, heavy metal ions, and some gases, for example NH_3.

Their absorbent properties can be used to remove toxic pollutants, both organic and inorganic, from the environment, to store and separate gases, and as photocatalysts. The lower density of COFs means that if all the pores are filled, they will have a higher weight % of pollutant or gases such as CO_2 than MOFs.

7.5 OTHER POROUS SOLIDS

7.5.1 Mesoporous Aluminosilicates

You saw in Section 7.2 that zeolites were aluminosilicates, but these are not the only porous solids of this type.

In 1992, scientists at Mobil Research and Development Corporation developed a family of silicate and aluminosilicate materials (M41S) that had pores in the mesoporous range. The mesoporous material that has received most attention so far is MCM-41 (Mobil Crystalline Materials). It has a highly ordered hexagonal array of uniformly sized mesopores and can have a huge surface area of 1200 $m^2\ g^{-1}$. It is made by a templating technique, where the silicate or aluminosilicate walls of the mesopores, instead of forming around a single molecule or ion, form around an assembly of molecules known as a micelle. In a solution of silicate or silicate and aluminate anions, cationic long-chain alkyl trimethylammonium surfactants ($[CH_3(CH_2)_n(CH_3)_3N^+]\ X^-$) form rod-like micelles (Figure 7.25), with the hydrophobic tails clustering together inside the rods and the cationic heads forming the outside; these silicate/aluminate ions form a cladding around the micelles. The silicate-coated

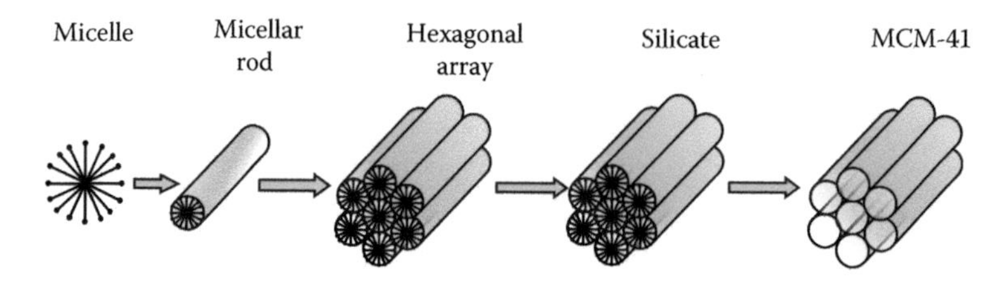

FIGURE 7.25 A possible preparative route for the formation of MCM-21 by liquid crystal templating.

micelles pack together along the axes of the rods, earning the synthesis technique the name **liquid crystal templating**; under hydrothermal conditions, this mesoporous structure precipitates out of the solution. The calcination of the filtered solid in air at temperatures up to 700°C removes the template and produces the mesoporous solid. As we would expect, the alkyl chain length determines the size of the pores; where $n = 11$, 13, and 15, the pore diameters are 300, 340, and 380 nm, respectively.

7.5.2 Clays

Another type are the **smectite clays**. Clays are often found as components of soils and sediments, and the Mars Rover has even found clays on Mars. The smectite clays have a basic layer structure, which is shown in Figure 7.26; it consists of parallel

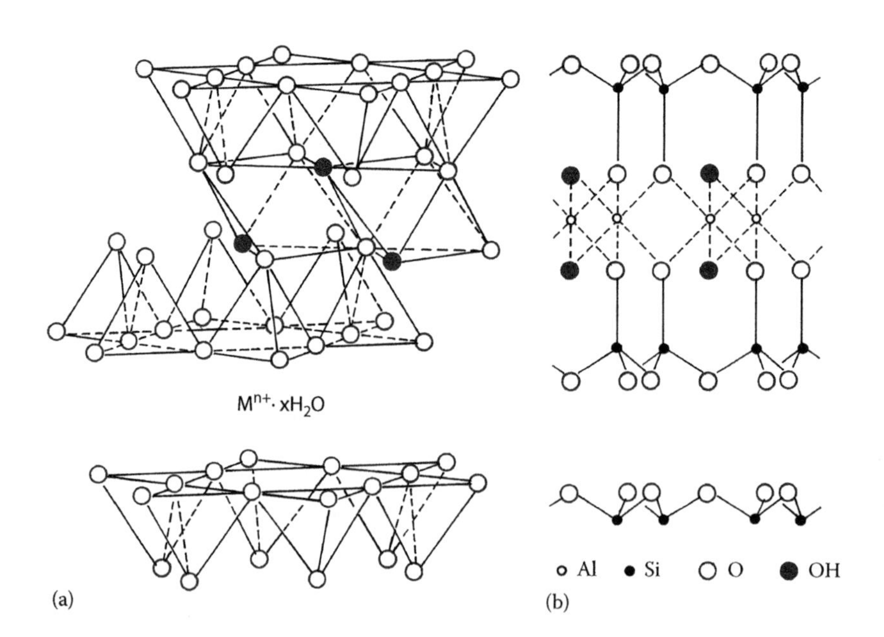

FIGURE 7.26 Idealised structure of the layers in a smectite clay, (a) showing only the oxygen/hydroxyl framework and (b) also showing the aluminum and silicon positions.

layers of tetrahedral silicate [SiO_4] sheets and octahedral aluminate [$Al(O,OH)_6$] sheets.

The silicate layers contain [SiO_4] tetrahedra, each linked through three corners to form an infinite layer. The aluminate layers contain a plane of octahedrally coordinated edge-linked aluminium ions, sandwiched between two inward-pointing sheets of corner-linked [SiO_4] tetrahedra. This three-sheet layer structure is typical of smectite clays.

If there is no replacement of Si or Al, then the layers are electrically neutral, producing the mineral pyrophyllite. In the smectite clays, different structures are formed because the substitution of silicon and aluminium by metal ions can take place in both the tetrahedral and the octahedral layers. The resulting negative charge is distributed on the oxygens of the layer surface, and any charge balance is restored by interlayer cations (usually Na^+ or Ca^{2+}); for instance, in montmorillonite, approximately one-sixth of the Al^{3+} ions have been replaced by Mg^{2+}, whereas in beidellite, about one-twelfth of the Si^{4+} have been replaced by Al^{3+}. The peculiar layer structure of these clays gives them cation exchange and intercalation properties that can be very useful. Molecules such as water and polar organic molecules such as glycol can easily intercalate between the layers and cause the clay to swell.

Clays have been used as catalysts, but one of the problems with using clays as catalysts is that at high temperatures (>200°C) they start to dehydrate and the interlayer region collapses.

7.5.3 Periodic Mesoporous Organosilicas

Periodic mesoporous organosilicas (PMOs) were first reported in 1999 and are composed of silsesquioxanes which form Si-C bridges, $(RO)_3Si\text{-}R'\text{-}Si(OR)_3$, the organic spacer group determining the pore size of the structure. Research has been partly driven by the novelty of the structures, and by their potential usefulness as materials in the type of functions already discussed above, but also for chromatography columns and silicone oils and resins. Their potential advantage lies both in their porosity and also in the organic group(s) that are now actually an intrinsic part of the framework structure and not just grafted on as in the modified frameworks.

A large number of these poly(trialkyloxysilyl) structures have been made with different structures and pore sizes although not many precursors have been found that are suitable as the frameworks tend to collapse when the template is removed. The synthetic method is similar to that used for the aluminosilicates, in that a silsesquioxane is mixed with a surfactant to form self-assembling micelles which are then aged and heated to extract the template. The synthesis has many factors that can be varied apart from the starting silicon-precursor: temperature, pH, surfactant, ageing time and temperature, and so on. Of the precursor material, $(RO)_3Si\text{-}R'\text{-}Si(OR)_3$, the R group is usually $-OC_2H_5$, but the R′ group has varied, based on for instance methane, ethane, benzene, thiophene, and even ferrocene: some examples are: $-(CH_2)_n-$; $-(CH{=}CH)-$; $-(C{\sim}C)-$; *o/p* $-(C_6H_4)_n-$; $-(CH_2CH_2Si(OEt)_2CH_2CH_2)$; $-(S/O\text{-}(C_6H_4)_2)-$; $-(Fe\text{-}(C_5H_4)_2)-$.

7.6 SUMMARY

1. This chapter looked at solids containing cavities and channels. Such solids are useful for heterogeneous catalysis, gas absorbance, and gas storage.
2. Microporous structures have pores of less than 2 nm in diameter and mesoporous structures pores of between 2 and 50 nm.
3. Zeolites are microporous aluminosilicates with pore sizes in the range 0.3–1 nm. There are over 60 naturally occurring zeolites, but only seven occur in large quantities. Over 200 artificial zeolites have been synthesized.
4. Metal organic frameworks contain metal ions or metal clusters linked by organic molecules. They have attracted interest because they have can have larger surface areas than zeolites. Most are microporous but some examples are mesoporous or even macroporous. They have many potential uses but many are thermally and chemically unstable.
5. Covalent organic frameworks contain light elements covalently bonded. They show promise as absorbers and in sensors.
6. Other porous solids include mesoporous organosilicas and aluminosilicates, and clays.

QUESTIONS

1. Zeolite A has a single peak in the ^{29}Si MAS NMR spectrum at 89 ppm and has an Si/Al ratio of 1. Comment on these observations. (Si MAS NMR of zeolites is discussed in Chapter 2.)
2. A series of zeolites were synthesised with an increasing Si/Al ratio. It was found that the catalytic activity also increased until the ratio was about 15: 1, after which it declined. Suggest a reason for this behaviour.
3. Zeolite A (Ca form), when loaded with platinum, has been found to be a good catalyst for the oxidation of hydrocarbon mixtures. However, if the mixture contains branched-chain hydrocarbons, these do not react. Suggest a possible reason.
4. Both ethene and propene can diffuse into the channels of a particular mordenite catalyst used for hydrogenation. Explain why only ethane is produced.
5. Explain why, when 3-methyl pentane and *n*-hexane are cracked over zeolite A (Ca form) to produce smaller hydrocarbons, the percentage conversion for 3-methyl pentane is less than 1%, whereas that for *n*-hexane is 9.2%.
6. When toluene is alkylated by methanol with a ZSM-5 catalyst, an increase in the crystallite size from 0.5 to 3 μm approximately doubles the amount of para-xylene produced. Suggest a possible explanation.
7. The infrared stretching frequency of the hydroxyl associated with the Brønsted sites in decationised zeolites falls in the range of 3600–3660 cm^{-1}. As the Si/Al ratio in the framework increases, this frequency tends to decrease. What does this suggest about the acidity of the highly siliceous zeolites?

8. What property of the organic linkages can be used to alter the pore size in metal organic frameworks (MOFs)?
9. Removing water molecules attached to Cu^{2+} in the MOF HKUST-1 promoted catalytic activity for the isomerization of terpenes. What property useful for catalysis does the removal of water produce?
10. Which aspects of MOFs impede their commercial adoption?
11. How are ZIFs (zeolitic imidazole frameworks) related to zeolites?
12. What type of bonding holds the framework together for zeolites, MOFs, and COFs?
13. List the types of solid discussed in this chapter which can be mesoporous.

8 Optical Properties of Solids

8.1 INTRODUCTION

Perhaps the most well-known example of solid-state optical devices is the laser. Of interest to the solid state chemist are two types of laser, typified by the ruby laser and the gallium arsenide laser. Because laser light is more easily modulated than light from other sources, it is used for sending information; light traveling along optical fibres replacing electrons traveling along wires in, for example, telecommunications. To transmit light over long distances, the optical fibres must have particular absorption and refraction properties, and the development of suitable substances has become an important area of research.

Light emitting diodes (LEDs) are used for displays, including those on digital watches and scientific instruments, and are currently replacing traditional light bulbs. Solar cells produce electricity from sunlight and are employed on individual houses and in larger arrays. Another important group of solids are the light-emitting solids, known as phosphors, which are used on plasma television screens and in white light LEDs.

Very broadly speaking, there are two situations that have to be considered in explaining the above-mentioned devices in which light is absorbed or emitted. In the first, which is relevant to the ruby laser and phosphor coatings in white light LEDs, the light is emitted by an impurity ion in a host lattice. We are concerned here with what is essentially an atomic spectrum modified by the lattice. We also look briefly at colour centres which are also due to impurities but have spectra characteristic of the host lattice. In the second case, which applies to photovoltaic cells, LEDs, and the gallium arsenide laser, the optical properties of the delocalised electrons in the bulk solid are important.

Some devices, such as optical fibres and 'cloaks of invisibility', depend on the refractive properties of materials; therefore, following on from discussions of devices that depend on the absorption/emission of light, the process of refraction is considered.

Photonic crystals and metamaterials differ from the other examples in that their characteristic properties are a result of their construction rather than their composition. We shall look briefly at photonic crystals. Thin film photonic crystals are used as reflective coatings on lenses. Photonic crystals are also found in nature. Finally, we consider the unusual optical properties of metamaterials, a class of solids that has attracted much attention due to their potential use in devices to render objects invisible.

8.2 INTERACTION OF LIGHT WITH ATOMS

When an atom absorbs a photon of light of the correct wavelength, it undergoes a transition to a higher energy level. To a first approximation in many cases, we can think of one electron in the atom absorbing the photon and being excited. The electron will only absorb the photon if the energy of the photon matches that of the energy difference between the initial and final electronic energy levels, and if certain rules, known as **selection rules,** are obeyed. In light atoms, the electron cannot change its spin and its orbital angular momentum must change by one unit; in terms of quantum numbers: $\Delta s = 0$, and $\Delta l = \pm 1$; there is no restriction on the changes of the principal quantum number. One way of thinking about this is that the photon has zero spin and one unit of angular momentum. The conservation of spin and angular momentum then produces these rules. For a sodium atom, for example, the 3s electron can absorb one photon and go to the 3p level. The 3s electron will not, however, go to the 3d ($\Delta l = +2$) or 4s ($\Delta l = 0$) level. Figure 8.1 illustrates allowed and forbidden transitions for Ca.

However, the spin and the orbital angular momenta are not entirely independent and coupling between them allows forbidden transitions to occur, although the probability of an electron absorbing a photon and being excited to a forbidden level is much less than the probability of it being excited to an allowed level. Consequently, spectral lines corresponding to forbidden transitions are less intense than those corresponding to allowed transitions.

An electron that has been excited to a higher energy level will sooner or later return to the ground state. It can do this in several ways. The electron may simply emit a photon of the correct wavelength at random, some time after it has been excited. This is known as **spontaneous emission**. Alternatively, a second photon may come along and instead of being absorbed, may induce the electron to emit. This is known as **induced** or **stimulated emission** and plays an important role in the action of lasers. The emitted photon in this case is in phase with, and traveling in the same direction as, the photon inducing the emission; the resulting beam of light is said to be **coherent**. Finally, the atom may collide with another atom losing energy in the process or giving energy to its surroundings in the form of vibrational energy. These are examples of **nonradiative transitions**. Spontaneous and stimulated emissions obey the same selection rules as absorption. Nonradiative transitions have different rules. In a crystal, as in a molecule, the atomic energy levels and the selection rules are modified. As an example, let us take an ion with one d electron outside a closed shell (Ti^{3+}, for example). This will help us understand the ruby laser.

In the free ion, the five 3d orbitals all have the same energy. In a crystal, these levels are split; for example, if the ion occupied an octahedral hole, the 3d levels would be split into a lower, triply degenerate (t_{2g}) level and a higher, doubly degenerate (e_g) level. This is shown in Figure 8.2.

An electronic transition between these levels is now possible. In the free ion, a transition from one d level to another involves zero energy change, so it would not be observed even if it were allowed. In the crystal, the transition involves a change in energy, but it is still forbidden by the selection rules. Lines corresponding to such transitions can, however, be observed, albeit with low intensity, because the crystal

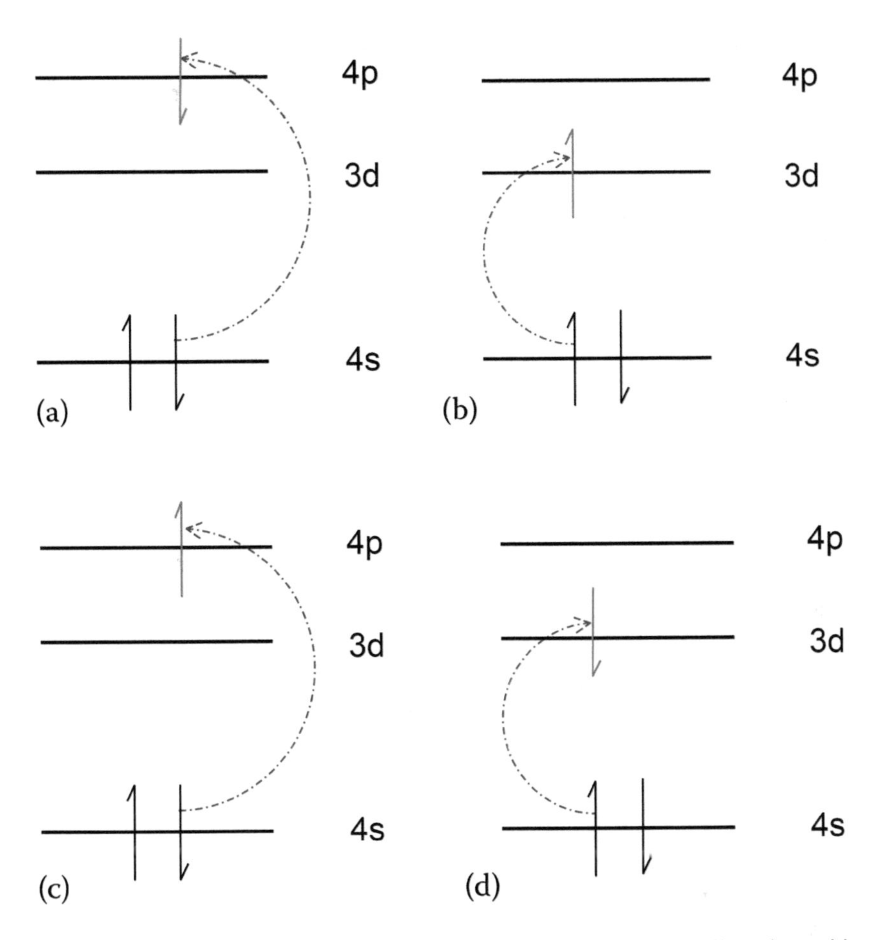

FIGURE 8.1 Allowed and forbidden atomic transitions for Ca. (a) an allowed transition, (b) a forbidden transition with $\Delta l \neq 1$, (c) a forbidden transition with $\Delta s \neq 0$, and (d) a forbidden transition in which $\Delta l \neq 1$ and $\Delta s \neq 0$.

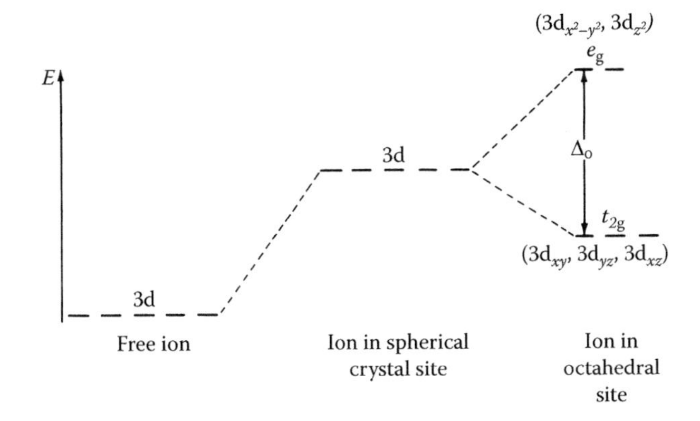

FIGURE 8.2 Splitting of d levels in an octahedral site in a crystal.

vibrations mix different electronic energy levels. Thus, the 3d levels may be mixed with the 4p levels, giving a small fraction of 'allowedness' to the transition. Figure 8.3 shows an absorption band due to a transition from t_{2g} to e_g for the ion Ti^{3+}, which has one d electron. (This band is, in fact, two closely spaced bands because the excited state is distorted from a true octahedron and the e_g level is further split into two.)

We shall see now the role played by a similar forbidden transition in the operation of the ruby laser.

8.2.1 Ruby Laser

Ruby is corundum (one form of Al_2O_3, see Chapter 1) with 0.04%–0.5% Cr^{3+} ions as an impurity replacing the aluminium ions. The aluminium ions, and hence the chromium ions, occupy distorted octahedral sites. Thus, as discussed above, the 3d levels of the chromium ion will be split. Cr^{3+} has three 3d electrons, and, in the ground state, these occupy separate orbitals with parallel spins. When light is absorbed, one of these electrons can undergo a transition to a higher-energy 3d level; this is similar to the transition discussed in the previous section, but with three electrons it is necessary to consider the changes in the electron repulsion as well as the changes in orbital energies. When the changes in the electron repulsion are included, we find that there are two transitions at different energies, which correspond to the jump from the lower to the higher 3d level.

Having absorbed light and undergone one of these transitions, the chromium ion could now simply emit radiation of the same wavelength and return to the ground state. However, in ruby, there is a fast, radiationless transition in which the excited electron loses some of its energy and the crystal gains vibrational energy. The chromium ion is left in a state in which it can only return to the ground state by a forbidden transition in which an electron changes its spin. Such a transition is doubly forbidden because it also breaks the rule that forbids 3d ↔ 3d, so it is even less likely to occur than the original absorption process. The states involved are shown

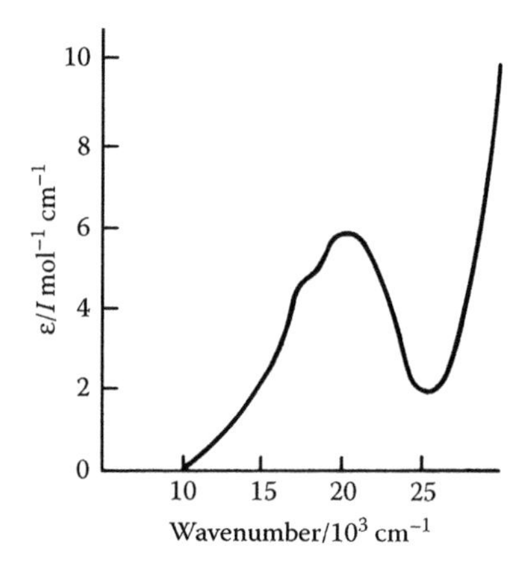

FIGURE 8.3 The t_{2g} to e_g transition of Ti^{3+}. The band is, in fact, two overlapping bands. This is due to a further splitting of the e_g levels.

schematically in Figure 8.4. The Cr^{3+} ions absorb light and go to states 3 and 4. They then undergo a radiationless transition to state 2. Because the probability of spontaneous emission for state 2 is low, and there is no convenient nonradiative route to the ground state, a considerable population of state 2 can build up. When eventually (about 5 msec later) some of the ions in state 2 return to the ground state, the first few spontaneously emitted photons interact with other ions in state 2 and induce these to emit. The resulting photons will be in phase and travelling in the same direction as the spontaneously emitted photons and will induce further emission as they travel through the ruby. In the laser, the ruby is enclosed by a reflecting cavity so that the photons are reflected back into the crystal when they reach the edge. The reflected photons induce further emission and by this means, an appreciable beam of coherent light is built up. The mirror on one end can then be removed and a pulse of light emitted. The name 'laser' is a reflection of this build-up of intensity. It is an acronym standing for **L**ight **A**mplification by **S**timulated **E**mission of **R**adiation. (Similar devices producing coherent beams of microwave radiation are known as masers.)

A typical arrangement for a pulsed ruby laser is shown in Figure 8.5. A high-intensity flash lamp excites the Cr^{3+} ions from level 1 to levels 3 and 4. The lamp can lie alongside the crystalline rod of ruby or can be wrapped around it. At one end of the reflective cavity surrounding the ruby crystal is a Q switch, which switches from being reflective to transmitting the laser light and can be as simple as a rotating segmented mirror, but it is usually a more complex device.

Ruby was the first material used for solid-state lasers, but now several other crystals are employed. The crystals used need to contain an impurity with an energy level such that a return to the ground state is only possible by a forbidden transition in the infrared, visible, or near ultraviolet. It must also be possible to populate this level through an allowed (or at least less forbidden) transition. Research has tended to concentrate on transition metal ions and lanthanoid ions in various hosts, because these ions have suitable transitions of the right wavelength. Some examples are given

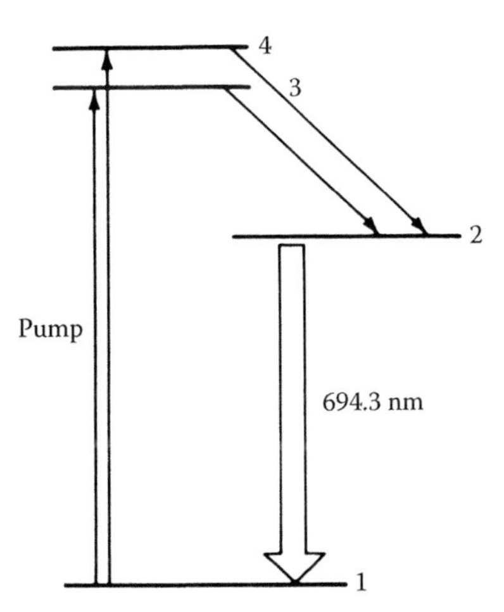

FIGURE 8.4 The states of the Cr^{3+} ion involved in the ruby laser transition.

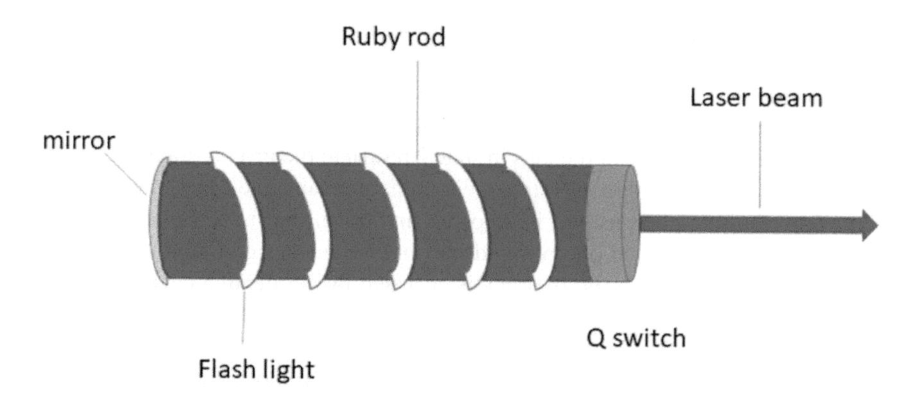

FIGURE 8.5 A sketch of a ruby laser.

in Table 8.1 with the wavelength of the laser emission. Note that the majority involve lanthanoid ions.

8.2.2 Phosphors in LEDs

Phosphors are solids that absorb energy and re-emit it as light. As in the lasers that we have just described, the emitter is usually an impurity ion in a host lattice. However, for the general uses of phosphors, it is not necessary to produce intense, coherent beams of light, and the emitting process is spontaneous rather than induced. There are many applications of phosphors, for example, the colours of plasma television screens are produced by phosphors that are bombarded with electrons from a beam. Phosphors are also used to produce white light in LED bulbs. The light emitting diode in white light LED bulbs actually produces radiation in the blue/violet region. Some of this radiation is absorbed by the phosphor coating and re-emitted as yellow, green, or red light to produce a white light. The yellow, red, or green light is produced by an impurity ion, often a lanthanoid ion, in a host lattice. Host lattices include alkaline earth aluminates, yttrium aluminium garnet (YAG), nitrides, and

TABLE 8.1
Examples of Crystals Used as Lasers

Ion	Host	Wavelength emitted/nm
Ti^{3+}	Sapphire	650–1100 (tunable)
Nd^{3+}	Fluorite (CaF_2)	1046
Sm^{3+}	Fluorite	708.5
Ho^{3+}	Fluorite	2090
Nd^{3+}	Calcium tungstate ($CaWO_4$)	1060
Nd^{3+}	Yttrium vanadate (YVO_4)	1064
Nd^{3+}	Yttrium aluminum garnet (YAG, $Y_3Al_5O_{12}$) (Nd/YAG laser)	1064

silicates. Ce^{3+}-doped YAG, for example, is a common phosphor producing a broad band of light due to a transition from 5d to 4f centered in the yellow region of the spectrum. The light from the LED is absorbed by an electron in a 4f orbital which is excited to an excited vibrational state of a 5d level. The electron then loses energy, moving to the vibrational ground state of the 5d level. From there it emits yellow and green light and returns to the 4f level (Figure 8.6).

In phosphors and in the ruby laser, light is absorbed and emitted by electrons localized on an impurity site. In the next section, we look at another example of color produced by different types of defects such as anion vacancies.

8.3 COLOUR CENTRES

The first observation of color centres was probably by Goldstein at the end of the nineteenth century. Goldstein noted changes in color of alkali halide crystals when cathode rays were aimed them. However these centers were not studied extensively until the work of Pohl in the 1930s. The color was thought to be associated with a defect, known then as a **Farbenzenter** (color centre), now abbreviated to **F-centre**. Since then, it has been found that many forms of high-energy radiation (UV, X-rays, neutrons) will cause F-centers to form. The color produced by the F-centres is always characteristic of the host crystal, for instance, NaCl becomes deep yellowy–orange, KCl becomes violet, and KBr becomes blue–green.

Subsequently, it was found that F-centres can also be produced by heating a crystal in the vapor of an alkali metal: this gives a clue to the nature of these defects. The excess alkali metal atoms diffuse into the crystal and settle on cation sites; at the same time, an equivalent number of anion-site vacancies are created and

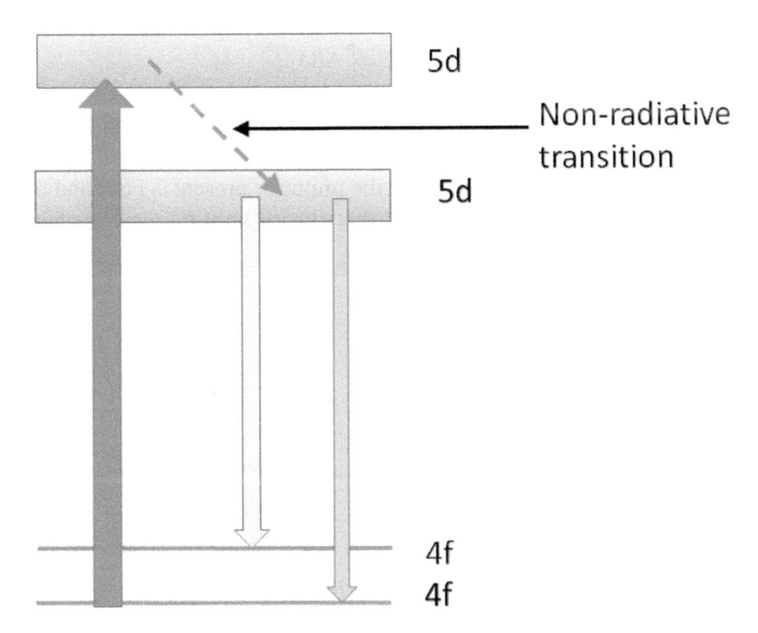

FIGURE 8.6 The phosphor process for Ce^{3+}in YAG.

ionisation gives an alkali-metal cation with an electron trapped at the anion vacancy (Figure 8.7). In fact, it does not even matter which alkali metal is used; if NaCl is heated with potassium, the color of the F-centre does not change because it is characteristic of the electron trapped at the anion vacancy in the host halide. Work with ESR has confirmed that F-centres are indeed unpaired electrons trapped at vacant lattice (anion) sites.

The trapped electron provides a classic example of an 'electron in a box' (Chapter 4). A series of energy levels are available for the electron, and the energy required to transfer from one level to another falls in the visible part of the electromagnetic spectrum—hence the color of the F-centre. There is an interesting natural example of this phenomenon: the mineral fluorite (CaF_2) can be found in Derbyshire, UK, where it is known as 'Blue John', and its beautiful blue-purple coloration is due to the presence of F-centres.

Many other color centres have now been characterised in alkali-halide crystals. The **H-centre** (halogen centre) is formed by heating, for instance, NaCl in Cl_2 gas. In this case, a $[Cl_2]^-$ ion is formed and it occupies a single anion site (Figure 8.7b). F-centres and H-centres are perfectly complementary—if they meet, they cancel one another out.

Another interesting natural example of color centres lies in the color of smoky quartz and amethyst. These semiprecious stones are basically crystals of silica (SiO_2) with some impurity present. In the case of smoky quartz, the silica contains a little aluminum impurity. The Al^{3+} substitutes for the Si^{4+} in the lattice and the electrical neutrality is maintained by H^+ present in the same amount as Al^{3+}. The color centre arises when ionising radiation interacts with an $[AlO_4]^{5-}$ group, liberating an electron that is then trapped by H^+:

$$[AlO_4]^{5-} + H^+ = [AlO_4]^{4-} + H.$$

The $[AlO_4]^{4-}$ group is now electron-deficient and can be considered to have a 'hole' trapped at its centre. This group is the color centre, absorbing light and producing the smoky color. In crystals of amethyst, the impurity present is Fe^{3+}, and on irradiation $[FeO_4]^{4-}$ color centres are produced that absorb light to give the characteristic purple coloration.

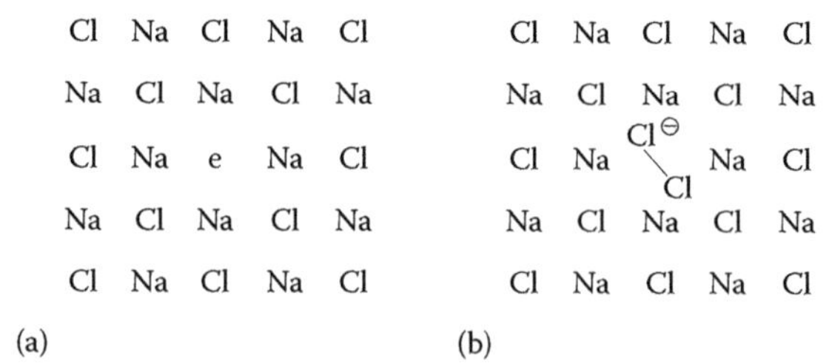

Cl Na Cl Na Cl
Na Cl Na Cl Na
Cl Na e Na Cl
Na Cl Na Cl Na
Cl Na Cl Na Cl
(a)

Cl Na Cl Na Cl
Na Cl Na Cl Na
Cl Na $Cl^{\ominus}$–Cl Na Cl
Na Cl Na Cl Na
Cl Na Cl Na Cl
(b)

FIGURE 8.7 (a) The F-centre, an electron trapped on an anion vacancy. (b) H-centre.

8.4 ABSORPTION AND EMISSION OF RADIATION IN CONTINUOUS SOLIDS

For continuous solids, the absorption and the emission of radiation involve transitions between energy bands rather than between discrete atomic energy levels. Radiation falling on an insulator or a semiconductor is absorbed by electrons in delocalised bands, in particular those near the top of the valence band, causing these electrons to be promoted to the conduction band. In metals, electrons can be promoted from the partially occupied conduction band to higher energy bands. Because there are many closely packed levels in an energy band, the absorption spectrum is not a series of lines as in atomic spectra, but a broad peak with a sharp threshold close to the band gap energy. The absorption spectrum of GaAs, for example, is shown in Figure 8.8.

Transitions to some levels in the conduction band are more likely than transitions to other levels. This is because transitions between levels in bands, like those between atomic energy levels, are governed by selection rules. The spin selection rule still holds; when promoted, the electron does not change its spin. However, the orbital angular momentum rules are not appropriate for energy bands, and the rule governing change in the quantum number l is replaced by a restriction on the wave vector (**k**). As seen in Chapter 4, the energy levels in a band are characterised by the wave vector, the momentum of the electron wave being given by $\mathbf{k}\hbar$. The momentum of a photon with a wavelength in the infrared, visible, or ultraviolet region is very small compared with that of the electron in the band, so conservation of momentum produces the selection rule for transitions between bands: an electron cannot change its wave vector when it absorbs or emits radiation. Thus, an electron in the valence band with wave vector ($\mathbf{k}_i$) can only undergo allowed transitions to levels in the conduction band that also have the wave vector ($\mathbf{k}_i$). In some solids, for example, GaAs, the level at the top of the valence band and that at the bottom of the conduction band have the same wave vector. There is then an allowed transition at the band gap energy. Such solids are said to have a **direct band gap**.

For other semiconductors, for example silicon, the direct transition from the top of the valence band to the bottom of the conduction band is forbidden. These solids are said to have an **indirect band gap**. Illustrations of band structures for solids with direct and indirect band gaps are given in Figure 8.9. Note that in these

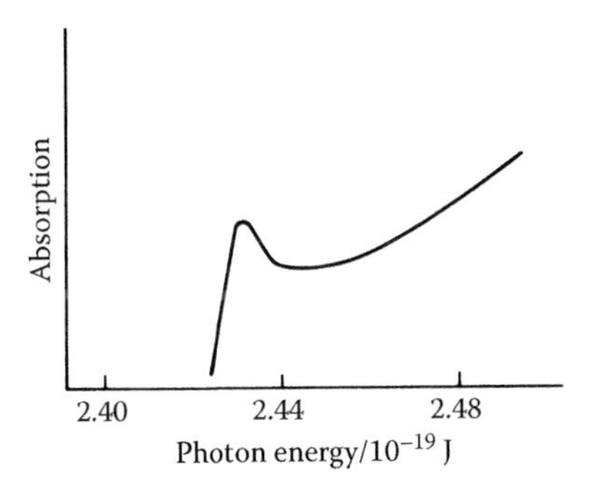

FIGURE 8.8 The absorption spectrum of GaAs.

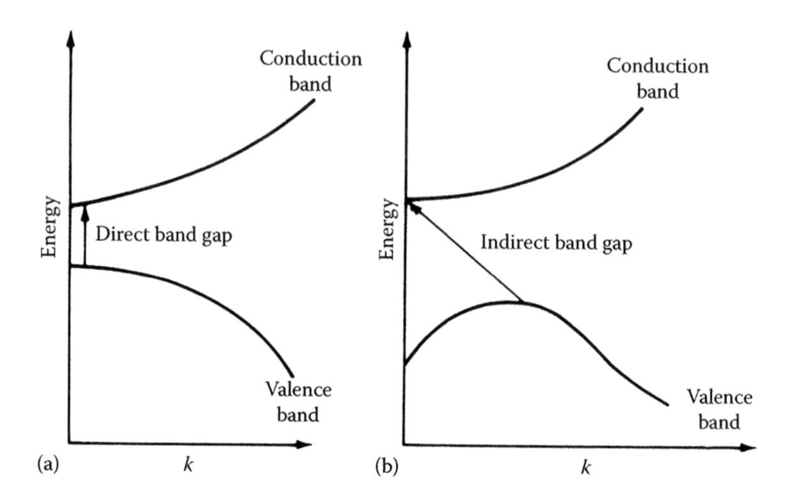

FIGURE 8.9 A sketch of the energy bands for (a) a solid with a direct band gap and (b) a solid with an indirect band gap. Note that in this diagram the horizontal axis is k, not the density of states. In this representation, a band is shown as a line going from 0 to the maximum value of k occurring for that band.

diagrams, the energy for a band in one direction is plotted against the wave number (k). Similar diagrams were given in Chapter 4, but with the energy plotted against the density of states.

The simple free electron model might suggest that the lowest energy orbital in any band is that with $k=0$. Figure 8.10, however, illustrates two combinations of orbitals that will have $k=0$ for a chain because all the atomic orbitals are combined in phase. The combination of p orbitals is obviously antibonding and would be expected to have the highest energy in its band; the combination of s orbitals is bonding and would have the lowest energy in its band. If the p band lay below the s band, a transition between these levels would be allowed and would correspond to a direct transition across the band gap. In real solids, the highest and the lowest levels in the bands will contain contributions from different types of atomic orbital and it becomes difficult to predict whether a band gap will be direct or indirect.

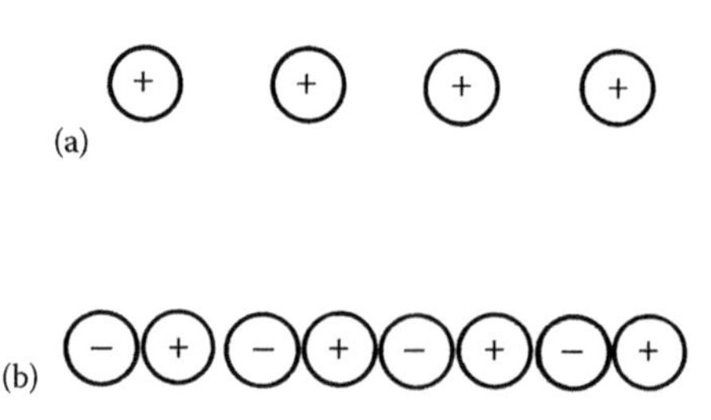

FIGURE 8.10 A row of (a) s orbitals and (b) p orbitals, both with $k=0$.

One consequence of an indirect band gap is that an electron in the bottom level of the conduction band has only a small probability of emitting a photon and returning to the top of the valence band. This is of importance when selecting materials for some of the applications we are going to consider.

Transitions across the band gap are also responsible for the appearance of many solids. Because a solid is very concentrated, the probability is very high that a photon with energy that corresponds to an allowed transition is absorbed. Many such photons will therefore be absorbed at or near the surface of the solid. These photons will then be re-emitted in random directions so that some will be reflected back towards the source of the radiation and some will travel further into the solid. Those traveling into the solid stand a very good chance of being reabsorbed and then re-emitted, again in random directions. The net effect is that the radiation does not penetrate the solid, but is reflected by its surface. If the surface is sufficiently regular, then solids that reflect visible radiation appear shiny. Thus, silicon, whose band gap is at the lower end of the visible region and has allowed transitions covering most of the visible wavelengths, appears shiny and metallic. Many metals have strong transitions between the conduction band and a higher-energy band, which lead to their characteristic metallic sheen. Some metals, such as tungsten and zinc, have a band gap in the infrared region, and transitions in the visible region are not as strong. These metals appear relatively dull. Gold and copper have strong absorption bands due to the excitation of d band electrons to the s/p conduction band. In these elements, the d band is full and lies some way below the Fermi level (Figure 8.11).

The reflectivity peaks in the yellow part of the spectrum, and blue light and green light are less strongly absorbed; hence, the metals appear golden. Very thin films of gold appear green because the yellow light and red light are absorbed and only the blue and green are transmitted. Insulators typically have band gaps in the ultraviolet region and, unless there is a localised transition in the visible region, appear colourless.

In Chapter 4 we introduced a *p–n* junction. We now describe several devices that depend on the properties of *p–n* junctions.

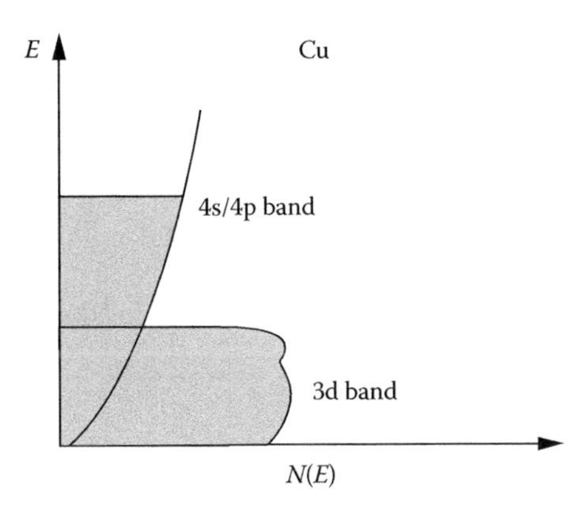

FIGURE 8.11 The band structure of copper.

8.4.1 Gallium Arsenide Laser

When an electric field causes conduction electrons to move across the *p*–*n* junction, the resulting situation is one in which the population in the conduction band is greater than the thermal equilibrium population. An excess of electrons in an excited state is an essential feature of lasers, and there are several semiconductor lasers based on the *p*–*n* junction. The best known of these is the gallium arsenide laser.

The **gallium arsenide laser** actually contains a layer of GaAs sandwiched between layers of *p*-type and *n*-type gallium aluminium arsenide ($Ga_{1-x}Al_xAs$). As shown in Figure 8.12, the band gap of gallium aluminium arsenide is larger than that of gallium arsenide.

An electric field applied across the *p*–*n* junction produces an excess of electrons in the conduction band of the gallium arsenide. These electrons do not drift across into the gallium aluminium arsenide layer because the bottom of the conduction band in this layer is higher in energy, and the electrons would therefore need to gain energy in order to move across. The excess conduction band electrons are therefore constrained to remain within the GaAs layer. Eventually, one of these electrons drops down into the valence band, emitting a photon as it does so. This photon induces other conduction band electrons to return to the valence band and thus a coherent beam of light begins to build up. As in the ruby laser, the initial burst of photons is reflected back by mirrors placed at the ends, thereby inducing more emission. Eventually, a beam of infrared radiation is emitted.

Several such lasers have been developed, most of them based on III–V compounds—compounds of In, Ga, and/or Al with As, P, or Sb. It is possible to

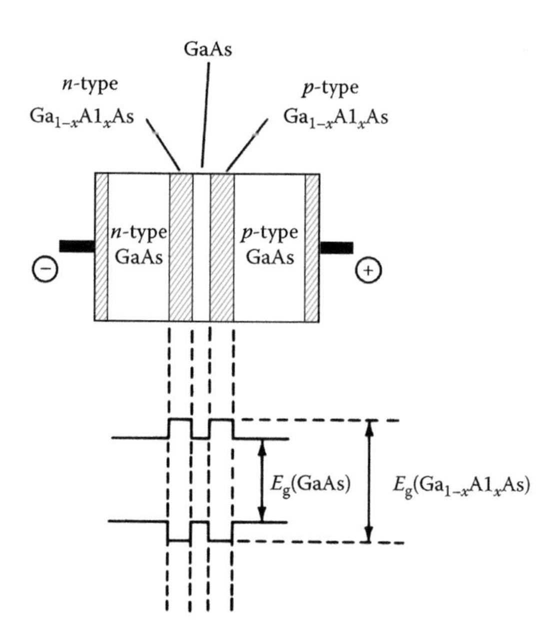

FIGURE 8.12 The arrangement of the different semiconductor regions in a GaAs laser. The band gap profile is shown below.

manufacture materials with band gaps over the range 400–1300 nm by carefully controlling the ratios of the different elements.

Red semiconductor lasers are used, among many other things, for reading DVDs. A DVD consists of a plastic disc coated with a highly reflective aluminium film and is protected against mechanical damage by a layer of polymer. The original recording is split up into a number of frequency channels and the frequency of each channel is given as a binary code (i.e., a series of 0's and 1's). This is converted to a series of pits in tracks on the disk, spaced approximately 1.6 μm apart. A laser is focused on the disk and reflected onto a photodetector. The pits cause some of the light to be scattered, thereby reducing the intensity of the reflected beam. The signal read by the photodetector is read as 1 when there is a high intensity of light and as 0 when the intensity is reduced by scattering. The binary code is thus recovered and can be converted back to sound or pictures.

8.4.2 Quantum Wells: Blue Lasers

Blue lasers allow a higher resolution, and hence a higher density of optical storage of information on devices. The earliest blue lasers were based on ZnSe, but their lifetime proved too short for commercial applications. Lasers based on gallium nitride (GaN), first demonstrated in 1995, have proved to have greater lifetimes. In these lasers, the photons are produced in quantum wells rather than in a bulk semiconductor.

The active region of GaN lasers consists of GaN containing several thin layers (3–4 nm thick) of indium-doped GaN, $In_xGa_{1-x}N$. The addition of indium reduces the band gap within the thin layers, so that the bottom of the conduction band is at a lower energy than that of the bulk GaN. The electrons in this conduction band are effectively trapped because they need to gain energy from an external source to pass into the conduction band of the bulk GaN. Figure 8.13 schematically shows the conduction band for a series of thin layers of $In_xGa_{1-x}N$ in GaN.

The trapped electrons behave like particles in a box (Chapter 4), but with finite energy walls to the box. Such boxes are **quantum wells**. Within the well, the electron energy is quantised and the spacing of the energy levels depends on the (energy) depth and the (spatial) width of the well. The depth of the wells is controlled by the extent of the doping, that is, the value of x. Although the electrons in the wells have insufficient energy to surmount the energy barrier to reach the next well, there is a probability that they will move to a similar energy level in the next well through **quantum mechanical tunneling** (see Chapter 2). The electrons traveling from the bulk GaN enter a high level in the first well (Figure 8.13). From this level, the electrons can emit a photon and go to a lower level or can tunnel through to the next well. Tunneling is faster than photon emission, so that the population of the higher levels builds up. There are now more electrons in the higher levels than in the lowest levels, that is, there is a population inversion—the requirement for laser action.

The active region containing the quantum wells is sandwiched between layers of n-doped and p-doped GaN and aluminium-doped GaN, $A1_yGa_{1-y}N$, which provide the electrons entering the quantum well and keep them confined to the active region. All these layers are built up on a substrate, for example, sapphire. The ends of the

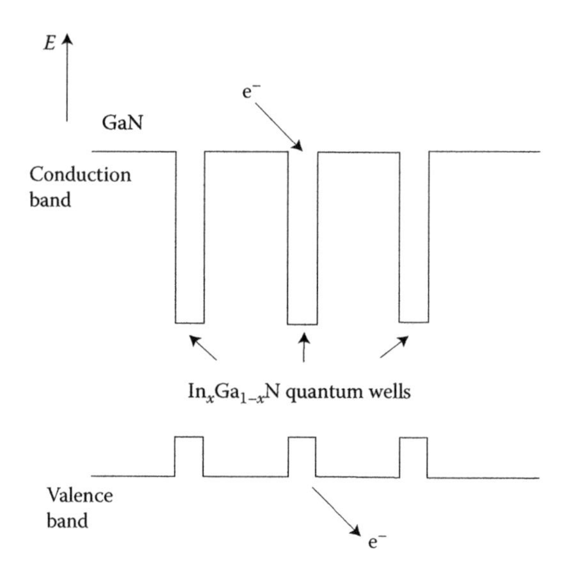

FIGURE 8.13 An energy level diagram for the conduction band in a series of thin layers of $In_xGa_{1-x}N$ in GaN.

whole device are etched or cleaned to form a partial mirror that reflects the emitted photons, allowing a coherent beam to build up.

Blu-ray disks are DVD-like devices using blue lasers to read the disk. However, the blue laser used is generally a diode-pumped solid-state laser in which light from an LED is used to excite ions in a solid-state laser, such as Nd/YAG, and the frequency of the radiation emitted by the laser is doubled by passing it through a solid such as lithium borate.

8.4.3 Light-Emitting Diodes

LEDs are widely used in homes, car headlights, and displays. Like transistors, they are based on the *p–n* junction, but the voltage applied across the *p–n* junction in this case leads to the emission of light. Figure 4.15 showed a *p–n* junction in a semiconductor such as GaAs. The band structure shown in Figure 4.15 was for the junction in the dark and with no electric field applied. Now suppose that an electrical field is applied so that the *n*-type semiconductor is made negatively charged relative to the *p*-type (i.e., in the reverse direction to the applied voltage in transistors, Chapter 4). Electrons will then flow from the *n*-type to the *p*-type. An electron in the conduction band moving to the *p*-type side can drop down into one of the vacancies in the valence band on the *p*-type side, emitting a photon in the process. This is more likely to happen if the transition is allowed, so that semiconductors with direct band gaps are usually used in such devices. To use the LED as a display, for example, it is then wired into a circuit so that an electric field is applied across the parts making up the required letters or numbers. Different colours can be produced by using semiconductors of differing band gap. GaP produces red light, but by mixing in various

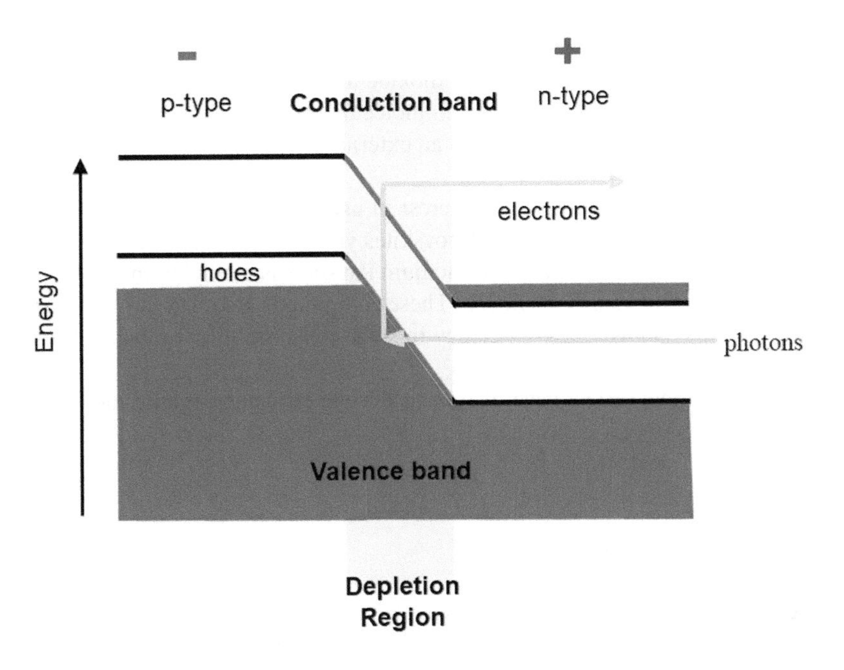

FIGURE 8.14 The effect of light on a p–n junction.

proportions of aluminium to form $Ga_{1-x}Al_xP$, green or orange light can be produced. Blue light is produced using InGaN sandwiched between GaN layers.

It should be noted that semiconductors with indirect band gaps are used for LEDs, but in these cases, the impurity levels play an important role. Thus, GaP is used although it has an indirect band gap. Silicon is not suitable because there is a nonradiative transition available to electrons at the bottom of the conduction band and these electrons donate thermal energy to the crystal lattice rather than emitting light when they return to the valence band. However p–n junctions in silicon are responsible for the production of electricity from sunlight in the most common solar (photovoltaic) cells.

8.4.4 Photovoltaic (Solar) Cells

Solar cells have become widely used as a source of clean energy. They rely on the absorption of sunlight by a p–n junction (Section 4.4.3). Electrons in the valence band of the depletion zone are promoted to the conduction band. The promoted electron moves into the n-type region. An electron from the p-type region then moves into the depletion zone to replace the promoted electron (Figure 8.14). The promoted electron and the vacancy are now separated in space, so that recombination does not occur. From the n-type region conduction band electrons can travel through an external circuit.

The original solar cells were based on a single crystal of silicon with p-doped regions and n-doped regions. Much research has gone into materials that are more efficient and/or cheaper. Polycrystalline or amorphous silicon have been used to make solar cells as have other semiconductors such as cadmium telluride, CdTe, gallium arsenide, GaAs, and copper indium gallium selenide. Dye-sensitised solar cells use dyes absorbed onto titanium dioxide to gather sunlight. The dye is excited to a higher energy level and the excited molecule transfers electrons to the titanium dioxide. Electrons then travel through an external circuit. The dye is regenerated by a reducing agent.

Currently (2019) there is much interest in using hybrid perovskites. These have the formula AMX_3, but unlike the perovskites you have met previously, A is a complex cation, usually a substituted ammonium ion such as methylammonium, $N(CH_3)H_3^+$, M is lead or tin, and X a halogen. These compounds are relatively easy to manufacture and have shown high efficiency in solar cells. Stability problems limit their adoption commercially.

Solar cells (and LEDs) have also been developed using organic molecules and polymers.

8.5 CARBON-BASED CONDUCTING POLYMERS

8.5.1 Discovery of Polyacetylene

The extension of the ideas of delocalised electrons in conjugated organic molecules led to a suggestion that a conjugated polymer might be an electrical conductor or semiconductor. The initial attempts to make polyacetylene and similar solids, however, resulted in short-chain molecules or amorphous, unmeltable powders. Then, in 1961, Hatano and co-workers in Tokyo managed to produce thin films of polyacetylene, by polymerising ethyne (acetylene). Polyacetylene exists in two forms, *cis* and *trans*, as shown in Figure 8.15, of which the *trans* form is the more stable. Ten years after Hatano's work, Shirakawa and Ikeda made films of *cis*-polyacetylene, which could be converted into the *trans* form. They achieved this by directing a stream of

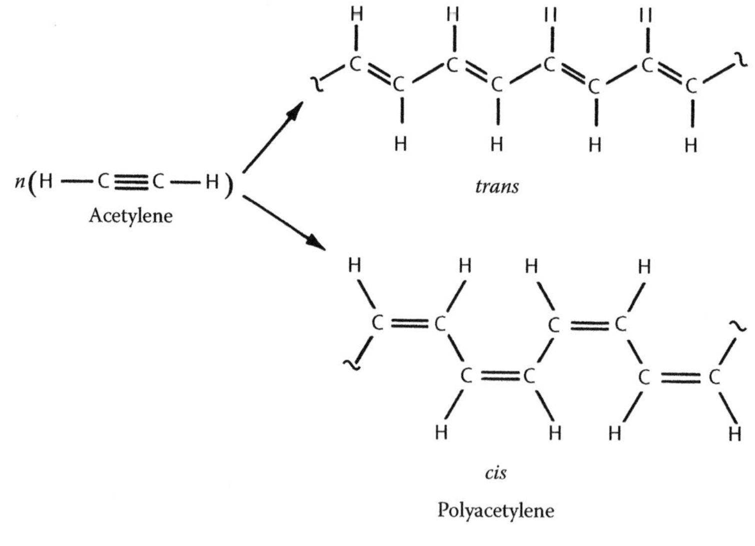

FIGURE 8.15 *cis*- and *trans*-polyacetylene.

ethyne gas onto the surface of a Ziegler–Natta catalyst (a mixture of triethyl aluminum and titanium tetrabutoxide). To make a large film, the catalyst solution can be spread in a thin layer over the walls of a reaction vessel (Figure 8.16) and then ethyne gas is allowed to enter. The polyacetylene produced in this way has a smooth shiny surface on one side and is a sponge-like structure. It can be converted to the thermodynamically more stable *trans* form by heating. The conversion is quite rapid above 370 K. After conversion, the smooth side of the film is silvery in appearance, becoming blue when the film is very thin.

A way of improving the conductivity was found when Shirakawa visited McDiarmid and Heeger in Pennsylvania later in the 1970s. This led to these three scientists being awarded the Nobel Prize for Chemistry in 2000. The Americans had been working on smaller conjugated molecules to which they added an electron acceptor in order to make them conducting. It was a natural step to try this approach with polyacetylene. If an electron acceptor such as bromine is added to polyacetylene forming $\left[(\mathrm{CH})^{\delta+}\,\mathrm{Br}_{\delta}^{-}\right]_n$, its conductivity is greater than that of the undoped

FIGURE 8.16 Polyacetylene film with metallic luster. (From M. Goh, S. Matsushita and K. Akagi, 'From Helical Polyacetylene to Helical Graphite: Synthesis in the Chiral Nematic Liquid Crystal Field and Morphology-Retaining Carbonisation'. *Chem. Soc. Rev.* (2010) **39**, 2467, by Chemical Society (Great Britain). Reproduced with permission of Royal Society of Chemistry via Copyright.)

material. Other examples of dopants that can oxidise polyacetylene are I_2, AsF_5, and $HClO_4^-$. The conductivity of polyacetylene is also increased by dopants that are electron donors. For example, the polymer can be doped with alkali metals to give, for example, $\left[Li_\delta^+(CH)^{\delta-}\right]_n$. The effect of these dopants is shown in Figure 8.17, where it can be seen that the conductivity rises from 10^{-3} S m^{-1} to as much as 10^5 S m^{-1} using only small quantities of dopant.

Polyacetylene is very susceptible to attack by oxygen. The polymer loses its metallic luster and becomes brittle when exposed to air. However, other conjugated polymers were found. Polypyrrole, polythiophene, polyaniline, polyphenylvinylene, and others (Figure 8.18) are conjugated polymers whose bonding and conductivity are similar to those of polyacetylene. These polymers are, however, less sensitive to oxygen and by attaching suitable side chains, can be made soluble in nonpolar organic solvents and are therefore easier to process.

As for polyacetylene, the conductivity of these polymers is sensitive to doping. This is exploited in polypyrrole gas sensors, which are based on the variation of conductivity of a thin polymer film when exposed to gases such as NH_3 and H_2S. Doped conducting polymers can also be used as a metallic contact in organic electronic devices.

8.5.2 Bonding in Polyacetylene and Related Polymers

In small conjugated alkenes such as butadiene with alternate double and single bonds, the π electrons are delocalised over the molecule. If we take a very long conjugated olefin, we might expect to obtain a band of π levels, and if this band were partly occupied, we would expect to have a one-dimensional conductor. Polyacetylene is

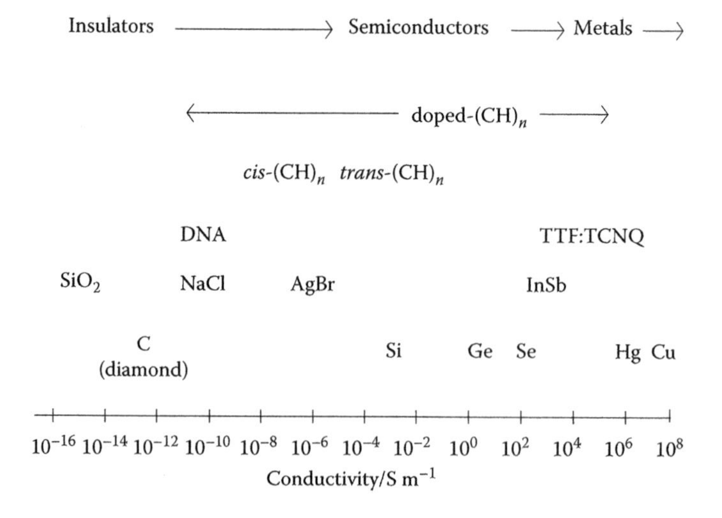

FIGURE 8.17 The conductivites of undoped and doped polyacetylene (CH)$_n$ compared with the values for some of the better-known insulators, semiconductors, and metals.

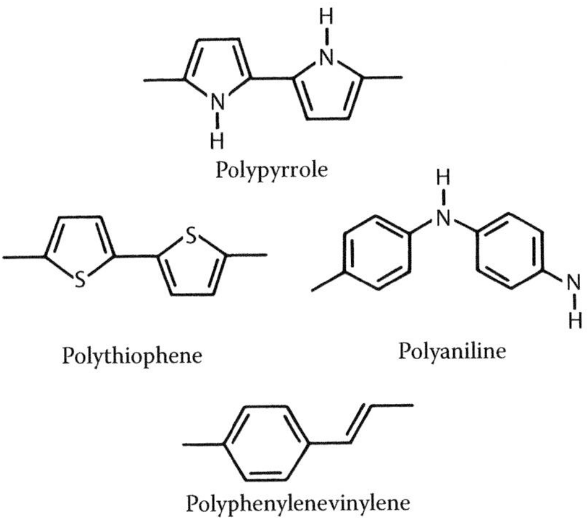

Polypyrrole

Polythiophene

Polyaniline

Polyphenylenevinylene

FIGURE 8.18 Repeating units of some conducting polymers.

just such a conjugated long-chain polymer. Now, if polyacetylene consisted of a regular, evenly spaced chain of carbon atoms, the highest-occupied energy band, the π band, would be half-full and polyacetylene would be an electrical conductor. In practice, polyacetylene shows only modest electrical conductivity, comparable with semiconductors such as silicon: the *cis* form has a conductivity of the order of 10^{-7} S m^{-1} and the *trans* form, 10^{-3} S m^{-1}. The crystal structure is difficult to determine accurately, but diffraction measurements indicate that there is an alternation in the bond lengths of about 6 pm. This is much less than would be expected for truly alternating single and double bonds (C–C, 154 pm in ethane; C=C, 134 pm in ethene). Nonetheless, this does indicate that the electrons tend to localise in double bonds rather than be equally distributed over the whole chain.

What, in fact, is happening is that two bands, bonding and antibonding, form with a band gap where nonbonding levels would be expected. There are just enough electrons to fill the lower band. As shown in Figure 8.19, this leads to a lower energy than the half-full single band.

This splitting of the band is an example of **Peierls' theorem**, which asserts that a one-dimensional metal is always electronically unstable with respect to a

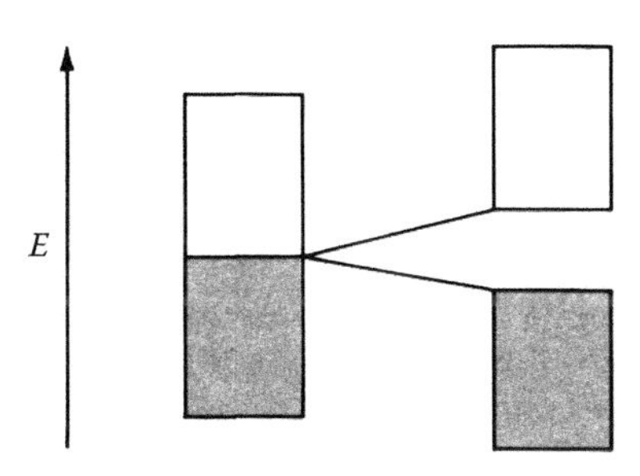

FIGURE 8.19 The band gap in polyacetylene produced by the alternation of long and short bonds.

nonmetallic state—there is always some way of opening an energy gap and creating a semiconductor. Thus, although a very simple bonding picture of such solids would suggest a half-filled band and metallic conductivity, the best that can be expected is a semiconducting polymer.

If an electron acceptor is added, it takes the electrons from the lower π bonding band. The doped polyacetylene now has holes in its valence band and, like *p*-type semiconductors, has a higher conductivity than the undoped material. Electron donor dopants add electrons to the upper π band, making this partly full, thereby producing an *n*-type semiconductor.

8.5.3 Organic LEDs and Photovoltaic Cells

Polymer LEDs are being developed for flat panel displays in mobile phones, laptops, and televisions. A typical polymer LED is built up on a glass substrate. At the bottom is a layer of a semitransparent metallic conductor, usually ITO, which acts as an electrode (Figure 8.20). On top of this is a layer of an undoped conjugated polymer, such as polyphenylenevinylene, and on top of that is an easily ionised metal, such as Ca, Mg/Ag, or Al. The metal forms the second electrode. To act as an LED, a voltage is applied across the two electrodes such that the ITO layer is positively charged. At the ITO/polymer interface, the electrons from the polymer move into the ITO layer, attracted by the positive charge. This leaves gaps in the lower energy band. At the same time, the electrons from the negatively charged second electrode move into the polymer and travel towards the ITO layer. When such an electron reaches a region of the polymer where there are vacancies in the lower energy band, it can jump down to this band, emitting light as it does so. The wavelength of the light, as in semiconductor LEDs, depends on the band gap. Polyphenylenevinylene LEDs emit green/yellow light, polyphenylene-based LEDs emit blue light, and some polythiophenes emit red light.

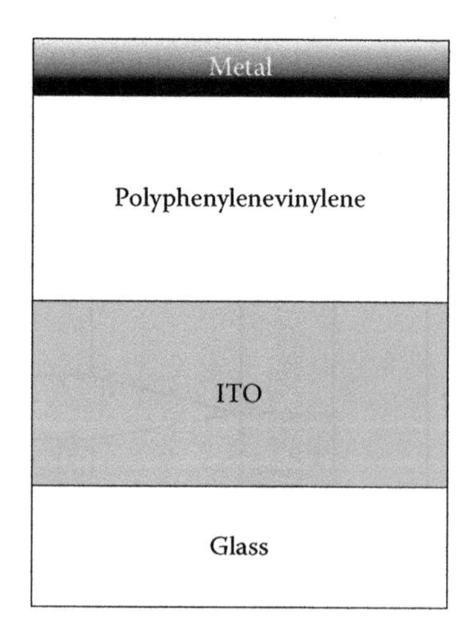

FIGURE 8.20 A section through a polymer LED.

Organic LEDs (OLED) are used in screens for TVs and other devices. However, in many cases these are based on a semiconducting organic molecule or an organometallic chelate rather than a conducting polymer.

8.6 REFRACTION

When light travels from a different medium through a solid, its velocity changes. A ray of light travelling at right angles to the surface of the solid will pass straight through it, but all other rays change direction. The size of the change in the velocity and hence the angle through which the radiation bends depend both on the material and the wavelength of the radiation. When electromagnetic radiation passes from a medium with one refractive index, say n_1, to a medium with a different refractive index, say n_2, the radiation bends according to Snell's law, $n_1\sin\theta_1 = n_2\sin\theta_2$, where θ_1 and θ_2 are the angles between the radiation and the normal to the interface. This is illustrated in Figure 8.21.

To see why the refractive index is altered by composition, the atomic origin of the refractive index is considered briefly. Electromagnetic radiation has associated with it an oscillating electric field. Even when the radiation is not absorbed, this field has an effect on the electrons in the solid. If we think of an electrical field applied to an atom, we can imagine the electrons pulled by the field so that the atom is no longer spherical. The applied field produces a separation of the centres of positive and negative charges, that is, it induces an electric dipole moment. (A molecule in a solid may also have a permanent electric dipole moment produced by an unequal distribution of bonding electrons between the nuclei, but this is present in the absence of an applied field.) The oscillating field of the radiation can be thought of as pulling the

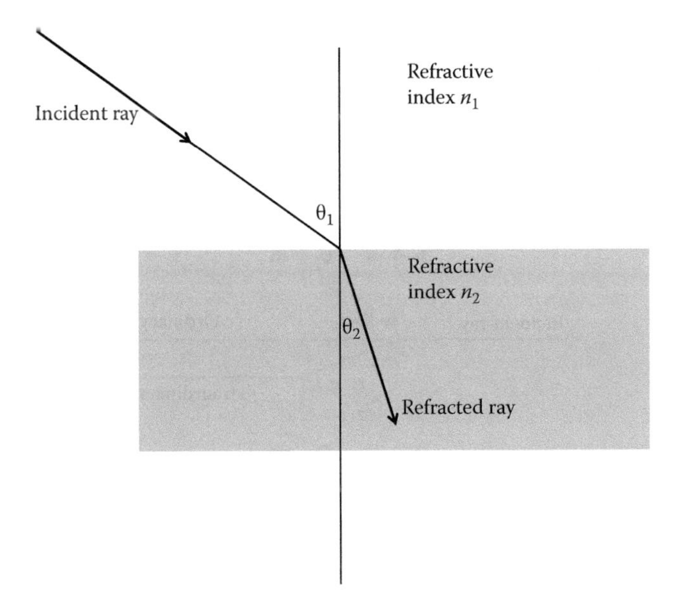

FIGURE 8.21 Diffraction of a ray of electromagnetic radiation when passing from one medium to another of a different refractive index.

electrons alternately one way and then the other way. The amount of response from the electrons depends on how tightly bound the electrons are to the nucleus. This property is called the **polarisability** and is higher for large ions with low charge, for example, Cs^+ and I^-, than for small, highly charged ions such as Al^{3+}. If there is a high concentration of polarisable ions in a solid, then the radiation will be slowed down more and the refractive index of the material increases. Adjusting the refractive index by adding carefully selected impurities is useful in a number of applications. For example, controlling the refractive index of glass is very important when making lenses for telescopes, binoculars, and cameras. Lead ions (Pb^{2+}) are highly polarisable and are used to produce glass of a high refractive index.

8.6.1 Calcite

Calcite is the most stable polymorph of calcium carbonate ($CaCO_3$). Single crystals are transparent and display the interesting optical property of **birefringence**. Birefringent materials such as calcite have different polarisabilities in the directions of different crystal axes and hence different refractive indices for light polarised perpendicular to these axes. Calcite has a particular refractive index along a unique axis and a differing refractive index along the directions perpendicular to this axis. The unique axis is known as the **optical axis**. When light is passed through the material, it splits into two beams traveling at different speeds due to the different refractive indices. One beam obeys Snell's law and is known as the **ordinary ray**. The other is the **extraordinary ray**. The effect is illustrated in Figure 8.22, which shows the crystal structure of calcite; the optical axis is perpendicular to the planes of carbonate ions $\left(CO_3^{2-}\right)$. The other two axes contain the planes. Thus, for calcite, the polarisability is not equal in all directions due to the asymmetry of the crystal structure; components in two directions are equal, but differ from that in the third direction, the optical axis. The ordinary rays are polarised in the plane perpendicular

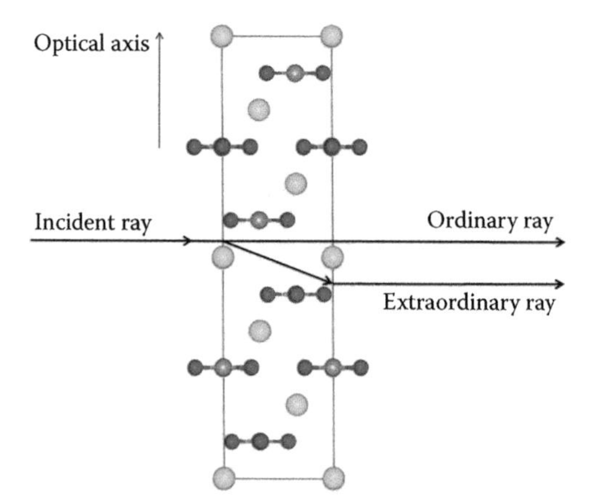

FIGURE 8.22 A ray of electromagnetic radiation passing into a birefringent crystal. The ray is entering perpendicular to the surface, which contains the optical axis, and the ordinary ray travels straight through.

to the optical axis. The extraordinary rays are polarised in the plane parallel to the optical axis. If the beam enters at right angles to the surface, the ordinary ray is not deflected, but the extraordinary ray is deflected.

The interaction of the extraordinary ray with anisotropic oscillations in the crystal causes the ray to be propagated in a direction that is not at right angles to the wavefront and this causes it to deviate from the ordinary ray. Birefringence can only occur for crystals displaying asymmetry. It is not observed with cubic crystals, which are identical in all three directions, unless an asymmetrical stress is applied. Interestingly, the effect of stress is employed by engineers in testing structures. A model of the planned structure is made from clear plastic and then viewed through crossed polaroids. Where there are stresses in the structure, colours are seen.

One way of producing polarised light is to use a Nicol prism. The Nicol prism exploits the splitting by calcite of a beam of light into two rays to produce polarised light. Two crystals of calcite are glued together using Canada balsam cement that has a refractive index between that of the ordinary ray and that of the extraordinary ray. When light hits the prism at certain angles, the ordinary ray is totally internally reflected at the boundary, but the extraordinary ray continues into the second crystal. Thus, the light emerging from the second crystal is polarized.

At the end of 2010, the birefringence of calcite was used independently by two groups to produce a **cloak of invisibility** that worked at optical wavelengths; an object behind such a 'cloak' cannot be seen. George Barbastathis's group at the Singapore-MIT Alliance for Research and Technology Centre produced a 2-D cloak that worked under water, and a group led by John Pendry at Imperial College, London, and Shuang Zhang at the University of Birmingham, UK, has produced one that hides an object in air. The former optical cloak was made of two prisms of calcite glued together such that the optical axes of both are at 30° to the interface, as shown in Figure 8.23.

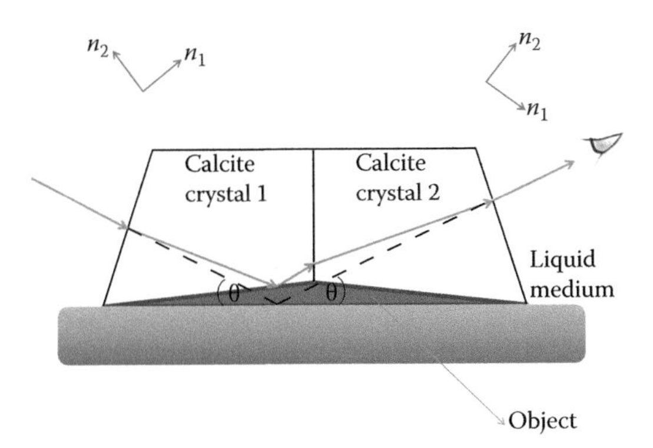

FIGURE 8.23 An optical cloak made from two crystals of calcite. The path of the light is shown by the blue trace. The ray exits in the direction it would have if reflected off the surface in the absence of object and cloak, as shown by the dashed trace. (Adapted from Zhang, B. L., Luo, Y. A., Liu, X. G., and Barbastathis, G.,(2011) Macroscopic invisibility cloak for visible light, *Phys. Rev. Lett.*, **106**, 033901.)

Polarised light enters one prism is refracted and then reflected by the object. The prisms are joined together at such an angle that the ray reflected from the object is refracted so that it appears to the observer that the light has simply been reflected off the base. Thus, both the object *and* the cloak are invisible.

8.6.2 Optical Fibres

Optical fibres are used to transmit light in the way that metal wires are used to transmit electricity. For example, a telephone call can be sent along an optical fibre in the form of a series of light pulses from a laser. The intensity, the time between pulses and the length of the pulse, can be modified to convey the contents of the call in coded form. In order to transmit the information over useful distances (of the order of kilometres), the intensity of the light must be maintained so that there is still a detectable signal at the other end of the fibre. Thus, much of the art of making commercial optical fibres lies in finding ways of reducing energy loss as the beam travels along the fibre.

The first requirement is that the laser beam keeps within the fibre. Laser beams diverge less than conventional light beams so that using laser light is of help, but even so, there is a small tendency for the beam to stray outside the fibre. Therefore, fibres are usually constructed with a variable refractive index across the fibre. The beam is sent down a central core. The surrounding region has a lower refractive index than the core so that light deviating from a straight path is totally internally reflected and hence remains in the core. This is illustrated in Figure 8.24.

The refractive index can be varied by adding selected impurities. In the case illustrated in Figure 8.24, the totally internally reflected rays travel a longer path than those that travel straight along the core. This will lead to a pulse being spread out in time. One way to keep the pulses together is to use very narrow cores, so that essentially all the light travels the straight-ray path. An alternative is the variable refractive index core: the lower the refractive index, the faster is the speed of light. Thus, if the outer parts of the core have a lower refractive index, then the reflected light moves faster and this compensates for the longer path length.

There are bound to be some imperfections in the fibre and these are another source of energy loss. The imperfections cause scattering of the light of a type known as Rayleigh scattering. Rayleigh scattering does not cause any change in the wavelength of the light, only in its direction. The amount of scattering depends on $(1/\lambda^4)$, where λ is the wavelength, so there is much less scattering for longer wavelengths. Even

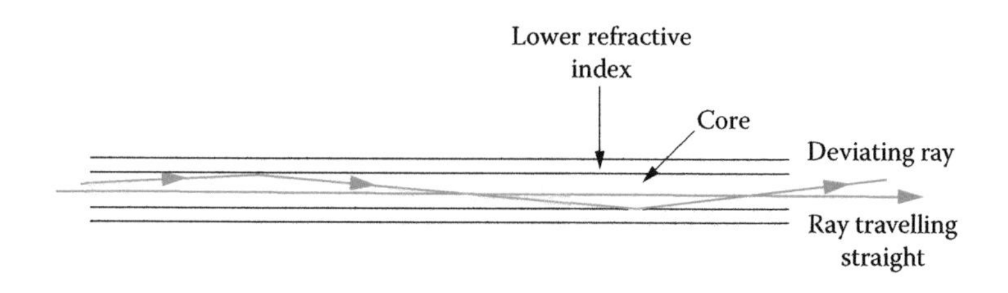

FIGURE 8.24 Rays of light traveling along an optical fiber.

the reduction in going from blue to red light is significant and is responsible for the color of the sky. To reduce Rayleigh scattering, the lasers employed for optical fibre systems usually emit long-wavelength infrared radiation.

A third source of energy loss is absorption of light by the fibre. In a fibre several kilometres long, a very small amount of impurity can give rise to substantial absorption. We can get an idea of this by looking at a sheet of window glass edge-on. Instead of being clear, the glass appears green. This is due to absorption by the Fe^{2+} ions in the glass. A windowpane is only about half a metre across, so that we can see that in a fibre of a few kilometres in length, there would be considerable loss due to such absorption. In a glass, the spectrum of the impurity ions is similar to that in a crystal, but because the ions occupy several different types of site in a glass, the absorption bands are wider; each site giving rise to a band at a slightly different wavelength. In a 3-km-long optical fibre operating at 1300 nm in the near infrared, the intensity of the Fe^{2+} absorption is still such that a concentration of two parts in 10^{10} would reduce the amount of radiation by half. The materials for optical fibres must therefore be very pure. One reason why silica has been widely used is because high-purity silicon tetrachloride, developed for the semiconductor industry, is commercially available as a starting material to make impurity-free SiO_2.

Metal ions are not the only source of absorption, however. Using infrared radiation means that there is likely to be loss due to absorption by molecular vibrations. In silica glass, the structure may contain dangling SiO bonds that easily react with water to form –OH bonds. The vibrational frequencies of –OH bonds are high and close to the frequencies used for transmission. It is important, therefore, to exclude water when manufacturing silica optical fibres. Even when water is excluded and there are no –OH bonds, the absorption by vibrational modes cannot be neglected. Si–O bonds vibrate at lower frequencies than O–H bonds, so the maximum in the absorption does not interfere. However, the SiO absorption is very strong and the peak tails into the region of the transmission frequency. There has been some research into substances with lower vibrational frequencies than silica, particularly fluorides, but as yet such substances are not economically viable, being difficult to manufacture and more expensive than silica.

There are still losses in the fibres developed for commercial use; nonetheless, these have been reduced to a point where they are used for delivering cable TV and broadband computer services. In the next section, we look at new materials that have been developed, which may form the basis of integrated optical circuits. These materials are photonic crystals.

8.7 PHOTONIC CRYSTALS

Photonic crystals have been hailed as the optical equivalent of semiconductors. The idea of such crystals was first developed in the 1980s by Eli Yablonovitch at Bell Communications Research.

A photonic crystal consists of a periodic arrangement of two materials of different refractive indexes. At each boundary between the two materials, light, or other electromagnetic radiation, will refract and partly reflect. The beams from the different interfaces will reinforce or cancel each other out, depending on their relative phases. Whether two beams will be in phase is determined by the wavelength of the radiation, its direction of travel, the refractive index of the photonic crystal materials, and the particular periodic arrangement. For certain wavelengths of radiation, refractive indices, and spacings of the materials, complete cancellation in all directions can occur so that such wavelengths are not transmitted by the crystal. The range of forbidden wavelengths is known by analogy with semiconductors as the **photonic band gap**. Figure 8.25 shows how a forbidden wavelength can occur for a one-dimensional arrangement—a row of slabs of dielectric material.

As the light reaches each slab, some is reflected due to the change in the refractive index. For the correct spacing of the slabs, the reflected rays from each slab are in phase with each other but out of phase with the incident light. For such wavelengths, the incident and the reflected rays cancel each other (Figure 8.25a).

The original photonic crystal was produced by accurately drilling holes a millimetre in diameter in a block of material with a refractive index of 3.6. This crystal had a photonic band gap in the microwave region. Similar structures with band gaps in the microwave and radio regions are being used to make antennae that direct radiation away from the heads of mobile phone users. Producing photonic crystals with band gaps at shorter wavelengths—infrared and visible—is less straightforward.

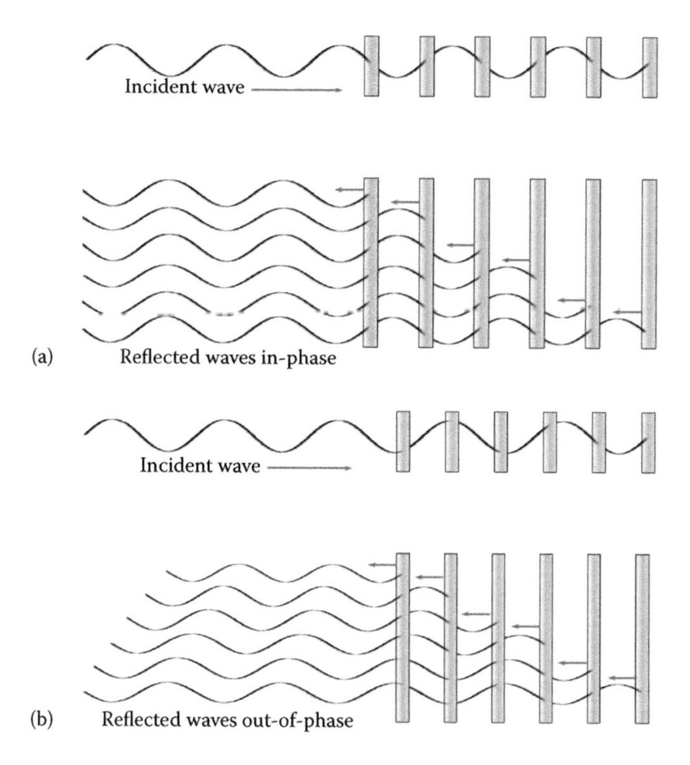

FIGURE 8.25 A row of dielectric slabs. Light traveling through can interfere destructively with the reflected light, giving rise to forbidden wavelengths. In (a) the reflected waves are all in phase with one another and out of phase with the incident wave and destructive interference occurs. In (b) the waves are reflected with slightly different phases and the incident light travels through.

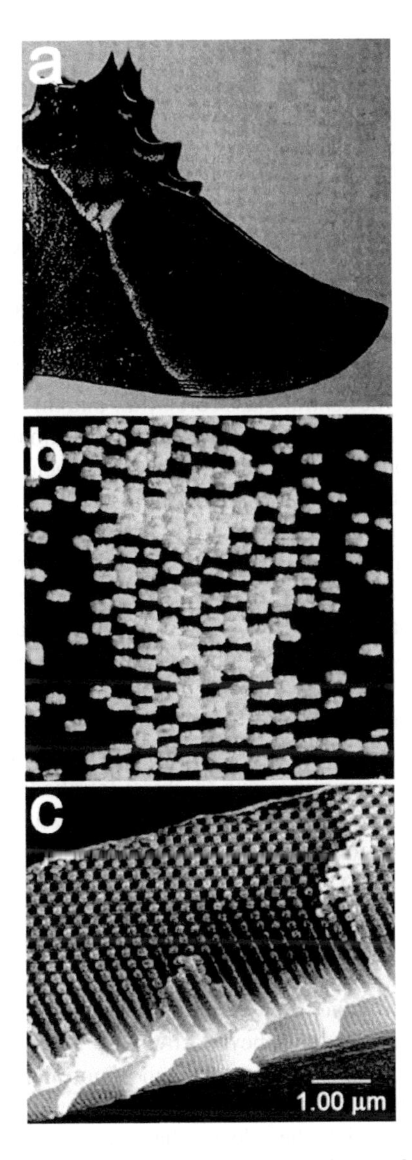

FIGURE 8.26 (a) The wing of the Kaiser-I-Hind butterfly. (b) Optical microscope image showing the scales on the wing, the black ground scales and the green cover scales containing the photonic crystals. (c) SEM image of a cross-section of one of the green scales showing the periodic structure. (Reprinted from Argyros, 2002, A. et al., 'Electron Tomography and Computer Visualization of a Three-Dimensional "photonic" Crystal in a Butterfly Wing-Scale', *Micron*, **33**, 483–487, Copyright 2002 with permission from Elsevier.)

The hole size or lattice spacing needed for photonic crystals is roughly equal to the wavelength of the radiation divided by the refractive index. A GaAs laser produces radiation of wavelength 904 nm. To use this radiation with a photonic crystal composed of silica with a refractive index of 1.45 would require a spacing of 623 nm or 0.623 μm. Machining holes of this size is impractical, and the production of regular periodic structures with spacings of this order is a technical problem that needs to be overcome before integrated optical circuits and other devices can be manufactured. One approach to this problem is to suspend spheres (usually of silica) with a diameter of less than 1 μm in a colloidal suspension or hydrogel. The spheres arrange themselves into a close-packed structure.

Two-dimensional photonic materials—materials in which light is blocked within a plane but transmitted perpendicular to the plane—are useful as optical fibres. An ingenious method of constructing such fibres is to pack a series of hollow capillary tubes around a central glass core. The structure is then heated and drawn until it is only a few micrometres thick. The central core is now surrounded by a periodic array of tubes of the right diameter to have a photonic band gap in the near infrared. It is also possible to replace the central glass core by air and this enables very-high-power laser signals to be transmitted along the fibre without damage to the fibre material.

Other potential applications of photonic crystals include crystals with rows of holes removed to guide radiation around sharp bends (something that cannot be attained with conventional optical fibres), nanoscopic lasers formed from thin films, ultrawhite pigment formed from a regular array of submicron titanium dioxide particles, radiofrequency reflectors for magnetic resonance imaging (MRI), and LEDs.

Photonic crystals have only been studied in the laboratory for a few decades, but there are naturally occurring examples, the best-known being the gemstone opal. Opals consist of tiny spheres of silica arranged in a face-centered cubic structure. These are thought to have formed from colloidal silica solutions and the color depends on the size of the spheres. The green color on the scale of the wings of the butterfly *Teinopalpus imperialis* is due to chiral tetrahedral units in a monoclinic structure. Figure 8.26 shows part of the butterfly wing, a microscope image of the green scales, and a SEM image showing the structure.

8.8 METAMATERIALS

Metamaterials like the photonic crystals just discussed, owe their properties of interest to their structure rather than their composition. They are composite materials, engineered to have properties beyond those of naturally occurring materials. They are often periodic, but you should note an important difference from the photonic materials described in the previous section. Metamaterials behave optically like a homogeneous material; they do not have a photonic band gap but they do have a refractive index that describes the bulk metamaterial. In photonic solids, the refractive index is characteristic of the materials from which the structure is made.

Great interest in these materials has centered around their unusual optical, electric, and magnetic properties, in particular their negative refractive index. In 1968, Victor Veselago put forward the possibility of a negative refractive index, but no natural materials have been found to possess this property. In general, the refractive index (n) is given by $n = \pm (|\varepsilon_r||\mu_r|)^{1/2}$, where ε_r and μ_r are the relative permittivity and permeability of the material, respectively, at the wavelength of interest. For the majority of materials, ε_r and μ_r are positive and n takes the positive square root. Some materials, such as metals and ferroelectrics, have negative permittivity. Some ferrites have negative permeability. These two groups also have positive refractive indices.

However, if both ε_r and μ_r are negative, then it can be shown that the negative square root has to be taken, meaning that the refractive index is negative. The first report of a negative refractive index material was not until 2000. This was a metamaterial with a negative refractive index in the microwave region.

The difference between materials with positive and negative refractive indices is illustrated in Figure 8.27. In Figure 8.27a, radiation passes from a medium, for example, air with a refractive index n_1 to a medium with a refractive index n_2, where $n_2 > n_1$. The radiation approaches and is bent through an angle θ_2. In Figure 8.27b, the second medium has a refractive index $-n_2$ and is bent through the angle $-\theta_2$. That is, the ray in Figure 8.27b is bent through an angle of the same magnitude as in Figure 8.27a, but in the opposite direction.

The refractive index is related to the velocity of radiation in a medium. One of the consequences of a negative index of refraction is that the phase velocity of the radiation is negative, that is, the change of phase of the wave travels in the opposite direction to the direction of propagation (and of the direction of energy transfer). This leads to some unusual properties such that the Doppler effect is reversed with radiation travelling towards the observer being shifted to longer wavelengths (red shift). A useful consequence of the negative refractive index is that a lens made of such a material is not subject to the diffraction limit of ordinary lenses, so a higher resolution can be achieved. Light from objects whose distance apart is less than half the wavelength of the light contains components that decay exponentially: **evanescent waves**. Lenses made of negative refractive index materials increase the amplitude of the evanescent waves. When they emerge from the lens, the waves decay again such that the amplitude at the image plane is equal to the original amplitude when the wave leaves the object. Thus, the evanescent wave component is not lost and all the light from the object is collected. Such lenses have thus been dubbed superlenses. In 2005, a superlens effect at optical frequencies was observed for a thin slab of silver.

The application that made metamaterials famous, however, was their use in 'cloaks of invisibility'. The apparent invisibility rests on the presumption that

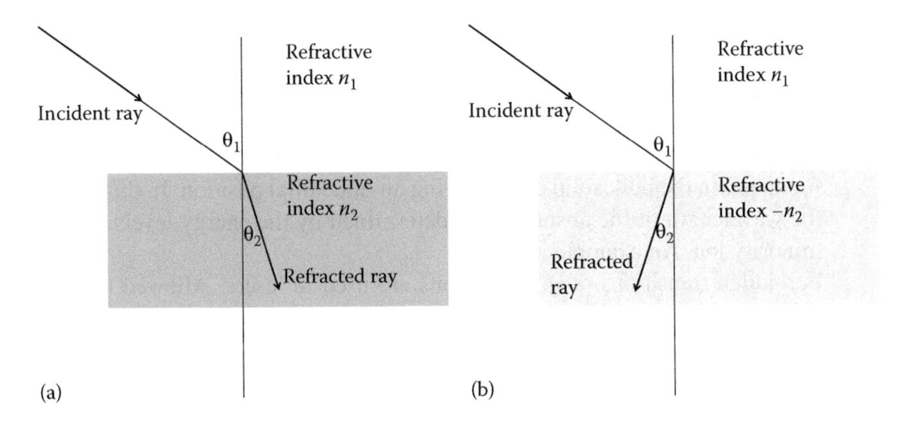

FIGURE 8.27 (a) Radiation passing from a medium with a refractive index n_1 to one with a refractive index n_2. (b) Radiation passing from a medium with a refractive index n_1 to one with a refractive index $-n_2$.

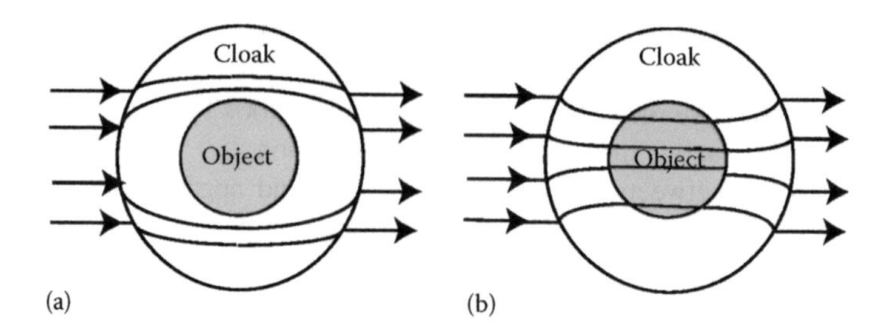

FIGURE 8.28 Two ways of cloaking. (a) Coordinate transformation and (b) cancellation of scattering.

electromagnetic radiation travels in straight lines. There are a number of methods of achieving a situation in which radiation is deviated in the vicinity of an object and led to continue parallel to the original beam on the other side of the object. The radiation then appears to have come straight through the object. We shall mention two methods, illustrated in Figure 8.28. In the first, coordinate transformation (Figure 8.28a), the radiation is diverted around the object, by varying the refractive index of the cloak. The skill in designing such a cloak lies in calculating the exact transformation needed at every point.

An alternative method is to select a cloak with a varying refractive index so that any scattering from the object is cancelled (Figure 8.28b). In this case, the radiation passes through the object but both the object and the cloak are invisible.

Both these methods depend on engineering a material with a refractive index varying in a precise way. Metamaterials have the advantage that they can be tuned to do this by varying the physical structure of a material. The first example of a cloak of invisibility, reported in 2006, used an array of metallic 'split ring' resonators. This device shielded a copper cylinder at a range of microwave frequencies. Researchers at the University of Texas have demonstrated cloaks to suppress scattering at microwave frequencies in all directions from a cylinder.

8.9 SUMMARY

1. The colour of solids can be due to the presence of minority ions substituting for an ion in the bulk solid or occupying an interstitial position. In this case, the wavelength of the absorption is determined by the energy levels of the minority ion. An example is ruby.
2. Forbidden transitions of minority ions are used in lasers. Allowed transitions are used in phosphors.
3. F-centres are produced when alkali halides are irradiated with high-energy photons, e.g., X-rays. The colour is determined by energy levels of electrons at positions of vacancies and can be described by a particle in a box model.
4. The absorption and emission of radiation can be determined in continuous solids by the band structure of the solid. Examples are semiconductors such as gallium arsenide and metals.

5. The refractive index of materials is very important for applications such as lenses and optical fibres. The refractive index depends on the polarizability of the atoms/ions in the solid. Highly polarizable ions such as Pb^{2+} lead to an increase in refractive index.
6. Photonic crystals contain a periodic arrangement of two materials of different refractive index. Some wavelengths of radiation are not transmitted by such crystals. The range of these wavelengths is known as the photonic band gap.
7. Metamaterials are also engineered and often periodic in arrangement. However, the refractive index of these materials is a property of the structure rather than the material from which they are made.

QUESTIONS

1. In the oxide MnO, the Mn^{2+} ions occupy octahedral holes in an oxide lattice. The degeneracy of the 3d levels of the manganese ion are split into two, as for Ti^{3+}. The five d electrons of the Mn^{2+} ions occupy separate d orbitals and have parallel spins. Explain why the absorption lines due to transitions between the two 3d levels are very weak for Mn^{2+}.
2. Figure 8.29 shows the energy levels of Nd^{3+} in yttrium aluminium garnet ($Y_3Al_5O_{12}$), which are involved in the laser action of this crystal. Describe the processes that occur when the laser is working.
3. Figure 8.6 showed transitions between energy levels of Ce^{3+} when it is acting as a phosphor in Ce^{3+}/YAG. Describe the processes occurring.
4. Figure 8.30 shows two bands of a semiconductor. Is the band gap of this solid direct or indirect?

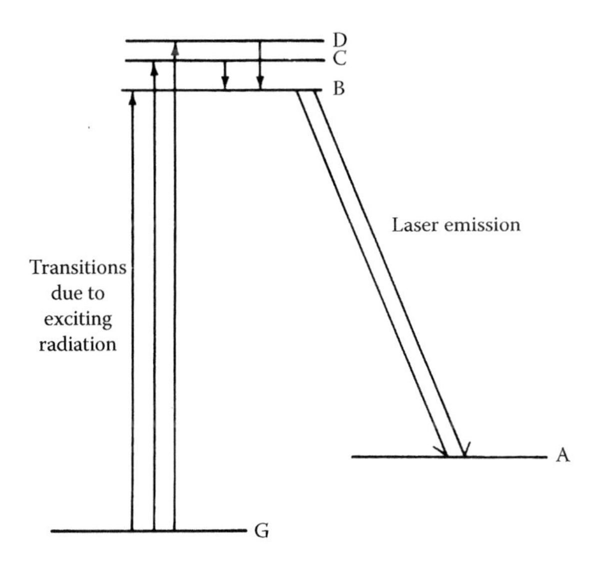

FIGURE 8.29 Energy levels of Nd^{3+}in yttrium ion garnet.

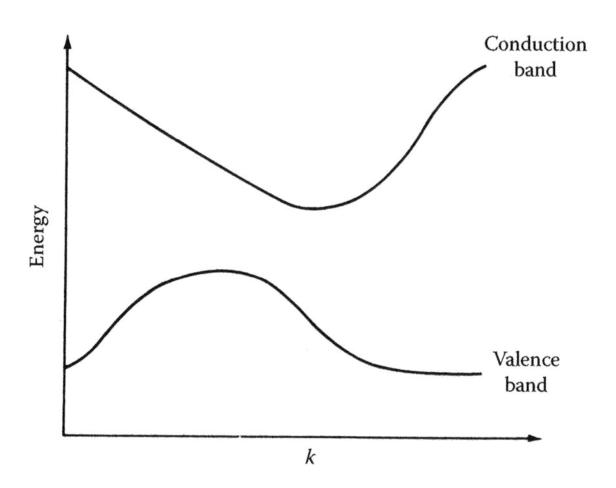

FIGURE 8.30 A plot of energy levels vs k for a solid.

5. A defect in which an interstitial iodide ion and a lattice iodide ion form a dimer I_2^- is found in the lead halide perovskite $CH_3NH_3PbI_3$. Is this an F-centre or an H-centre?
6. How would you expect the formation of colour centres to affect the density of the crystal?
7. Explain why silicon is used for solar cells but not for LEDs.
8. Which orbitals would you expect to combine to form a delocalised band in polyphenylenevinylene?
9. Would doping of polyacetylene with (a) Rb or (b) H_2SO_4 give an n-type or a p-type conductor?
10. When polyacetylene is doped with chloric(VII) acid ($HClO_4$), part of the acid is used to oxidise the polyacetylene and part is used to provide a counter anion. The oxidation reaction for the acid is given in Equation 8.1. Write a balanced equation for the overall reaction.

$$ClO_4^- + 8H^+ + 8e^- = Cl^- + 4H_2O \qquad (8.1)$$

11. Undoped polythiophene is red. What does this suggest about the size of the conduction band/valence band energy gap in this polymer? Would you expect a polythiophene LED to emit a red light?
12. Tetragonal calomel, whose empirical formula is HgCl, is birefringent, whereas alkali metal halides such as NaCl are not birefringent. The refractive index of the ordinary ray of calomel is higher than that of NaCl. From a consideration of the molecular and crystal structures of calomel and sodium chloride, explain these observations.
13. An optical fiber has a silica core and a doped silica surround. Why would B_2O_3 or P_2O_5 be a suitable dopant?
14. It has been proposed that the coloration of peacock feathers is due to the presence of photonic structures. By changing the periodicity of these structures, different colors are produced. Explain why.

9 Magnetic and Electrical Properties

9.1 INTRODUCTION

One consequence of the closeness of atoms in a solid is that the properties of the individual atoms or molecules can interact cooperatively to produce effects not found in fluids. In a piece of iron used as a magnet, for example, the magnetism of the iron atoms aligns to produce a strong magnetic effect. This is known as ferromagnetism. Other cooperative magnetic effects lead to a cancelling (antiferromagnetism) or a partial cancelling (ferrimagnetism) of the magnetism of different atoms. Ferromagnets and ferrimagnets have many commercial applications, from compass needles and watch magnets to hard-disk read heads and computer memory devices.

Cooperative effects are not confined to magnetism; similar effects can occur in the response of a crystal to mechanical stress and to electric fields. The electrical analogue of ferromagnetism is the ferroelectric effect, in which the material develops an overall electrical polarisation, that is, a separation of charge. Ferroelectric materials are important in the electronics industry as capacitors (for storing charge) and transducers (e.g., in the conversion of ultrasound to electrical energy). Ferroelectric crystals are a subclass of piezoelectric crystals, which have commercial uses of their own. For example, quartz watches use piezoelectric quartz crystals as oscillators.

This chapter looks at the types of materials that display cooperative magnetic and dielectric properties, and in the final section, we discuss multiferroics—materials that display more than one type of cooperative property, electric and magnetic.

To start, we consider the weaker magnetic effects that can be found in all types of matter.

9.2 MAGNETIC SUSCEPTIBILITY

A magnetic field produces lines of force that penetrate the medium to which the field is applied. These lines of force show up, for example, when we scatter iron filings on a piece of paper covering a bar magnet. The density of these lines of force is known as the **magnetic flux density**. In a vacuum, the magnetic field and the magnetic flux density are related by the permeability of free space, μ_0

$$B = \mu_0 H \tag{9.1}$$

If a magnetic material is placed in the field, however, it can increase or decrease the flux density. Diamagnetic materials reduce the density of the lines of force, as shown in Figure 9.1, whereas paramagnetic materials increase the flux density.

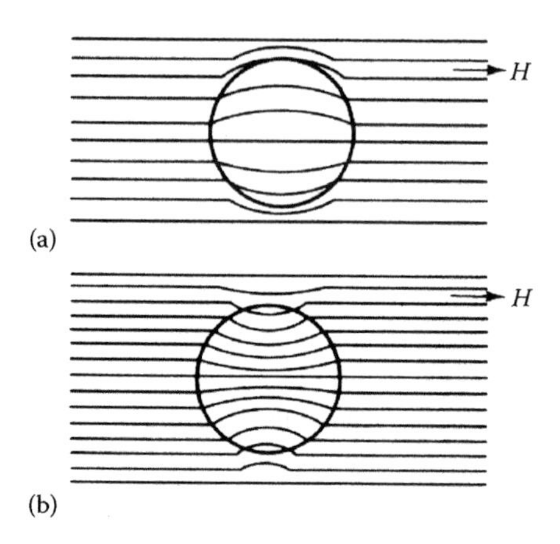

FIGURE 9.1 The flux density in (a) a diamagnetic and (b) a paramagnetic sample.

The field of the sample in the applied field is known as its **magnetisation** (**M**). The magnetic flux density is now given by

$$B = \mu_0 (H + M) \quad (9.2)$$

The magnetisation is usually discussed in terms of the **magnetic susceptibility** (χ), where $\chi = M/H$, or the permeability (μ_r), where $\mu_r = 1 + \chi$.

Diamagnetism is present in all substances, but is very weak, so it is not normally observed if other effects are present. Diamagnetism is produced by the circulation of the electrons in an atom or a molecule. Atoms or molecules with closed shells of electrons are diamagnetic. Unpaired electrons, however, give rise to **paramagnetism**. Simple paramagnetic behaviour is found for substances such as liquid oxygen or transition metal complexes, in which the unpaired electrons on different centres are isolated from each other. In a magnetic field, the magnetic moments on different centres tend to align with the field and hence with each other, but this is opposed by the randomising effect of thermal energy, and in the absence of a field, the unpaired electrons on different centres are aligned randomly. The interplay of the applied field and thermal randomisation leads to the temperature dependence described by the **Curie law**:

$$\chi = \frac{C}{T} \quad (9.3)$$

where χ is the magnetic susceptibility, C is a constant known as the Curie constant, and T is the temperature in Kelvin.

Different temperature dependence is observed when there is cooperative behaviour. The changeover from independent to cooperative behaviour is associated with a characteristic temperature. In **ferromagnetism**, electron spins on different atoms/molecules are aligned, via spin-spin coupling, in the absence of a magnetic field. The susceptibility of the solid above the Curie temperature, T_C, is now given by the Curie–Weiss law:

$$\chi = \frac{C}{(T - T_C)} \tag{9.4}$$

where T_C is the **Curie temperature.** For **antiferromagnetism**, the temperature dependence in the paramagnetic region is of the form:

$$\chi = \frac{C}{(T + T_N)} \tag{9.5}$$

where T_N is the **Néel temperature**. These two behaviours are illustrated in Figure 9.2.

Ferrimagnetism has a more complicated form of temperature dependence, with ions on different sites having different characteristic temperatures. The characteristics of the various types of magnetism that we shall be concerned with are given in Table 9.1.

9.3 PARAMAGNETISM IN METAL COMPLEXES

In solids containing metal complexes such that the unpaired electrons on the different metal atoms are effectively isolated, the susceptibility can be discussed in terms of magnetic moments. The isolated metal complex can be thought of as a

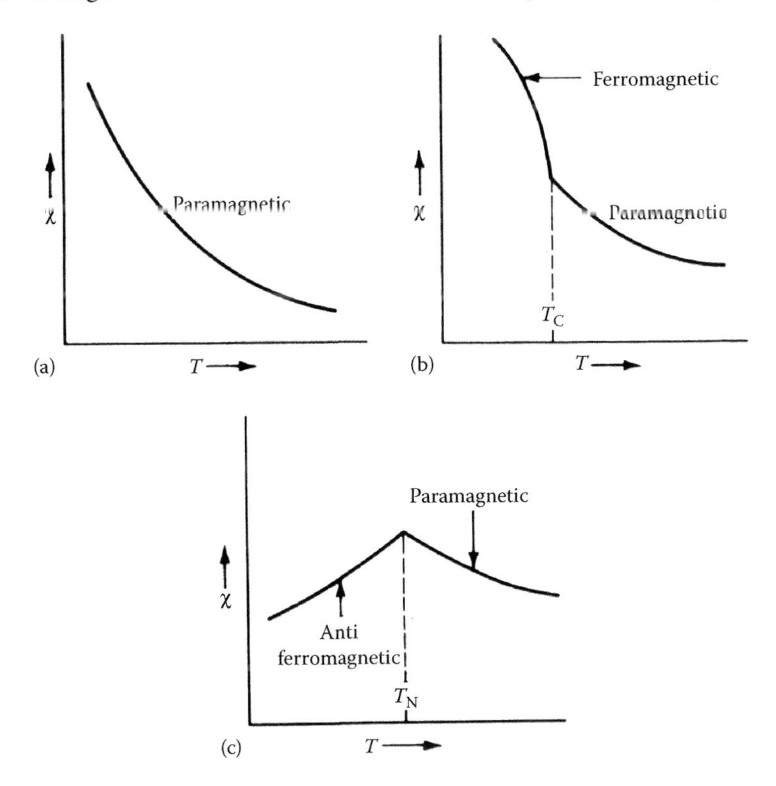

FIGURE 9.2 The variation of magnetic susceptibility with temperature for (a) a paramagnetic substance, (b) a ferromagnetic substance, and (c) an antiferromagnetic substance.

TABLE 9.1
Characteristics of the Types of Magnetism

Type	Sign of χ	Typical χ Value (calculated Using SI units)	Dependence of χ on H	Change of χ with increasing temperature	Origin
Diamagnetism	–	$-(1–600)\times10^{-5}$	Independent	None	Electron charge
Paramagnetism	+	0–0.1	Independent	Decreases	Spin and orbital motion of electrons on individual atoms
Ferromagnetism	+	$0.1–10^{-7}$	Dependent	Decreases	Cooperative interaction between
Antiferromagnetism	+	0–0.1	May be dependent	Increases	magnetic moments of individual atoms
Pauli paramagnetism	+	10^{-5}	Independent	None	Spin and orbital motion of delocalized electrons

small magnet. Each complex in a solid will produce its own magnetic field due to the unpaired electrons. If the solid consists of one type of complex, then each complex produces a magnetic field of the same magnitude. However, thermal motion causes the orientation of these fields to be random. In the Curie law (Equation 9.3), the temperature dependence is the result of this thermal motion, but the constant (C) gives us information on the value of the magnetic field, known as the **magnetic moment** (μ) of the complex. The magnetic moment, unlike the susceptibility, does not normally vary with temperature. The dimensionless quantity χ in Equation 9.2 is the susceptibility per unit volume. To obtain the size of the magnetic field due to an individual complex, χ is divided by the specific gravity to give the susceptibility per unit mass of the sample and then multiplied by the relative molecular mass to obtain the molar susceptibility χ_m. Assuming that each complex has a fixed magnetic moment (μ) and that the orientation is randomised by thermal motion, it can be shown that χ_m is proportional to μ^2, as in Equation 9.6.

$$\chi_m = \frac{N_A \mu_0}{3kT} \mu^2 \tag{9.6}$$

where N_A is Avogadro's number, k is Boltzman's constant, μ_0 is the permeability of free space, and T is the temperature in Kelvin. In SI units, χ_m is in cubic meters per mole ($m^3\ mol^{-1}$) and the magnetic moment is in joules per tesla ($J\ T^{-1}$). It is usual to quote μ in Bohr magneton (BM or μ_B), where one **Bohr magneton** has a value of $9.274 \times 10^{-24}\ J\ T^{-1}$.

The magnetic moment (μ) is a consequence of the angular momentum of the unpaired electrons. The electrons possess both spin and orbital angular momenta. For the first-row transition elements, the contribution from the orbital angular momentum is greatly reduced or 'quenched' as a consequence of the lifting of the fivefold degeneracy of the 3d orbitals. In complexes of these elements, the magnetic moment is often close to that predicted for spin angular momentum only:

$$\mu_s = g\sqrt{S(S+1)} \tag{9.7}$$

where g is a constant with a value of 2.00023 for a free electron and μ_s is in Bohr magnetons.

The value of S depends on the number of unpaired electrons. Table 9.2 gives the values of S and μ_s for the possible numbers of unpaired 3d electrons.

The contributions from the orbital angular momentum cause deviations from these values. For complexes containing heavier metal ions, the interaction of the spin and the orbital angular momenta is greater. For the lanthanoids, the magnetic moment depends on the total angular momentum of the electrons (**J**), not just or mainly on the spin angular momentum. The total angular momentum (**J**) is the vector sum of the orbital (**L**) and the spin (**S**) angular momenta.

$$\mathbf{J} = \mathbf{L} + \mathbf{S} \tag{9.8}$$

TABLE 9.2

Values of S and μ_S for Unpaired 3d Electrons

No. unpaired electrons	Spin quantum number (S)	Magnetic moment (μ_s) in Bohr magneton
1	1/2	1.73
2	1	2.83
3	3/2	3.87
4	2	4.90
5	5/2	5.92

If **J** is the quantum number for the total electronic angular momentum, then the magnetic moment is given by

$$\mu = g\sqrt{J(J+1)} \tag{9.9}$$

Where

$$g = 1 + \frac{J(J+1) + S(S+1) - L(L+1)}{2J(J+1)}$$

This can give rise to large magnetic moments, especially for shells that are more than half-full. For example, Tb^{3+} with an f^8 configuration has a magnetic moment of 9.72 μ_B from Equation 9.9.

We now turn to solids for which the magnetism is not due to isolated spins but to cooperative effects involving the entire crystal.

9.4 FERROMAGNETIC METALS

When discussing the electrical conductivity of metals, we described them in terms of ionic cores and delocalised valence electrons. The core electrons contribute a diamagnetic term to the magnetic susceptibility, but the valence electrons can give rise to paramagnetism or a cooperative effect.

In filling the conduction band, we have implicitly put electrons into energy levels with paired spins. Even in the ground state of simple molecules, such as O_2, however, it can be more favourable to have the electrons in different orbitals with parallel spins than in the same orbital with paired spins. This occurs when there are degenerate or nearly degenerate levels. In an energy band, there are many degenerate levels and many levels very close in energy to the highest occupied level. It might well be favourable to reduce electron repulsion by having the electrons with a parallel spin singly occupying the levels near the Fermi level. To obtain a measurable effect, however, the number of parallel spins would have to be comparable with the number of atoms; 10^3 unpaired spins would not be noticed in a sample of 10^{23} atoms. Unless the density of states is very high near the Fermi level, a large number of electrons would have to be promoted to high energy levels in the band in order to achieve a measurable number of unpaired spins. The resulting promotion energy would be too great to be compensated for by the loss in electron repulsion. In the wide bands of the simple metals, the density of states is comparatively low, so that in the absence of a magnetic field, few electrons are promoted.

When a magnetic field is applied, the electrons acquire an extra energy term due to the interaction of their spins with the field. If the spin is parallel to the field, then its magnetic energy is negative; that is, the electrons are at lower energy than they were in the absence of a field. For an electron with a spin antiparallel to the field, it is now worthwhile to go to a higher-energy state and change the spin, as long as the promotion energy is not more than the gain in magnetic energy. This will produce a measurable imbalance of electron spins aligned with and against the field and hence the solid will exhibit paramagnetism. This type of paramagnetism is known as **Pauli paramagnetism**, and it is a very weak effect, giving a magnetic susceptibility much less than that due to isolated spins and comparable in magnitude to diamagnetism.

For a very few metals, however, the unpaired electrons in the conduction band can lead to ferromagnetism. In the whole of the periodic table, only iron, cobalt, nickel, and a few of the lanthanoids (Gd, Tb) possess this property. So, what is it about these elements that confers this uniqueness on them? It is not their crystal structure; they each have different structures and their structures are similar to those of other non-ferromagnetic metals. Iron, cobalt, and nickel; however, all have a nearly full, narrow 3d band.

The 3d orbitals are less diffuse than the 4s and 4p orbitals; that is, they are concentrated nearer the atomic nuclei. This leads to less overlap so that the 3d band is a lot narrower than the 4s/4p band. Furthermore, there are five 3d orbitals, so that for a crystal of N atoms, $5N$ levels must be accommodated. With more electrons and a narrower band, the average density of states must be much higher than in the ns/np band. In particular, the density of states near the Fermi level is high. In this case, it is energetically favourable to have substantial numbers of unpaired electrons at the cost of populating higher-energy levels. Thus, these elements have large numbers of unpaired electrons even in the absence of a magnetic field. For iron, for example, in a crystal of N atoms, there are up to $2.2N$ unpaired electrons, all with their spins aligned parallel. Note the contrast with a paramagnetic solid containing transition metal complex ions, where each ion may have as many as five unpaired electrons, but in the absence of a magnetic field, electrons on different ions are aligned randomly.

Ferromagnetism thus arises from the alignment of the electron spins throughout the solid, and this occurs for partially filled bands with a high density of states near the Fermi level. The 4d and 5d orbitals are more diffuse than the 3d orbitals and produce wider bands, so that ferromagnetism is not observed in the second-row and third-row transition elements. The 3d orbitals themselves become less diffuse across the transition series and lower in energy. In titanium, the valence electrons are in the 4s/4p band with a low density of states and, at the other end of the row in copper, the 3d band has dropped in energy so that the Fermi level is also in the 4s/4p band. Thus, it is only at the middle of the series that the Fermi level is in a region of high density of states. Schematic band diagrams for Ti, Ni, and Cu are given in Figure 9.3. The occupied levels are indicated by shading.

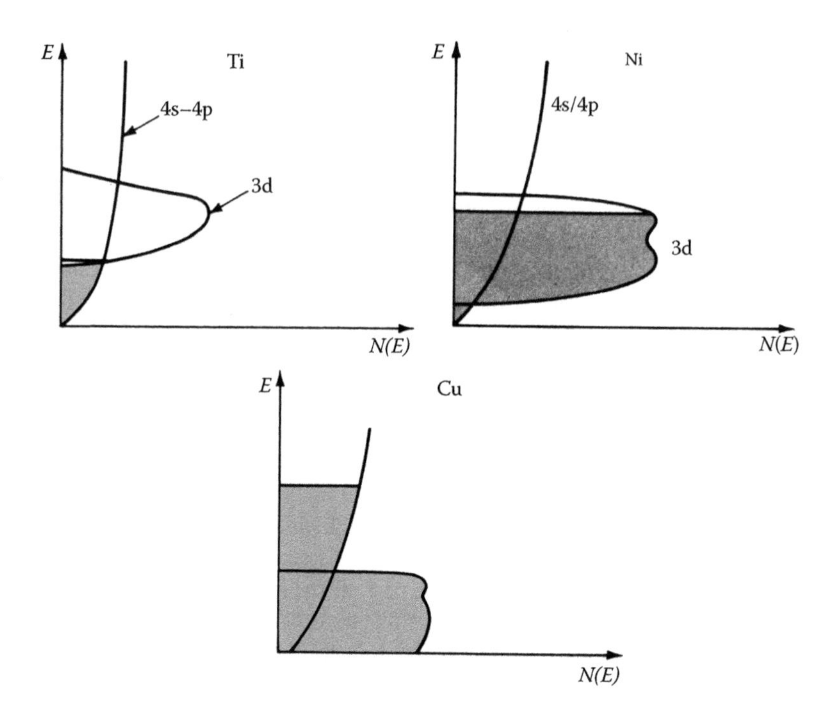

FIGURE 9.3 Schematic energy level diagrams for Ti, Ni, and Cu. The shading represents the occupied energy levels.

The high density of states found in the 3d bands of Fe, Co, and Ni also leads to a reduction of the mean free path of the electrons in this band. This causes a decrease in their mobility and hence in the electrical conductivity of these elements, compared with simple metals and copper, where the conduction electrons are in the s/p band.

The pure elements are not always suitable for applications requiring a metallic ferromagnet and many ferromagnetic alloys have been produced. Some of these contain one or more ferromagnetic elements and among these, alloys of iron, cobalt, and nickel with the lanthanoids, for example, $SmCo_5$ and $Nd_2Fe_{14}B$, have produced some of the most powerful permanent magnets known. In the lanthanoid alloys, f electrons contribute to the magnetism. Potentially, this could lead to a very high magnetisation because there are seven f orbitals, and so a maximum possible magnetisation corresponding to seven electrons per atom. The theoretical maximum magnetisation for the transition metals is five electrons per atom, as there are only five d orbitals. In practice, this maximum is never reached. In the pure lanthanoid metals, the overlap of f orbitals is so small that they can be regarded as localized. In the ferromagnetic lanthanoids, the magnetism is produced by delocalised d electrons. The interaction between these d electrons and the localised f electrons causes the alignment of the d and f electrons in order to reduce the electron repulsion. Thus, the f electrons on different atoms are aligned through the intermediary of the d electrons. In alloys, the f electrons can align via the transition metal d electrons, and although not all of the d and f electrons are aligned, it can be seen that high values of the magnetisation could be achieved. It is not surprising, then, that it is these transition metal/lanthanoid alloys that are the most powerful magnets. Other alloys can be made from nonmagnetic elements, such as manganese, and in these, the overlap of the d orbitals is brought into the range necessary for ferromagnetism, by altering the interatomic distance from that in the element.

The usefulness of a particular ferromagnetic substance depends on factors such as the size of the magnetisation produced, how easily the solid can be magnetised and demagnetised, and how readily it responds to an applied field. The number of unpaired electrons determines the maximum field, but the other factors depend on the structure of the solid and the impurities it contains, as discussed in the next section.

9.4.1 Ferromagnetic Domains

A drawback to the previous explanation may have occurred to you. If all 2.2N electrons are aligned in any sample of iron, why are all pieces of iron not magnetic? The reason for this is that our picture only holds for small volumes (typically 10^{-24} to 10^{-18} m^3) of metal within a crystal called **domains**. Within each domain, all the spins are aligned, but the different domains are aligned randomly with respect to each other.

It is actually possible to see these domains through a microscope on the polished surface of a crystal (Figure 9.4). What then causes these domains to form?

The spins tend to align parallel because of short-range **exchange interactions** stemming from electron–electron repulsion, but there is also a longer-range **magnetic dipole interaction** that tends to align the spins antiparallel. If you consider building up a domain starting with just a few spins, initially the exchange interactions dominate and, so the spins all lie parallel. As more spins are added, an individual spin will be subjected to an increasing magnetic dipole interaction. Eventually, the magnetic dipole interaction overcomes the exchange interaction and the adjacent piece of crystal has its spins aligned antiparallel to the original domain. Thus, within

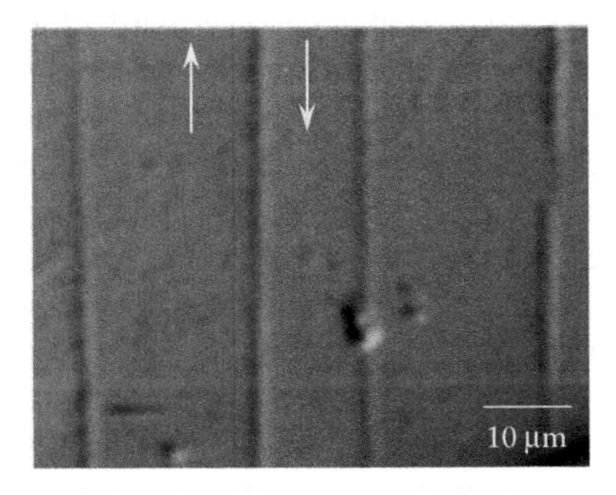

FIGURE 9.4 Domains in iron. (Reprinted with permission from Bathany, C., Le Romancer, M., Armstrong, J.N. and Chopra, H.D. *Phys. Rev. B* 82 184411 2010, Copyright 2010 by the American Physical Society. http://link.aps.org/PRB/v82/e184411.).

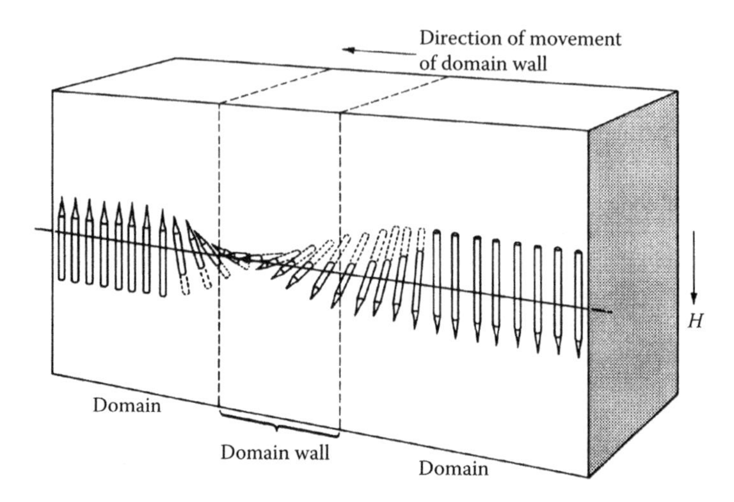

FIGURE 9.5 The movement of a domain wall. The broken lines show the domain wall, which separates two domains with the magnetic moments lined up in opposite directions. The moments twist to align with the applied field (H), and the wall moves in the direction of the arrow shown.

domains, exchange forces keep the spins parallel, whereas the magnetic dipole interaction keeps the spins of different domains aligned in different directions.

When a magnetic field is applied to a ferromagnetic sample, all the domains tend to line up with the field. This alignment can be accomplished in two ways. First, a domain of correct alignment can grow at the expense of a neighbouring domain. Between the two domains is an area of finite thickness, known as the domain wall. The changeover from the alignment of one domain to that of the next is gradual within the wall. When the magnetic field is applied, the spins in the wall nearest the aligned domain alter their spins to line up with the bulk of the domain. This causes the next spins to alter their alignment. The net effect is to move the wall of the domain further out, as shown in Figure 9.5. This process is reversible; the spins return to their former state after the magnetic field is removed.

If impurities or defects are present, it becomes harder for a domain to grow; there is an activation energy to aligning the spins through the defect, and therefore a larger magnetic field is required. Once the domain has grown past the defect, however, it cannot shrink back once the magnetic field is removed because this will also need an energy input. In this case, the solid retains its magnetisation. The amount retained depends on the number and the types of defects. Thus, steel (which is iron with a high impurity content) remains magnetic after the field is removed, whereas soft iron, which is much purer, retains hardly any magnetisation.

The second mechanism of alignment, which only occurs in strong magnetic fields, is when the interaction of the spins with the applied field becomes large enough to overcome the dipole interaction and entire domains of spins change their alignment simultaneously. The two mechanisms are compared in Figure 9.6.

The magnetic behaviour of different ferromagnetic substances is shown by their **hysteresis curves**. This is a plot of the magnetic flux density (B) against applied magnetic field (H). If we start with a nonmagnetic sample in which all the domains

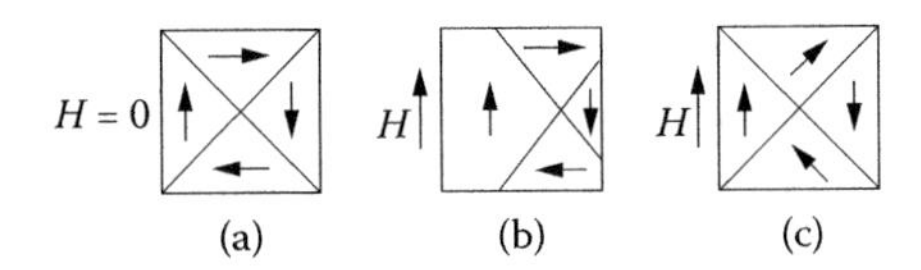

FIGURE 9.6 Magnetisation processes according to the domain model: (a) unmagnetised, (b) magnetised by domain growth, and (c) magnetised by domain rotation (spin alignment).

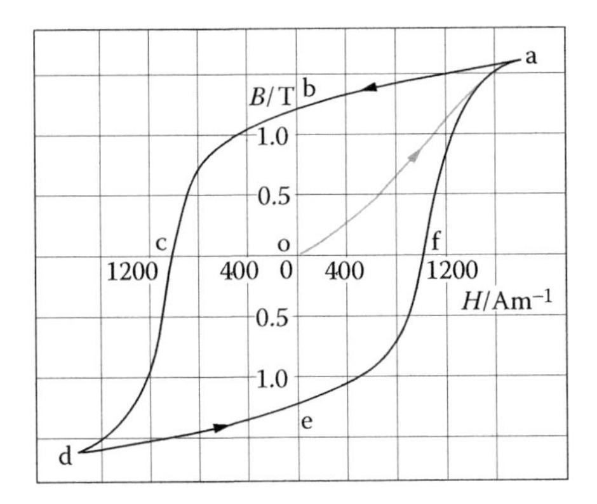

FIGURE 9.7 A B–H curve for a typical hard steel.

are randomly aligned, then in the absence of a magnetic field, B and H are zero. As the field is increased, the flux density also increases. The plot of B against H is shown in Figure 9.7. Initially, the curve is like '*oa*', which is not simply a straight line because the magnetisation is increasing with the field. At point '*a*', the magnetisation has reached its maximum value—all the spins in the sample are aligned. When the applied field is reduced, the flux density does not follow the initial curve. This is due to the difficulty of reversing processes where domains have grown through crystal imperfections. A sufficiently large field in the reverse direction to provide the activation energy for realignment through the imperfection must be applied before the magnetisation process can be reversed. At the point '*b*', therefore, where H is zero, B is not zero because there is still a contribution from M. The magnetisation at this point is known as the **remanent magnetisation**. The field that needs to be applied in the reverse direction to reduce the magnetisation to zero is the **coercive force** and is equal to the distance '*oc*'. If the coercive force is large, then the material is said to have high **coercivity**.

9.4.2 Permanent Magnets

Substances used as permanent magnets need a large coercive force, so that they are not easily demagnetised, and preferably should have a large remanent magnetisation. These substances have fat hysteresis curves and are referred to as hard magnets. They are often made from alloys of iron, cobalt, or nickel that form with small crystals and include nonmagnetic areas so that domain growth and shrinkage are difficult. The magnets for electronic watches, for example, are made from samarium/cobalt alloys. The best known of these alloys is $SmCo_5$, which has a coercive force of 6×10^5 A m^{-1} compared with 50 A m^{-1} for pure iron. The most widely used lanthanoid magnet is an alloy of neodymium, iron, and boron. Such magnets are used in MRI (magnetic resonance imaging) machines, wind turbine generators, smartphones, and even jewellery. The alloy has a tetragonal structure which makes it magnetise preferentially along a particular crystal axis and resist a change of direction, leading to a high coercivity. The coercive force varies with the precise composition but is of the order of 9×10^5 A m^{-1} and the remanent magnetisation is 1–1.5 T compared to 1.3 T for 99.5% iron.

9.4.3 Magnetic Shielding

Shielding is used either to protect electronic components, for example magnetic lenses in electron microscopes or components of electronic devices, from external magnetic fields or to confine magnetic fields from powerful magnets such as those in MRI machines. One way of doing this is to use a soft magnetic material. The shield acts by providing an alternative path for the magnetic field lines. The shield must be easy to magnetise and have a strong response to the magnetic field. This means it should have a low coercivity and high permeability, that is, a thin, steep hysteresis curve. Widely used materials are nickel–iron alloys. One example is mu-metal., which is 80% Ni, 15.5% Fe, and 4.5% Mo. The alloy is heated to high temperatures to increase the size of the magnetic domains. Most ferromagnetic materials undergo small changes in their shape and dimensions when a magnetic field is applied. This effect can be used in transducers to produce ultrasound waves and is responsible for the hum from transformers. Mu-metal is interesting because its shape and dimensions are virtually unchanged when a magnetic field is applied.

9.5 FERROMAGNETIC COMPOUNDS: CHROMIUM DIOXIDE

Chromium dioxide crystallises with a rutile structure (see Chapter 1) and is ferromagnetic with a Curie temperature of 392 K. CrO_2 has metal 3d orbitals that can overlap to form a band. In chromium dioxide, however, this band is very narrow, and, like Fe, Co, and Ni, chromium dioxide displays ferromagnetism. The dioxides later in the row have localised 3d electrons (e.g., MnO_2) and are insulators or semiconductors. TiO_2 has no 3d electrons and is an insulator. VO_2 has a different structure at room temperature and is a semiconductor. However, it does undergo a phase transition to a metal at 340 K, when it becomes Pauli paramagnetic. Therefore, chromium dioxide occupies a unique position among the dioxides, similar to that of iron, cobalt, and nickel among the first-row transition metals, in which the dioxides of the elements to the left have wide bands of delocalised electrons and the elements to the right have dioxides with localised 3d electrons. Because the metal atoms are further apart in the dioxides than in the elemental metals, the narrow bands that give rise to ferromagnetism occur earlier in the row than for the metallic elements.

9.6 ANTIFERROMAGNETISM: TRANSITION METAL MONOXIDES

These oxides have already been discussed in Chapter 5, and you may remember that they all had the sodium chloride structure, but had varying electrical properties. In this section, we shall see that their magnetic properties are equally varied. In TiO and VO, the 3d orbitals are diffuse and form delocalised bands. These oxides are metallic conductors. The delocalised nature of the 3d electrons also determines the magnetic nature of these compounds and, like the simple metals, they are Pauli paramagnetic. MnO, FeO, CoO, and NiO have localised 3d electrons and are paramagnetic at high temperatures. On cooling, however, the oxides become antiferromagnetic. In antiferromagnetism, the spins on the different nuclei interact cooperatively but in such a way as to cancel out the magnetic moments. Antiferromagnetic materials therefore show a drop in magnetic susceptibility at the onset of cooperative behaviour, as we saw in Figure 9.2. The temperature that characterises this process is known as the Néel temperature (T_N). The Néel temperatures for this transition in MnO, FeO, CoO, and NiO are 122, 198, 293, and 523 K, respectively.

The appearance of cooperative behaviour suggests that the d electrons on different ions interact, but the electronic properties were explained by assuming that the d electrons are localised. So how do we reconcile these two sets of properties?

The magnetic interaction in these compounds is thought to arise indirectly through the oxide ions; a mechanism known as **superexchange**. In a crystal of, say, NiO, there is a linear Ni–O–Ni arrangement. The d_{z2} orbital on the nickel can overlap with the $2p_z$ on oxygen, leading to partial covalency. The incipient NiO bond will have the d_{z2} electron and a $2p_z$ electron paired. The oxide ion has a closed shell, so there is another $2p_z$ electron, which must have the opposite spin. This electron forms a partial bond with the next nickel, so the d_{z2} on this nickel pairs with the 2p electron of opposite spin. As shown in Figure 9.8, the net result is that adjacent nickel ions have opposed spins.

The alternating spin magnetic moments in antiferromagnets such as NiO can be observed experimentally using neutron diffraction. Because neutrons have a magnetic moment, a neutron beam used for diffraction responds not only to the nuclear positions but also to the magnetic moments of the atoms. X-rays, on the other hand, have no magnetic moment and respond only to the electron density and hence to the atomic positions. The structure of NiO as determined by X-ray diffraction is a simple

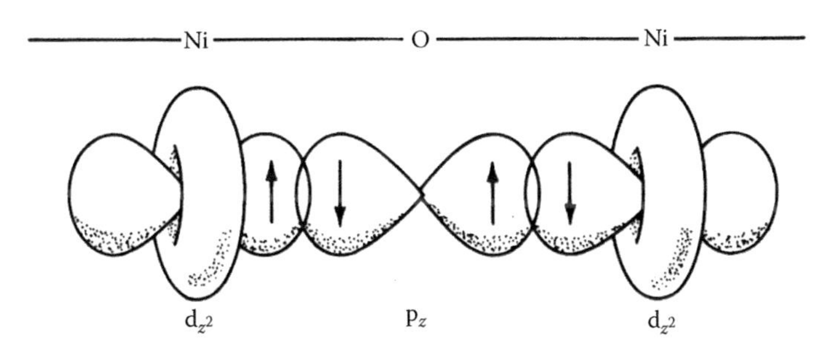

FIGURE 9.8 The overlap between the Ni d_z2 orbitals and the O p_z orbitals in NiO.

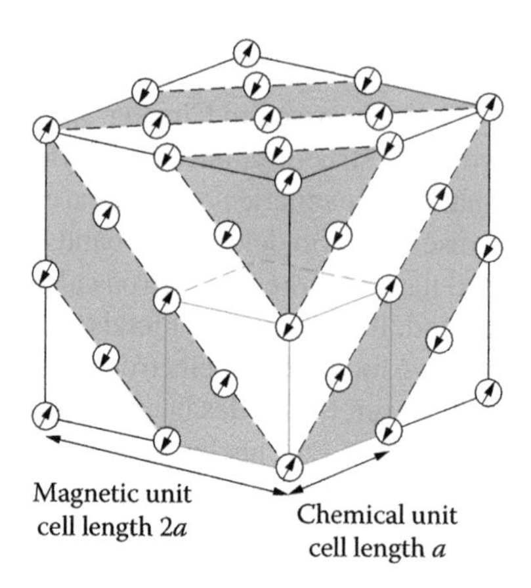

FIGURE 9.9 The magnetic unit cell of NiO with the crystallographic or chemical unit cell indicated in color.

NaCl structure. When the structure is determined by neutron diffraction, however, extra peaks appear that can be interpreted in terms of a magnetic unit cell, which is twice the size of the unit cell determined by X-ray. The positions of the nickel ions in this cell are shown in Figure 9.9. The normal crystallographic unit cell is bounded by identical atoms. The magnetic unit cell is bounded by identical atoms with an identical spin alignment. The shading indicates layers of nickel ions parallel to the body diagonal of the cube. The spins of all the nickel ions in a given layer are aligned parallel, but antiparallel to the next layer.

The arrangement of spins in NiO is not the only one found in antiferromagnets. Figure 9.10 shows different types of antiferromagnetic spin ordering for a cubic system. Overall each type has an equal number of up and down spins. However it is only in type G that each spin is aligned antiparallel to all its neighbours. In A-type antiferromagnetism, there are planes of feromagnetically coupled spins with alternate planes of opposite spin. In C-type, there are columns of ferromagnetically coupled spins antiferromagnetically coupled to adjacent columns.

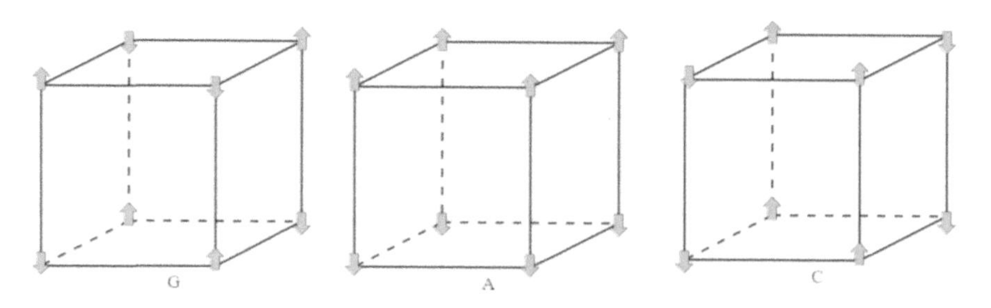

FIGURE 9.10 Different antiferromagnetic arrangements of spins in a cubic lattice (M. Fraser).

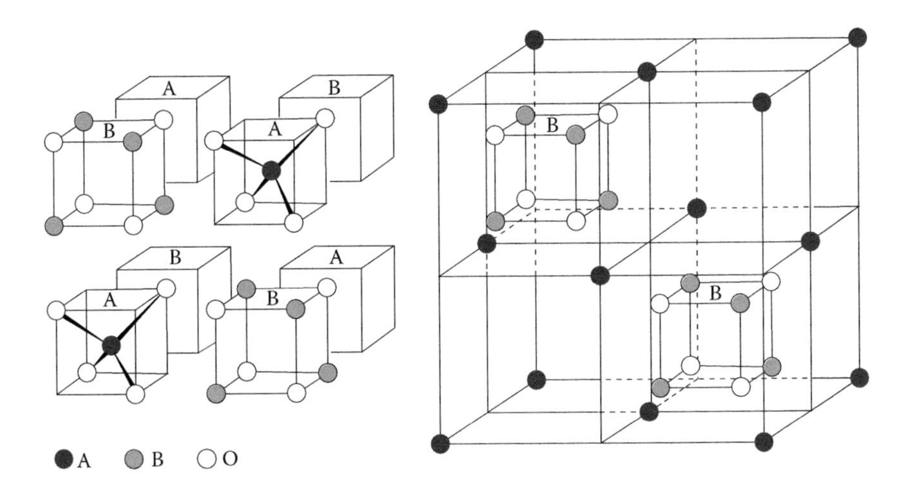

FIGURE 9.11 The spinel structure AB_2O_4.

9.7 FERRIMAGNETISM: FERRITES

The name ferrite was originally given to a class of mixed oxides having an inverse spinel structure (see Chapter 1) and the formula MFe_2O_4, where M is a divalent metal ion. The term has been extended to include other oxides, not necessarily containing iron, which have similar magnetic properties.

The spinel (AB_2O_4) structure (Figure 1.43) is a common mixed oxide structure, typified by the spinel itself ($MgAl_2O_4$), in which the oxide ions form a face-centered cubic close-packed array. For an array of N oxide ions, there are N octahedral holes and $2N$ tetrahedral holes; the divalent A ions (Mg^{2+}) occupy one-eighth of the $2N$ tetrahedral sites and the trivalent B ions (Al^{3+}) occupy half of the octahedral sites, $A^{II}_{tet}\left(B^{III}_{oct}B^{III}_{oct}\right)O_4$.

Figure 9.11 breaks this complex structure down into eight octants of two kinds, A and B, shown on the left of the diagram. In the inverse spinel structure, the oxide ions have the same cubic close-packed arrangement, but the divalent metal ions now occupy octahedral sites, and the trivalent ions are equally divided among tetrahedral and octahedral sites, $A^{III}_{tet}\left(B^{II}_{oct}B^{III}_{oct}\right)O_4$.

Using Figure 9.11 to describe the inverse spinel ferrite structure (MFe_2O_4): half of the Fe^{3+} ions occupy tetrahedral positions in the A-type octants, together with the corners and face-centres of the unit cell, and the other half of the Fe^{3+} ions, together with the M^{2+} ions, occupy the octahedral sites in the B octants.

The ions on octahedral sites interact directly with each other and their spins align parallel. The ions on octahedral sites also interact with those on tetrahedral sites, but in this case, they interact through the oxide ions and the spins align antiparallel, as in NiO.

In ferrites (MFe_2O4), the Fe^{3+} ions on tetrahedral sites are therefore aligned antiparallel to those Fe^{3+} ions on octahedral sites, so that there is no net magnetisation from these ions. The divalent M ions, however, if they have unpaired electrons, tend

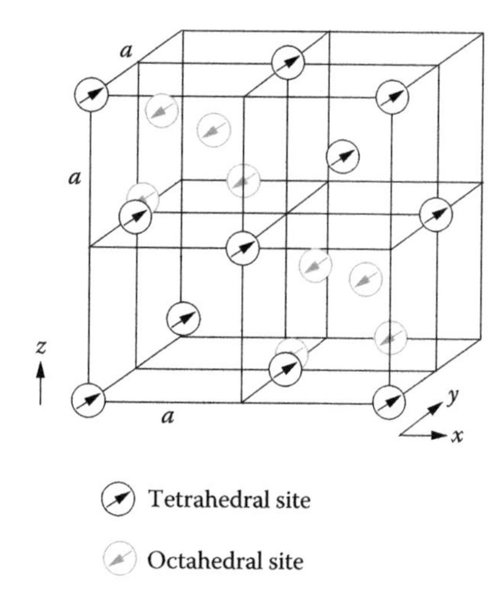

FIGURE 9.12 The magnetic structure of a ferrimagnetic inverse spinel.

to align their spins parallel to those of Fe^{3+} on the adjacent octahedral sites and hence with those of other M^{2+} ions. This produces a net ferromagnetic interaction for ferrites in which M^{2+} has unpaired electrons. The magnetic structure of such a ferrimagnetic ferrite is shown in Figure 9.12.

In magnetite (Fe_3O_4), the divalent ions are also iron (Fe^{2+}), and the interaction between the ions on the adjacent octahedral sites is particularly strong. One way of looking at the electronic structure of this oxide is to consider it as an array of O^{2-} ions and Fe^{3+} ions with the electrons that would have made half of the Fe ions divalent, delocalised over all the ions on octahedral sites. The Fe^{3+} ions have five 3d electrons, all with parallel spins. Since there can only be five 3d electrons of one spin on any atom, the delocalised spin must have the opposite spin. Being delocalised, it must also have the opposite spin to the 3d electrons on the next Fe ion. Hence, the two ions must have their spins aligned, and these spins must be aligned with those of all the other Fe ions on octahedral sites. Delocalisation will be less for other ferrites. The delocalisation of the electrons means that Fe_3O_4 is a good conductor.

Magnetite is the ancient lodestone used as an early compass. Ferrites have also found use as memory devices in computers, as magnetic particles on recording tapes, and as transformer cores.

9.7.1 Magnetic Strips on Swipe Cards

Magnetic strips are the brown or black strips found on many plastic cards, such as debit and credit cards. They were first used on the London Underground in the 1960s and were introduced into credit cards in the 1970s. They consist of small crystals of a magnetic material in a resin. A material commonly used for this is barium ferrite ($BaFe_2O_4$). Data such as your name and account number are encoded in the strip by altering the direction of the magnetisation of some crystals. Each character is represented by a set of crystals, some magnetised in one direction to represent 1 and others magnetised in the opposite direction to represent 0. When the card is swiped, the direction of the magnetisation is detected. High-coercivity materials, such as barium ferrite, are used as they retain their magnetisation so that the data are not accidentally erased.

9.8 SPIRAL MAGNETISM

When discussing ferromagnetism, ferrimagnetism, and antiferromagnetism, we have assumed that the electron spin angular momentum can only point in two directions (spin up and spin down).

In many solids, however, the spins are at an angle to the direction of the net magnetisation, they are canted with each spin pointing in a direction slightly different from that of its neighbour. This tilting arises from the spin–orbit coupling interaction of the spin angular momentum with the orbital angular momentum of the ion to give a spin–spin coupling termed the **Dzyaloshinskii–Moriya interaction.** This introduces a contribution to the magnetic moment at right angles to the direction of alignment of the spin-only magnetic moment. The spin direction changes by the same angle between each pair of spins to give a helical or a spiral effect and gives rise to **spiral magnetism**. There are various types of spiral arrangements (Figure 9.13). In the screw arrangement (a), successive spins point in different directions in a plane perpendicular to the wave vector. The cycloidal arrangement (b) is similar, except that the spin direction changes in a plane including the wave vector, rather like the change in direction of a particular spoke of a bicycle wheel as the wheel travels along the road. In the longitudinal and transverse conical (c and d) arrangements, the spins behave as though they were precessing from one ion to another. In longitudinal conical arrangements, the precession is around an axis coincident with the wave vector. In transverse conical arrangements, the precession is around an axis perpendicular to the wave vector.

Note that the net magnetisation cancels out for the screw and cycloidal arrangements, making solids with such arrangements effectively antiferromagnetic. For the

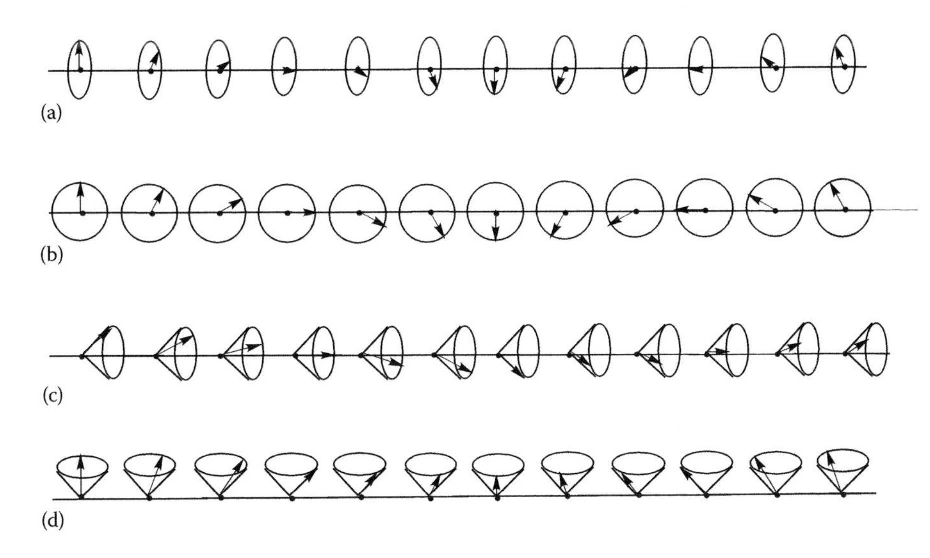

FIGURE 9.13 The spiral spin arrangements: (a) screw, (b) cycloidal, (c) longitudinal conical, and (d) transverse conical.

conical arrangements, there is a net ferromagnetic effect. Spiral magnetism can lead to the coupling of the magnetism with polarisation, producing materials that are both ferroelectric and either ferromagnetic or antiferromagnetic. Such materials are examples of multiferroics and are discussed further in Section 9.12.

9.9 GIANT, TUNNELLING, AND COLOSSAL MAGNETORESISTANCE

9.9.1 Giant Magnetoresistance

Magnetoresistance is the change in the electrical resistance of a material as a result of applying a magnetic field to the sample. In 1988, it was found that for certain metallic multilayer materials, the application of a magnetic field of strength 1–3 mT caused the resistivity to drop to half or less of its zero-field value. This phenomenon was termed **giant magnetoresistance (GMR)**. Within a decade, commercial devices such as computer hard-disk read heads based on GMR were available.

GMR is observed in metallic magnetic multilayers, which are stacks of nanometer-thick layers of different metals, in particular, alternate layers of a ferromagnetic metal and a nonmagnetic metal, for example, Fe/Cr/Fe... or Co/Cu/Co.... Within a single ferromagnetic layer, all the spins are aligned, but the coupling between the adjacent ferromagnetic layers depends on the thickness of the intervening nonmagnetic layer. For certain thicknesses, the adjacent ferromagnetic layers are coupled antiferromagnetically (Figure 9.14b) and this increases the resistivity. However, when a magnetic field is applied, this aligns the layers in a ferromagnetic manner (Figure 9.14a) and there is a dramatic drop in electrical resistance.

GMR can be understood if we realise that in metals, spin-up and spin-down electrons conduct electricity independently and that electrical resistance occurs when the electrons are scattered by the material. The scattering is very different for spin-up and spin-down electrons. This means that the resistivities are different for the two spin states: the scattering is strong for spin-down electrons, which are antiparallel to the magnetisation direction, but is low for spin-up electrons, which are parallel to the magnetisation direction.

In a normal metal, there are the same number of spin-up and spin-down electrons, but in a ferromagnet, such as Fe, there are more electrons in which the spin is parallel to the direction of the magnetisation; these are the spin-up electrons (majority-spin electrons) and they are scattered only weakly. The minority, spin-down electrons are antiparallel to the direction of the magnetisation and they are scattered strongly. The resistance of a single layer of a ferromagnet thus depends on both resistances; we can think of the two conduction channels as resistances in parallel and the overall resistance is dominated by the low resistance of the majority-spin electrons.

Now look at the situation where we have two layers of ferromagnetic material separated by a nonmagnetic conducting layer. In (a), where there is an applied field and the layers are aligned parallel, the majority, spin-up, electrons pass through both layers with only weak scattering. The spin-down electrons are strongly scattered by both layers. The resistance will be dominated by the majority-spin electrons and is therefore low. In (b), the ferromagnetic layers are antiparallel. The spin-up electrons will be strongly scattered by the first antiparallel layer, but will then pass easily through the second layer. The minority, spin-down, electrons will pass easily

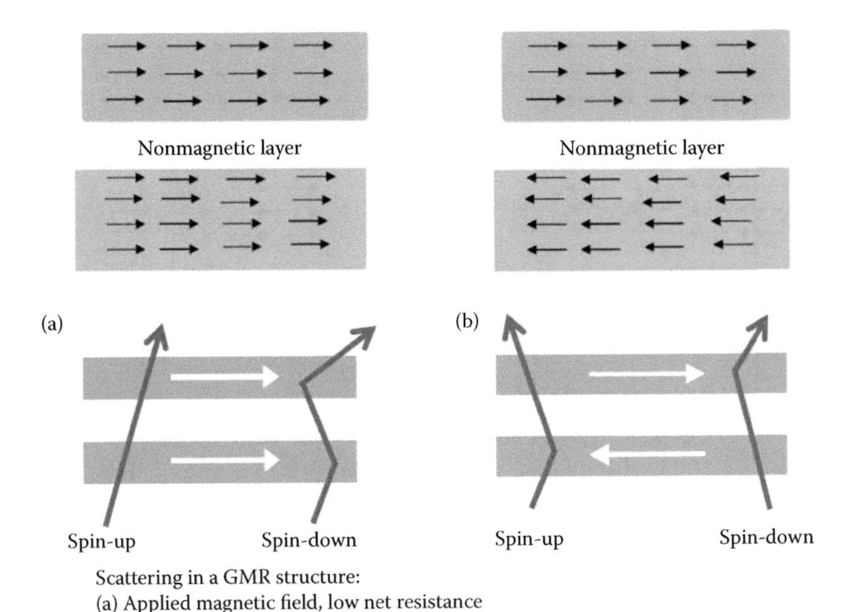

FIGURE 9.14 Two magnetic layers separated by a nonmagnetic layer showing the electron path, with (a) the spins in the two magnetic layers aligned ferromagnetically and (b) the spins in the two magnetic layers aligned antiferromagnetically.

through the first layer but will be strongly scattered by the second layer, so both categories of electrons will experience strong scattering in one or other of the layers. Hence, the overall resistance of this multilayer structure is high.

We now see why the application of a magnetic field, which switches the coupling from antiferromagnetic to ferromagnetic, leads to a large drop in resistance.

To understand why the electrons in one state conduct differently from the other state, we need to return to band theory. Consider the separate energy bands for spin-up and spin-down electrons. For d^8 metallic iron electrons, the majority-spin species, spin-up, the 3d bands are full, but for the spin-down electrons, the 3d band is only partially full. Because the spin-up 3d band is full, the spin-up current carriers occupy a higher-energy s/p band, whereas the spin-down current carriers are in the 3d band. The mean free path of electrons (the average distance that electrons travel through the solid without being scattered) in the s/p band is greater than in the 3d band, so the spin-up electrons carry more current than the spin-down electrons.

9.9.2 Tunnelling Magnetoresistance

A related phenomenon, **tunnelling magnetoresistance (TMR)**, superseded GMR as the basis of the action of hard-disk read heads. TMR read heads can produce a larger signal and are easier to make into smaller devices.

TMR, like GMR, occurs for layered structures, but here the two ferromagnetic layers are separated by a very thin insulating layer—of a few nanometres—that allows

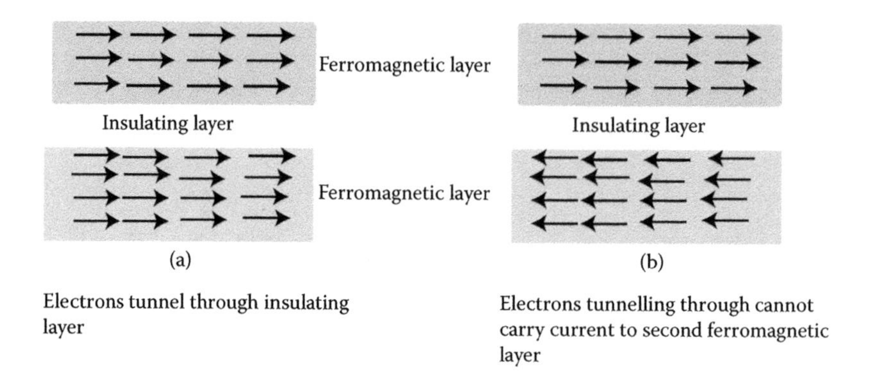

FIGURE 9.15 (a, b) Tunnelling magnetoresistance.

electrons to flow from one ferromagnetic layer to the next by quantum mechanical tunnelling through the insulating oxide layer. When the magnetisations are parallel, it is more likely that the electrons can tunnel through than when they are antiparallel. A junction made of this material can thus be switched between high- and low-resistance states.

TMR was discovered in 1975 using a sandwich of Co/Ge/Fe, which showed a 14% difference in the resistance between the two states; however, subsequently, amorphous alumina (Al_2O_3) came to be the favoured material for the insulating layer; Al_2O_3 forms a very dense thin layer with few vacancies, so no gaps in the insulating layer, and has shown differences in resistance of up to 70%.

More recently, MgO has been used as the oxide layer. In the case of Fe/MgO/Fe, only wave functions of a particular symmetry can tunnel through efficiently. If the ferromagnetic material is chosen so that the bands at the Fermi level are of this symmetry for the majority carrier spins, but not for the minority carrier spins, then there is a sharp increase in resistance when the two ferromagnetic layers are aligned antiparallel (Figure 9.15). Theory has predicted that there could be a difference of several thousand percent between the two states, and at room temperature, ~600% has been observed and even higher at very low temperatures.

9.9.3 Hard-Disk Read Heads

In hard-disk read heads, one ferromagnetic layer has its spin orientation fixed by coupling to an antiferromagnetic layer (Figure 9.16). A second ferromagnetic layer separated from the first ferromagnetic layer by a nonmagnetic metal in the case of GMR, or an insulator in the case of TMR, is free to change its spin orientation when a field is applied.

As the read head moves over the hard disk, the magnetic fields on the disk cause the spins in the second layer to align either parallel or antiparallel to those in the first layer. The information on the hard disk is coded as a series of 0's and 1's corresponding to the different orientations of the magnetic field on the disk, and these give rise to a high or low current in the read head. A similar principle is used for magnetic random access memory chips (MRAM).

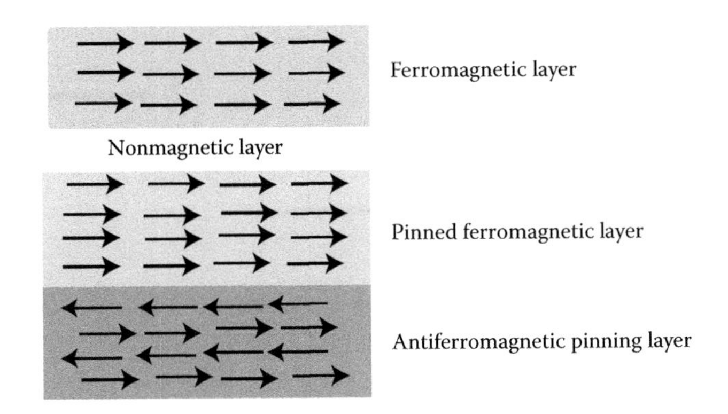

FIGURE 9.16 Two ferromagnetic layers as in a magnetic hard-disk read head showing pinning by an antiferromagnetic layer.

9.9.4 Colossal Magnetoresistance: Manganites

In 1993, **colossal magnetoresistance (CMR)** was observed for certain compounds such as doped manganite perovskites (e.g., $La_{1-x}Ca_xMnO_3$). In these compounds, a change in electrical resistance of orders of magnitude is observed, but large magnetic fields of the order of several tenths of a tesla (i.e., a hundred times stronger than those that produce GMR) or larger are needed.

Doped manganite perovskites exhibiting CMR have the general formula $Ln_{1-x}M_xMnO_3$, where Ln represents a lanthanoid element and M is a divalent metal such as Cu, Cr, Ba, or Pb. The trivalent Ln ions and divalent M ions occupy the A sites in the perovskite structure (Figures 1.45 and 9.17) and they have 12-fold coordination to oxygen. The Mn ions occupy the octahedral B sites. Of the manganese ions,

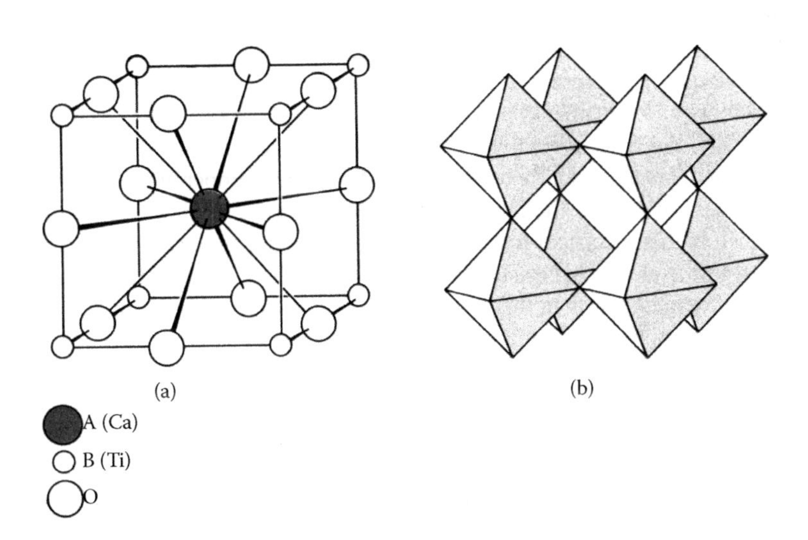

FIGURE 9.17 The perovskite structure (a) A-type unit cell, (b) showing octahedral environment of B ions.

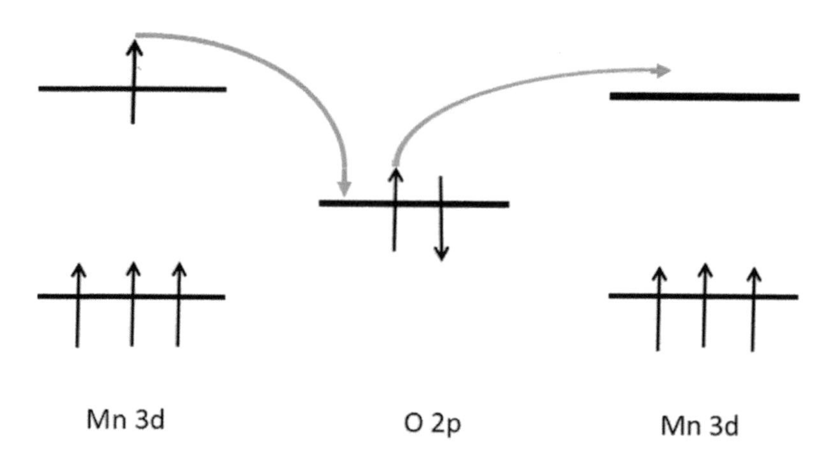

FIGURE 9.18 Double exchange.

(1–x) are Mn^{3+} and x are Mn^{4+}. The environment of the $Mn^{3+}(d^4)$ ions is distorted due to the Jahn–Teller effect, and for small values of x, a cooperative distortion occurs.

As in other transition metal oxides such as NiO, the spins on the transition metal ions are coupled via the oxide ions. The Mn^{4+} ions couple antiferromagnetically with each other via superexchange as in NiO. The coupling of the Mn^{3+} ions varies and can be ferromagnetic or antiferromagnetic. The Mn^{3+} ions couple to the Mn^{4+} ions in a process known as **double exchange**. In this, a simultaneous hop of an electron from an Mn^{3+} ion to an O 2p orbital and from an O 2p orbital to an Mn^{4+} ion takes place (Figure 9.18). This produces ferromagnetic coupling.

Double exchange is strong in manganites with $x \approx \frac{1}{3}$, and these manganites form ferromagnetic phases at low temperatures. These phases can be described in terms of band theory. The e_g orbitals on manganese and the O 2p orbitals combine to form two bands, one for spin-up electrons and one for spin-down electrons. Unusually, there is a large band gap between the two bands, so that at low temperatures, one band is empty and the other is partly full. (Such solids are referred to as **half metals** because only one type of spin is free to carry an electrical current.) The ferromagnetism disappears at the Curie temperature. Above this temperature, the e_g electrons on manganese are better thought of as localised and the solid is paramagnetic and has a much higher electrical resistance. As the ferromagnetic phase approaches the Curie point, the electrical resistance rises as the thermal energy starts to overcome the double exchange. It is in this region that the manganites exhibit CMR. A strong magnetic field applied to the manganite realigns the spins, restoring the half-metallic state and thus decreasing the resistivity.

9.10 ELECTRICAL POLARISATION

Although solids consist of charged particles (nuclei and electrons), a solid has no overall charge. For most solids, there is also no net separation of positive and negative charges; that is, there is no net dipole moment. Even if a solid is composed of molecules with permanent dipole moments, for example ice, the molecules are generally arranged in such a way that the unit cell of the crystal has no net dipole moment, so the solid has none. If such a solid is placed in an electric field, then a field is induced in the solid which opposes the applied field. This field arises from two sources, a distortion of the electron cloud of the atoms or molecules and the slight movement of the atoms themselves. The average dipole moment per unit volume that is induced in the solid is the electrical polarisation (P), and is proportional to the field (E).

$$P = \varepsilon_0 \chi_e E \tag{9.10}$$

where ε_0 is the permittivity of free space ($= 8.85 \times 10^{-12}$ F m^{-1}) and χ_e is the (dimensionless) dielectric susceptibility. For most solids, the electric susceptibility lies between 0 and 10. The susceptibility is often determined experimentally by determining the capacitance of an electric circuit with and without the solid present. The ratio of these two capacitances is the relative permittivity or dielectric constant of the solid, ε_r.

$$\frac{C}{C_0} = \varepsilon_r \tag{9.11}$$

where C is the capacitance in farads in the presence of the solid, and C_0 is that in the absence of the solid. The dielectric constant (ε_r) is related to the dielectric susceptibility by

$$\varepsilon_r = 1 + \chi_e \tag{9.12}$$

If the experiment is performed using a high-frequency alternating electric field, then the atoms cannot follow the changes in the field and only the effect due to electron displacement is measured. Electromagnetic radiation in the visible and the ultraviolet regions provides such a field, and the refractive index of a material is a measure of the electron contribution to the dielectric constant. Substances with high dielectric constants also tend to have high refractive indices, and therefore we will find that similar types of materials will be considered in the following sections as were discussed at the end of Chapter 8 in connection with optical devices.

Although most solids do not have a dipole moment in the absence of an electric field, the classes of solids that do have a dipole moment are commercially important and therefore form the subject matter of the next two sections.

9.11 PIEZOELECTRIC CRYSTALS: A-QUARTZ

A **piezoelectric crystal** is one that develops an electrical voltage when subjected to mechanical stress, for example if pressure is applied to it, and conversely develops strain when an electric field is applied across it. The application of an electric field causes a slight movement of atoms in the crystal so that a dipole moment develops in the crystal. Most piezoelectric crystals must be made up from units that are non-centrosymmetric; that is, they do not possess a centre of symmetry. (There are a few crystals of high symmetry that cannot be piezoelectric because of other symmetry elements that they possess.)

α-Quartz is based on SiO_4 tetrahedra. The structure is similar to β-quartz (Figure 7.1) but in α-quartz the tetrahedra are distorted so that each unit has a net dipole moment. However, these tetrahedra are arranged in such a way that normally the crystal does not have an overall polarisation.

External stress changes the Si–O–Si bond angles between the tetrahedra so that the dipole moments no longer cancel and the crystal has a net electrical polarisation. The effect in α-quartz is small; the output electrical energy is only 1% of the input strain energy, whereas for Rochelle salt (another commercially used piezoelectric crystal), the ratio of output energy to input energy is 81%. α-Quartz, however, is useful in applications where an oscillator of stable frequency is needed, such as in quartz watches. An electric field causes distortion of quartz and if an alternating electric field is applied, the crystal vibrates. When the frequency of the electric field matches the natural vibration frequency of the crystal, then resonance occurs and a steady oscillation is set up with the vibrating crystal feeding energy back to the electrical circuit. The importance of α-quartz in devices such as watches is due to the fact that for cuts along some crystal planes, the natural frequency of the crystal is independent of the temperature, so the crystal will oscillate at the same frequency, and the watch will keep time, however hot or cold the day is.

For other applications, such as ultrasonic imaging, it is more important that the conversion of mechanical energy to electrical energy is high.

Some piezoelectric crystals are electrically polarised in the absence of mechanical stress; one example is gem-quality tourmaline crystals. Normally, this effect is unnoticed because the crystal does not act as the source of an electric field. Although there should be a surface charge, this is rapidly neutralised by the charged particles from the environment and from the crystal itself. However, the polarisation decreases with increasing temperature, and this can be used to reveal the polar nature of the crystal. If tourmaline is heated, its polarisation decreases and it loses some of its surface charges. On rapid cooling, it has a net polarisation and will attract small electrically charged particles such as ash. Such crystals are known as **pyroelectric**, and ferroelectric crystals are a special subclass of pyroelectric crystals.

9.12 FERROELECTRIC EFFECT

Ferroelectric crystals possess domains of different orientation of electrical polarisation that can be reorientated and brought into alignment by an electric field. Among the most numerous ferroelectrics are the perovskites (Chapter 1), of which a classic example is barium titanate ($BaTiO_3$). This substance has a very large dielectric constant (around 1000) and is widely used in capacitors. Above 393 K, $BaTiO_3$ has a cubic structure, as in Figures 1.45 and 8.16, with Ba^{2+} ions in the centre, Ti^{4+} ions at the cube corners, and an octahedron of O^{2-} ions around each titanium ion. The Ti^{4+} ion is small (75-pm radius), so there is room for it to move inside the O_6 cage. At 393 K, the structure changes to a tetragonal structure in which the Ti atom moves off-centre along a Ti–O bond. At 278 K, further change occurs in which the Ti atom moves off-centre along a diagonal between two TiO bonds, and at 183 K a rhombohedral phase is formed in which there is distortion along a cube diagonal. The three distortions are shown in Figure 9.19. This figure magnifies the effect: the Ti atom is moved about 15 pm off-centre.

In these three phases, the TiO_6 octahedra have a net dipole moment. To illustrate how ferroelectricity arises, we use the tetragonal structure (Figure 9.19a) as

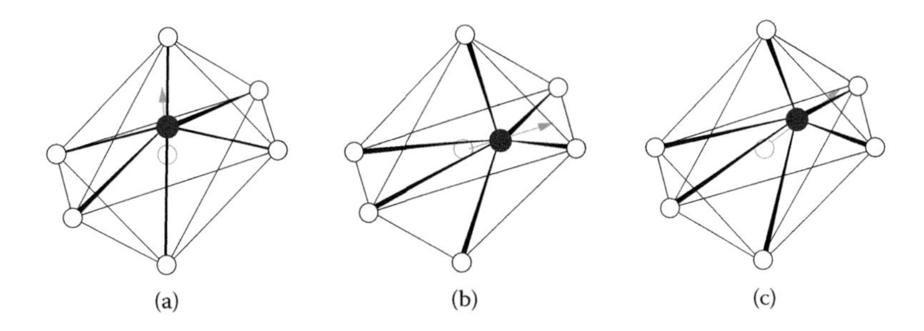

FIGURE 9.19 Distortions of TiO_6 octahedra in (a) the tetragonal structure, (b) orthorhombic structure, and (c) rhombohedral structure of barium titanate.

an example. If all the Ti atoms were slightly off-center in the same direction, then the crystal would have a net polarisation. Like ferromagnets, however, ferroelectrics have domains within which there is a net polarisation, but different domains have their polarisation in different directions, thus giving a net zero polarisation. In the tetragonal phase of $BaTiO_3$, the Ti atom can be off-centre in six directions along any one of the Ti–O bonds. As a result, neighbouring domains have polarisations that are either at 90° or at 180° to each other (Figure 9.20).

A domain can be of the order of 10^{-5} m or even more. Several methods can be used to obtain pictures of the domains. Figure 9.21 shows a thin slice of barium titanate under a polarising microscope in which different domains can clearly be seen. Note the sharpness of the domain boundaries.

When an external electric field is applied, the domains that are favorably aligned grow at the expense of others. As with ferromagnetics, the response to the field exhibits hysteresis; the polarisation grows until the whole crystal has its dipoles aligned, and

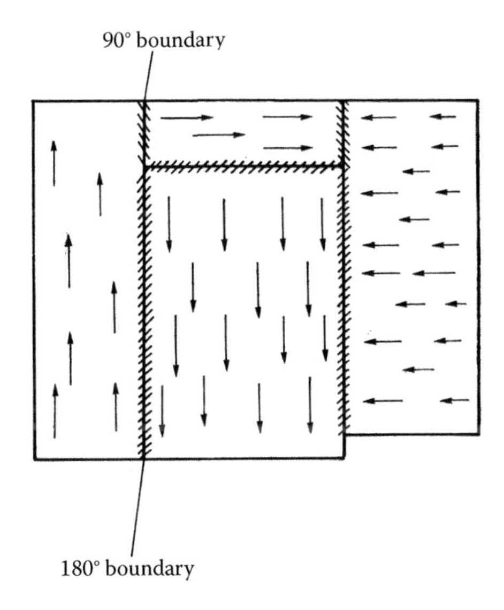

FIGURE 9.20 A sketch of the domains in barium titanate showing 90° and 180° boundaries.

FIGURE 9.21 A photograph of a thin slice of barium titanate taken under a polarizing microscope, showing domains of different polarisation. (From Guinier, A. and Julien, R., *The Solid State from Superconductors to Superalloys,* 3rd ed., Oxford University Press/International Union of Crystallography, Oxford, Figure 2.9, p. 67, 1989. Reproduced by permission of Oxford University Press.)

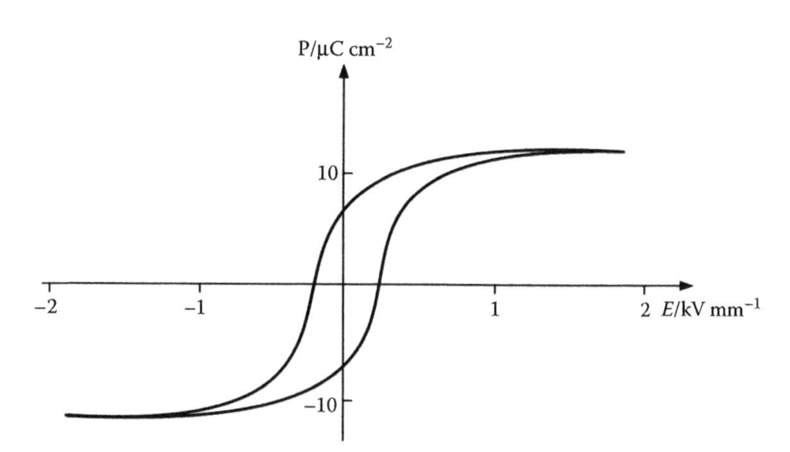

FIGURE 9.22 A plot of the polarisation versus the applied electrical field for barium titanate of grain size 2 × $^{10-5}$ to 1 × $^{10-4}$ m.

this polarisation remains while the field is reduced to zero, only declining as a field of opposite polarity is applied. Figure 9.22 shows a hysteresis curve for barium titanate.

The dielectric susceptibility and dielectric constant of ferroelectric substances obey the Curie law dependence on temperature:

$$\varepsilon_r = \varepsilon_\infty + \frac{C}{(T - T_c)} \tag{9.13}$$

Where ε_∞ is the permittivity at optical frequencies and T_C is the Curie temperature for ferroelectricity. The origin of the Curie temperature in barium titanate is quite easy to see because the Curie temperature is the temperature (393 K) at which barium titanate undergoes a phase transition to a cubic structure. Above the Curie temperature, the structure has a centre of symmetry and no net dipole moment. The dielectric constant is still high because the atoms can be moved off-centre by an applied electric field, but the polarisation is lost as soon as the field is removed.

In $PbZrO_3$, which also has a perovskite structure, the offset atoms are arranged alternately in opposite directions. This produces an antiferroelectric state. $PbZrO_3$ with some zirconium replaced by titanium gives the widely used ferroelectric material PZT ($PbZr_{1-x}Ti_xO_3$).

Remarkably, some polymers are ferroelectric. Polyvinylidene fluoride ($(-CH_2-CF_2-)_n$) as a thin film (approximately 25 μm thick) is used in shockwave experiments to measure stress. The pressure range over which this polymer operates is an order of magnitude larger than that of quartz.

In the β-phase of polyvinylidene fluoride (PVDF), the $-CF_2$ groups in a polymer chain all point in the same direction (Figure 9.23), so that a dipole moment is produced. Within a domain, the $-CF_2$ groups on different chains are aligned. When an electric field is applied, the chains rotate through 60° increments to align with an adjacent domain.

9.12.1 Multilayer Ceramic Capacitors

Capacitors are used to store charge. An electric field is applied to induce charge in the capacitor. The capacitor then remains charged until a current is required. To be useful in modern electronic circuits, for computers, spacecraft, television sets, and many other applications, a capacitor must be small. In order to retain a high capacitance, that is, to store a large amount of electrical energy whilst remaining small, a material needs a high permittivity. Hence, barium titanate with its very high permittivity has proved invaluable for this purpose. Pure barium titanate ($BaTiO_3$) has a high permittivity (of about 7000) close to the Curie temperature, but this rapidly drops with temperature to the room temperature value of 1–2000. While this is still high, for electronic circuits it would be useful to retain the higher value so that the size of the capacitor could be reduced. For some applications, it is also necessary for the permittivity to be constant with temperature over a range of 180 K, from 218 to 398 K. Barium titanate with some of the titanium substituted by zirconium or tin has a Curie temperature closer to room temperature and a flatter permittivity versus temperature curve. Further improvements can be made by partially substituting the barium ions. Materials made in this way consist of several phases mixed together, each with a different Curie temperature, and it is this that gives rise to the flatter

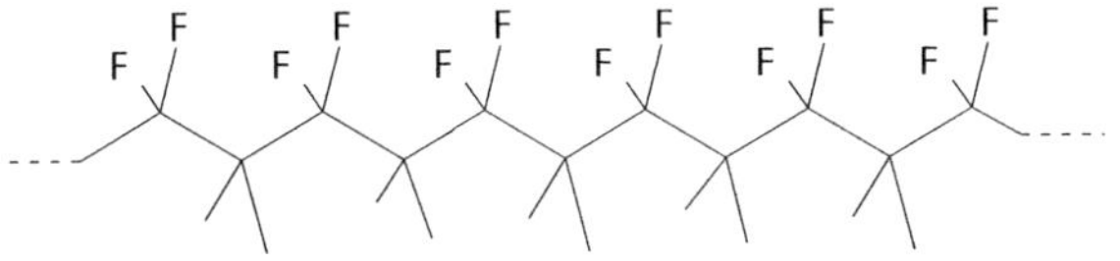

FIGURE 9.23 The alignment of $-CF_2$ groups in β-polyvinylidene fluoride.

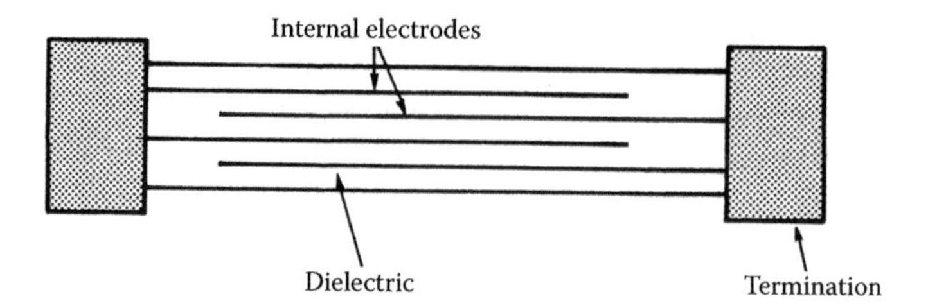

FIGURE 9.24 A section through a multilayer capacitor.

permittivity versus temperature curve. Another factor that affects the dielectric properties of barium titanate is the grain size. On the surface of the tetragonal crystals, the structure is cubic, so that for small particles with a large surface-to-volume ratio, there is a high proportion of cubic material. This leads to a higher room temperature permittivity, but a smaller permittivity at the Curie temperature. For very small particles, there is no ferroelectric effect; therefore, it is important to produce grains of a suitable size to give the properties needed. To manufacture multilayer capacitors, barium titanate of a suitable grain size and appropriately doped, is interleaved with conducting plates (see Figure 9.24). This enables one device to be used in place of several single disc capacitors in parallel.

Barium titanate is one example of a ferroelectric material. Other oxides with the perovskite structure are also ferroelectric, for example, lead titanate and lithium niobate. One important set of such compounds, used in many transducer applications, are the mixed oxides PZT ($PbZr_{1-x}Ti_xO_3$). These, like barium titanate, have small ions in O_6 cages which are easily displaced. Other ferroelectric solids include hydrogen-bonded solids such as KH_2PO_4 and Rochelle salt ($NaKC_4H_4O_6.4H_2O$), salts with anions that possess dipole moments such as $NaNO_2$ and copolymers of PVDF (polyvinylidenedifluoride).

9.13 MULTIFERROICS

Multiferroics are materials that possess two or more cooperative properties, such as ferromagnetism, ferroelectricity, and ferroelasticity. In practice, it is generally used to refer to materials that are ferroelectric and ferromagnetic or antiferromagnetic. The interest in these materials stems from the possibility of applying a magnetic field to alter the charge or applying a voltage to change the spin. Possible applications include computer memory that can be laid down electrically but read magnetically, and electronic devices with 4-state logic. When writing to memory using ferroelectrics, charged plates are placed on either side of the ferroelectric material, making the atoms move up or down and giving a 0 or 1 as with magnetic memory. Writing using an electric field rather than a magnetic field has advantages, including using less energy. Reading back using electric fields results in the memory being wiped. However, if changing the polarisation changes the magnetic state, then the memory can be read by detecting the magnetism, a process that is nondestructive. Other possible applications include magnetic sensors and energy harvesting.

We shall concentrate on single-phase multiferroics, but currently (2019) multiferroic composites (where the two ordered states are in separate phases within the material) have shown more promise.

Single-phase multiferroics can be divided into two broad classes: **Type I multiferroics** possess both magnetic and electric cooperative properties, but the two arise from different features of the structure and are weakly coupled; **Type II multiferroics** have strongly coupled magnetic and electrical cooperative effects.

9.13.1 Type I Multiferroics: Bismuth Ferrite

One of the most studied examples of Type I multiferroics is $BiFeO_3$, bismuth ferrite. The interest in this compound was sparked by the publication of a paper in *Science* in 2003, reporting the results from Ramamoorthy Ramesh's group at the University of Maryland. The measurements showed that thin films of $BiFeO_3$ had a large remanent polarisation. Bismuth ferrite has a Néel temperature of 643 K and a ferroelectric transition temperature of 1100 K, so it is multiferroic at room temperature. It adopts a perovskite structure with Bi–O and Fe–O layers. The spins on the Fe^{3+} ions are coupled antiferromagnetically. The ferroelectric effect is due to the alignment of the lone pairs of the Bi^{3+} ions, giving a cooperative offset arrangement.

However, because the polarisation and magnetisation are due to different ions, there is very little interaction between the two and the effect of a magnetic field on polarisation or of an electric field on magnetisation is very small.

Although the magnetoelectric coupling is small, bismuth ferrite does have potential applications. It has been shown that it can act as the insulating thin film in a tunneling magnetoresistance sandwich when placed between ferromagnetic (La, Sr) MnO_3 and Co metal. A possibility is that it could be used to modify the TMR through the application of an electric field.

9.13.2 Type II Multiferroics: Terbium Manganite

The first Type II multiferroics to be studied were the terbium manganites, $TbMnO_3$ and $TbMn_2O_5$. $TbMnO_3$ adopts a perovskite structure. Polarisation is induced by spiral magnetism. For $TbMnO_3$, the spins on the manganese ions are in the cycloidal arrangement below 28 K. $TbMnO_3$ is an example of a lanthanoid manganite, which is the parent compound of colossal magnetoresistance materials. The Mn^{3+} ions are subjected to Jahn–Teller distortion in their surroundings. This encourages the ordering of the e_g orbitals in the *ab* plane and leads to spins being ferromagnetically coupled along the *a* and *b* axes and antiferromagnetically coupled along the *c* axis (Figure 9.25).

As the radius of the lanthanoid ion decreases, the distortion increases and the distance between Mn^{3+} ions becomes smaller. This enables ferromagnetic coupling of Mn^{3+} ions through space to compete with the antiferromagnetic Mn–O–Mn coupling.

The spins become frustrated and adopt a modulated spin ordering. Below 28 K, this modulation takes the form of a cycloidal arrangement of spins in the *bc* plane, and this, in turn, leads to the development of polarisation along the *c* axis. A

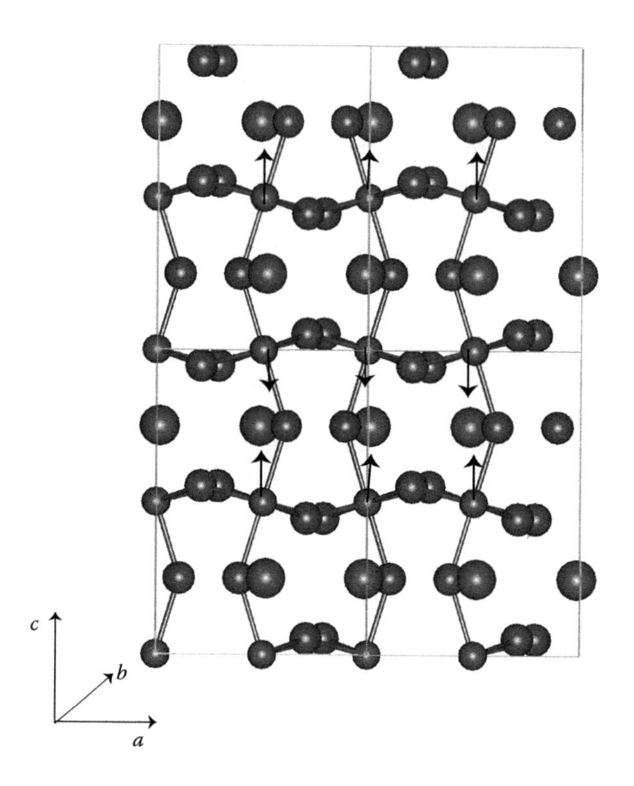

FIGURE 9.25 The structure of $TbMnO_3$ showing *a*, *b*, and *c* axes.

magnetic field of about 5 T applied along the *b* axis was found to cause the electrical polarisation of $TbMnO_3$ to change direction from along the *c* axis to along the *a* axis.

There are a number of other solids in which cycloidal spin arrangements produce a magnetoelectric effect. Examples are CuO, $MnWO_4$, and $Ni_3V_2O_8$.

Solids with cycloidal spin arrangements are not the only solids to show Type II multiferroicity. It can be demonstrated that in a cubic crystal, the screw and longitudinal conical arrangements do not give rise to electrical polarisation, but the transverse conical arrangement does. The transverse conical arrangement is interesting because it gives rise to a solid that is both ferromagnetic and ferroelectric. An example of such a solid is $CoCr_2O_4$, which has a spinel structure. The inverse Dzyaloshinskii–Moriya or spin-current model predicts the value of the polarisation to be

$$P = a\sum_{ij} e_{ij} \times (S_i \times S_i) \tag{9.14}$$

where e_{ij} is the unit vector connecting spins S_i and S_j and *a* is a constant dependent on the spin–orbit coupling constant, the spin exchange, and the spin–lattice interaction.

Although the screw and longitudinal conical arrangements do not give rise to polarisation, it is possible for an applied magnetic field to alter the spin arrangement such that the solid becomes ferroelectric. One such example is the hexaferrite ($Ba_2Mg_2Fe_{12}O_{22}$; BMFO). This has a longitudinal conical spin arrangement below 50 K. The spins are arranged as though precessing around the *c* axis. By applying a magnetic field at an angle to the *c* axis, the cones are tilted so that a component perpendicular to the *c* axis is produced. The solid then becomes polarised.

A screw arrangement of spiral magnetism can lead to a magnetoelectric interaction if the symmetry of the crystal is lowered. The oxides AMO_2, where A=Cu, Ag and M=Cr, Fe, have a structure in which each ion forms a triangular sublattice. The Cr and Fe ions prefer to couple antiferromagnetically to the adjacent ion of that type. Taking a triangle of Fe ions, there is a problem with this. If one ion has an opposite spin to both its neighbours, then the two neighbouring ions must have parallel spins, but there is a preference for them to have opposite spins. The spins are said to be **frustrated** because there is no way each spin can be of opposite spin to both its neighbours.

For $CuFeO_2$, the adoption of a screw arrangement of spins to avoid this frustration has been shown to lead to the appearance of ferroelectricity. The triangles of the Fe ions lie perpendicular to the *c* axis and polarisation develops parallel to this axis.

9.14 SUMMARY

1. Diamagnetism is the weak magnetic field produced in all materials by the circulation of electrons. Other types of magnetism can exist in addition to this.
2. Substances with isolated unpaired electrons, such as many metal complexes and liquid oxygen, are paramagnetic as a result of electron spin magnetic moments. Metals with sparsely populated conduction bands also exhibit paramagnetism.
3. Ferromagnetism arises when all the spins in the sample are aligned with each other. In antiferromagnetism half the spins are spin up and half spin down. Ferrimagnetism occurs when there are spins of both orientations but the numbers of each are unequal, giving rise to a net magnetic moment. Ferrites are examples of ferrimagnetic materials.
4. Spins can be aligned at an angle to the direction of the net magnetisation, they are canted with each spin pointing in a direction slightly different from that of its neighbor. This is spiral magnetism.
5. Applying a magnetic field to some materials can give rise to a drop in electrical resistance. This is particularly noticeable for metallic magnetic multilayers (giant magnetoresistance), multilayers including a very thin insulating layer (tunneling magnetoresistance), and doped manganite perovskites (colossal magnetoresistance).
6. Piezoelectric crystals develop a voltage under mechanical stress and are distorted by the application of an electric field.
7. Ferroelectric crystals possess domains of different orientation of electrical polarisation that can be reorientated and brought into alignment by an electric field. A classic example is barium titanate.
8. Materials may possess more than one cooperative effect. These are known as multiferroics. This chapter considered multiferroics that were ferroelectric and either ferromagnetic or antiferromagnetic. In Type I multiferroics

the two properties are weakly if at all coupled. In Type II multiferroics the two properties are strongly coupled. Multiferroics can be single compounds or composites of two or more materials.

QUESTIONS

1. Calculate the spin-only magnetic moments for (a) Fe^{2+} in a complex where the maximum number of d electrons are unpaired, (b) Fe^{3+} in a complex where the maximum number of d electrons are unpaired, (c) Fe^{2+} in a complex where the minimum number of d electrons are unpaired, (d) Fe^{3+} in a complex where the maximum number of d electrons are unpaired.
2. Ho^{3+} complexes have 4 unpaired f electrons. $L=6$ and $J=8$. Use Equation 9.9 to calculate the magnetic moment of such a complex.
3. Although manganese is not ferromagnetic, certain alloys such as Cu_2MnAl are ferromagnetic. The Mn–Mn distance in these alloys is greater than in manganese metal. What effect would this have on the 3d band of manganese? Why would this cause the alloy to be ferromagnetic?
4. The compound EuO has an NaCl structure and is paramagnetic above 70 K, but magnetically ordered below it. Its neutron diffraction patterns at high and low temperatures are identical. What is the nature of the magnetic ordering?
5. ZnV_2O_4 has a spinel structure at low temperatures. What type of magnetism would you expect it to exhibit?
6. In transition metal pyrite disulfides (MS_2), the M^{2+} ions occupy octahedral sites. If a d band is formed, it will split into two as in the monoxides. Consider the information on some sulfides given below and decide whether the 3d electrons are localised or delocalised, which band the electrons are in if delocalised, and in the case of semiconductors, between which two bands the band gap of interest lies.

   MnS_2 antiferromagnetic ($T_N=78$ K), insulator, above T_N paramagnetism fits five unpaired electrons per manganese
   FeS_2 diamagnetic, semiconductor
   CoS_2 ferromagnetic ($T_C=115$ K), metal

7. In hydrogen-bonded ferroelectrics, the Curie temperature and permittivity alter when deuterium is substituted for hydrogen. What does this suggest about the origin of the ferroelectric transition in these compounds?
8. Pure $KTaO_3$ has a perovskite structure but is not ferroelectric or antiferroelectric. Replacing some K ions with Li, however, produces a ferroelectric material. Explain why the substitution of Li might have this effect.
9. $ScFeO_3$ (in which both ions are present as M^{3+} ions) is a Type I multiferroic. The Sc^{3+}ion occupies an off-centre position leading to the ferroelectricity of this solid. Why would you expect this to be decoupled from the magnetic ordering?
10. In the composite multiferroic $La_{0.7}Sr_{9.3}MnO_3/BaTiO_3$, what are the roles of the two components?

10 Superconductivity

10.1 INTRODUCTION

In 1908, Kamerlingh Onnes succeeded in liquefying helium, and this paved the way for many new experiments to be performed on the behaviour of materials at low temperatures. For a long time, it had been known from conductivity experiments that the electrical resistance of a metal decreased with temperature. In 1911, Onnes was measuring the variation of the electrical resistance of mercury with temperature when he was amazed to find that at 4.2 K the resistance suddenly dropped to zero. He called this effect **superconductivity** and the temperature at which it occurs is known as the **(superconducting) critical temperature** (**T_C**). One effect of the zero resistance is that there is no power loss in an electrical circuit made from a superconductor. Once an electrical current is established, it shows no discernible decay for as long as the experimenters have been able to watch. More than 30 elements can be made superconducting at ambient pressure (Table 10.1) and even more, including some non-metals such as C and O, can be made superconducting at higher pressures.

In 1957, Bardeen, Cooper, and Schrieffer published their theory of superconductivity, known as the **BCS theory**. This theory explained the properties of the superconducting materials known at the time. However these materials were only superconducting at very low temperatures and required cooling with liquid helium.

In 2001, a temperature of 40 K for the onset of superconductivity was observed in MgB_2. Although this temperature is not high enough for liquid nitrogen cooling, it is technologically important because electrical cryocoolers are now available that can reduce temperatures to 30 K, so that devices made from this material would not require liquid helium cooling. In 2015 and 2019, even higher critical temperatures were reported for hydrides under pressure. These compounds could be accounted for by BCS theory.

In 1986, Bednorz and Muller discovered a barium-doped lanthanum copper oxide that became superconducting at 35 K; the first of the so-called 'high-temperature' superconductors. This is a complex oxide unlike the metals and alloys discovered previously and did not fit the BCS theory.

Their findings were thought to be so important that only a year later they were awarded the Nobel Prize in Physics.

The discovery of this barium-doped lanthanum copper oxide led to a flood of new high-temperature superconductors, some of which were superconducting above the boiling temperature of nitrogen, 77 K.

The past three decades have also seen the discovery of other unexpected types of superconductors. Considerable interest has been raised by the announcement in 2008 of a high-temperature superconductor that was iron-based rather than copper-based.

Another unexpected discovery was of superconductors that are also ferromagnets, a combination of properties that is incompatible with BCS theory.

TABLE 10.1

Elements That Are Superconducting at Atmospheric Pressure

H																	He
Li	Be											B	C	N	O	F	Ne
Na	Mg											Al	Si	P	S	Cl	Ar
K	Ca	Sc	Ti	V	Cr	Mn	Fe	Co	Ni	Cu	Zn	Ga	Ge	As	Se	Br	Kr
Rb	Sr	Y	Zr	Nb	Mo	Tc	Ru	Rh	Pd	Ag	Cd	In	Sn	Sb	Te	I	Xe
Cs	Ba	La	Hf	Ta	W	Re	Os	Ir	Pt	Au	Hg	Tl	Pb	Bi	Po	At	Rn
Fr	Ra	Ac	Rf	Db	Sg	Bh	Hs	Mt	Ds	Rg	Cn	Nh	Fl	Mc	Lv	Ts	Og
			Ce	Pr	Nd	Pm	Sm	Eu	Gd	Tb	Dy	Ho	Er	Tm	Yb	Lu	
			Th	Pa	U	Np	Pu	Hm	Cm	Bk	Cf	Es	Fm	Md	No	Lr	

Key:

Ti Superconducting at atmospheric pressure.

Fe Magnetically ordered metal.

We shall look at all these exciting developments; however, we start by looking at the properties of superconductors.

10.2 PROPERTIES OF SUPERCONDUCTORS

10.2.1 Electrical Conductivity

For a long time, it had been known from conductivity experiments that the electrical resistance of a metal decreased with temperature, but resistance was still present even at very low temperatures. The defining characteristic of superconducting materials is that the electrical resistance drops to zero at the superconducting critical temperature, T_C. This effect is illustrated for tin in Figure 10.1. One effect of the zero resistance is that there is no power loss in an electrical circuit made from a superconductor. Once an electrical current is established, it shows no discernible decay for as long as the experimenters have been able to watch. This makes superconductors useful as electromagnets, as high magnetic fields can be produced without the loss of energy as heat which arises from electrical resistance. Such magnets are used in NMR machines including for magnetic resonance imaging (MRI).

Contrary to what you might have expected, a superconductor will have high resistance at room temperature. Indeed, the best room-temperature electronic conductors—silver and copper—do not superconduct at all. Superconductors do not have low electrical resistance above the superconducting critical temperature, T_C.

10.2.2 Magnetic Properties of Superconductors

Meissner and Ochsenfeld found that when a superconducting material is cooled below its critical temperature (T_C), it expels all magnetic flux from within its interior (Figure 10.2a): the magnetic flux (B) is thus zero inside a superconductor. Since $B = \mu_0 H(1+\chi)$, when $B = 0$, χ must equal –1, that is, superconductors are perfect diamagnets. If a magnetic field is applied to a superconductor, the magnetic flux is

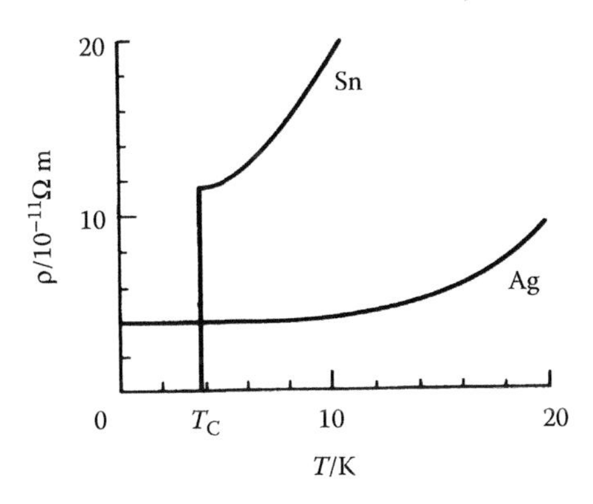

FIGURE 10.1 A plot of resistivity (ρ) versus temperature (T) showing the drop to zero at the critical temperature (T_C) for a superconductor and the finite resistance of a normal metal at absolute zero.

excluded (Figure 10.2b) and the superconductor repels a magnet. This is shown in Figure 10.3, where a magnet is seen floating in midair above a superconductor.

It is also found that the critical temperature (T_C) changes in the presence of a magnetic field. A typical plot of T_C against an increasing magnetic field is shown in Figure 10.4, where we can see that as the applied field increases, the critical temperature drops.

It follows that a superconducting material can be made nonsuperconducting by the application of a large enough magnetic field. The minimum value of the field strength required to bring about this change is called the **critical field strength** (**H_C**), its value depending on the material in question and on the temperature. Type

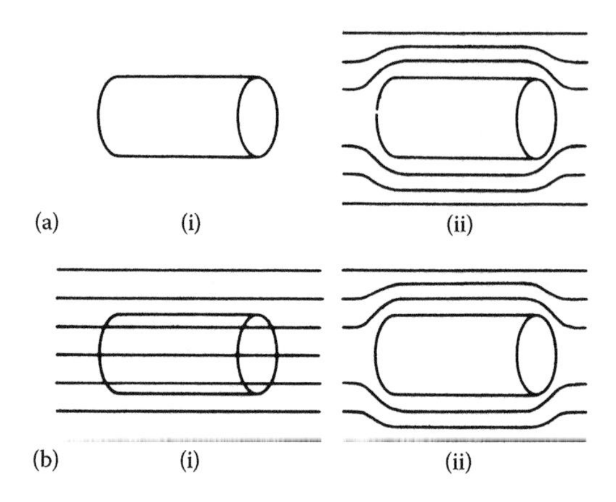

FIGURE 10.2 (a) (i) A superconductor with no magnetic field. When a field is applied in (ii), the magnetic flux is excluded. (b) (i) A superconducting substance above the critical temperature (T_C) in a magnetic field. When the temperature drops below the critical temperature (ii), the magnetic flux is expelled from the interior. Both are called Meissner effects.

FIGURE 10.3 A permanent magnet floating over a superconducting surface. (From Darren Peets, UBC Superconductivity Group. With permission.)

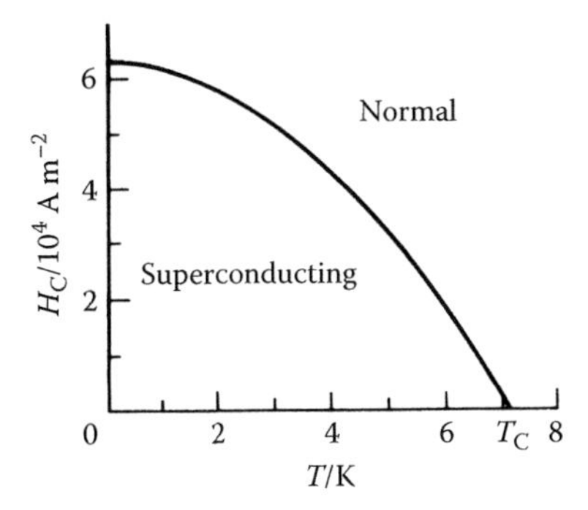

FIGURE 10.4 The variation of the critical field strength (H_C) for lead. Note that H_C is zero when the temperature (T) equals the critical temperature (T_C)

I superconductors, which include most of the pure metal superconductors, have a single critical field. Type II superconductors, which include alloys and the high-T_C superconductors, allow some penetration of the magnetic field into the surface above a critical field, H_{C1}, but do not return to their nonsuperconducting state until a higher field, H_{C2}, is reached. Type II superconductors tend to have higher critical temperatures than Type I superconductors.

The magnetic field does not have to be applied externally. A superconductor acting as an electromagnet will cease to be superconducting when the magnetic field produced exceeds the critical value. Since the field depends on the current passing through the material, this means that there is a **critical current**, above which the superconductivity is destroyed. This is known as the **Silsbee effect**. The size of the critical current is dependent on the nature and geometry of the particular sample.

10.2.3 BCS Theory of Superconductivity

This section attempts to give a qualitative picture of the ideas involved and also to give some familiarity with the terminology.

Physicists worked for many years to find a theory that explained superconductivity. To begin with, it looked as though the lattice played no part in the superconducting mechanism because X-ray studies showed that there was no change in either the symmetry or the spacing of the lattice when superconductivity occurred. However, in 1950, an **isotope effect** was first observed: for a particular metal, the critical temperature was found to depend on the isotopic mass (M) such that

$$T_C \alpha \frac{1}{\sqrt{M}}. \tag{10.1}$$

It is well known that the frequency (ν) of vibration of a diatomic molecule is given by

$$\nu = \frac{1}{2\pi}\sqrt{\frac{k}{\mu}}, \tag{10.2}$$

where μ is the reduced mass of the molecule and k is the force constant of the bond. We can see that a vibration also changes frequency on isotopic substitution such that the frequency (ν) is proportional to $(1/\sqrt{\text{mass}})$, suggesting to physicists that superconductivity was in some way related to the vibrational modes of the lattice and not just to the conduction electrons. The vibrational modes of a lattice are quantised, as are the modes of an isolated molecule: the quanta of the lattice vibrations being called **phonons** (see Chapter 4).

Frohlich suggested that there could be a strong phonon–electron interaction in a superconductor, leading to an attractive force between two electrons that is strong enough to overcome the Coulomb repulsion between them. Very simply, the mechanism works like this: as a conduction electron passes through the lattice, it can disturb some of the positively charged ions from their equilibrium positions, pushing them together and giving a region of increased positive charge density. As these oscillate back and forth, a second electron passing this moving region of increased positive charge density is attracted to it. The net effect is that the two electrons have interacted with one another, using the lattice vibration as an intermediary. Furthermore, the interaction between the electrons is attractive because each of the two separate steps involved an attractive Coulomb interaction.

It is the scattering of conduction electrons by the phonons that produces electrical resistance at room temperature. Superconductors have high resistance above the critical temperature because of their strong electron–phonon interactions. At low temperatures, it is predominantly the scattering by lattice defects that gives electrical resistance. **BCS theory** predicts that under certain conditions, the attraction between two conduction electrons due to a succession of phonon interactions can slightly exceed the repulsion that they exert directly on one another due to the Coulomb interaction of their like charges. The two electrons are thus weakly bound together, forming the so-called **Cooper pair**. It is these Cooper pairs that are responsible for superconductivity.

Cooper pairs are weakly bound, with typical separations of 10^6 pm for the two electrons. They are also constantly breaking up and reforming (usually with other partners). Thus, there is an enormous overlap between different pairs, and the pairing is a complicated dynamic process. The ground state of a superconductor, therefore, is a 'collective' state, describing the ordered motion of large numbers of Cooper pairs. This state can be written as a complex order parameter $|\Psi|e^{i\varphi}$ where φ is the phase of the superconducting wave function. For BCS superconductors, φ is independent of direction in k-space and the order parameter is said to be s-wave. (Note that this does not imply that s orbitals are involved.) When an external electrical field is applied, the Cooper pairs move through the lattice under its influence. However, they do so in such a way that the ordering of the pairs is maintained: the motion of each pair is locked to the motion of all the others, and none can be individually scattered by the lattice. Because the pairs cannot be scattered by the lattice, the resistance is zero and the system is a superconductor.

The BCS theory shows that there are several conditions that have to be met for a sufficient number of Cooper pairs to be formed and superconductivity to be achieved. It is beyond the scope of this book to go into this in any depth: suffice it to say that the electron–phonon interaction must be strong and that low temperature favours pair formation—hence, high-temperature superconductors were not predicted by the BCS theory. The relatively high critical temperature of MgB_2 is thought to be due to the high vibrational frequencies associated with the light boron atoms and the strong interaction between the electrons and lattice vibrations. Evidence for the involvement of B atom vibrations comes from the observation that T_C is increased by about 1 K when ^{11}B is replaced by ^{10}B. Higher critical temperatures have been reported for hydrogen-containing materials under high pressure. In 2015 a Russian group found that under 150 GPa pressure, hydrogen sulfide, H_2S, is a superconductor with a critical temperature of 203 K. In 2019, groups in Germany and the United States reported critical temperatures of 250–260 K for a lanthanide hydride under 170–185 GPa pressure. (To give some idea of the magnitude of these pressures, atmospheric pressure at the Earth's surface is around 100 kPa.) These materials are compatible with BCS theory.

10.3 HIGH-TEMPERATURE SUPERCONDUCTORS

Until 1986, the highest temperature found for the onset of superconductivity was 23.3 K for a compound of niobium and germanium (Nb_3Ge). However in 1986, Bednorz and Müller reported finding superconductivity in an oxide material at a temperature of 35 K. The compound that prompted their initial paper has been shown to be $La_{2-x}Ba_xCuO_4$, where $x = 0.2$, with a structure based on that of K_2NiF_4, a perovskite-related layer compound. Figure 10.5 shows the structure of the parent compound

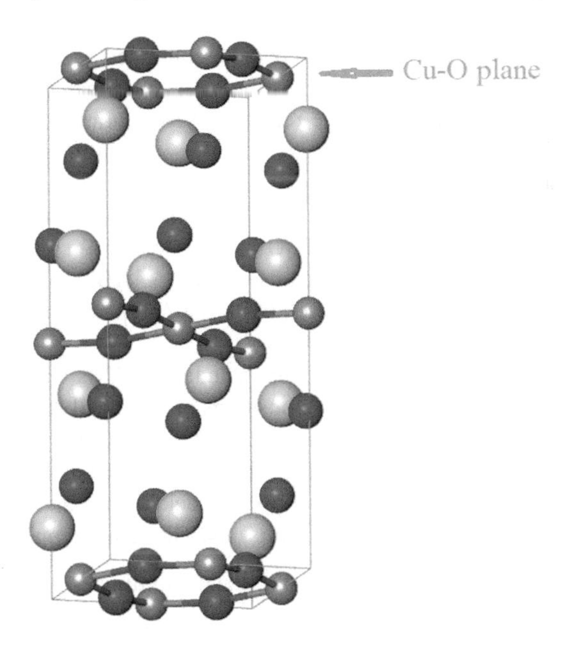

FIGURE 10.5 The structure of La_2CuO_4. Large blue spheres, La; pink spheres, Cu; red spheres, O.

La_2CuO_4. Barium substitutes for some lanthanum ions, but the superconductivity is observed in the CuO_2 layers.

Soon, the idea was born that it might be possible to raise the temperature even further by substitution with different metals. Using this technique, it was Chu's group in Houston, Texas, that finally broke through the liquid–nitrogen temperature barrier, with the superconductor that is now known as '1-2-3': this superconductor replaces lanthanum with yttrium and has the formula $YBa_2Cu_3O_{7-x}$. The onset of superconductivity for 1-2-3 occurs at 93 K. Since then, many other high-T_C superconductors with CuO_2 planes have been reported. These are referred to as **cuprate superconductors**. The highest critical temperature at ambient pressure discovered so far is 135 K for a doped $HgBa_2Ca_2Cu_3O_{8+\delta}$.

Other types of superconducting compounds have since been identified. An unusual superconductor that is structurally related to the cuprates was identified in 1994: Sr_2RuO_4 has a crystal structure almost identical to La_2CuO_4.

In 2008, Hideo Hosono reported superconductivity with $T_C = 26$ K in fluorine-doped LaFeAsO. This prompted researchers around the world to search for iron-based superconductors.

Other materials have also been shown to display superconductivity, for example alkali metal-intercalalated fullerides (anions of Buckminsterfullerene and related materials; see Chapter 11). These have low critical temperatures, the highest being 38 K for Cs_3C_{60}.

10.3.1 Cuprate Superconductors

The first high-T_C superconductors discovered contained weakly coupled copper oxide (CuO_2) planes. Since then, many other high-T_C superconductors with such planes have been reported. These are referred to as cuprate superconductors. The highest critical temperatures are found for cuprates containing a Group 2 metal (Ca, Ba, Sr) and a heavy metal, such as Tl, Bi, or Hg. The structures of all the cuprate superconductors are based on or related to the perovskite structure. The perovskite structure is named after the mineral $CaTiO_3$ (see Chapter 1): many oxides of the general formula ABO_3 adopt this structure (also fluorides [ABF_3] and sulfides [ABS_3]). The so-called perovskite A-type unit cell (with the A-type atom in the centre of the cell) was shown in Chapter 1 (Figure 10.6).

The crystal structure of the 1-2-3 superconductor, $YBa_2Cu_3O_{7-x}$, is shown in Figure 10.7. Figure 10.7a shows only the positions of the metal atoms: we can see the strong similarity between this and the structure of $La_{2-x}Ba_xCuO_4$. If, as before, we discuss it in terms of the perovskite structure ABO_3, where B = Cu, the central section is now an A-type perovskite unit cell and above and below it are also A-type perovskite unit cells with their bottom and top layers missing. This gives copper atoms at the unit cell corners and on the unit cell edges at fractional coordinates $\frac{1}{3}$ and $\frac{2}{3}$. The atom at the body centre of the cell (i.e., at the centre of the middle section) is now **yttrium**. The atoms at the centres of the top and bottom cubes are **barium**.

If, in this structure, all three sections were based exactly on perovskite unit cells, we would expect to find the oxygen atoms in the middle of each cube edge

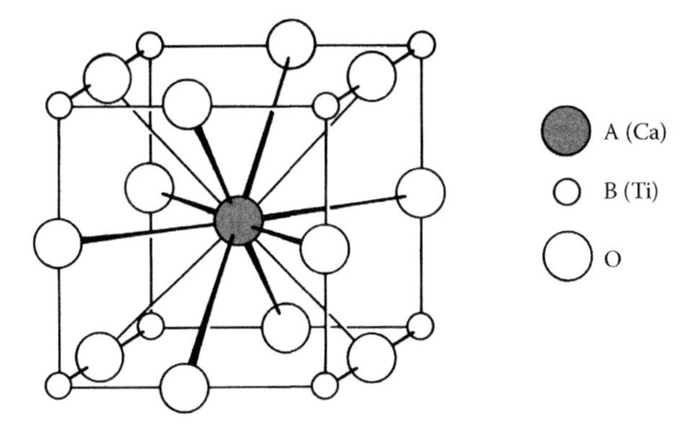

FIGURE 10.6 The A-type unit cell of the perovskite structure for compounds ABO_3, such as $CaTiO_3$.

(Figure 10.7b), giving an overall formula of $YBa_2Cu_3O_9$. This formula is improbable because it gives an average oxidation state for the three copper atoms of $\frac{11}{3}$, implying that the unit cell contains both Cu(III) and Cu(IV), which is unlikely because Cu(IV) complexes are extremely rare. The unit cell, in fact, contains only approximately seven oxygen atoms ($YBa_2Cu_3O_{7-x}$): when $x=0$, the oxygen atoms on the vertical edges of the central cube are not there and there are also two missing from both the top and the bottom faces (Figure 10.7c). A unit cell containing seven oxygen atoms has an average copper oxidation state of 2.33, indicating the presence of Cu(II) and Cu(III) in the unit cell (but no longer Cu(IV)).

In the 1-2-3 structure (when $x=0$), the yttrium atom is coordinated by 8 oxygens and the barium atoms by 10 oxygens. The oxygen vacancies in the 1-2-3 superconductor create sheets and chains of linked copper and oxygen atoms running through the structure (this is shown slightly idealised in the diagrams as, in practice, the copper atoms lie slightly out of the plane of the oxygens): the copper is in fourfold square–planar or fivefold square–pyramidal coordination (Figure 10.7d). The superconductivity is found in directions parallel to the copper planes, which are created by the bases of the Cu/O pyramids and are separated by layers of yttrium atoms. If this superconductor is made more deficient in oxygen, at $YBa_2Cu_3O_{6.5}$ ($x=0.5$), the superconducting critical temperature (T_C) drops to 60 K and at $YBa_2Cu_3O_6$ the superconductivity disappears. The oxygen is not lost at random, but goes from specific sites, gradually changing the square–planar coordination of the Cu along the c direction into the twofold linear coordination characteristic of Cu^+ and the arrangement of the copper and oxygen atoms in the base of the pyramids is not affected. However, when the formula is $YBa_2Cu_3O_6$, all the square–planar units along the c direction have become chains containing Cu^+ and the pyramid bases contain only Cu^{2+}; the unpaired spins of the Cu^{2+} are aligned antiparallel and the compound is antiferromagnetic. It is not until the oxygen content is increased to $YBa_2Cu_3O_{6.5}$ that the antiferromagnetic properties are destroyed and the compound becomes a superconductor. It is thought that this compound contains copper in all three oxidation states—I, II, and III. $YBa_2Cu_3O_7$ contains Cu(II) and Cu(III) both in the sheets

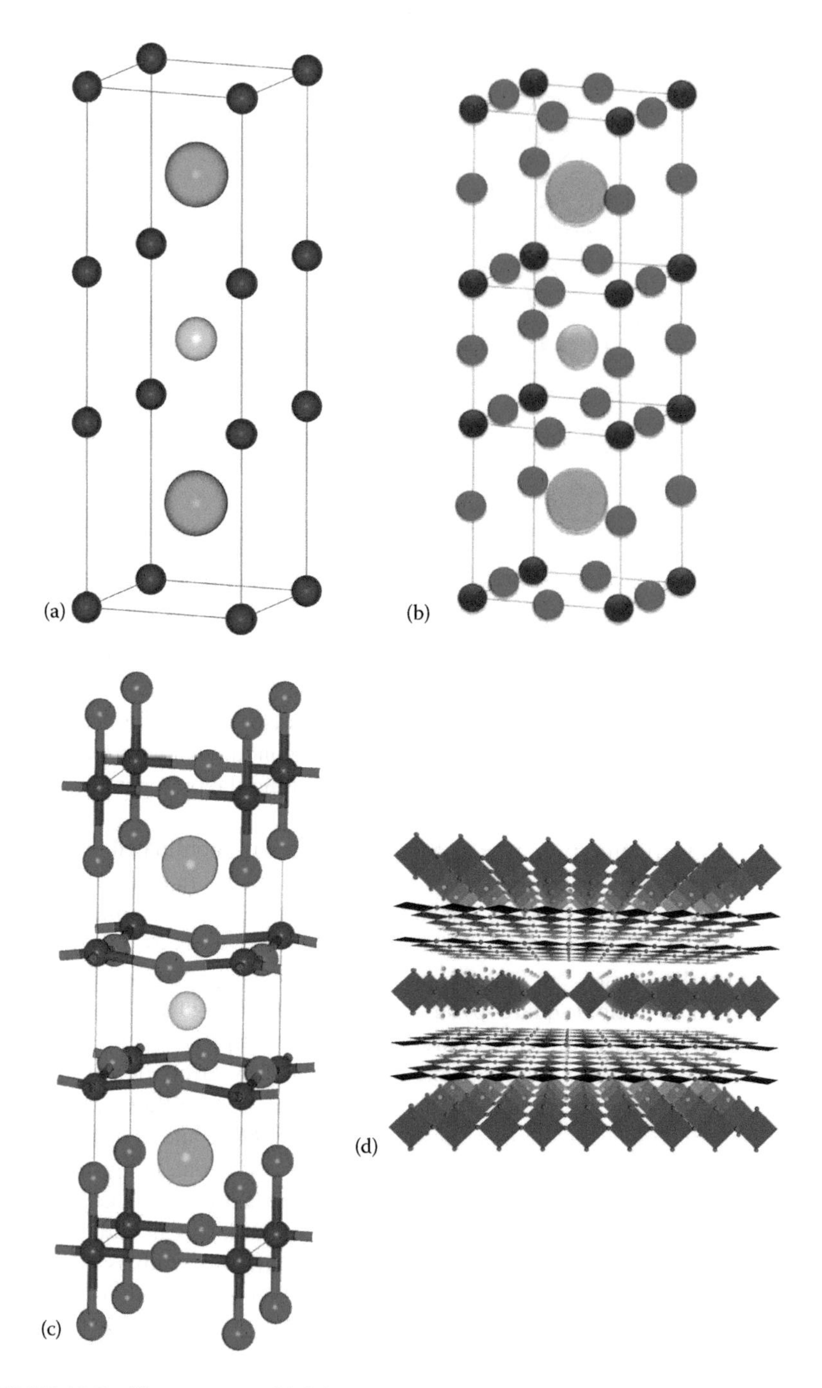

FIGURE 10.7 The structure of 1-2-3: (a) the metal positions; (b) an idealised unit cell of the hypothetical $YBa_2Cu_3O_9$, based on three perovskite A-type unit cells; (c) an idealised structure of $YBa_2Cu_3O_{7-x}$; and (d) the structure of $YBa_2Cu_3O_7$, showing copper–oxygen planes formed by the bases of the pyramids, with the copper–oxygen diamonds in between.

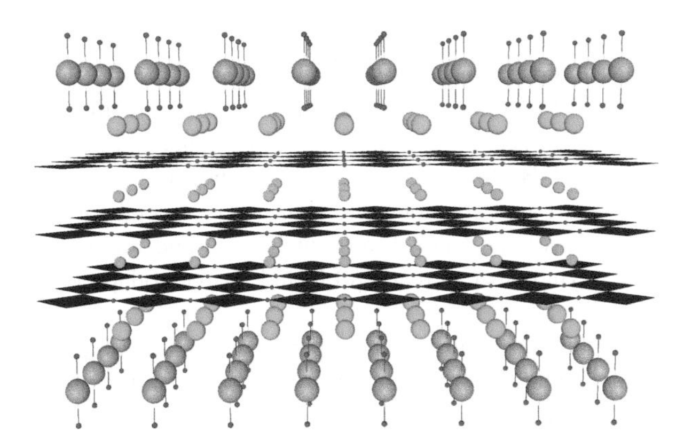

FIGURE 10.8 The structure of $HgBa_2Ca_2Cu_3O_8$ showing copper–oxygen diamonds between the Ca layers and the copper–oxygen layers forming the bases of the pyramids. The apices of the pyramids are in the barium–oxygen layers. In the superconductor $Hg_{0.8}Tl_{0.2}Ba_2Ca_2Cu_3O_{8+0.33}$, one-fifth of the Hg^{2+} ions are replaced by Tl^{3+} ions and there are additional oxygen ions in the mercury layer.

and in the chains. Clearly, the oxidation state of the copper atoms in the structure (and thus their bonding connections and bond lengths) is extremely important in determining both whether superconductivity occurs at all and the temperature below which it occurs (T_C).

The Hg-containing cuprate superconductors, of which more than 40 are known and include the material with the highest known (in 2019) T_C at ambient pressure, are also perovskite-based structures. The structures of these solids fall into one of three classes, 1201 based on $HgBa_2CuO_4$, 1212 based on $HgBa_2CaCu_2O_6$, and 1223 based on $HgBa_2Ca_2Cu_3O_8$. The Hg layers in the parent compounds contain no oxygen and are poor superconductors. In 1993, it was found that annealing these compounds in oxygen resulted in oxygen atoms being inserted into the $\left(\frac{1}{2},\frac{1}{2},0\right)$ position in the Hg layer and that $HgBa_2CuO_{4+\delta}$ was superconducting. Under pressures above 10 GPa, $HgBa_2Ca_2Cu_3O_{8+\delta}$ has a critical temperature greater than 150 K. By replacing some of the Hg atoms with metal atoms in higher oxidation states, compounds with extra oxygen in the Hg layer are formed, which are more stable at normal pressures. The highest critical temperature at the time of writing is for the 1223 structure with one-fifth of the Hg atoms replaced by Tl: $Hg_{0\cdot 8}Tl_{02}Ba_2Ca_2Cu_3O_{8+0.33}$. The structure of this compound is shown in Figure 10.8.

The unit cell contains four CuO_2 layers with Cu in square–pyramidal coordination and with the apices of the pyramids alternately pointing up and down. Ba occupies sites close to the apices of these pyramids and the Ca positions are just above or below the bases. At the top and bottom of the cell, between two layers of Ca atoms are CuO layers with the Cu in square–planar coordination. In the centre of the cell is the Hg layer. Overdoping with Tl gives a metallic compound with poor superconducting properties.

In 1-2-3, $La_{2-x}Ba_xCuO_4$, and the mercury cuprates, the average oxidation state of the Cu is greater than 2 and, as a result, positive holes are formed in the valence bands. The charge is carried by the positive holes and, as a consequence, such materials are known as ***p*-type superconductors**. In 1-2-3 and the mercury cuprates, positive holes are formed by adding oxygen, but there are other methods. In $SrCuO_2F_{2+\delta}$, superconductivity is induced by the insertion of fluorine, and in the collapsed oxycarbonates, such as $TlBa_2Sr_2Cu_2(CO_3)O_7$, superconductivity can be produced by shearing so that a shift appears along one plane every *n* octahedra (where *n* is typically 3–5), which leaves the CuO_2 layers unchanged and mixes the TlO and CO_3 layers. T_C for the oxyhalides can reach 80 K and for the collapsed carbonates, T_C is in the range of 60–77 K.

Until 1988, all the high-temperature superconductors that had been found were *p*-type and it was assumed by many that this would be a feature of high-temperature superconductors. However, some ***n*-type superconductors** have also been discovered, where the charge carriers are electrons: the first to be found was based on the compound Nd_2CuO_4 with small amounts of the three-valent neodymium substituted by four-valent cerium—$Nd_{2-x}Ce_xCuO_{4-y}$, where $x \sim 0.17$ (samarium, europium, or praseodymium can also be substituted for the neodymium). Other similar compounds have since been found based on this structure where the three-valent lanthanide is substituted by, for example, four-valent thorium in $Nd_{2-x}Th_xCuO_{4-y}$. The superconductivity occurs at $T_C \leq 25$ K for these compounds.

It seems clear that in cuprate superconductors, the superconductivity takes place in the CuO_2 planes and that the other elements present and the spacings between the planes change the superconducting transition temperature—exactly how is not yet understood.

An unusual superconductor that is structurally related to the cuprates was identified in 1994: Sr_2RuO_4 has a crystal structure almost identical to La_2CuO_4, the parent compound of the high-T_C superconductor 1-2-3 (see Figure 10.5). However, whereas La_2CuO_4 is an antiferromagnetic insulator, Sr_2RuO_4 is metallic and, below about 1.5 K, it is superconducting. Interest in this compound, despite its low T_C, is due to the presence of ferromagnetism as well as superconductivity. Conventional superconductors are diamagnetic and, previously, diamagnetism had been thought of as a necessary condition for superconductivity. The ruthenium in Sr_2RuO_4 is formally present as Ru(IV), giving an electronic configuration of $4d^4$. The Ru is in the centre of an octahedron of oxygen and so the d bands are split into t_{2g} and e_g. The four 4d electrons partly occupy the t_{2g} band. It is the electrons close to the highest-occupied levels of this band that are responsible for the superconductivity.

10.3.2 Iron Superconductors

Due to the idea that magnetism and superconductivity were believed to be mutually exclusive, the use of a magnetic ion to form a superconducting layer was generally avoided. Then, in 2008, Hideo Hosono reported superconductivity with $T_C = 26$ K in fluorine-doped LaFeAsO. This prompted researchers around the world to search for iron-based superconductors. Critical temperatures for these superconductors are not as high as for cuprate superconductors. However some have critical temperatures attainable by electric cryocooling. They have high critical fields and are easier to form into wires. The structure of the parent compound, SmFeAsO, is shown in Figure 10.9.

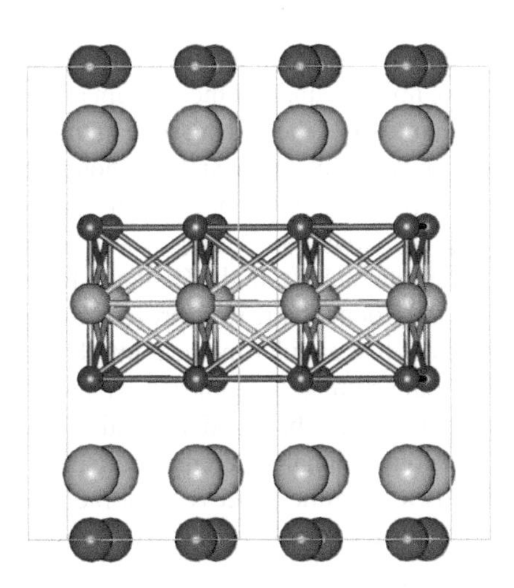

FIGURE 10.9 The structure of SmFeAsO. The Fe and As atoms form a closely packed layer with the Fe atoms at the centre. Above and below this layer are Sm ions and above and below these are oxide ions.

Like the cuprates, the iron-based superconductors form a layer structure. The characteristic layer consists of Fe arranged on a square lattice with P, As, Se, or Te tetrahedrally coordinated to the iron above and below the plane. These layers are either simply stacked or interspersed with layers of Group 1, Group 2, or lanthanoid oxyfluorides, or with a perovskite-type structure. Five structural classes have been identified. The Fe ions form the conducting layer, and the highest critical temperatures so far are for solids containing As.

The parent compounds, such as LaFeAsO and Fe_2Se_2, contain iron in a formal oxidation state of +2. It is perhaps surprising, therefore, that most of the superconductors found contain dopants that result in the addition of electrons to the Fe layer. The electrons are supplied in a variety of ways—substitution of fluoride for oxide, oxygen vacancies, insertion of Group 1 or 2 metals in between layers to give M^{2+} or M^+ ions, or substitution of cobalt on iron sites. The addition of electrons to the iron layer will, of course, reduce the oxidation state of Fe. This together with the high coordination number of the Fe, which includes coordination to adjacent Fe as well as to P, As, Se and Te, suggests that the superconducting layer is more like an alloy than an ionic structure.

There are also some Fe-based superconductors in which doping withdraws electrons from the iron layer, for example, by the substitution of Group 2 metals on lanthanoid sites.

10.3.3 Theory of High-T_C Superconductors

The general consensus is that, in common with the conventional superconductors, high-temperature superconductors contain Cooper pairs (Section 10.2.3). However, there is a difference in the pairs formed. In conventional superconductors, the Cooper pairs couple to give a total angular momentum of zero and are described by an order parameter of s-wave symmetry. In high-temperature cuprate superconductors, the Cooper pairs couple to give nonzero angular momentum and are described by an order parameter of d_{x2-y2}-wave symmetry. The pairing state in the BCS theory does not need to be caused by electron–phonon interaction, and it is thought that pairing in the d state is due to antiferromagnetic spin fluctuations or electron–magnon coupling. Such a possibility had been examined by Kohn and Luttinger, who found that a weak collective residual attraction could be generated from the Coulomb repulsion between electrons, but only if the electrons in the Cooper pairs were prevented from close encounters. This could be achieved if the pairs had angular momentum, that is, if they were described by p-or d-waves. Their initial estimate of T_C in such systems was very low, but it was hoped that higher temperatures might be found in metals with strong spin fluctuations. As in the BCS theory for conventional superconductors, the coupling is a collective property and the pairing is a dynamic process. Spin-mediated coupling is far more effective in bringing about superconductivity in two dimensions than in three, and this may explain why high T_C superconductivity is found in quasi–two-dimensional solids such as the cuprates. Experiment suggests that the quasi–two-dimensional organic superconducting polymers are also d-wave superconductors.

The presence of d-wave symmetry has been demonstrated in an elegant experiment using scanning tunnelling microscopy (STM, see Chapter 2). Collaborators at Berkeley and Tokyo, led by Davis and Uchida, used STM to measure differential tunnelling conductance on a specimen of bismuth strontium calcium copper oxide with zinc atoms replacing a small number of copper atoms. Differential tunnelling conductance is proportional to the density of states available to electrons tunnelling from the sample into the microscope's tip. This should be zero within a defined energy above the highest-occupied levels, known as the superconducting symmetry gap. For d-wave symmetry, the energy gap is zero in certain directions resembling the nodes of d_{x2-y2} orbitals. The Zn atoms broke the superconducting pairs at their positions so that a conductance peak appeared at the positions of the Zn atoms. Emanating from the Zn positions were lines of relatively high conductance in the direction of the nodes expected for the d-wave function.

The iron-based superconductors are also thought to contain Cooper pairs, and there is some evidence from isotope studies that phonons are involved. As for the cuprate semiconductors, the solids are antiferromagnetic above T_C and the onset of superconductivity may be linked to the quashing of magnetic ordering. NMR has shown that these superconductors are singlet, $S=0$. The wave function of the Cooper pair must be antisymmetric with respect to the exchange of electrons. For singlet Cooper pairs, exchange leads to a change in spin, so the order parameter must be even (s-, d-, etc.). Measurements on some solids suggest an s-wave symmetry. However, the type of s-wave symmetry is not that of the conventional superconductors because there is evidence of phase change of the s-wave. The nature of the mechanism, however, is still a matter of debate.

Ferromagnetic superconductors are thought to have Cooper pairs that are formed by magnetic interactions as in high-T_C superconductors, but with their spins aligned, giving a triplet state. For a triplet pair, both electrons have the same spin, so the order parameter must be odd. It is generally accepted that the ferromagnetic superconductors have p-wave order parameters.

10.4 USES OF HIGH-TEMPERATURE SUPERCONDUCTORS

Metal and alloy superconductors are used to form magnets, for example, for use in NMR spectrometers. Other uses include superconducting quantum interference devices (SQUIDS), which are loops of wire containing Josephson junctions (constructed from superconducting layers) and can measure very small voltages, currents, and magnetic fields. However these have to be cooled to liquid helium temperatures, which is expensive; therefore using high-temperature superconductors would reduce costs in these and other applications. However, the cuprates are brittle ceramic materials that are not easily formed into wires and the superconductivity in these materials is anisotropic. Progress has been slow in overcoming these problems.

Electric motors using high-temperature superconductors, and magnets containing such materials, have been developed. In 1997, a high-T_C superconducting magnet was installed in the beamline of a carbon-dating van der Graaf accelerator at the Institute for Geological and Nuclear Sciences in Wellington, New Zealand. In 2018, the magnet in a working wind turbine generator in Denmark was replaced by a superconducting magnet. This had a composite tape with the superconductor gadolinium–barium–copper-oxide (a cuprate superconductor $GdBa_2Cu_3O_{7-x}$ with a structure similar to that of 1-2-3) on a steel ribbon.

10.5 SUMMARY

1. In a superconducting phase of a material, the electrical resistance is 0 and all magnetic flux is expelled from the interior.
2. Metallic elements that displayed magnetic ordering (ferro- or antiferromagnetic) were found not to form a superconducting phase.
3. Superconductivity was initially found in metals and alloys at low temperatures and the properties were explained by a theory put forward by Bardeen, Cooper, and Schrieffer (BCS theory).
4. BCS theory attributes the superconductivity to strong phonon–electron coupling giving rise to coupled Cooper pairs of electrons which constantly break up and reform with new partners.
5. Some compounds containing light atoms (for example MgB_2, H_2S under very high pressure) are superconductors at higher temperatures. These can be explained using BCS theory.
6. Bednorz and Muller found that an oxide material, a cuprate with a structure based on that of perovskite, had a superconducting phase that was present at higher temperatures than the metals and alloys previously studied.
7. In cuprate superconductors, superconductivity occurs in planes of Cu and O atoms. These materials do not obey BCS theory.
8. Cuprate superconductors are brittle ceramics and thus difficult to form wires from.
9. Iron-based superconductors also form a layer structure. The characteristic layer consists of Fe arranged on a square lattice with P, As, Se, or Te tetrahedrally coordinated to the iron above and below the plane. The superconducting phase in these materials occurs at lower temperatures than for the cuprates, but is easier to form into wires.

QUESTIONS

1. One condition for superconductors obeying BCS theory with high critical temperatures is the existence of high-frequency phonons. Why might hydrides fulfill this condition?
2. Calculate the average oxidation state of Cu in the mercury cuprate $Hg_{0.8}Tl_{0.2}Ba_2Ca_2Cu_3O_{8.33}$. Assume Tl is present as Tl^{3+} and Hg as Hg^{2+}.
3. Calculate the average oxidation state of Fe in the superconductor $SmFeAsO_{0.5}F_{0.5}$ and comment on the value you obtain. Assume As is present as As^{3-}.
4. Using Figures 10.8 and 10.9, compare the structures of cuprate and iron-based superconductors, in terms of broad features of similarity and difference between the two classes.
5. Complete the following table, summarising the types of superconductivity.

Material	Symmetry of order parameter	Spin state
Metals and alloys		
Cuprates		
Iron-based superconductors		
Ferromagnetic superconductors		
Hydrides		

6. List the advantages and disadvantages of high-T_C cuprate superconductors compared to metallic superconductors.

11 Nanostructures

11.1 INTRODUCTION

In this chapter we consider structures that are between 0.2 and 100 nm in size in at least one dimension and in particular the effect that this has on their properties. The term nanostructures covers nanolayers with one nanosized dimension, nanowires with two dimensions, and nanoparticles, which are nanosized in three dimensions. A nanometre is 10^{-9} m or 10 Å. For comparison, the width of a human hair is in the range 15,000–200,000 nm.

Nanoparticles have a larger surface area to volume ratio than larger particles, affecting the way that they react with other particles and with each other. A 10 nm diameter nanoparticle has about 15% of the atoms on their surface, and by comparison, this drops to <1% for a bulk solid. In addition the energy levels associated with extended solids such as bands in metals no longer apply, and the electronic energy levels are more similar to the quantized levels found in individual atoms and molecules. This affects, for example, their conductivity and the way that they interact with light and other forms of energy.

We shall consider these properties and also synthesis of nanostructures. The methods and conditions of nanostructure manufacture are crucial because, as we shall see, the properties depend critically on the size and the shape of the particles produced. The synthetic techniques used are often so particular to each system that they make a subject in their own right, but we look at two man categories of method—bottom-up and top-down.

11.2 CONSEQUENCES OF THE NANOSCALE

11.2.1 Nanoparticle Morphology

The high surface-to-bulk ratio in nanostructures also means that the crystal structure is determined by a balance between bulk terms such as lattice energy, surface energy terms, and terms due to faults such as dislocations, as all these terms are now significant. This can lead to unusual crystal structures, such as thin films of body-centred cubic (*bcc*) copper, compared with the normal bulk structure, which is cubic closed packed (*ccp*).

Varying the condition of a film in chemical vapour deposition (CVD) can alter the morphology of the nanocrystals formed, and the techniques for producing specific morphologies could be very important in the production of catalysts, because different crystal faces have been found to catalyse very specific reactions.

Many nanomaterials can be made in different forms. In Section 11.3, we look at carbon, which can be found as diamond films, graphene, carbon black, fullerenes, and multi- and single-walled nanotubes. Inorganic materials can also form a variety of nanostructures; for example, ZnO can be formed as thin films, nanoparticles, nanotubes, nanowires, nanorods (Figure 11.1), nanorings, and nanoflowers.

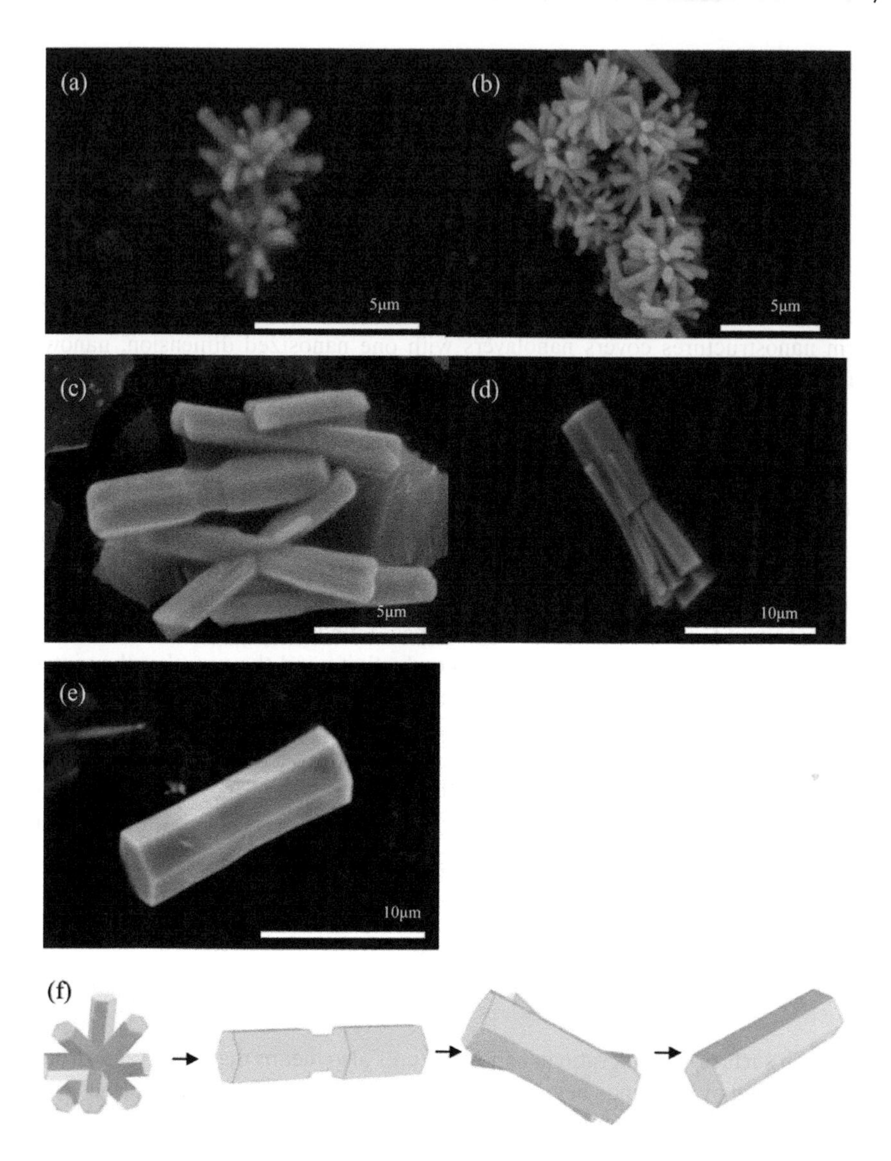

FIGURE 11.1 SEM images of ZnO nanorods reaction without NaOH at 100°C: (a) 3 h, (b) 6 h, (c) 12 h, (d) 24 h, (e) 48 h, (f) the schematic of the growth of ZnO nanorod. (Reprinted from Hsu, K.-C., Liao, J.-D., and Fu, Y.-S. 2013. 'Hydrothermal Synthesis of ZnO Nanorods Using the HMT Surfactant', *Integr. Ferroelectr*, Vol. 143 issue 1, 2013, Taylor & Francis Ltd. With permission. http//www.tanfonline.com.)

11.2.2 Electronic Structure

In Chapter 4, we discussed how the energy levels of a crystal could be obtained by thinking of a crystal as a very large molecule. For crystals of, say, micrometre dimensions, the number of energy levels is so large and the gap between them so small that we could treat them as essentially infinite solids with continuous bands of allowed energy. At the nanometrescale, we can still think of the particles as giant molecules, but a typical nanoparticle contains 10^2–10^4 atoms, very large for a molecule but not large enough to make an infinite solid a good approximation. The result is that in nanoparticles, we can still distinguish bands of energy, but the gaps between the bands may differ from those found in larger crystals and, within the bands, the energy levels do not quite form a continuum so that we can observe the effects due to the quantised nature of the levels within the bands. This is illustrated in Figure 11.2 for the nanosized crystals of a semiconductor (quantum dots).

The band diagram for a semiconductor is shown in Figure 11.2a. The bonding electrons are held in the lower valence band consisting of a continuum of many energy levels, and two electrons can occupy each energy level. The orbital density of states diagrams show that, in general, there is a lower density of states at the top and the bottom of the band and a higher density of states in the middle (indicated by shading). Above the valence band there is a conduction band, with a band gap separating the two bands. If the electrons are promoted from the valence band to the empty conduction band by supplying sufficient energy for them to jump the band gap, then the solid conducts.

As a crystal of a semiconductor becomes smaller in size, there are fewer atomic orbitals available to contribute to the bands. The orbitals are removed from each of the band edges until a point when the crystal is very small—a 'dot'—the bands are no longer a continuum of orbitals, but individual quantized orbital energy levels (Figure 11.2b). At the same time, this has the effect of increasing the band gap: as the size of the nanoparticles of most semiconductors decreases, so does the number of

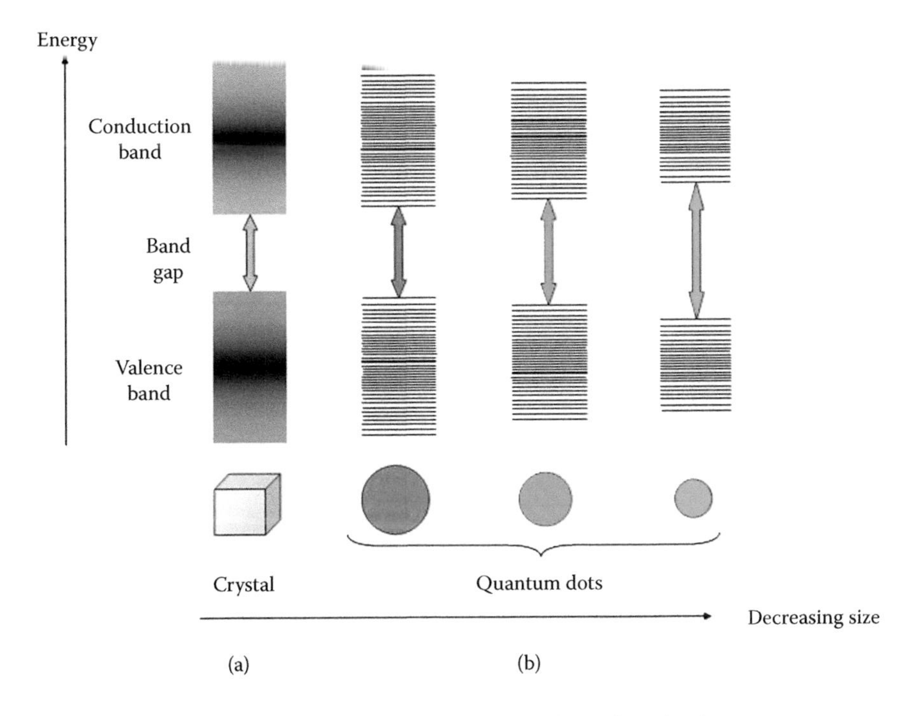

FIGURE 11.2 (a, b) The band gap of a semiconductor depends on its size.

orbital energy levels, but the band gap increases. The band gap in CdSe crystals, for example, is approximately 1.8 eV for crystals of diameter 11.5 nm, but it is approximately 3 eV for crystals of diameter 1.2 nm.

Quantum dots are nanometre scale in three dimensions, but structures that are only nanometre scale in two dimensions (quantum wires) or one dimension (quantum wells or films) also display interesting properties. The quantized nature of the bands in nanostructures can be seen in the density of states. Schematic, theoretical density of states diagrams for bulk material, quantum wells, quantum wires, and quantum dots are shown in Figure 11.3.

Figure 11.4 shows the density of states for a specific example, a single-wall carbon nanotube (SWCNT). The density of states for semiconducting carbon nanotubes is predicted to show sharp peaks (known as van Hove singularities) corresponding to specific energy levels. These can be seen in Figure 11.4 (S_{11} and S_{22}) and have been confirmed by scanning tunnel microscopy (STM, Chapter 2) experiments. In Figure 11.4a, there is a gap with zero density about the Fermi energy ($E=0$ on the figure). This is the band gap that we associate with semiconductors. For semiconducting carbon nanotubes, the band gap generally increases with decreasing diameter, but for particular nanotube structures, the band gap becomes zero and the nanotubes are then metallic conductors (Figure 11.4b), similar to graphite.

Electrical conductance in solids (other than ionic conductors) depends on the availability of delocalised orbitals close enough together in energy to form bands.

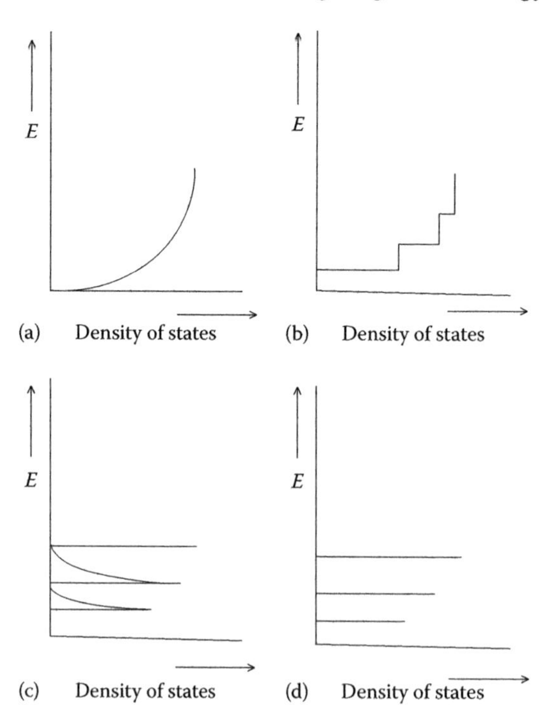

FIGURE 11.3 The theoretical density of states diagrams for (a) bulk material, (b) a quantum well, (c) a quantum wire, and (d) a quantum dot.

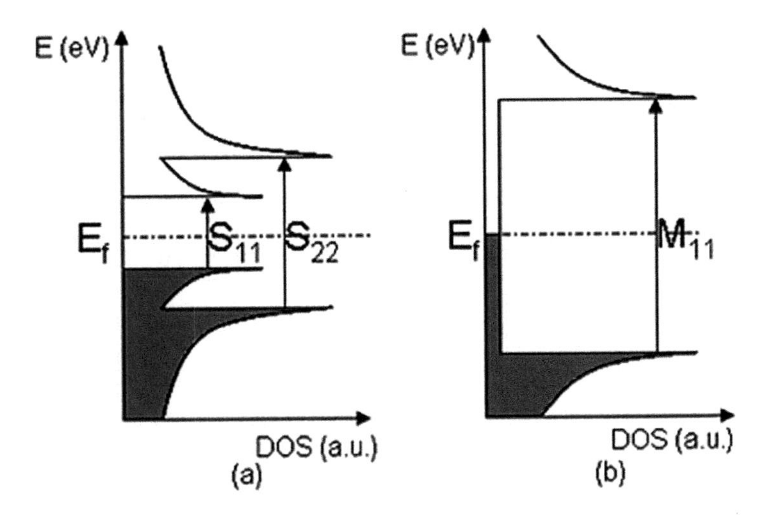

FIGURE 11.4 Energy diagram showing the density of states in semiconducting SWCNT (left) and metallic SWCNT (right) valence bands are grey while conduction bands are white. (From V. Gobba and D. M. Guldi, 'Carbon Nanotubes - Electronic/Electrochemical Properties and Application for Nanoelectronics and Photonics'. *Chem. Soc. Rev.* (2009) **38**, 165–184, by Chemical Society (Great Britain). Reproduced with permission of Royal Society of Chemistry via Copyright Clearance Center.)

When we have structures on the nanoscale where the levels at the top of occupied bands are not as close in energy as we have seen, then this can affect the conductance. Under certain conditions, electrical conductance through a nanostructure is quantised and increases in a stepwise fashion with increasing voltage. The electrons tunnel into the structure and fill the lowest empty quantised levels until all the levels below the highest-filled level providing electrons is full (Figure 11.5). If the thermal energy is insufficient to raise the electrons to the next discrete energy level, then no more electrons can tunnel in. The conductance thus drops until the voltage is increased to a value V, such that the energy $e \times V$ is sufficient to raise the energy of the electrons in the adjacent solid so that they can reach the next energy level, when the current increases again. Figure 11.6, for example, shows the increase in conductance with gate voltage along a quantum wire connecting two interfaces in a transistor.

11.2.3 Optical Properties

In this section, we consider two aspects of the interaction of light with nanostructures—the differences in the optical properties of bulk solids that arise because of the different energy levels of nanostructures, and the changes in the scattering properties.

The last section revealed that semiconductor band gaps for nanostructures vary with the size of the structure. The wavelength of light emitted when an electron in the conductance band returns to the valence band will therefore also vary. Thus, different colour fluorescence emission can be obtained from differently sized particles of the same substance. To produce fluorescence, light of greater photon energy than

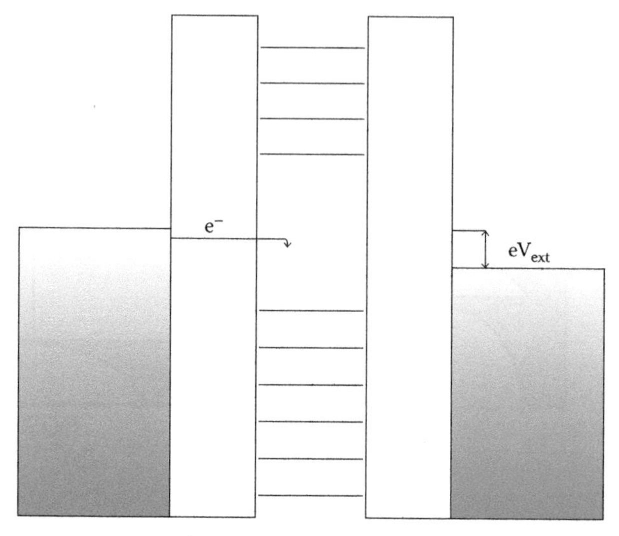

FIGURE 11.5 Electrons fill the lowest empty quantized levels until all the levels below the highest-filled level providing electrons are full.

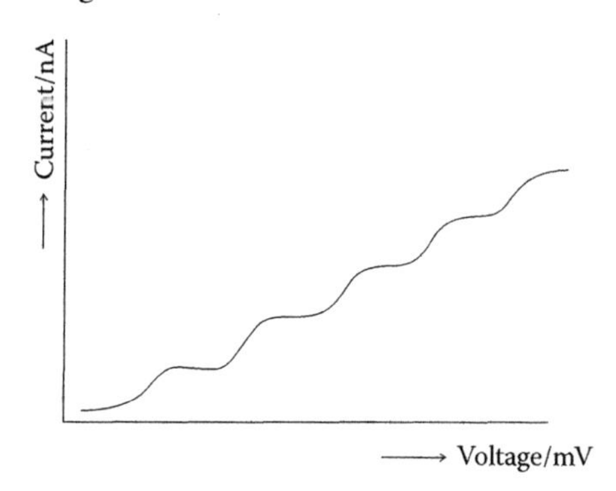

FIGURE 11.6 Schematic of the increase of current with gate voltage along a quantum wire connecting two interfaces in a transistor.

the band gap is shone onto the nanostructure. An electron is excited to a level in the conduction band, from where it reaches the lowest energy level in the conduction band through a series of steps by losing energy as heat. The electron then returns to the valence band, emitting light as it does so. This is illustrated for quantum dots (Section 11.5.2) in Figure 11.7.

Figure 11.7a depicts the irradiation of a large quantum dot (smaller band gap). It then decays into the valence band, emitting a photon of light, which is the colored fluorescence seen. If the smaller quantum dot (larger band gap) (Figure 11.7b) undergoes the same process, we can see that the photon emitted as it decays back into the valence band has more energy. From the Planck–Einstein equation, $E=h\nu$, the higher-energy photon will have the higher frequency and thus be closer to the blue

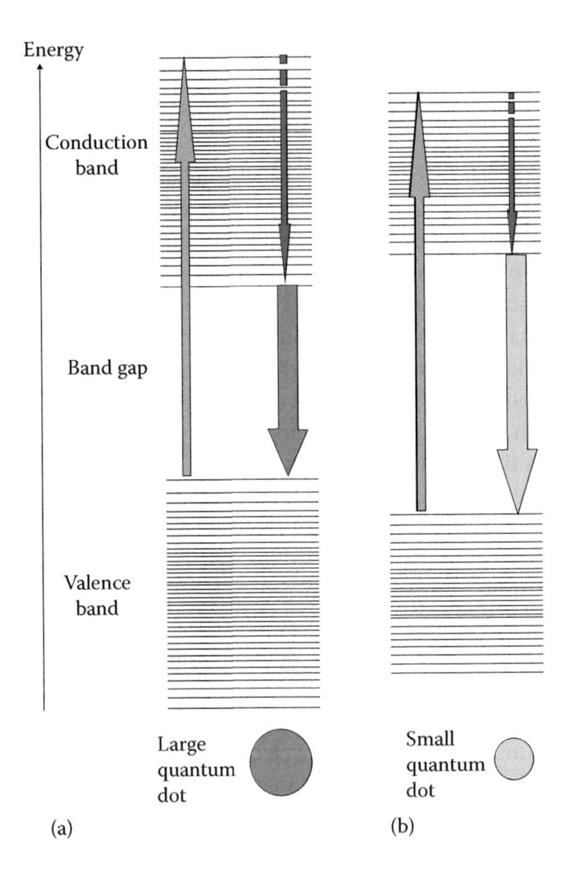

FIGURE 11.7 (a, b) The colour of the fluorescence from a nanosized particle depends on its dimensions.

end of the spectrum, accounting for the difference in color. The smaller the nanostructure, the larger is the band gap and hence the shorter is the wavelength of the emitted light. Thus, 5.5-nm-diameter particles of CdSe emit orange light, whereas 2.3-nm-diameter particles emit turquoise light.

In the absorption spectra of nanoparticles of CdSe and other semiconductors, not only can the shift in wavelength be observed, but there are also bands corresponding to absorption to discrete energy levels in the conduction band. For example, 11.5 nm diameter particles of CdSe have an absorption spectrum showing an almost featureless edge, but particles of diameter 1.5 nm show features resembling molecular absorption bands shifted about 200 nm to shorter wavelengths.

The colours produced by nanoparticles of gold (colloidal gold) have been used to colour glass since Roman times and have been studied since Faraday in the mid-nineteenth century. In metals, light interacts with the surface electrons and is reflected. The surface electrons in nanoparticles are induced by the light to oscillate at a particular frequency; absorption at this frequency gives rise to the colour. The oscillation frequency and hence the colour depend on the size of the particle. For example, whereas bulk gold is yellow, thin films of gold appear blue when light passes through them, but as the particle size is reduced, the wavelength of the absorbed light decreases and the film appears first red, then, for 3 nm diameter particles, orange. Oscillation frequencies in the visible region are only observed for Ag, Au, Cu, and the Group 1 metals.

Particles are known to scatter light as well as absorb it, and this produces the white or pale appearance of fine powders. The even-smaller nanosized particles, however, are transparent because the scattering efficiency is reduced. This effect has led to the use of nanoparticles in sunscreens and cosmetics. TiO_2 and ZnO have been used in sunscreens for many years because of their ability to absorb the UV radiation that is harmful to the skin. But TiO_2, in particular, also has a very high refractive index and therefore scatters light very efficiently; so efficiently, in fact, that it makes a very good white pigment and consequently, it is used in most white paint because of its high covering power. Nanoparticles of TiO_2 of about 50 nm diameter are transparent as they are too small to scatter the visible light, but they still absorb the harmful shorter-wavelength UV radiation, and therefore are now being used in sunscreens to get around this problem.

11.2.4 Magnetic Properties

In Chapter 9, the idea of ferromagnetic domains was introduced. Domains typically have dimensions of the order of 10–1000 nm. In nanocrystals, therefore, we can reach a situation where the domain size and the crystal dimensions are comparable. Such single-domain crystals have all the electron spins in the crystal aligned. Single-domain crystals of magnetite are found in certain bacteria (magnetotactic bacteria), which use the magnets to locate the sediment/water interface in muddy ponds.

In larger crystals, the main mechanism for magnetisation and demagnetisation is the rotation of the domain walls (Section 9.3.1). If the crystal size is reduced, as the single-domain region is approached, it becomes harder to demagnetise the crystal by applying a magnetic field. This is because the only possible mechanism is now the disruption of spin–spin coupling within a domain. If the particle size is decreased still further, however, the number of spins decreases and the force aligning them becomes weaker. Eventually, as this force gets weaker, and under the influence of temperature and in the absence of an applied magnetic field, the spins are able to flip over to their other orientation. Then, the crystal is no longer ferromagnetic but **superparamagnetic**. Superparamagnetism is different from normal paramagnetism because it occurs *below* the Curie temperature, and at this temperature the spins are able to flip to the opposite orientation, thereby reversing the direction of magnetisation. If the time between flips is short, when the magnetisation is measured it appears to be zero. Figure 11.8 is a plot of the magnetic field needed to demagnetise ferromagnetic particles (coercive force) as a function of the particle size. The radius D_C is the radius at which the particle becomes single domain. The values for D_C for the ferromagnetic transition metals and the ferrimagnetic Fe_3O_4 (magnetite) are given in Table 11.1.

In the superparamagnetic state, if a magnetic field is applied, above a critical temperature (T_B) known as the blocking temperature, the particle can have its spins aligned by a magnetic field and become magnetised. It behaves in a similar fashion

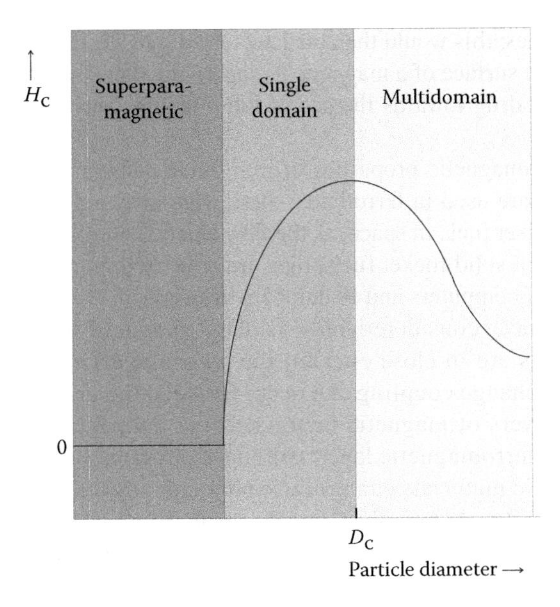

FIGURE 11.8 A plot of the magnetic field needed to demagnetize ferromagnetic particles (coercivity, H_C) as a function of the particle size. The particle becomes single domain at radius D_C.

TABLE 11.1
Single-Domain Radii for Selected Ferromagnetic Solids

Solid	D_C/nm
Fe	14
Co	70
Ni	55
Fe_3O_4	128

to paramagnetic materials, but with a much larger susceptibility and magnetic moment than those observed for paramagnetism. Below the blocking temperature, there is insufficient kinetic energy to overcome the energy barrier to reorientation of the spins. The blocking temperature varies with the strength of the applied field. In the superparamagnetic state, the particles, like paramagnetic solids, do not show hysteresis.

Superparamagnetic iron oxide nanoparticles are used to enhance magnetic resonance images (MRI scans) acting as contrast agents.

There are other promising medical uses for these magnetic nanoparticles. If they can be functionalised appropriately, say, with an antibody, then they could also be injected into the bloodstream and used to seek out markers of disease or infection such as cancer sites; this would then be detectable by MRI. If a drug molecule could be attached to the surface of a magnetic nanoparticle, then a magnetic field could be used to draw the drug towards the site of the infection, thereby targeting the drug more efficiently.

The superparamagnetic properties of iron oxide nanoparticles when suspended in hydrocarbons are used in ferrofluids; these were invented by NASA as a way to control liquid rocket fuels in space, as the flow can be controlled by magnetic fields. With the advent of solid rocket fuels, they are now used mainly as bearing seals in the hard drives of computers and as dampers in speakers.

So far, we have considered only isolated magnetic nanoparticles. When nanosized grains are in close contact, the magnetic effects can differ. This is because spin exchange coupling can occur between the grains. In Chapter 9, we saw that thin layers of magnetic material could have the spin state pinned by an adjacent antiferromagnetic layer. Exchange coupling between grains of hard and soft magnetic materials can produce nanocomposites with high remanence and high coercivity, that is, they remain strongly magnetic when the applied magnetic field is reduced to zero and need a large magnetic field to demagnetise them.

11.2.5 Mechanical Properties

The properties of solids, such as hardness and plasticity, have long been known to vary with the grain size. For example, down to micrometre-sized grains, the resistance to plastic deformation increases with decreasing grain size. However, as the grain size is reduced still further, the resistance levels off or even decreases. The effect has been attributed to a different mechanism for plastic deformation in nanosized grains. In larger crystals, deformation is mainly governed by the movement of dislocations (Chapter 5) in the crystal. Such movements are inhibited by grain boundaries. As the grain size is reduced, the ratio of grain boundary to bulk grain increases, so deformation becomes harder. Eventually, as the grain size reaches 5–30 nm, the movement of the dislocations becomes negligible, but deformation through atoms sliding along the grain boundaries becomes favourable. This latter mechanism is aided by a large ratio of the grain boundary to the bulk grain, so softening increases as the grain size is reduced.

11.2.6 Melting Temperature

Nanocrystals show a melting temperature depression with decreasing size. Melting of nanocrystals can be observed in a transmission electron microscope, and gold nanoclusters have been found to melt at 300 K compared with a melting temperature of 1338 K for elemental gold in bulk. Three-nanometer CdS crystals melt at about 700 K in a vacuum compared with 1678 K in bulk. The surface energy becomes an increasing factor as the size of a crystal becomes smaller; thermodynamics predicts a lowering of the melting temperature if the surface energy of the solid is higher than that of the liquid.

11.3 NANOSTRUCTURAL CARBON

11.3.1 Carbon Black

Carbon black is made by the vapour-phase incomplete pyrolysis of hydrocarbons to produce a fluffy fine powder consisting of nanoparticles of amorphous carbon. The method of preparation produces particles with few impurities in contrast to powders of similar appearance such as soot. Worldwide, about 8 million tonnes a year is produced. It is used as a reinforcing agent in rubber products such as tyres, as a black pigment in printing inks, paints and plastics, in photocopier toner, and in electrodes for batteries and brushes in motors. Adding carbon black to tyres improves the tensile strength and resistance to wear of the tyres, with smaller particles (around 20 nm) having a greater effect than larger ones.

11.3.2 Graphite

Graphite is, of course, a very familiar substance, with many uses. The lead in lead pencils is graphite, and finely divided forms of graphite are used to absorb gases and solutes. Its absorption properties find a wide range of applications from gas masks to decolouring food. Graphite formed by the pyrolysis of oriented organic polymer fibres is the basis of carbon fibres. Graphite is also used as a support for several industrially important catalysts. Its electronic conductivity is exploited in several industrial electrolysis processes where it is used as an electrode. Crystals of graphite are, however, only good conductors in two dimensions, and it is this two-dimensional aspect that we are concerned with here.

Crystals of graphite contain layers of interlocking hexagons, as shown in Figure 11.9. Figure 11.9 shows the most stable form in which the layers are stacked ABAB, with the atoms in every other layer directly above each other. There is another form that also contains layers of interlocking hexagons, but the layers are stacked ABCABC.

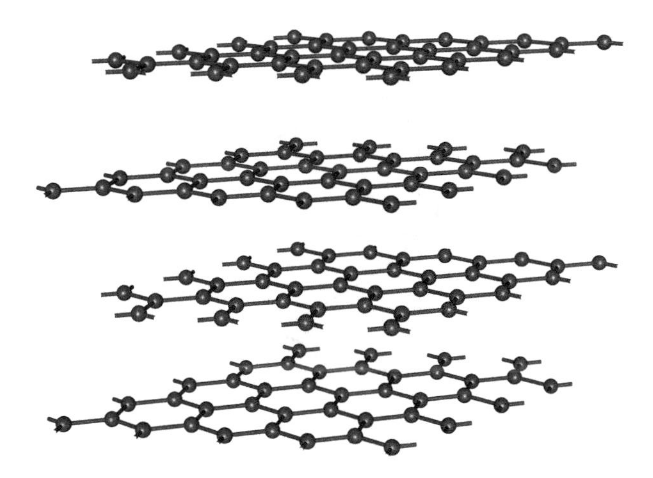

FIGURE 11.9 Hexagonal structure of graphite layers.

Each carbon can be thought of as having three single bonds to the neighboring carbons. This leaves one valence electron per carbon in a p orbital at right angles to the plane of the layer. These p orbitals combine to form delocalised orbitals that extend over the whole layer. If the layer contains *n* carbon atoms, then *n* orbitals are formed, and there are *n* electrons to fit in them. Thus, half of the delocalised orbitals are filled. Were the orbitals to form one band, this would explain the conductivity of graphite very nicely, since there would be a half-filled band confined to the layers. The situation is close to this, but not quite as simple. The delocalized orbitals, in fact, form two bands, one bonding and one antibonding. The lower band is full and the upper band empty. Graphite is a conductor because the band gap is zero, and thus the electrons are readily promoted to the upper band. The band structure for graphite is shown in Figure 11.10. Because the density of states is low at the Fermi level, the conductivity is not as high as that for a typical metal. It can, however, be increased, as we will now see.

11.3.3 Intercalation Compounds of Graphite

Because the bonding between the layers in graphite is weak, it is easy to insert molecules or ions into the spaces between the layers. The solids produced by reversible insertion of such guest molecules into lattices are known as intercalation compounds. Since the 1960s, attention has been paid to the intercalation compounds as of possible importance as catalysts and as electrodes for high-energy-density batteries.

Many layered solids form intercalation compounds, but graphite is particularly interesting because it forms compounds with both the electron donors and the electron acceptors. Among the electron donors, the most extensively studied are the alkali metals. The alkali metals enter the graphite between the layers and produce strongly coloured solids in which the layers of carbon atoms have moved further apart. For example, potassium forms a golden compound (KC_8) in which the interlayer spacing

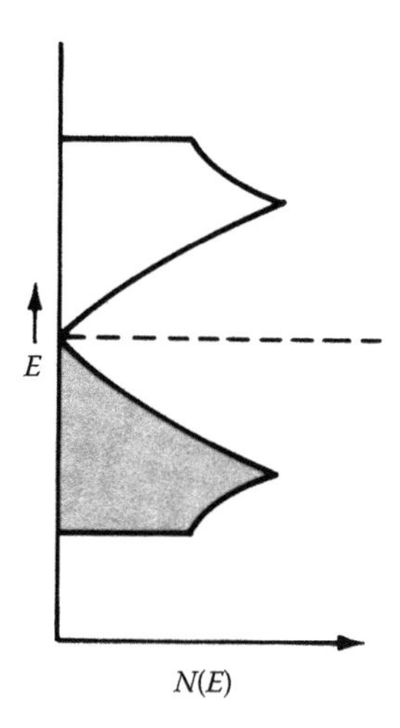

FIGURE 11.10 The band structure of graphite.

is increased by 200 pm (Figure 11.11). The potassium donates an electron to the graphite (forming K^+) and the conductivity of the graphite now increases because it has a partially full antibonding band. In Chapter 6, you saw that the anode of lithium-ion batteries was an intercalate of Li in graphite, typically C_6Li.

In 1841, the first intercalation compound was made, containing sulfate, an electron acceptor. Since then, many other electron acceptor intercalation compounds have been made with, for example, NO_3^-, CrO_3, Br_2, $FeCl_3$, and AsF_5. In these compounds, the graphite layers donate electrons to the inserted molecules or ions, thus producing a partially filled bonding band. This increases the conductivity, and some of these compounds have electrical conductivity approaching that of aluminium.

In graphite, the current is carried through the layers by delocalised p electrons. Layered structures in which the current is carried by d electrons are common in transition metal compounds, such as the disulfides of Ti, Zr, Hf, V, Nb, Ta, Mo, and W, and mixed lithium metal oxides.

11.3.4 Graphene

The 2010 Nobel Prize in Physics was awarded to Geim and Novoselov for producing, isolating, analysing, and identifying graphene. Graphene consists of a single layer of carbon atoms, essentially a single layer of graphite; the first truly two-dimensional solid (Figure 11.12).

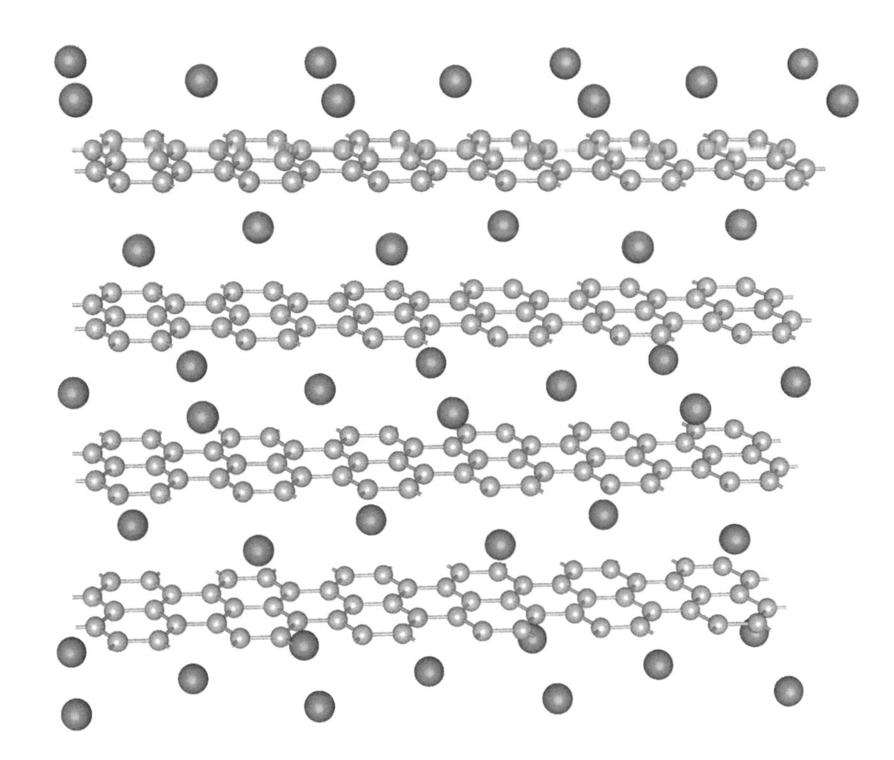

FIGURE 11.11 The structure of KC_8. K, blue; C, grey.

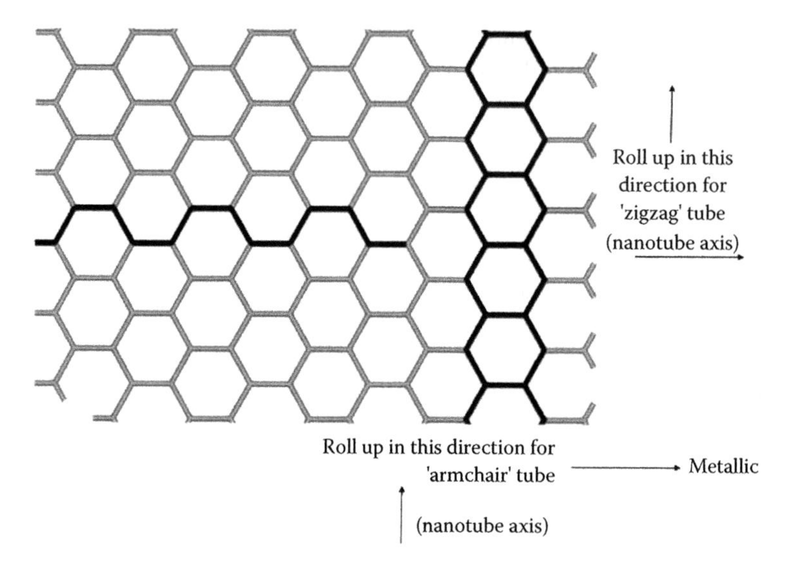

FIGURE 11.12 The structure of graphene. The sheet can be rolled up in the directions of the two arrows, forming distinct types of nanotubes (see Section 11.3.7).

The electronic structure of graphene had been predicted theoretically in 1947, but it had proved difficult to be isolated and identified. In 2004, Geim and Novoselov extracted layers of graphene from graphite using Scotch tape and transferred them to a silicon substrate. The thinnest layers were characterised using atomic force microscopy (AFM). Some of these proved to be the single-layer graphene, and they were able to measure its electrical properties.

Since then, it has become possible to make large sheets of graphene. Graphene is stronger than steel (its breaking strength is 200 times that of steel), but it is very flexible. Its electronic properties are also unusual. It is, for example, a transparent electrical conductor and could replace indium tin oxide (ITO) in touch screens and solar panels. Graphene conducts electrons with a constant velocity of about 10^6 m s^{-1}. In addition, it has high thermal conductivity. The density of states plot for graphene shows zero density at the Fermi level with the density of states rising rapidly as the energy increases or decreases (Figure 11.13).

Graphene's properties make it potentially useful in a wide range of applications. Examples are membranes for water filtration, targeted drug delivery, weatherproofing, ultrasensitive sensors, batteries, and supercapacitors. However at the time of writing (2019), it has proved difficult to manufacture graphene on the scale needed for industry and this has hindered the exploitation of its properties.

Recently, interesting properties have been found when two layers of graphene are stacked (**bilayer graphene**). If the two layers are not exactly matched but one is twisted through a small angle, then a set of 'flat' bands appear in the density of states more like those of strongly correlated metals such as Ni (Chapter 4). In addition, it was found that the electron velocity was much smaller than in graphene. Predictions of the offset angle at which the electron velocity would be zero were published in 2010. In 2018, a team at Massachusetts Institute of Technology found that such twists

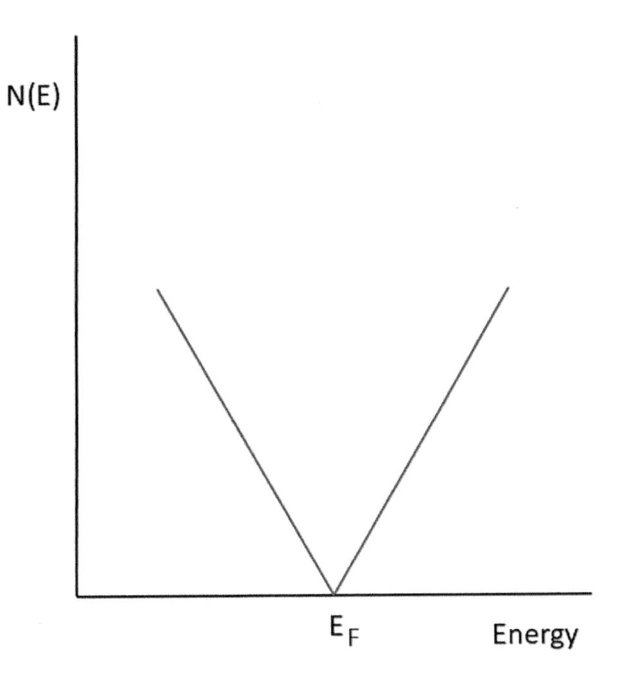

FIGURE 11.13 Sketch of density of states for graphene close to the Fermi level (E_F).

could produce insulating or superconducting states depending on the angle. These states were similar to the corresponding phases of cuprate superconductors. Perhaps more surprising, with some angles, teams in the United States and Spain have found evidence that the bilayer is a ferromagnetic insulator.

11.3.5 Graphene Oxide

Oxidation of graphite using a strong acid plus an oxidant gives yellow **graphene oxide**. The oxidation results in expansion of the gap between the graphite layers and the addition of functional groups such as carbonyl, epoxide, and hydroxide to the carbon atoms. The layers will separate in aqueous solution with sonification, and can be reduced to black graphene by thermal, chemical, or electrochemical methods. This has been suggested as a possible method for manufacturing graphene.

Buckminsterfullerene and carbon nanotubes, which we go on to discuss in the next sections, were discovered earlier and are related to graphene in that they can be thought of as being formed by rolling up sheets of graphene.

11.3.6 Buckminsterfullerene

This polymorph of carbon was only discovered in 1985 by Sir Harry Kroto at the University of Sussex while he was looking for carbon chains. It is made by passing an electric arc between two carbon rods in an inert atmosphere of helium, forming an arc discharge and thus heating the electrodes to >1700°C; many different fullerenes can be extracted from the resulting soot, C_{60} and C_{70} being the major products. Kroto was awarded the 1996 Nobel Prize in Chemistry along with two American researchers, Robert F. Curl Jr. and Richard E. Smalley.

The crystals of **buckminsterfullerene** are not necessarily nanoparticles, but we have included fullerenes here because they are related to carbon nanotubes, which we discuss in the next section.

The molecule has the formula C_{60} and has the same shape as a soccer ball—a truncated icosahedron; it takes its name from the engineer and philosopher Buckminster Fuller, who discovered the architectural principle of the hollow geodesic dome that this molecule resembles (a geodesic dome was built for EXPO 67 in Montreal). The structure is shown in Figure 11.14.

The molecule is extremely symmetrical with every carbon atom in an identical environment; it consists of 12 pentagons of carbon atoms joined to 20 hexagons. The carbon–carbon distances between the adjacent hexagons is 139 pm, and where a hexagon joins a pentagon, the carbon–carbon distance is 143 pm—similar to that found in the graphite layers. Indeed, we can think of this structure as a carbon sheet with graphite-type delocalised bonding, which bends back on itself to form a polyhedron. Other fullerenes are also known, such as C_{70}, and they all have 12 pentagons of carbon atoms linked to different numbers of hexagons. The C_{60} molecules, commonly called buckyballs, pack together in a cubic close-packed array in the crystals. The use of fullerenes has been suggested for electronic devices and even in targeted cancer therapy. One of the interesting features is the formation of salts with the alkali metals, known as buckides or fullerides. For instance, potassium buckide (K_3C_{60}) has a *ccp* array of buckyballs, with all the octahedral holes and tetrahedral holes filled by potassium atoms: this is a metallic substance, which becomes superconducting below 18 K. Other alkali metal buckides become superconducting at even higher temperatures.

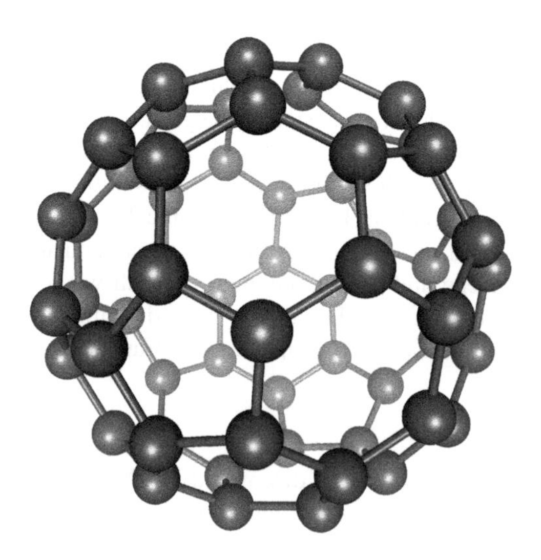

FIGURE 11.14 The structure of buckminsterfullerene (C_{60}).

11.3.7 Carbon Nanotubes

In 1991, Sumio Iijima discovered nanotubes. Similar in structure to buckyballs, they consist of sheets of graphene that roll up to form cylinders. Initially multilayered **(multi-walled nanotubes; MWNTs)**, they were subsequently made single-walled. **Single-walled nanotubes (SWNTs)** (Figure 11.15) have diameters of about 1 nm and lengths of the order of 10^{-6} m.

There are several common methods for synthesising CNTs, which are nowadays manufactured in very large quantities (hundreds of tonnes). They can either be grown in situ or synthesised separately and then incorporated into the matrix. When the arc discharge method is used for producing CNTs, a metal catalyst, such as Fe, Ni, or Co, is incorporated into one of the electrodes; the addition of diamond powder to the catalyst increases the yield. The laser ablation method is very similar to the arc discharge, only now a high-power laser is used to vaporise a graphite pellet containing the metal catalyst. CVD techniques are usually used for producing nanotubes when growing them in position. In this method, carbon-containing gases are deposited onto a substrate that contains a metal catalyst. If the catalyst has been deposited in nanoparticle arrays, then forests of vertically aligned nanotubes can be formed (Figure 11.16). This method has the advantage of using much lower temperatures than the laser and arc discharge techniques, usually between 700°C and 900°C.

The tubes are capped at each end by half of a fullerene-type structure. They have remarkable properties and are being used for a variety of high-performance applications. They can adsorb 100 times their volume of hydrogen, so they could be developed as a safe storage medium for hydrogen for fuel cells. They have a remarkable tensile strength, 100 times greater than steel, but are very light, with half the density of aluminium, and are being used in high-performance sports equipment such as tennis rackets. Although strong, they are also very flexible. Using AFM, they can be bent into a variety of shapes, including springs. Amroy Europe Oy (a spin-off company of the University of Jyväskylä, Finland), for example, produces epoxy resins reinforced with nanotubes that are widely used for example for sports equipment and wind turbine blades.

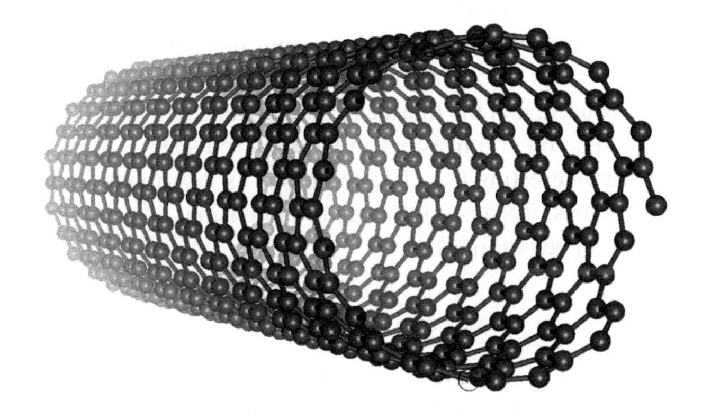

FIGURE 11.15 A computer simulation of a single-walled carbon nanotube.

FIGURE 11.16 A scanning electron microscope image of a forest of carbon nanotubes (CNTs) grown at the University of Virginia Nanoscale Energy Transport Laboratory. Each individual CNT is around 50 µm in length and 100 nm in diameter. (LeighAnn Larkin, Quang Pham, Carina Cerri, and Christopher Saltonstall Advisor: Dean Pamela Norris, University of Virginia, Charlottesville, VA. 'A Forest of Carbon Nanotubes', https://www.nano.gov/node/1797. Courtesy of Nano.gov.)

Carbon nanotubes (CNTs) are good conductors of heat. This property plus their small size enables them to act as thermal interface materials filling gaps in the interface between two materials in devices such as transistors.

Nanotubes can be made as conductors and semiconductors, and one use has been found in car bumpers, where they not only provide strength but also prevent the build-up of static electricity.

The electronic properties of SWNTs depend on the direction in which the graphene sheet rolls up. In the structure of graphene shown in Figure 11.12, notice that the line joining the rows of hexagons in the vertical direction is a simple zigzag, but those at right angles to this, the so-called 'armchair lines', join the rows. Tubes that roll up along the armchair direction—the so-called **armchair nanotubes**—always exhibit metallic levels of conductivity. If graphite sheets roll up along the zigzag lines, with the armchairs along the axis of the tube (**zigzag nanotubes**) or if the sheets roll up along any other direction except the zigzag or the armchair lines, forming *helical* nanotubes, a band gap is introduced, and the tubes can be semiconducting. It is this property that has allowed the formation of transistors and diodes from carbon nanotubes.

One chemically interesting use of nanotubes is as 'test-tubes'. Solids can be prepared inside nanotubes. Because of the small dimensions of the tube, unusual crystalline structures can be obtained.

11.4 NONCARBON NANOPARTICLES

Nanoparticles have been known and used for centuries; think, for instance, of the pigments used to colour stained glass and ceramic glazes and of the colloidal gold particles used for making 'ruby' glass, known since Roman times. Within the modern era, fumed silica—small particles of silica—has been added to solids and liquids to improve their flow properties since the 1940s.

At the simplest level, nanoparticles of hard substances are useful as polishing powders which are able to give very smooth, defect-free surfaces. Indeed, 50-nm nanoparticles of cobalt tungsten carbide are found to be much harder than the bulk material and therefore can be used to make cutting and drilling tools that will last longer.

Nanoparticles of zirconium oxide (ZrO_2) are being used in UV-cured dental fillings. They give strength and are transparent to visible light, but are opaque to X-rays. Rare-earth oxides can also be used.

The large surface area of nanoparticles lends increasing dominance to the behaviour of the atoms on the surface of the particles; in catalysis, this is exploited to improve the rate of production in commercial processes and in the structure of electrodes, to improve the performance of batteries and fuel cells (see Chapter 6).

The interactions between these surface atoms and a surrounding matrix determine the properties of high-performance nanocomposites.

11.4.1 Fumed Silica

This very pure form of silica is made by reacting $SiCl_4$ with an oxyhydrogen flame. The resulting SiO_2 particles, with dimensions of 7–50 nm, have an amorphous structure. The particles have silanol (Si–OH) groups on their surface, and these hold the particles together by hydrogen bonding, forming chain-like structures. When added to liquids, this three-dimensional network traps the liquid and increases the viscosity, but when the thickened liquid is subsequently brushed out or sprayed, the liquid and any trapped air are released. When the shear force is removed, the liquid thickens again. This property is known as **thixotropy** and is very useful in paints for preventing the settling of pigments and for improving the flow properties of paints, coatings, and resins. It is also used to improve the flow properties of powdered solids, such as pharmaceuticals, cosmetics, cement, inks, and abrasives, helping to prevent caking.

11.4.2 Quantum Dots

A **quantum dot** or **nanocrystal** is defined as a crystal of a semiconductor which is a few nanometres in diameter, typically containing only 10^2–10^4 atoms. As we saw in Section 11.2, quantum dots exhibit quantum size effects in their optical properties. Figure 11.17 shows samples of quantum dots of CdSe of varying size prepared by an undergraduate at Pacific Lutheran University, WA.

Cadmium chalcogenide quantum dots (CdS_xSe_{1-x}, CdTe) are commercially available. The shape of the dots is improved by the growth of a thin layer of zinc sulfide (ZnS) on the surface. How are the unusual properties of quantum dots being exploited?

If quantum dots are made reliably to a particular size, they can be used to make LEDs (Chapter 8) of a very pure colour. This purity of colour makes them suitable for use as pixels in colour displays. Samsung marketed QLED (quantum dot light

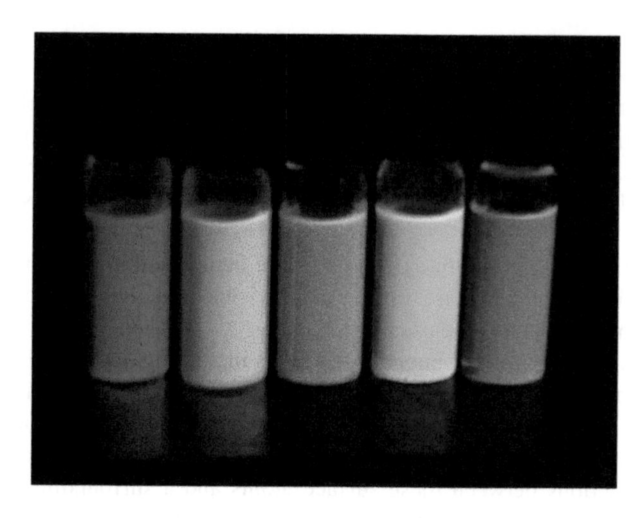

FIGURE 11.17 CdSe quantum dots of varying sizes. (Preparation and photograph by John Stiller.)

emitting diode) TVs in 2017. The screens of these have a quantum dot layer to give bright colours combined with a blue LED layer to give a purer white light.

The light-emitting properties of quantum dots mean that they can be used as fluorescent probes in biological systems, where they can have many advantages in replacing conventional organic fluorescent probes. Quantum dots have been used both in vivo and in vitro; if they can be attached to a biological molecule of interest, such as an antibody or a protein, they can be used to follow the reactions of that molecule. They have the advantages of being stable, can be excited by broadband excitation but emit a narrow band of frequencies of high intensity, and are available in many colours; because the colour emitted by each size of dot is pure, several different colour dots can be used at the same time to track different processes.

11.4.3 Metal Nanoparticles

The colours produced by nanoparticles of gold (colloidal gold) have been used to colour glass since Roman times and have been studied since Faraday in the mid-nineteenth century. Particles of around 30 nm diameter absorb light in the blue-green region and the scattered red light gives the glass its colour. Larger particles absorb light of longer wavelength and the scattered light becomes more blue.

Gold nanoparticles are also used in electronic devices, for example as electrical conductors, and for medical purposes such as detection of biomarkers for heart disease and in home pregnancy kits.

Silver nanoparticles have been found to have very good antibacterial action and are being used to impregnate bandages; they are also used to impregnate socks and are added to underarm deodorants, as the antibacterial action kills the bacteria responsible for the unpleasant smells.

Cobalt clusters embedded in silver, display giant magnetoresistance (GMR, Chapter 9) with an increase in resistance of up to 20%, and these are used for magnetic recording and data storage.

11.5 OTHER NONCARBON NANOSTRUCTURES

A variety of nanostructures have been synthesized from inorganic materials. **Nanofilms** can be used simply as very thin coatings. For example, nanoscale coatings to protect and enhance modern plastic spectacle lenses have been developed, including self-assembling top coatings for non-reflective lenses to protect the anti-reflective layer from dirt, dust, and skin oils, and a super-hard coating of carbides to protect from scratching.

The electronic and magnetic properties of nanolayers are important in devices formed from more conventional electronic materials. We have already discussed quantum-well lasers (Chapter 8) and colossal magnetoresistance (CMR) devices that are used for hard-disk read heads (Chapter 9). Quantum-well lasers may be an important component of light-based computers.

Multi-walled nanotubes (MWNTs) have been synthesised from inorganic compounds, and some, such as WS_2 nanotubes, are routinely synthesised in bulk quantities. Nanotubes are often free of bulk defects such as dislocations and grain boundaries, and this is responsible for their efficiency in potential applications.

TiO_2 nanotubes have been investigated because of their potential application in dye-sensitised solar cells (Chapter 8). The band gap of TiO_2 nanotubes is close to that of nanosheets, but is higher than that of bulk anatase. TiO_2 nanotubes are more efficient at transporting charge than nanoparticles, and this is attributed to a reduction of the resistance due to the grain boundaries.

To prevent nanotubes clumping or to make them soluble, groups have been attached to the surface of the nanotubes. For example, WS_2 nanotubes have been made water-soluble by interaction with Ni^{2+} coordinated to 2,2′,2″-nitrilotriacetic acid ($N(CH_2COOH)_3$).

Nanowires have been prepared from many materials, for instance, from semiconductors such as Si and Ge, metals, and oxides particularly ZnO. In Section 11.2, we described the electrical properties of nanowires and we saw that conductance along such wires as a function of voltage could be stepped. In nanowires of less than 100 nm diameter, scattering at the surface is the dominant mechanism for electrical resistance. Stepped magnetisation hysteresis loops have been observed for iron nanowires. As for nanotubes, the mechanical properties of nanowires indicate that they are free of bulk defects.

Figure 11.1 showed nanorods of ZnO prepared by hydrothermal synthesis under different conditions. Other structures include nanoribbons and nanowhiskers.

11.6 SYNTHESIS OF NANOMATERIALS

Nanoparticles can be the result of the synthetic methods described in Chapter 3.

However, increasingly ingenious methods of making nanostructures have been developed and we describe some of them here. The methods of synthesis of nanoparticles are often divided into **top-down** and **bottom-up syntheses**. The meaning of these terms is becoming increasingly blurred, but the original meaning of 'top-down' referred to making products in bulk and then manipulating them into the form required, for example, casting, recrystallising, and etching. 'Bottom-up' was coined to refer to the building up of a structure, atom by atom, in a controlled fashion, so that each atom is placed in its specific place and the structure is built with no defects.

11.6.1 Top-Down Methods

Top-down methods for nanoparticles apply to taking an already synthesised bulk structure and rendering it into nanosized particles.

Mechanical grinding, such as **high-energy milling**, can be used to form nanoparticles from bulk solids.

This technique has been successfully used to produce:

- Fine metal powders, such as Fe, Co, Ni, Cu, and Ag, for example, $NiCl_2+2Na=Ni+2NaCl$, produces 5-nm Ni particles embedded in NaCl. A top-down milling approach without the diluent gives particles that are several thousand times larger.
- Oxides, such as Al_2O_3, ZrO_2, Cr_2O_3, SnO_2, and ZnO, for example, $ZrCl_4+2CaO=ZrO_2+2CaCl_2$.
- Sulfides, such as ZnS and CdS, for example, $CdCl_2+Na_2S=CdS+NaCl$, produces 4 nm particles of the CdS II–VI semiconductor.

Microemulsions can be used to synthesize nanoparticles. When preparing nanoparticles, the simultaneous control of several factors is important: particles need to be both in the nanometre range and homogeneous, and they have to be prevented from growing larger, and from agglomeration. Reactions in microemulsions have proved to be very efficient for controlling these conditions in nanoparticle preparation. A microemulsion is defined as a dispersion of droplets of either oil in water or water in oil, of a diameter <100 nm; the droplets are stabilized by an interfacial film of surfactant molecules. A cosurfactant such as an alcohol or amine may also be used. When the droplets consist of a dispersion of aqueous droplets in an organic solvent, inverted micelles (also known as inverse and reverse micelles) are formed so that the hydrophilic heads of the surfactant lie in the aqueous core of the droplets and the hydrophobic tails are in the organic solvent (Figure 11.18).

In a typical preparation, two separate microemulsions of aqueous solutions containing the reactant salts are prepared in oil (A and B in Figure 11.18) and these are then mixed together. As the droplets collide, they fuse to make a larger micelle with which the salts interact. This larger micelle is less stable and breaks back down into two droplets containing the product that has precipitated out.

This micelle method has been used to prepare a range of useful nanoparticles including: metals; oxides such as CeO_2 used in fuel cells and gas sensors; titanates and zirconates, such as $SrTiO_3$ and $SrZrO_3$, which are important dielectric oxides used in electronic applications (Chapter 9); and magnetic particles.

11.6.2 Bottom-Up Methods: Manipulating Atoms and Molecules

The term bottom-up is now often used to mean structures that are built from constituent atoms, ions, or molecules or that self-assemble. An example that we covered earlier (Section 3.8.1) is chemical vapour deposition.

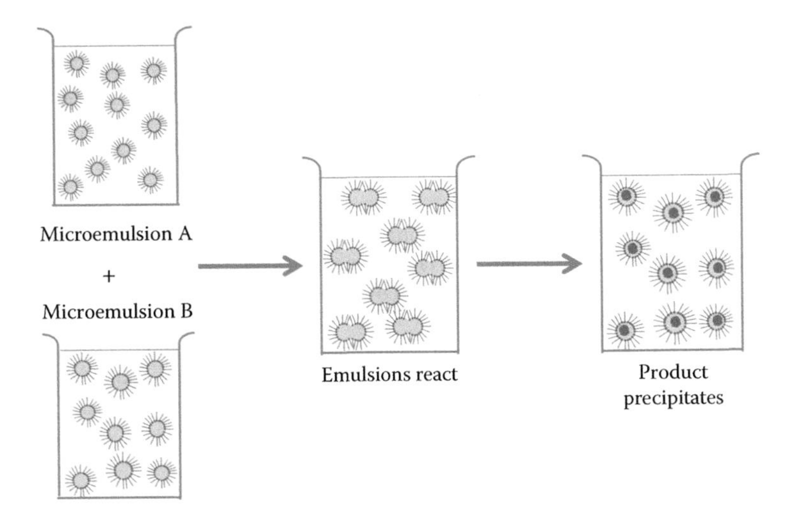

FIGURE 11.18 Microemulsion synthesis, showing the mixing and interaction of two microemulsions, A and B, to give a solid precipitate.

Another example of bottom-up synthesis is the use of scanning probe tips to move atoms into specific sites. One of the drivers for this research has been the production of ever-smaller electronic circuits. Traditionally, integrated circuits have been assembled using techniques in which a pattern is laid down on the surface of a crystal and the uncovered parts are treated to form a differently doped semiconductor, an insulating oxide, or a metallic connector. The patterns are produced by photolithography, that is shining light through a mask onto a light-sensitive chemical on a surface. The spacing of the components using this technique is limited by diffraction effects to about 100 nm. To produce integrated circuits in which the components are nanometre scale requires new techniques. It is possible to use variations on the existing approach by, for example, using UV radiation, X-rays, or electron beams in place of visible light. Another approach is to use scanning probe microscopy (SPM; see below) to print nanoscale patterns onto a substrate. For example, if a current is passed through the probe tip to raise its temperature, it can soften a thermoplastic polymer in the immediate vicinity of the tip, producing a small indentation. A completely different approach is to build up the required nanoscale structures, atom by atom, or molecule by molecule.

The operation of the **scanning tunnelling microscope** was explained in Chapter 2. In addition to giving us a picture of the surface and any molecules on it, we noted then that STM can also be used to move atoms and molecules across a surface and to make molecules react. If the STM tip is brought closer to the surface than usual, the force between it and the atoms/molecules on the surface increases. Repulsive forces between the tip and the surface (produced by reversing the sign of the voltage) can push molecules away from the tip across the surface. If this force is attractive, moving the tip across the surface pulls the molecule or atom along after it. Molecules can also be changed by the STM in single-molecule chemical reactions when the STM tip is used to channel electrons into a molecule. The energy can both dissociate bonds and provide the activation energy for bond formation.

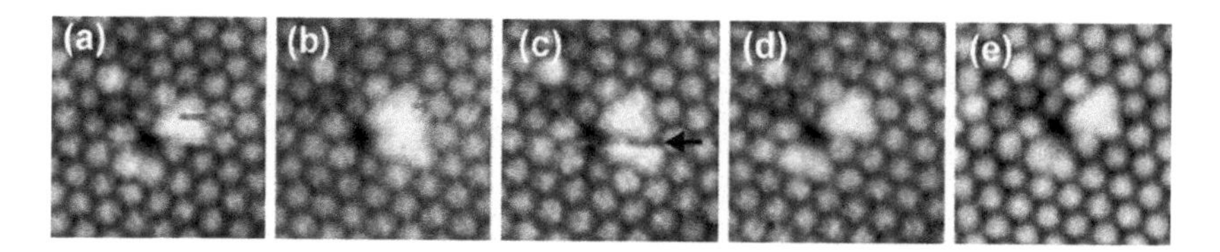

FIGURE 11.19 Series of successively recorded images (a)–(e) showing the tip-induced diffusion of a water molecule around a surface oxygen vacancy recorded at 80 K. The black arrow in (c) marks the line for the tip-induced movement. (Image size 3×3 nm^2. Acquisition parameters $f_0 = 150\ 936$ Hz, $A = 8:7$ nm, $k = 24:6$ N m^{-1}, $\Delta f = 41:6$ Hz.) ('Manipulation of Individual Water Molecules on CeO2(111)', S. Torbrügge *et al* 2012 *J. Phys.: Condens. Matter* **24** 084010 doi:10.1088/0953-8984/24/8/084010 © IOP Publishing. Reproduced with permission. All rights reserved.)

Atomic force microscopy (AFM) (see Chapter 2), like STM, relies on a very sharp tip, but in this case the tip is brought close enough to the surface such that the intermolecular forces between the tip and the surface can be measured. Nonconducting and semiconducting solids form essential components in circuits and in order to make nanocircuits on an atomic scale, the ability to manoeuver single atoms into place is an essential tool. An AFM tip has been used to move a single water molecule on a CeO_2 (*111*) surface simply by lowering and raising the tip (Figure 11.19).

By attaching a particular molecule to the tip, the AFM can be made responsive to certain molecules or groups of molecules and not others. AFM is mostly used for the study of biological molecules, and modified tips have been used for extracting particular molecules, for example, proteins from cell membranes.

By scanning AFM tips along a surface or tapping them, it is possible to induce chemical reactions along a line or to form a pattern of dots. One example is the application of a voltage to an AFM tip to oxidise a silicon semiconductor with lines of oxide of nanometre dimensions. The silicon surface is cleaned of any oxide on the surface with dilute hydrofluoric acid (HF) and then a negative voltage is applied to the AFM tip. The width of the SiO_2 structures produced is of the order of tens of nanometres. Some water vapour must be present in the surrounding atmosphere.

In **dip-pen nanolithography** using a molecular 'ink', the molecules are placed on an AFM tip and are delivered to a substrate surface via a water meniscus; examples include nanoscale patterns of thiols on gold and of silanes on silicon. The tip is loaded either by dipping in a solution and allowing to dry, or by vapour deposition. Under ambient conditions, a water meniscus naturally forms between the tip and the sample, due to the humidity of the air, so that when the tip is close to the substrate, the molecules are delivered to the surface via the meniscus and form a self-assembled monolayer. Surface tension holds the tip at a fixed distance from the surface, as it moves across it. The size of the deposit is dependent on the tip radius of curvature, the relative humidity, which controls the size of the water droplet, and any diffusion of the molecules across the surface. Nanoscale structures of 50 nm can be generated. The technique has the advantage that the same instrument can be used to both build the nanostructure and image it. Huge arrays of AFM tips linked together can be used (more than 200×200); one tip is positioned slightly further (0.4 nm) above the surface than the others and is used for imaging and hence guidance on moving the array across the surface. The other tips all lay down identical structures.

11.6.3 Synthesis Using Templates

Nanoparticles and nanowires can be made using template methods. There are broadly three main ways of doing this and these are summarised in a scheme for the formation of one-dimensional structures in Figure 11.20:

- Porous solids and membranes
- Organic and biological templates
- Using a preexisting nanostructure

In the first method, porous solids that possess nanosized cavities or channels are impregnated with a solution of the reactants so that the cavities are filled. Sufficient heat is then applied for the reaction to take place within the cavities, thus restricting

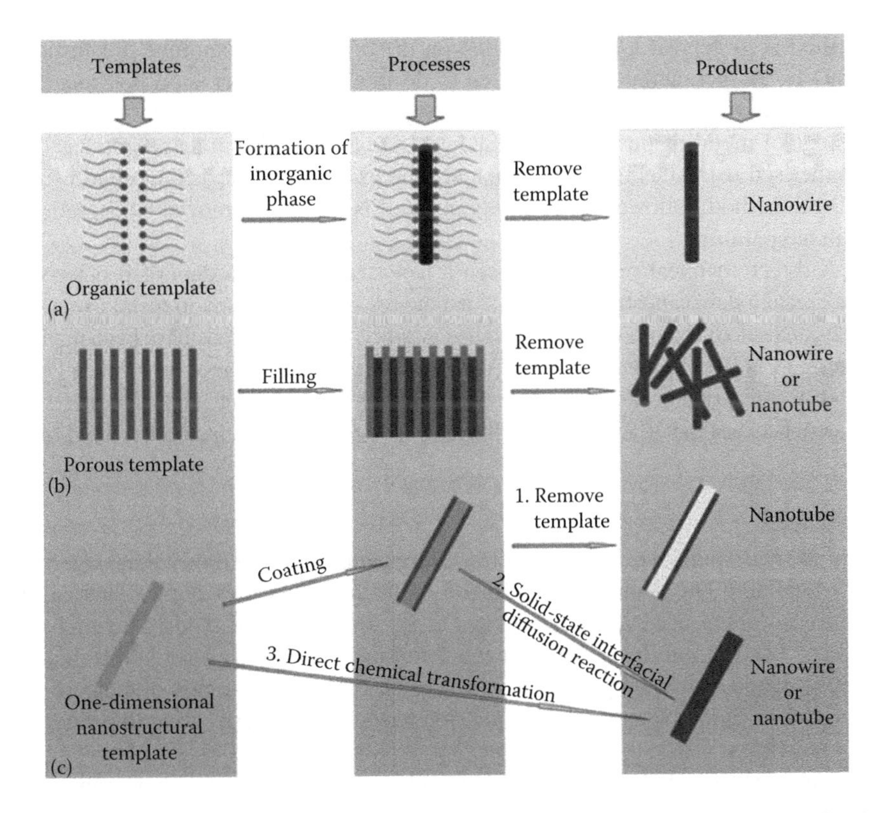

FIGURE 11.20 Schematic illustrations showing three different template processes for the synthesis of one-dimensional inorganic nanostructures: (a) cylindrical micelle, (b) porous membrane, and (c) preexisting nanostructure templates. (Liang, H-W., et al.: *Advanced Materials*. 2010, **22**, 3925. Copyright Wiley-VCH Verlag GmbH & Co. KGaA. Reprinted with permission.)

the size of the product. The final step removes the original template scaffold by dissolving (silica-based scaffolds) or reaction (oxidation of carbon-based scaffolds). If the pores of the template scaffold interconnect, then an inverse replica structure of the template will be formed. If the cavities are discrete, as in some zeolites, for instance, then nanoparticles result.

Porous solids such as silica gels, mesoporous silica, and zeolites (see Chapter 7) can be used for the templates. Mesoporous silica has a long-range order with specific pore sizes and geometry and has been used to form nanostructures or nanoparticles of such diverse solids as mesoporous carbon, metal oxides, sulfides, nitrides, spinels, and phosphates. Mesoporous carbon, made by this method, can then itself be used as a template to form metal oxides.

Organic surfactants are used to make self-assembling micelles or inverted micelles, which encapsulate the inorganic material. Cylindrical micelles will form nanowires or nanorods. The reaction takes place within the micelle, which is subsequently removed to leave the nanoparticle or wire. Gold and metal oxide nanowires have been made this way.

A preformed nanostructure can be used as a template in several ways. One method is to deposit a new compound on the outside of a nanotube, for instance. CVD or sol–gel methods can be used to do this and an outer shell is formed, the template is then dissolved or oxidised away to leave a hollow core. For example, a range of metal oxide nanotubes, SiO_2, Al_2O_3, and TiO_2, among others, has been synthesised using CNTs as the template. The oxide precursor is deposited using a sol–gel method, followed by calcination. The CNT is then removed by oxidation at high temperature.

A direct chemical reaction between a nanostructure and a deposited outer shell can create a new nanostructure. Metal nanowires can be oxidised to metal oxide, sulfides, and selenide nanowires, for instance. Metal carbide nanorods have been made by reacting CNTs with metal halides. Complex oxide nanotubes have been made by the solid state reaction between two oxides. Al_2O_3 was deposited onto ZnO nanowires and heated; because the diffusion rate of the ZnO is faster, $ZnAl_2O_4$ nanotubes are formed.

11.7 SAFETY

Due to their small size, nanoparticles can reach parts of the body that larger particles cannot. They can cross the blood–brain barrier, for example. This can be beneficial when used to deliver drugs to the brain but could also have adverse effects. Particles in the air of diameter less than 2500 nm can reach the smallest tubes, alveoli, in the lungs, and particles of this size are currently of concern as a component of atmospheric pollution. This has led to concerns about the safety of nanoparticles.

Nanoparticles of ZnO or TiO_2 are found in some sunscreens, deodorants, face cleansers, and moisturisers. Some antifungal and antimicrobial creams and shampoos contain silver nanoparticles.

So how safe are these products? To answer this, we need to bear in mind not just the toxicity, but also the amount transferred to the body, where in or on the body the nanoparticles are, and whether over time, they move to different places in the body. For readily soluble materials several scientific committees have concluded that the toxicity will be generally the same as for larger particles.

One recent (2018) estimate is that 90% of the nanoparticles in these products are on the skin, 6% are inhaled (for example via powders and sprays) and 4% are ingested (for example via toothpaste and lip balm). There is little evidence for these particles moving through the outer layer of the skin, although there are as yet no long-term studies.

In the workplace, the most likely route of exposure is by inhalation and should be taken into account in risk assessment.

11.8 SUMMARY

1. Nano structures have at least one dimension between 0.2 and 100 nm.
2. Structures of this size may have different properties from those of the bulk material.
   a) Nanoparticles have a high surface-to-volume ratio and this can lead to the particles adopting unusual crystal structures. This can also affect the speed of chemical reaction, particularly catalysis. In addition it can reduce the melting point.
   b) The electronic structure of nanostructures differs from that of bulk solids. For the smaller particles, there can be discrete energy levels as for molecules. This leads to differences in several properties.
   c) Electrical conductance can be quantized, giving a stepwise increase in current with increasing voltage.
   d) Colours of nanoparticles can vary with size. Scattering of light is reduced and can make the particles transparent.
   e) Magnetic nanoparticles can be small enough to only contain one domain or exhibit superparamagnetism.
3. Many nanostructures of carbon are known. Carbon Black consists of nanoparticles. Graphene is a single layer of graphite. Bilayer graphene is two sheets of graphene brought close enough to interact. Oxidation of graphene produces graphene oxide in which the graphene layers have functional groups on. Carbon nanotubes (CNT) can be thought of as sections of graphene rolled round to form a tube. CNTs can be single-walled (SWCNT) or multi-walled (MWCNT). Fullerenes are small 3D structures formed of linked pentagons and hexagons of carbon atoms. The most famous is buckminsterfullerene C_{60}.
4. Nanostructures of many inorganic compounds and metals can be synthesized. They can take many forms such as nanotubes, nanorods, nanowires, nanoribbons, and nanoflowers.
5. Nanostructures can be synthesized using top-down methods or bottom-up methods. In top-down methods, larger particles are broken down. In bottom-up methods, the nanostructures are built up from scratch from the constituent ions, atoms, or molecules.

QUESTIONS

1. Calculate the surface-to-volume ratio of particles of diameter (a) 2 nm and (b) 10 nm. Assume the particles are spherical.
2. Magnetotactic bacteria build rows of nanoparticles of magnetite that consist of a single domain which they use to help them locate the bottom of muddy ponds. What advantage might forming particles of this size have for the bacteria?
3. QLED TV screens use quantum dots to provide the colour. What properties of quantum dots make them suitable for this?
4. What optical property of ZnO and TiO_2 makes them suitable for transparent sunscreen?
5. Sketch the band structure of graphite intercalated with (a) an electron donor and (b) an electron acceptor.
6. Graphene was initially made by lifting layers from graphite using adhesive tape. Is this a top-down or bottom-up synthetic method?
7. Fullerenes consist of 12 pentagons and a number of hexagons; 20 in the case of C_{60} and 25 in the case of C_{70}. What would be the formula of the smallest possible fullerene?

12 Sustainability

Mary Anne White

12.1 INTRODUCTION

This chapter introduces the important concepts concerning sustainability in solid-state chemistry, including sustainable goals and tools for sustainable approaches. As readers will realize, sustainability is much further reaching than just solid-state chemistry, but the emphasis here is a chemical point of view, with broader implications.

12.1.1 DEFINITION OF MATERIALS SUSTAINABILITY

The term 'sustainability' has become such a buzzword that a clear definition is required. In 1987, the UN Brundtland Commission defined sustainable development as meeting the needs of the present without compromising the ability of future generations to meet their own needs. We can adapt the UN definition to sustainable approaches to materials. Although sustainability goals can be promoted by government legislation and by and nongovernmental agencies, and by the close and often parallel relationship between business profits and sustainable approaches, here we consider sustainable goals in a materials chemistry context.

12.1.2 SUSTAINABLE MATERIALS CHEMISTRY GOALS

Sustainable approaches to materials chemistry allow development of new materials without compromising the needs of future generations. Some sociopolitical issues, such as ethical concerns related to mining of minerals, are difficult to quantify, but of underlying importance to the discussion that follows.

Foremost in the context of sustainability is the minimisation of production of CO_2 and other greenhouse gases (GHGs). The minimisation of GHG production includes all aspects of the materials development, from obtaining the elements from their natural source (e.g., mineral extraction), to synthesis of the material, to use of the material in a device, and ultimate fate of the material at the end of its useful lifetime. In addition, we can include consideration of the role of materials in reduction of GHGs, such as by use in a device that harvests or stores energy, replacing energy sources that would otherwise create more GHG emissions.

Energy itself is another major consideration for sustainable approaches. For example, if a material is used in a device that scavenges energy that would otherwise be wasted, we need to consider how much energy is required to make the material. Quantitative approaches to address this matter will be introduced in this chapter.

For many parts of the world, fresh water is in short supply. Many of the processes to produce materials use very large quantities of water. For example, typically about 200 L of water are required to produce 1 kg of a polymer. Again, as for GHG emissions and energy consumption, sustainable approaches to materials production should aim to minimise water consumption.

Although not always appreciated, even by chemists, not all elements are readily available in the quantities required for scale-up of materials to widespread use. Therefore, as we explore in more detail in this chapter, elemental availability is also a significant factor in sustainable materials chemistry approaches.

Before we introduce tools for sustainable approaches, it is useful to present an overview of material use in society, to have a context for the sustainable tools.

12.1.3 Materials Dependence in Society

In some sense, humanity has always been defined by its cutting-edge materials. Before about 3000 BC, during the Stone Age, a person with a sharp arrow prepared from a rock by flint knapping had a significant advantage over others. The Bronze Age, starting from about 3000 BC, was defined by the copper–tin alloy that has exceptional properties such as malleability and strength, giving better tools than stone. The Iron Age, starting about 1000 BC, allowed wider use of tools, as iron ore was more abundant than tin ore.

Our modern age of materials relies on a wide variety of materials, from polymers, to semiconductors, to rare-earth magnets. As described elsewhere in this book, the defining materials properties derive from the properties of the individual atoms (e.g., the role of *f*-electrons in magnetic materials), and of their aggregates (e.g., the role of dislocations in solid-state materials).

However, it is interesting to quantify the worldwide production of materials, to indicate the relative quantities of various commodity materials (Figure 12.1). Note that concrete accounts for about half of the worldwide production of materials. Therefore, if you really want to make a difference in, for example, GHG emissions or energy consumption in materials production—both areas in which concrete is near the top of any list—consider new, sustainable approaches to concrete production.

12.1.4 Elemental Abundances

It is a bit startling to realize that everything around us is made from the relatively small number of elements in the periodic table, about 90 of which play prominent roles in our lives. However, as Table 12.1 indicates, not all elements are very abundant on Earth.

In fact, several of the elements are thought to be in danger of depletion in the next 100 years, or even in the coming decades. A prime example is indium, which has been greatly depleted for use in computer screens as indium tin oxide, which has the (usually contradictory) properties of high electrical conductivity and high optical transparency. Indium is a good example to consider the role of recycling. A typical computer display screen contains only about 50 mg of indium. With such widespread use of displays however, about 50% of the current mining production of indium is

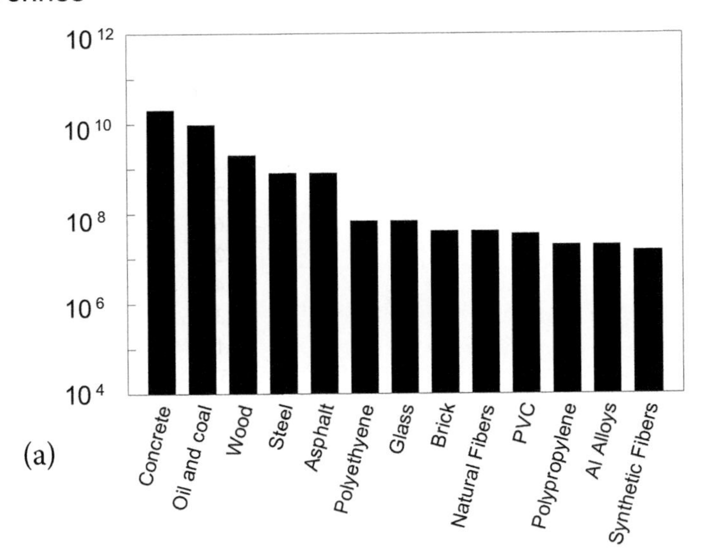

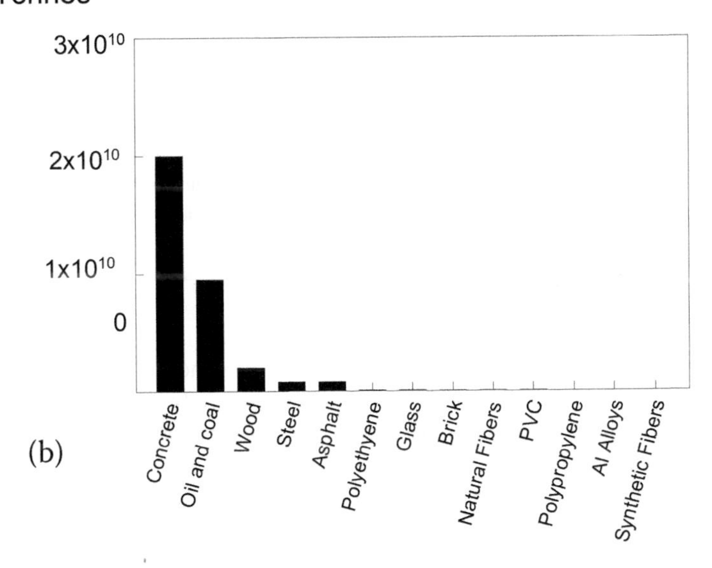

FIGURE 12.1 Annual worldwide production of materials, in tonnes (1 tonne = 10^3 kg), expressed (a) on a log scale to show all the major materials, and (b) on a linear to highlight the dominant materials. (Data from *Materials and the Environment*, 2nd ed., 2013, M. F. Ashby, Elsevier.)

TABLE 12.1
Natural Abundance of Elements in the Earth's Crust

Element	Natural abundance in the Earth's crust (%)
O	46.1
Si	28.2
Al	8.2
Fe	5.6
Ca	4.2
Na	2.4
Mg	2.3
K	2.1
Ti	0.57
H	0.14
Cu	0.005
B	0.001
In	2×10^{-5}
Ag	1×10^{-5}
Bi	2×10^{-6}
Ru	1×10^{-6}
He	5×10^{-7}
Au, Pt, Ir	1×10^{-7}
Te	8×10^{-8}
Rh	4×10^{-8}

Source: Data from *Encyclopedia Britannica, Inc.* Elements with lower abundance in the Earth's crust are not listed. As originally deduced by geochemist Frank W. Clarke (1847–1931) and extended by others.

used for display devices. The consequent dispersal of indium around the globe, with only a few milligrams in each display device, makes indium recycling impractical.

Several of the elements that are required for energy-related applications also are in low availability. Examples include Re, Te, Pt, Pd, He, Ag, Nd, La, and Dy. These are critical elements. For example, an electric car can have about 1 kg of Nd in its motor, and about 10 kg of La in its battery. Such large quantities indicate the importance of recycling.

The rare-earth elements deserve special emphasis. For example, a 1-MW wind turbine requires up to 1 tonne of rare-earth permanent magnets (Nd, Tb, and Dy), all quite scarce. Dy and Tb are predicted to run out in a few decades. They each have unique magnetic properties due to their specific electron configuration, making elemental substitution problematic.

Helium also requires particular mention. It has the unique property of an exceptionally low boiling point (4.2 K), and is used to maintain cryogenic temperatures, such as needed for NMR and MRI magnets. However, He is a low-density gas, and easily escapes the Earth's atmosphere, never to be here on Earth again. Therefore, we should treat helium as the precious commodity that it is, and forgo trivial uses, such as helium balloons.

We cannot feasibly create new quantities of elements, and therefore need to work within the confines of availability. However, there are tools to assess availability and provide directions for new materials development, as introduced later in this chapter.

12.1.5 Solid-State Chemistry's Role in Sustainability

Solid-state chemistry has already played a pivotal role in materials development, from new photovoltaics, to superconductors, and far beyond. With a few tools to quantify sustainability matters, solid-state chemistry can lead important new developments in a sustainable way.

One example is replacement of indium tin oxide with other materials that are free of indium. Recent developments based on carbon nanotubes and graphene allow carbon-based materials with the unusual property combination of high electrical conductivity and high optical transparency. Carbon's wide availability avoids the problem associated with indium.

Another area of importance is the enhanced used of Fe-based materials, for example to replace precious metals in catalysts. Even if the efficiency is lower, the availability of Fe is so much higher, and the resultant cost is so much lower, that substitution can be advantageous.

New, sustainable approaches to materials synthesis and processing are also active areas in which solid-state chemistry can be of fundamental importance. For example, green chemistry approaches (see Section 3.10 and also below), can greatly reduce GHG emissions, water and energy consumption, and material use.

12.1.6 Material Life Cycle

As consumers, we are now familiar with consideration of the end of the useful life of a product, such as repurposing or recycling. Those same principles apply to sustainability matters for solid-state chemistry. In addition, we need to consider the pre-product starting point, which is often the raw materials from which the product is made. In other words, we should consider the full life cycle.

The sustainability tools discussed below present several methods to quantify the relative burden on our planet to synthesize new materials, namely the concepts of embodied energy, the Herfindahl–Hirschman Index, and exergy. These concepts apply to the starting stage of the life cycle of a material, and are defined below.

At the end of the life of a material, we need to consider the best path forward. Options, in decreasing order of preference, include reuse, refurbishment, recycling, combustion, and landfill. The last should be the final resort. If the heat is captured, combustion can provide back some of the energy embodied in the material, but it can also give noxious byproducts, and might not be the best use of the material. Recycling is very appealing for some materials that are already in nearly pure form, such as aluminum from beverage containers, but highly impractical for widely dispersed materials such as indium in indium tin oxide coatings, as discussed above.

For high-value materials, such as battery components or rare-earth magnets, recycling is a viable and necessary path. Of course, if a material can be refurbished or reused, this is the best outcome.

The 'Cradle to Cradle' concept of ecologically intelligent design, initiated by architect William McDonough and chemist Michael Braungart, places the 'waste' of one process as the input material of another. This approach is an example of upcycling: transforming low-value waste into higher-value products. It is already in use in industry; for example, one major sneaker manufacturer has a line of shoes with an infinitely recyclable polyester outside and a biodegradable sole. Sustainable approaches often align with optimisation of the bottom line for businesses.

12.2 TOOLS FOR SUSTAINABLE APPROACHES

This section provides a brief introduction to several tools that a solid-state chemist can use to help select the most sustainable approaches and directions. For more details, see the Further Reading section at the end of this book.

12.2.1 Green Chemistry

The **12 principles of green chemistry** have already been elucidated in Section 3.10. These are approaches that can be used to make production of materials more sustainable. They can be categorised as reducing waste (principles: prevention, atom economy, less hazardous synthesis, renewable feedstocks, avoiding derivatisation, use of catalysts, real-time monitoring), lowering toxicity (safer chemicals, safer solvents, design for degradation, minimizing accidents), and reducing energy use (energy efficiency).

While most chemists view these principles as applying to synthesis, they also apply to processing of materials, and to synthesis steps that include processing. For example, aqueous solvents can be used to replace organic solvents, as a more sustainable approach. Another example is that many inorganic products are produced via high-temperature processes, requiring large material quantities, high-temperature furnaces, and high-value containers that can only be used once. More sustainable approaches can include microwave chemistry, or mechanochemical methods such as ball-milling. These approaches are all active areas of research.

12.2.2 Herfindahl–Hirschman Index (HHI)

The **Herfindahl–Hirschman Index (HHI)** is a concept from economics used to measure the monopoly of entities. Recently, considering elemental accessibility and supply risk (crustal abundance, global production, reserves, use, geopolitics, and other factors), HHI values have been assessed for each element (see Table 12.2). A low value of HHI means that the element is plentiful, taking into account present rate of production and accessible reserves.

Note that HHI values (Table 12.2) roughly correspond with abundance in the Earth's crust (Table 12.1) but also take into account other factors, such as accessibility of the elemental reserves, both geographically and politically. For example, Sn is

TABLE 12.2
Herfindahl–Hirschman Index (HHI) for Selected Elements. A low HHI Value Indicates That An Element is Plentiful

Element	HHI
O	(500)
Si	(1000)
Al	(1000)
Fe	1400
Ca	(1500)
Na	(500)
Mg	(500)
K	7200
Ti	1600
Cu	1500
B	2000
In	(2000)
Ag	1400
Bi	6000
Ru	(8000)
He	3900
Au	1000
Pt	(9100)
Ir	(9100)
Te	4900
Rh	(8000)

Source: Data from M. Gaultois et al., 'Data-Driven Review of Thermoelectric Materials: Performance and Resource Considerations', *Chem. of Mat.* **25** (2013): 2911–2920.

Note: These are HHI_r values, i.e., HHI values that include known reserves of the elements. For convenience, elements are listed in order of decreasing abundance in the Earth's crust (see Table 12.1), but that does not necessarily correlate with HHI order as HHI considers other factors besides abundance.

Note: Numbers in parentheses have higher uncertainty.

only the 50th most abundant element overall in the Earth's crust, and therefore not listed in Table 12.1, but it is more accessible than some elements, giving it an HHI of 1600, which is quite low (i.e., favorable) considering the low abundance.

The use of HHI allows rapid relative determination of elemental scarcity and supply risk, when considering solid-state chemistry approaches. For example, consider electronically conductive, optically transparent materials for display devices. Indium tin oxide (which is 35 mol% In, 61 mol% O, 4 mol% Sn) has an HHI (weighted by stoichiometry) of 1070, whereas carbon nanotubes have a more favourable HHI of only 500, due to the higher availability of carbon. Therefore, HHI can be used as a rapid screening tool for potential materials approaches. However, HHI is silent on how sustainable it might be to produce a material to use in a device, highlighting that many approaches are needed.

12.2.3 Embodied Energy

Another very useful tool to assess sustainability of a material is its **embodied energy**. This term encompasses the energy required to produce the product from raw materials, including extraction from ores or crops, transport, and refining. Examples of embodied energies for some materials are presented in Table 12.3. In addition, the CO_2 equivalency for such materials (i.e., GHG emission for all GHG expressed in as the equivalent amount of CO_2 that would give the same global warming effect) is presented. Note that embodied energy and CO_2 equivalent show similar trends.

From Table 12.3, it is apparent that energy-intensive materials processing leads to very high embodied energy, and high GHG emissions. Therefore, it is compelling to reuse or recycle such materials. Even materials with lower embodied energies per unit mass, such as concrete, are used in such large quantities that they have considerable value even after the end of their (first) useful life.

For example, the embodied energy of a plastic is nearly twice its enthalpy of combustion. So when plastics are burned, nearly half the energy is lost forever, even if we capture the heat. On the other hand, the energy required to recycle most plastics

TABLE 12.3

Embodied Energies and CO_2 Footprints of Selected Materials

Material	Embodied energy (MJ kg^{-1})	CO_2 equivalent (kg kg^{-1})
Concrete	1.1	0.1
Glass	10	0.8
Steel (virgin)	27	1.8
Lead	27	1.9
Brass	54	3.5
Copper	60	3.7
PVC	60	2.5
Polypropylene	80	3
Aluminum (43% recycled)	131	8
Aluminum (virgin)	210	12
Silver	1,500	100
Single crystal silicon for electronics	6,000	305
Platinum	270,000	15,000

Source: Data from *Materials and the Environment*, 2nd ed., by M. F. Ashby, Elsevier (2013).

is only a fraction of the embodied energy, and when we recycle plastic, we still make use of the embodied energy.

How we process a material greatly influences the embodied energy. An example of a process with very high embodied energy is the use of high-temperature furnaces and high-value containers that can only be used once. Another example of high embodied energy is materials made by chemical vapour deposition, due to the high energy costs in this method and the low yield.

Although not always easy to assess, embodied energy considerations can point to more sustainable approaches for solid-state synthesis and processing.

12.2.4 Exergy

In thermodynamic terms, **exergy** is the maximum quantity of useful work that can be provided by a system (such as during a Carnot cycle). The concept of exergy, defined as the minimum quantity of work required to produce a material in a specified state from common materials in the environment, can be used to guide sustainable approaches.

Exergy is the Gibbs energy of formation of a material with an altered definition of the standard state. This new standard state is referred to as the reference compound, and is a common material from a natural environment. (The usual Gibbs energy of formation is relative to the elements, but most elements do not occur in pure form in Nature.) In exergy, examples of common reference compounds include SiO_2, H_2O, Cl^-(aq), and PdO_2. Exergy is similar to embodied energy in that it can be destroyed.

Exergy, represented here as B, can be defined for a compound q in its standard state represented by the superscript ^{0}, as

$$B_q^0 = \Delta G_f^0 + \sum_i n_i B_i^0 \tag{12.1}$$

where compound q is made from elements i, each of stoichiometric number n_i, ΔG_f^0 is the standard state Gibbs energy of formation of the compound, and B_i^0 is the standard state exergy of each element i. Tables of elemental exergies exist (see the Further Reading section at the end of this book), and care must be taken in noting their reference state.

Consider the production of aluminum from the mineral sillimanite, Al_2SiO_5, which has a Gibbs energy of formation of −2441 kJ mol^{-1}. Based on Equation 12.1, and the exergies of Si (855 kJ mol^{-1}, with SiO_2 α-quartz as the reference state), O_2 (3.92 kJ mol^{-1}, with O_2 as the reference state), and Al_2SiO_5 (15.2 kJ mol^{-1}, with Al as the reference state), we can calculate that the standard state exergy of aluminum is 796 kJ mol^{-1}. It is readily apparent from this large, positive value that refinement of the mineral Al_2SiO_5 to produce pure aluminum is an energy-intensive process. (Note that this conclusion correlates with the results from Table 12.3, which show that the embodied energy of aluminum is very high, and that of recycled aluminium is also still quite high, even though considerable energy is required to recycle aluminum.)

With data from published tables of exergies (see the Further Reading section at the end of this book), an exergy calculation can be very informative concerning the thermodynamic minimum quantity of work energy required to produce a compound.

Note that exergy does not account for additional processing energy, or other energetic inputs, but it is simple to use and a useful guide.

12.2.5 Life Cycle Assessment

The tools presented above are relatively quick guides, but do not allow a full quantitative assessment of sustainability of approaches. A more quantitative methodology is a **life cycle assessment (LCA)**, also called life cycle analysis, which quantifies the environmental impacts imposed by the manufacture and use of a material, including its end of life. LCA is a very useful tool to compare two materials or devices, and the results, such as energy requirements, GHG emissions, and water consumption, can be used to inform decisions.

To define the scope of an LCA study, we must first define the boundaries. To do this, we should define the **functional unit** for the investigation, which is typically the amount of material (by mass or volume), or the number of final products, or the duration of use. Some examples of functional units could include: 100 kg of material; or the amount of material for a given device; or one device in its entirety; or the number of a particular devices to be required in a household over the course of a given number of years. The latter can allow ongoing comparisons between materials, processes, or devices.

Based on the selected functional unit, the next step in an LCA is an inventory analysis. In this step, the types and amounts of materials and their transport and processing are summarised, so that the environmental impact of each can be quantified. Typical matters that are quantified, using published LCA information and/or databases, are energy use, water consumption and emissions, materials consumption, GHG emissions, and solid waste. The quantified information generally includes production/manufacture of the material/device, and also the use stage, and possibly the end-of-life stage, depending on the functional unit. Other factors can be added, depending on the goals of the LCA.

The inventory analysis is then used to carry out an impact assessment for the functional unit. In many cases, this impact will be compared with a benchmark, or another process or material or device, to allow an informed decision on the best way forward.

For example, consider a comparative LCA of single-use plastic sandwich bags compared with reusable sandwich containers. A functional unit of five years' use, chosen as the expected lifetime for a reusable sandwich container, corresponds to 1200 sandwiches and plastic bags. Knowledge of the quantities of polymers used in each container type, and the processing methods and packaging, allows calculation of the energy consumption, water use, and GHG emissions. All three are 10–20 times higher for the plastic bags. However, taking into account materials transportation, use (including washing for the reusable container), and end-of-life processes, during the full five years of the functional unit, the plastic bags consume about twice the energy, but only half the water, and produce less than half the GHG emissions. Therefore, the answer to 'which is best?' depends on the rank of importance of energy use, water consumption, and GHG emissions in the local situation. This simple example shows the nuanced information that can be obtained in an LCA.

LCAs can quantitatively address questions such as the following: How much energy does it take to make a particular energy storage material and how long does it need to be used to recoup its embodied energy? Which of two processes to prepare a given material has lower water consumption? For two comparable materials for a given purpose, which has the lower GHG emission over the lifetime of the device in which it will be used? In the production of automobiles, which type of insulation foam (synthetic vs. bio-based) has the lower environmental impact? Is it better to recycle newsprint, or burn it and capture the heat?

However, an LCA does not cover all aspects of sustainability. For example, the impact of plastic bags that are loose in the environment, wreaking havoc with wildlife on land and sea, is not addressed in an LCA.

12.3 CASE STUDY: SUSTAINABILITY OF A SMARTPHONE

The smartphone is a useful device, and around the world we rely on its availability for many aspects of our lives. In itself, the smartphone is a wonder of materials science, bringing so many functions together so efficiently. However, of the approximately 40 elements that make up a smartphone, over half are in danger of shortages in the decades to come, due to scarcity. Again, this is an example of elemental dispersion, with more than 300 million smartphones discarded or replaced annually worldwide, each with very small quantities of precious elements. If each person used their phone longer than the present average of three years, the materials consumption could drop by considerably. And if manufacturers emphasised refurbishment, materials consumption could drop further.

The information and communications technology sector is responsible for 2% of the worldwide GHG emissions, and it is interesting to see the contribution of smartphones. A lifecycle assessment of smartphones has been carried out, including the impact of the network usage, using a functional unit of lifetime usage: three years of the smartphone device and its accessories for a representative usage scenario, including all raw materials acquisition (smartphone, packaging, battery), production (parts, integrated circuit, display, transportation, assembly, integrated circuit manufacture and distribution), use (smartphone and network), and various end-of-life scenarios. The results show that by far the largest contributor to GHG emissions is the production processes, especially integrated circuit production (~10% of smartphone GHG emissions). The total emissions for production of a smartphone are about 60 kg of CO_2 equivalent, which compares with an annual impact for their network use of about 20 kg of CO_2 equivalent. Therefore, for a smartphone used for 3 years before replacement, the overall annual GHG impact is about 40 kg of CO_2 equivalent, or about 1% of the GHG emissions from typical annual use of a combustion engine car. Note also that the production of one smartphone uses 3,000 to 10,000 L of fresh water, mostly for the extraction of the raw materials. On the other hand, in the United States the total energy consumption of consumer electronics such as computers and televisions has fallen considerably in recent years, due to the increased use of smartphones for those functions.

Therefore, the example of smartphones shows the important role of advances of solid-state materials in society, and the responsibility of researchers and consumers to consider the impacts of materials and devices, to sustainably guide future endeavors.

12.4 CONCLUDING REMARKS

The concept of sustainability in solid-state chemistry underlies the goal of future development of new materials without compromising the needs of future generations.

Solid-state chemists need access to tools to assess sustainability, including the principles of green chemistry. Concerning elemental availability, the range of abundance of different elements in the Earth's crust is important, and availability can be quantified via the Herfindahl–Hirschman Index (HHI). The concept of embodied energy quantifies the energy that goes into making the material or device from raw materials. Exergy quantifies the minimum quantity of work required to produce a chemical compound from common materials in the environment. Although these tools are useful and relatively easy to use, the concept of life cycle analysis provides a more though quantitative assessment of the environmental impacts imposed by the manufacture and use of a material or device.

Sustainability considerations can provide new directions for synthetic approaches, processing, and the use of novel materials in consumer products. Consideration of sustainability matters should be the first step along the path to improving our future lives through solid-state chemistry.

QUESTIONS

1. Considering sustainability alone, why is there such strong interest in carbon-based electronic devices?
2. Rare earths have a special role in magnetic materials, yet they are not very abundant in the Earth's crust, and are scarce when geopolitical aspects are taken into account. Why do we not just replace rare-earth elements with other elements that are more abundant?
3. The embodied energy of monocrystalline Si is 2800 MJ/m^2. If the Si layer in a solar cell is 200 microns thick, and a photovoltaic cell produces 1 W of power for every 10 g of Si, calculate the approximate payback time to recover the embodied energy of the Si. The density of Si is 2.3 g/cm^3.
4. Two competing cathode materials for lithium-ion batteries are $LiCoO_2$ and $LiFePO_4$. Their respective enthalpies of formation are −680 kJ mol^{-1} and −151 kJ mol^{-1}, and can be used to approximate their Gibbs energies of formation. Calculate the respective exergies of these two compounds, and comment on which would be the more favourable cathode material on this basis. The corresponding exergies of the elements in kJ mol^{-1} are as follows: Li: 393; Co, 313; O_2: 3.92; Fe, 374; P, 861.
5. Mechanochemical methods can be used to synthesize organic and metal-based solids without solvents, and at ambient temperature, by mechanical action of the components, such as in ball-milling. Such methods can replace traditional solvent-based synthesis, and also high-temperature synthesis of solid-state materials. Discuss advantages of mechanochemical methods over these traditional methods, in terms of the 12 principles of green chemistry, and also in terms of GHG emissions and water consumption.

Answers to Questions

ANSWERS FOR CHAPTER 1

1. No. Tetrahedral molecules do not have a centre of inversion. A twofold (C_2) axis of symmetry is coincident with the $\bar{4}$.
2. This cell has $a \neq b \neq c$, $\alpha = \beta = \gamma = 90°$. It therefore is an orthorhombic cell.
3. Hexagonal cells have $a = b \neq c$, $\alpha = \beta = 90°$, $\gamma = 120°$.
4. Four. There are $\left(6 \times \frac{1}{2}\right) = 3$ fluorine units at the centres of the faces, $\left(8 \times \frac{1}{8}\right) = 1$ at the corners, $(12 \times ¼) = 3$ on edge sites and 1 at the centre entirely enclosed by the cell. This is a total of 8. These eight are matched by four calcium units entirely enclosed in the cell.
5. The ideal perovskite cell is cubic. It is based on a primitive cell (P), has planes of symmetry perpendicular to *x*, *y*, and *z* (m). As a cubic cell it has a –3 axis along the body diagonal and it has planes of symmetry perpendicular to the face diagonals. Its space group is therefore Pm-3m.
6. a). The unit cell projection for NaCl is shown in Figure A.1a. This is a view looking down *c* on the unit cell.
   b). The unit cell projection for ZnS (zinc blende) is shown in Figure A.1b.
   c). The unit cell projection for perovskite is shown in Figure A.1c.
   d). The unit cell projection is shown for ReO_3 in Figure A.1d.
7. The unit cell projection for the D-type unit cell of perovskite is shown in Figure A.2.
8. For the spheres representing atoms in the cubic primitive cell to be just touching, they must have a radius of $a/2$ where a is the dimension of the unit cell. One-eighth of each sphere lies within the unit cell. As there are 8 spheres, the total occupied space is equal to the volume of one sphere $= (4\pi/3)(a/2)^3$. The total volume of the unit cell is a^3. The packing efficiency is thus $\pi/6 = 0.52$ or 52% to two significant figures.
9. The indices are B—$1\bar{1}$; C—01; D—21; E—$2\bar{1}$. (By choosing a different line, you may have come up with the following answers: $\bar{1}1$, $0\bar{1}$, $\bar{2}\bar{1}$ and $\bar{2}1$. These are equally valid answers.)
10. Figure 1.28b shows the *200* planes: the planes are parallel to *y* and *z* and divide *a* into two.
    Figure 1.28c shows the $11\bar{1}$ planes: they leave *a*, *b*, and *c* undivided.
    Figure 1.28d shows the *110* planes: the planes are parallel to *z* and leave both *a* and *b* undivided.
11. Assuming that anion–anion contact occurs as in Figure 1.46b, the iodide ion radius is $300/\sqrt{2}$ or 212 pm.

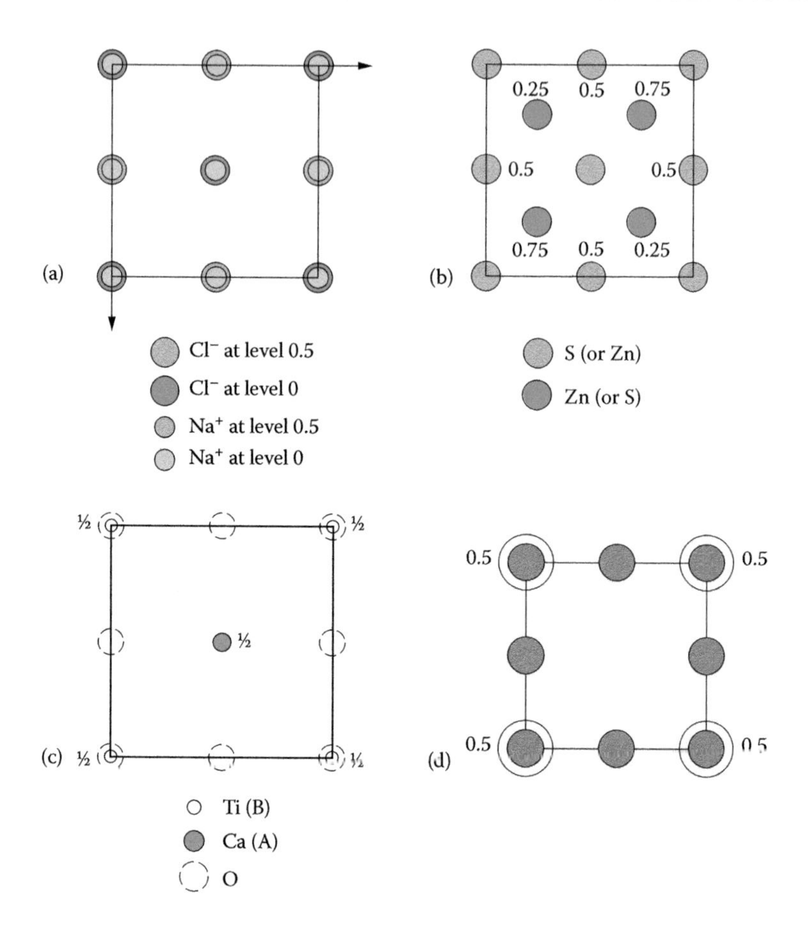

FIGURE A.1 (a) Plan of the NaCl cell, (b) plan of the ZnS cell, (c) plan of the ideal perovskite cell, and (d) plan of the ReO_3 cell. ((d) M. Fraser.)

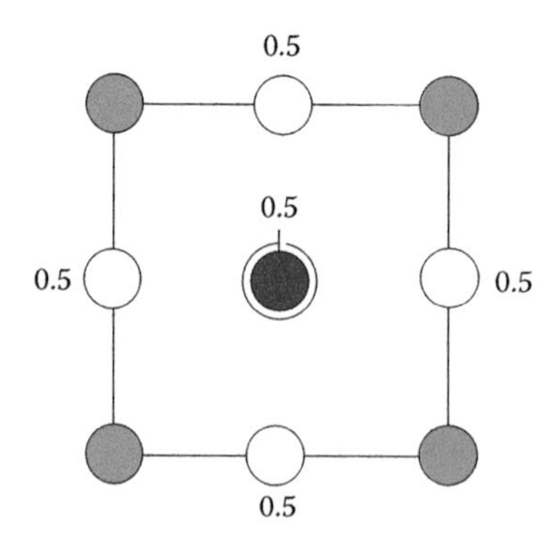

FIGURE A.2 Packing diagram for a perovskite B-type cell.

12. From the internuclear distance in NaI, $r(Na^+)=(323-212)$pm$=111$ pm. Then from the internuclear distance in NaF, $r(F^-)=(231-111)$pm$=120$ pm. The same procedure for RbI and RbF gives $r(Rb^+)=154$ pm and $r(F^-)=128$ pm.
13. Consider a cube, side r with a sphere on each vertex. This will give you 8 coordination. A sphere in the center of the cube must touch any 2 of the 8 spheres on vertices that lie at the ends of a body diagonal of the cube. These 2 spheres have radii $r_1=r/2$ and so together they occupy a distance of r along the body diagonal. The length of the body diagonal is $\sqrt{3}r=2\sqrt{3}r_1$. The diameter of the central sphere $2r_2$ is thus $2\sqrt{3}r_1-2r_1$ and the radius $r_2=(\sqrt{3}-1)r_1$. Thus $r_2/r_1=0.732$.
14. The Madelung constant, $A=-3.99$.

    There are seven ions in the structure in Figure 1.54: six cations and one anion. First calculate the contribution to the potential energy of interactions of the six cations with the central anion. Each cation is at a distance r_0 from the central ion.

$$E_1=-\frac{6e^2}{4\pi\varepsilon_0 r_0}$$

Each cation also interacts with a diametrically opposite cation (distance $2r_0$). There are three such interactions, so

$$E_2=+\frac{3e^2}{4\pi\varepsilon_0 2r_0}$$

Finally E_3 is calculated from the interactions between adjacent cations (distance $\sqrt{2}r_0$) of which there are twelve.

$$E_3=+\frac{12e^2}{4\pi\varepsilon_0\sqrt{2}r_0}$$

$$E=E_1+E_2+E_3$$

$$=-\frac{e^2}{4\pi\varepsilon_0 r_0}\left(6-\frac{3}{2}-\frac{12}{\sqrt{2}}\right)$$

$$=-\frac{Ae^2}{4\pi\varepsilon_0 r_0}$$

so,

$$A=6-\frac{3}{2}-\frac{12}{\sqrt{2}}$$

$$A=-3.99$$

15. The appropriate Born–Haber cycle is shown in Figure 1.50 and leads to the relationship in Equation 1.2:

$$\Delta H_f^{\ominus}(\text{MCl, s}) = \Delta H_{\text{atm}}^{\ominus}(\text{M, s}) + I_1(\text{M}) + \frac{1}{2}D_m(\text{Cl}-\text{Cl}) - E(\text{Cl}) + L(\text{MCl, s}) \quad (1.2)$$

Using the data in Table 1.20, we have

$$L = (-436.7 - 89.1 - 418 - 122 + 349)\ \text{kJ mol}^{-1}$$

$$= -717\ \text{kJ mol}^{-1}$$

For KCl, $Z_+ = 1$, $Z_- = 1$, $r_{K^+} = 152$ pm, $r_{Cl^-} = 167$ pm, $v = 2$, and using Equation 1.15, this gives $L = -677$ kJ mol^{-1}.

There is a discrepancy of 40 kJ mol^{-1} between the calculated value and the thermodynamic value.

16. An appropriate cycle is shown in Figure A.3.
We wish to calculate $[E(\text{O}) + E(\text{O}^-)]$, which we can do if $L(\text{MgO,s})$ can be calculated. Values of all the other terms in the cycle are known. The lattice energy relationship of Equation 1.15 is used:

Substituting $v = 2$, $Z+ = 2$, $Z- = 2$, $r+ = 86$, and $r^- = 126$, it gives

$$L = -4072\ \text{kJ mol}^{-1}$$

From the cycle in Figure A.3

$$E(\text{O}) + E(\text{O}^-) = -\Delta H_f^{\ominus}(\text{MgO, s}) + \Delta H_{\text{atm}}^{\ominus}(\text{Mg, s}) + (I_1 + I_2)$$

$$+ \frac{1}{2}D_m(\text{O}-\text{O}) + L(\text{MgO, s})$$

$$= (601.7 + 147.7 + 736 + 1452 + 249 - 4072)$$

$$= -885.6\ \text{kJ mol}^{-1}$$

$$E(\text{O}) = 141, \text{ so } E(\text{O}^-) = -885.6 - 141 = -1026.6\ \text{kJ mol}^{-1}$$

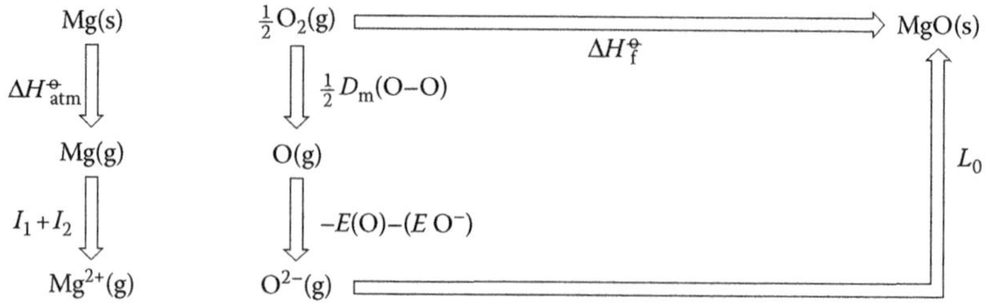

Mg(s) $\frac{1}{2}O_2(g)$ → $\Delta H_f^{\ominus}$ → MgO(s)
$\Delta H_{atm}^{\ominus}$ ↓ $\frac{1}{2}D_m(\text{O–O})$ ↓
Mg(g) O(g) L_0
$I_1 + I_2$ ↓ $-E(\text{O})-(E\ \text{O}^-)$ ↓
$Mg^{2+}(g)$ $O^{2-}(g)$

FIGURE A.3 Born–Haber cycle for magnesium oxide.

This implies that the enthalpy change for the addition of an electron to the O^- (g) anion is also endothermic:

$$O^-(g) + e^-(g) = O^{2-}(g); \quad \Delta H_m^{\ominus} = 1026.6 \text{ kJ mol}^{-1}$$

17. First we use Equation 1.15 to calculate a value for the lattice energy: For NH_4Cl, $v=2$, $Z_+=1$, $Z_-=1$, $r_+=151$ pm, and $r_-=167$ pm, giving

$$L = -679 \text{ kJ mol}^{-1}$$

From Figure 1.58:

$$P(NH_3,g) = -\Delta H_f^{\ominus}(NH_4Cl,s) + \Delta H_f^{\ominus}(NH_3,g) + \frac{1}{2}D(H-H) + I(H)$$

$$+\frac{1}{2}D(Cl-Cl) - E(Cl) + L(NH_4Cl,s)$$

$$= (314.4 - 46.0 + 218 + 1314 + 122 - 349 - 679) \text{ kJ mol}^{-1}$$

$$= 894.4 \text{ kJ mol}^{-1}$$

The addition of a proton to the ammonia molecule is an exothermic process:

$$NH_3(g) + H^+(g) = NH_4^+(g); \quad \Delta H_m^{\ominus} = -894.4 \text{ kJ mol}^{-1}$$

(The experimentally determined value is 871 ± 15 kJ mol^{-1}, quite good agreement!)

ANSWERS FOR CHAPTER 2

1. The spacings for these planes are $d_{100}=a$, $d_{110} = a/\sqrt{2}$, and $d_{111} = a/\sqrt{3}$ and so the reflections occur in the order *100*, *110*, and *111*.
2. Since the separation of planes decreases as $h^2+k^2+l^2$ increases, it follows that the value of sinθ increases as $h^2+k^2+l^2$ increases. The $h^2+k^2+l^2$ values are therefore: for *220*, 8; for *300*, 9 and for *211*, 6. The sequence of reflections is *211*, *220*, and *300*.
3. If the *100* and *110* reflections are absent, then the crystal is likely to be face-centered cubic.
4. In this case, the common factor of the first two lines is 0.05. Dividing all the $\sin^2\theta$ values by 0.05 gives the series 3, 4, 8, 11, 12, 16, 19, …, which is characteristic of face-centring.
   The common factor 0.05 is equal to $\lambda^2/4a^2$, and so

$$a^2 = \frac{\lambda^2}{4\times 0.05} = \frac{154.2^2}{0.2}$$

$$a = 345 \text{ pm}$$

5. The $\sin^2\theta$ values do not have the ratio 1: 2: 3: 4: 5: 6: 8, …, or 1: 2: 3: 4: 5: 6: 7: 8, …, so the cell is not primitive or body-centred. If it is face-centred, then the common factor A will be 0.0746−0.0560=0.0186. Dividing the $\sin^2\theta$ values by the common factor gives the $h^2+k^2+l^2$ values, and these are listed in below table rounded to the nearest integer. *hkl* only takes values where *hkl* are either all even or all odd, so the data fits a face-centred cubic structure.

Powder Diffraction Data for NaCl

θ_{hkl}	$\sin^2\theta$	$h^2+k^2+l^2$	*hkl*
13° 41′	0.0560	3	*111*
15° 51′	0.0746	4	*200*
22° 44′	0.1492	8	*220*
26° 56′	0.2052	11	*311*
28° 14′	0.2239	12	*222*
33°7′	0.2984	16	*400*
36°32′	0.3544	19	*331*
37° 39′	0.3731	20	*420*
42° 0′	0.4477	24	*422*
45° 13′	0.5036	27	*511, 333*
50° 36′	0.5972	32	*440*
53° 54′	0.6529	35	*531*
55°2′	0.6715	36	*600, 442*
59° 45′	0.7462	40	*620*

6. The mean value of the common factor A is 0.01875, and this is equal to $\lambda^2/4a^2$.
   This gives

$$a^2 = \frac{154.2^2}{4 \times 0.01875}$$

$$a = 563.1 \text{ pm}$$

7. If the mass of the unit cell contents is M and the unit cell volume is V, then the density, ρ, is given by

$$\rho = \frac{M}{V} = 2.17 \times 10^3 \text{ kg m}^{-3}$$

but

$$V = \left(563.1 \times 10^{-12}\right)^3 \text{ m}^3$$

the mass of one mole of NaCl=(22.99+35.45)×10^{-3} kg dividing by the Avogadro constant, we get that the mass of one formula unit of

$NaCl = \frac{(22.99 + 35.45) \times 10^{-3}}{6.022 \times 10^{23}}$ kg, and if there are Z formula units in one unit cell, then the mass of the unit cell contents is

$$M = \frac{Z(22.99 + 35.45) \times 10^{-3}}{6.022 \times 10^{23}} \text{kg}$$

so

$$\rho = 2.17 \times 10^3 = \frac{Z(22.99 + 35.45) \times 10^{-3}}{6.022 \times 10^{23} \times (563.1 \times 10^{-12})^3} \text{kg m}^{-3}$$

and rearranging gives

$$Z = \frac{2.17 \times 10^3 \times 6.022 \times 10^{23} \times (563.1 \times 10^{-12})^3}{(22.99 + 35.45) \times 10^{-3}}$$

$Z = 3.99$, or $Z = 4$ (rounding to the nearest whole number)

8.

$$\theta = 19.1°$$

$$\sin\theta = 0.327$$

From the Bragg equation

$$\lambda = 2d_{111} \sin\theta_{111}$$

so, $d_{111} = \frac{\lambda}{2\sin\theta_{111}} = \frac{154.2}{0.654} = 235.8$ pm

the d spacing between the *111* planes is $a / \sqrt{3}$

so,

$$a = \left(\sqrt{3} \times 235.8\right) \text{pm}$$

$$= 408.4 \text{ pm}$$

density, $\rho = \frac{M}{V}$

$$V = \left(408.4 \times 10^{-12}\right)^3 \text{m}^3$$

$$M = \frac{Z \times 107.9 \times 10^{-3}}{N_A}$$

where $Z = 4$, the number of atoms of silver in the unit cell, and N_A is the Avogadro constant.

So,

$$N_A = \frac{4 \times 107.9 \times 10^{-3}}{10.5 \times 10^3 \times \left(408.4 \times 10^{-12}\right)^3}$$

$$= 6.03 \times 10^{23}\ \text{mol}^{-1}$$

9.

$$\rho = \frac{M}{V}$$

$$3.35 \times 10^3 = \frac{Z\left(40.08 + 15.99\right) \times 10^{-3}}{6.022 \times 10^{23}} \div \left(481 \times 10^{-12}\right)^3\ \text{kg m}^{-3}$$

$$Z = \frac{3.35 \times 10^3 \times \left(481 \times 10^{-12}\right)^3}{9.311 \times 10^{-26}}$$

$$Z = 4$$

10.

$$\rho = \frac{M}{V}$$

$$8.5 \times 10^3 = \frac{Z\left(232.04 + 2\left(78.96\right)\right) \times 10^{-3}}{6.022 \times 10^{23}} \div \left(442.0 \times 761.0 \times 906.4 \times \left(10^{-12}\right)^3\right) \text{kg m}^{-3}$$

$$Z = \frac{8.5 \times 10^3 \times 3.049 \times 10^{-28}}{6.476 \times 10^{-25}}$$

$$Z = \frac{2.591 \times 10^{-24}}{6.476 \times 10^{-25}}$$

$$Z = 4$$

11. From Equation 2.7, we know that $\sin^2\theta_{hkl} = \lambda^2/4a^2\ (h^2 + k^2 + l^2)$.
A cubic close-packed structure has a face-centered unit cell. The first two reflections observed will therefore be *111* and *200*, with $h^2 + k^2 + l^2$ values of 3 and 4, respectively. Thus, $\sin^2\theta_{111} = 3\lambda^2/4a^2$ and $\sin^2\theta_{200} = 4\lambda^2/4a^2$, and $\lambda^2/4a^2 = \sin^2\theta_{200} - \sin^2\theta_{111} = 0.181 - 0.136 = 0.045$.

$$a = 363.5\ \text{pm}$$

In a close-packed structure where the atoms are considered to be in contact, the radius of an atom, r, is $\frac{1}{2}$ of the length of the body-diagonal.

$$r = \frac{1}{4}\sqrt{3}a$$

$$r = \left(\frac{1}{4} \times 1.732 \times 363.5\right) \text{pm}$$

$$r = 157.4 \text{ pm}$$

12. The scattering power is determined by the number of electrons possessed by each atom, that is, the atomic number. In increasing order, the atoms can be arranged as follows: H, O, F, Na, Cl, Co, Cd, Pt, Tl.
13. Both *111* and *222* are observed in P and F lattices, but *111* is not present for I. The *001* is not observed for F but would be present in a P unit cell. The Bravais lattice is thus F.
14.

Indexing

Peak number	2θ(°)	θ(°)	2d	d (Å)	d²	1/d²	(1/d²)/Z	h	k	l
1	18.52	9.26	9.57376	4.786880	22.914222	0.043641	1	1	0	0
2	25.80	12.90	6.90	3.45	11.90	0.08	1.9248250	1	1	0
3	32.16	16.08	5.56200	2.781002	7.7339745	0.129299	2.9628003	1	1	1
4	38.01	19.005	4.73071	2.365356	5.5949125	0.178733	4.0955460	2	0	0
5	42.13	21.065	4.28615	2.143078	4.5927864	0.217732	4.9891764	2	1	0
6	45.98	22.99	3.94438	1.972192	3.8895418	0.257099	5.8912394	2	1	1
7	54.23	27.115	3.38006	1.690033	2.8562132	0.350113	8.0225881	2	2	0
8	57.91	28.955	3.18216	1.591084	2.5315506	0.395014	9.0514570	2	2	1
9	61.09	30.545	3.03131	1.515657	2.2972190	0.435308	9.9747659	3	1	0
10	64.53	32.265	2.88583	1.442915	2.0820045	0.480306	11.005846	3	1	1

Z = common factor = 0.04364
System is cubic (absence of 7)
Taking peak 1, 1.5406 = (2a 0.161)/1
therefore a = b = c = 4.787 Å, angles are all = 90°

15. $Co(CO)_4$ is a tetrahedral molecule. The first coordination shell corresponds to four carbon atoms surrounding the Co at a distance of 177 pm. The second coordination shell refines to four oxygen atoms at a distance of 292 pm.
16. The peaks in the spectrum maximize at approximately −88, −93, −99, and −105 ppm. If you mark these values on the chart in Figure 2.34a, then you will see that the best correspondence is to the four linkages: $Si(OAl)_3(OSi)$, $Si(OAl)_2(OSi)_2$, $Si(OAl)(OSi)_3$, and $Si(OSi)_4$.
17. There is only one peak in the new spectrum at approximately −108 ppm. Reference to the chart in Figure 2.34a suggests that this is due to a $Si(OSi)_4$ environment. Clearly, the treatment with $SiCl_4$ has removed all the tetrahedral Al from the framework. The intensity measurements on this peak indicate a Si/Al ratio of 55.

18. The starting sample (a) clearly shows the presence of tetrahedral Al in the framework (peak at 61 ppm). After treatment with $SiCl_4$ (b), the amount of Al in the framework has been reduced considerably, but there is a very strong peak due to $[AlCl_4]^-$ (at 100 ppm) and also a peak due to octahedral aluminum at 0 ppm. The first washings (c) remove $Na^+[AlCl_4]^-$ from the sample, and repeated washing (d) also removes some of the octahedrally coordinated Al.
19. The relative molecular mass (RMM) of MnC_2O_4 is $(54.9+(2\times 12.0)+(4\times 16.0))=142.9$, and the RMM of H_2O is $(2\times 1.0)+16.0)=18.0$. 25 mg of $MnC_2O_4 \cdot xH_2O$ gives 20 mg of MnC_2O_4, so 20/142.9 moles MnC_2O_4 are produced from 25/X moles of MnC_2O_4 xH_2O, if X is the RMM of $MnC_2O_4 \cdot xH_2O$. Therefore, $X=(25\times 142.9)/20=178.6$. The molecular mass of water, $18x=178.6-142.9=35.7$ and $x=2$.

    The oxalate is thus the dihydrate, $MnC_2O_4 \cdot 2H_2O$.

    At higher temperatures, the oxalate decomposes to CO and CO_2 and MnO. Above 900°C, the MnO oxidises to Mn_3O_4.
20. The first exotherm at about 60°C coincides with a sharp weight loss and is due to dehydration of the ferrous sulfate. The second exotherm at 90°C does not coincide with any weight loss and must therefore be due to a phase change (this is the melting temperature). The third exotherm at about 600°C again coincides with weight loss and is due to decomposition.
21. ± 0.001 Å

ANSWERS FOR CHAPTER 3

1. This compound could be made from the elements in the correct stoichiometric proportions. The reactants would have to be very well mixed, and the reaction vessel would have to be closed to prevent the loss of the volatile sulfur (see SmS preparation).
2. Of the methods described in this chapter, the following are suitable: (a) sol–gel because the gel can be sliced thinly; CVD methods; (b) hydrothermal synthesis, vapour phase and molecular beam epitaxy; (c) vapour phase and molecular beam epitaxy; (d) sol–gel precursor method.
3. The compound given could be used as a precursor as it has a ratio of Cu:Cr of 1: 2. Heating would pyrolyse the ammonia and ammonium, which would be driven off as nitrogen oxides and water. The product would be more homogeneous than a ceramic product because the components are mixed at a molecular level. The temperature needed to pyrolyse the ammonia and ammonium is less than that required to enable the solid-state reaction to proceed at a reasonable rate.

    The precursor compound contains ammonia of solvation and was probably therefore precipitated from liquid ammonia.
4. β-TeI is metastable and contains Te in an unusual oxidation state. A method that employs temperatures of this order and can produce compounds in unusual oxidation states is hydrothermal synthesis.

5. Alumina is insoluble in water even under hydrothermal conditions but needs to be dissolved for the synthesis. However, alumina is amphoteric and will dissolve in alkali. OH^- ions from the alkali coordinate to the aluminium, forming aluminate ions $Al(OH)_4^-$, which are soluble.
6. At least one component of the reaction mixture must absorb microwave radiation strongly. Splitting the mixed oxides up into their component oxides that would be used as starting materials, we get $CaO + TiO_2$; $BaO + PbO_2$; $ZnO + Fe_2O_3$; $ZrO_2 + CaO$; $K_2O + V_2O_5$. PbO_2, ZnO, and V_2O_5 absorb microwaves strongly, and so $BaPbO_3$, $ZnFe_2O_4$, $Zr_{1-x}CaxO_{2-x}$, and KVO_3 are good candidates.
7. Since the reaction is exothermic, it is driven to the left by raising the temperature. Thus, the crystals grow at the hotter end of the tube.
8. Niobium is in a high oxidation state in lithium niobate; the addition of oxygen to the carrier would prevent any tendency for decomposition to a lower oxidation state plus oxygen. The hydrogen in the preparation of mercury telluride would act to reduce the tellurium (in HgTe, it is in an oxidation state of −2) and form hydrocarbons with the ethyl radicals.
9. The two methods are
(1) arsenic(III) chloride ($AsCl_3$) (boiling temperature 103°C) is used to transport gallium vapour to the reaction site where gallium arsenide is deposited in layers. The reaction involved is

$$2Ga(g) + 2AsCl_3(g) = 2GaAs(s) + 3Cl_2(g)$$

(2) the reaction of trimethyl gallium ($Ga(CH_3)_3$) with the highly volatile and toxic arsine (AsH_3)

$$Ga(CH_3)_3(g) + AsH_3(g) = GaAs(s) + 3CH_4(g).$$

For method 1, chlorine gas is produced as a byproduct. It could be recycled and reacted with arsenic to produce $AsCl_3$. It could however simply become a waste material (Principle 1). The production of chlorine means that not all materials used are incorporated in the product (Principle 2). Chlorine gas is harmful to health. At 1–3 ppm, it produces mild irritation, but the toxicity increases with concentration and at 1000 ppm, it is lethal within minutes. Chronic low-level exposure may lead to impairment of the lung. Gallium is considered mildly toxic, but can cause irritation when inhaled. $AsCl_3$ is carcinogenic and toxic when inhaled or ingested (Principle 3).

For method 2, Methane is produced as a waste product. Methane contributes to global warming (Principle 1). Carbon and hydrogen are not present in the final product (Principle 2). Arsine (AsH_3) is toxic as stated. $Ga(CH_3)_3$ can cause burns to the eye, skin, and respiratory tract (Principle 3).

For full details of toxicity of these substances and how to handle them, you need to consult safety data sheets.

ANSWERS FOR CHAPTER 4

1. For the Fermi level in sodium, $E = 3.2$ eV and the mass of an electron is 9.11×10^{-31} kg. With 1 eV $= 1.602 \times 10^{-19}$ J, this gives

$$3.2 \times 1.602 \times 10^{-19} = \frac{1}{2} \times 9.11 \times 10^{-31} \times v^2$$

$$v = \left(\frac{2 \times 3.2 \times 1.602 \times 10^{-19}}{9.11 \times 10^{-31}} \right)^{\frac{1}{2}}$$

$$= 1.1 \times 10^6 \text{ m s}^{-1}$$

2. 10^{-12} m^3 of metal contains 1740×10^{-12} kg of magnesium, but one atom of magnesium weighs $(24/6.022 \times 10^{23})$ g $= (24 \times 10^{-3})/(6.022 \times 10^{23})$ kg. So there are $(1740 \times 10^{-12}) \times (6.022 \times 10^{23})/(24 \times 10^{-3})$ atoms of magnesium $= 4.4 \times 10^{16}$ atoms.
3. a) $N = 10^{-12} \times (2 \times (9.11 \times 10^{-31}) \times 7.1 \times (1.602 \times 10^{-19}))^{3/2} / (3\pi^2 \times (1.055 \times 10^{-34})^3) = 8.7 \times 10^{16}$
   b) 8.7×10^{22}
   c) 0.87

   Each level can take two electrons, and a crystal of N atoms of magnesium has $2N$ electrons to fill the band. As you can see, the agreement between the number of filled levels predicted by this very simple theory and the number needed to accommodate the available electrons is very good for crystals. For example, the answer to question 2 shows there are 4.4×10^{16} magnesium atoms in a volume of 10^{-12} m^3 and these will supply 8.7×10^{16} electrons. Comparing the answers to (a), (b), and (c) shows how the number of states increases with volume.
4. These examples are analogous to the case of magnesium discussed in the text. s-p mixing of the type discussed in the text occurs and the resultant combined band remains incompletely filled and these elements are metallic conductors. Remember that in band theory, conductivity is always associated with the presence of partially filled bands.
5. The band gap lies in the energy range for visible photons, and so the photons of visible light can promote electrons from the valence band to the conduction band. Electrons in both bands can then conduct electricity.
6. Si, Ge.
7. (a) n-type, (b) neither, (c) p-type, (d) p-type, and (e) p-type.
8. (a) n-type (b) p-type since boron has three valence electrons, unlike carbon's four (c) p-type since Li has only one valence electron, unlike carbon's four (d) n-type since now the Li is not replacing (substituting for) a carbon atom but sitting in a non-lattice position and it can donate its valence electron: the outer 2s electron can be excited into the empty conduction level.

9. Carborundum, like silicon and germanium, has $4N$ valence electrons for a crystal of N atoms. The tetrahedral diamond structure will be favored because all $4N$ electrons will then be in bonding orbitals, and the energy is lower than in the higher coordination structure.
10. In general, compounds with broader d bands will be metallic. Broad bands will tend to occur for elements at the beginning of the transition series. Going from Ti to V to Mn the bands would be expected to narrow as the 3d-orbitals contract and 3d orbital overlap decreases, and the compounds become less metallic.

ANSWERS FOR CHAPTER 5

1. $n_S \approx Ne^{-\Delta H_S/2RT}$ where $\Delta H_S = 200$ kJ mol^{-1} and R = 8.314 J K^{-1} mol^{-1}

Schottky Defect Concentration in MX Compound at Various Temperatures

Temperature/°C	Temperature/K	n_s/N	n_s/mol^{-1}
27	300	3.87×10^{-18}	2.33×10^{6}
227	500	3.57×10^{-11}	2.15×10^{13}
427	700	3.45×10^{-8}	2.08×10^{16}
627	900	1.57×10^{-6}	9.45×10^{17}

2. Rearrange Equation 5.4 and take logs:

$$\frac{n_S}{N} \approx e^{-\Delta H_S/2RT}$$

$$\ln\frac{n_S}{N} = -\frac{\Delta H_S}{2RT}, \text{ or}$$

$$\log\frac{n_S}{N} = -\frac{\Delta H_S}{2.303 \times 2RT}$$

A plot of *log* n_s/N against $1/T$ gives a straight line plot passing through the origin. The slope of this graph, $-[\Delta H_s/(2.303 \times 2R)]$, gives a value of 183.4 kJ mol^{-1} for ΔH_S. Dividing by Avogadro's number, 6.022×10^{23}, gives the enthalpy of formation of one Schottky defect, 3.045×10^{-19} J.

$$E = -\left(4.302 \times 10^{-19}\text{ J}\right)\left(\frac{8 \times (-1)}{0.43} + \frac{6 \times 2}{0.5}\right)$$

$$= -\left(4.302 \times 10^{-19}\text{ J}\right)\left(5.395\right)$$

$$= -2.32 \times 10^{-8}\text{ J}$$

3. The anion at the body-centre of the unit cell in Figure 5.3c is surrounded by six anions at distance ($a/2$) and by four cations at a distance of $0.43a$. The interstitial site at the body-centre of the unit cell in Figure 5.3a is surrounded by *eight* anions at a distance of $0.43a$ and by *six* cations at a distance of $a/2$. for the normal anion site:
   $r = 0.43 \times 537 \times 10^{-12}$ m and $Z = +2$, for interaction with four cations.
   $r = 0.5 \times 537 \times 10^{-12}$ m and $Z = -1$, for interaction with the six anions.

$$E = -\left(\frac{2.31 \times 10^{-28}\ \text{J m}}{537 \times 10^{-12}\text{m}}\right)\left(\frac{4 \times 2}{0.43} + \frac{6 \times (-1)}{0.5}\right)$$

$$= -\left(4.302 \times 10^{-19}\ \text{J}\right)\left(6.605\right)$$

$$= -2.84 \times 10^{-18}\ \text{J}$$

for the interstitial site:

$$E = -\left(4.302 \times 10^{-19}\ \text{J}\right)\left(\frac{8 \times (-1)}{0.43} + \frac{6 \times 2}{0.5}\right)$$

$$= -\left(4.302 \times 10^{-19}\ \text{J}\right)\left(5.395\right)$$

$$= -2.32 \times 10^{-8}\ \text{J}$$

The energy of defect formation is the difference in energy between the two sites and so is given by:

$$\left(-2.32 \times 10^{-18}\ \text{J}\right) - \left(-2.84 \times 10^{-18}\ \text{J}\right) = 5.2 \times 10^{-19}\ \text{J}$$

The experimental value for fluorite is given in Table 5.1 as 4.49×10^{-19} J. This calculation gives a very good level of agreement considering that we have ignored all the more distant interactions, internuclear repulsion, lattice vibrations, and lattice relaxation!

4. One possible equation is

$$3\text{MgO(s)} + 2\text{Fe}_{\text{Fe}}^{\text{x}} = 2\text{Mg}_{\text{Fe}}^{\bullet} + \text{Mg}_{\text{int}}^{\prime\prime} + \text{Fe}_2\text{O}_3\text{(s)}$$

In which two Mg ions substitute for two Fe ions and the charge is balanced by an Mg ion entering an interstitial site.

See if you can think of other possibilities.

5. Unit cell volume is $(428.2\ \text{pm})^3 = 7.8513 \times 10^{-29}\ \text{m}^3$. Mass of contents for iron vacancies:

$$\left[(4 \times 55.86 \times 0.910) + (4 \times 16.00)\right] / \left(N_{\text{A}} \times 10^3\right)\ \text{kg giving a density of}\ 5.653 \times 10^{-3}\ \text{kg m}^{-3}$$

Mass of contents for oxygen interstitial:

$$\left[(4 \times 55.85) + (4 \times 16.00 \times 1/0.910)\right] / \left(N_{\text{A}} \times 10^3\right)\ \text{kg giving a density of}\ 6.213 \times 10^{-3}\ \text{kg m}^{-3}$$

Comparing these theoretical values with the experimental value, we see that the evidence supports an iron vacancy model.

6. Let the atomic mass of X be x. The mass of the AgX unit cell in kg is thus:

$$\frac{(4\times 107.868)+4x}{6.0220\times 10^{26}}\ \text{kg}$$

The volume of the unit cell is:

$$\left(577.5\times 10^{-12}\right)^3\ \text{m}^3 = 1.926\times 10^{-28}\ \text{m}^3$$

From the density and the volume, we calculate the mass of the unit cell to be:

$$6.477\times 10^3\ \text{kg m}^{-3}\times 1.926\times 10^{-28}\ \text{m}^3 = 1.247\times 10^{-24}\ \text{kg}$$

Equating the two values for the mass of the unit cell gives:

$$\frac{(4\times 107.868)+4x}{6.0220\times 10^{26}}\ \text{kg} = 1.247\times 10^{-24}\ \text{kg}$$

Solving for x, $431.5+4x=751.2$
$x=79.93$, and so X is bromine.

7. From Table 5.6, we see that as the Fe:O ratio decreases, the unit cell volume also decreases; this is the trend we would expect to see as more vacancies are introduced. If the interstitial model were correct, as the Fe:O ratio decreases, the number of interstitial oxygen rises and we would expect to see a slight increase in lattice parameter.

8. (a) The central section has *two* Fe^{3+} ions in tetrahedral sites and *seven* vacancies, so it is known as a 7: 2 cluster. (b) There are 32 oxide anions. There are seven octahedral vacancies and two Fe^{3+}_{tet} interstitial ions, so there will be a total of 27 Fe cations. (There are *two* Fe^{3+} ions and *six* Fe_{oct} ions enclosed within the cluster. The outer layer will be the same as the Koch–Cohen cluster with $\left(8\times\frac{1}{8}\right)=1\ Fe_{oct}$ at the corners, $\left(12\times\frac{1}{4}\right)=3\ Fe_{oct}$ at the midpoints of the edges, and $\left(30\times\frac{1}{2}\right)=15\ Fe_{oct}$ on the faces; this makes 27 Fe ions in total). The formula would be $Fe_{27}O_{32}$. (c) The 32 oxide ions provide 64 negative charges to be balanced. The two tetrahedral Fe^{3+} ions reduce this to 58 to be balanced by the ions in octahedral positions.
Setting up simultaneous equations, we know that if x is the number of Fe^{2+} ions and y the number of Fe^{3+}, then

$$x+y=25$$

And adding up the charges, we get

$$2x+3y=58$$

Solving gives $x=17$ and $y=8$.

- **Titanium vacancies:** there are 8 at the corners $\left(8\times\frac{1}{8}\right)=1$ and $\left(2\times\frac{1}{2}\right)=1$ on cell faces.
- **Titanium ions:** cell edges, $\left(4\times\frac{1}{4}\right)=1$. Cell faces, $\left(8\times\frac{1}{2}\right)=4$ on the top and bottom. There are 5 ions contained within the cell boundary, making 10 in total.

  The titanium stoichiometry of the unit cell is obviously representative of the whole structure: of the 12 sites, 10 are occupied and 2 are vacant. This is also true for oxygen:
- **Oxygen vacancies:** cell faces, $\left(4\times\frac{1}{2}\right)=2$
- **Oxide ions:** cell faces, $\left(8\times\frac{1}{2}\right)=4$. Cell edges, $\left(8\times\frac{1}{4}\right)=2$. There are 4 ions contained within the cell boundary, making 10 in all.

10. Taking the titanium positions first: there are vacancies at all eight corners $\left(8\times\frac{1}{8}\right)=1$ and 1 vacancy in the centre of the cell, making 2 vacancies in all. There are no Ti^{2+} ions on cell edges. There are Ti^{2+} ions on only two of the faces—the top and bottom, which have 4 each—contributing $\left(2\times4\times\frac{1}{2}\right)=4$ There are 4 Ti^{2+} ions enclosed within the cell. Four edges have O^{2-} ions giving $\left(4\times1\times\frac{1}{4}\right)=1$; only the top and bottom faces have O^{2-} ions, and these have 5 each contributing $\left(2\times5\times\frac{1}{2}\right)=5$; there are 4 O^{2-} ions enclosed within the cell, making 10 in all. The overall content of this unit cell is thus, Ti_8O_{10}, which corresponds to $TiO_{1.25}$.
11. The structure has Ti vacancies—every fifth Ti is missing. For every absent Ti^{2+} ion, there must be two Ti^{3+} present or one Ti^{4+}.
12. The effects of sharing are shown in Figure A.4.
13. Figure A.5 shows the shear structure with a unit cell added. Within the boundary, there is one group of four edge-sharing octahedra and four $[WO_6]$ octahedra. The formula is thus $W_4O_{11}+4WO_3=W_8O_{23}$.
14. Nonstoichiometric ZnO is an *n*-type semiconductor. Aluminium ions entering the structure of ZnO have a charge of +3. If the Al^{3+} substitutes for Zn^{2+} and the crystal maintains its stoichiometry, oxygen will be lost during the reaction. The electrons made available from the oxide ions becoming oxygen molecules will remain in the structure to effect the necessary charge compensation, thus enhancing the *n*-type semiconduction. An equation for the reaction is:

$$xAl_2O_3+2(1-x)ZnO=2Al_xZn_{1-x}O+\frac{1}{2}xO_2$$

where we have used 1: 1 ZnO for simplicity.

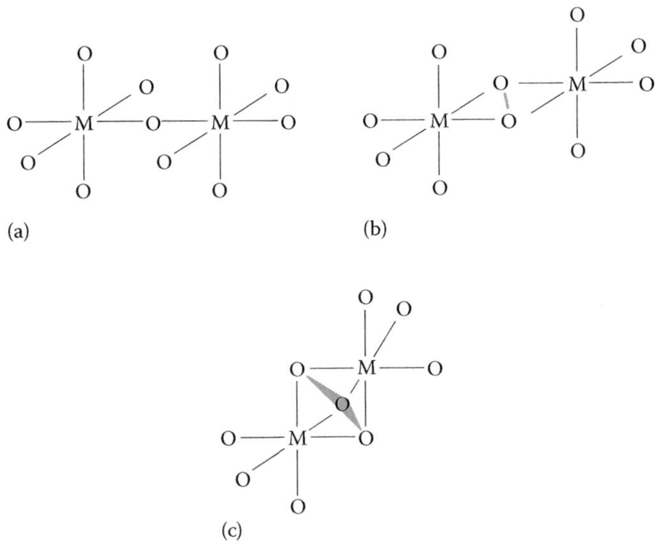

(a) (b) (c)

FIGURE A.4 (a) Sharing a corner M_2O_{11}, (b) sharing an edge M_2O_{10}, (c) sharing a face M_2O_9.

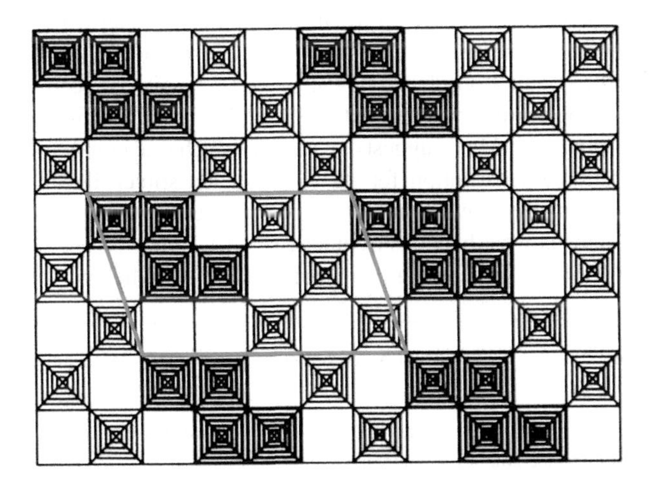

FIGURE A.5 Unit cell of W_8O_{23}.

ANSWERS FOR CHAPTER 6

1. The path is shown in Figure A.6. The anion jumps from the tetrahedral site, through a trigonal position on the common triangular face, and into the octahedral site at the body-centre.
2. Both structures contain Ag^+ ions in tetrahedral sites, but in γ-AgI, half of the tetrahedral holes are occupied, so there is only *one* vacant equivalent site per Ag^+ ion, whereas in α-AgI, there are *five*. Both structures contain vacant octahedral and trigonal sites. In γ-AgI, the octahedral holes lie at the body-centre and at the midpoint of each edge. The trigonal sites lie on the

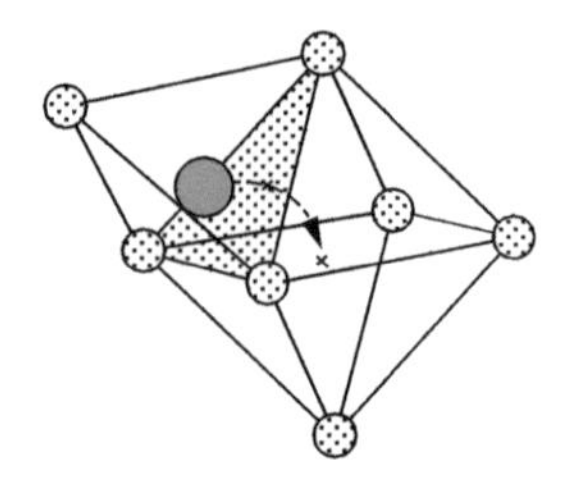

FIGURE A.6 Path taken by an anion moving from one tetrahedral site to an octahedral site.

triangular faces of the octahedra, where adjacent octahedra and tetrahedra join (Figure A.7). Both structures have the same monovalent ions and a polarisable anion.

The jump from one tetrahedral site to another in γ-AgI could take either of the routes sketched out in Figure A.7, which first pass through a trigonal face and then through an octahedral hole. Alternatively (not shown), if the ion passed through only the top half of the octahedron, it would be five coordinate before passing through another three-coordinate face. All these routes would be of higher energy than the pathways we looked at for α-AgI.

3. I^-, S^{2-}, Se^{2-}, and Te^{2-} are all polarizable anions.
4. Li is a light element and therefore difficult to locate using X-ray diffraction.
5. In undoped β-alumina, the excess Na^+ ions are balanced by additional oxygen in the conduction plane. Beyond a certain point, it seems likely that these will begin to block the motion of the Na^+ ions, as observed. By contrast, doping with Mg^{2+} suggests an alternative charge compensation mechanism: simple substitution for Al^{3+} ions in the spinel-like blocks allows extra Na^+ ions into the conduction planes without the need for oxygen interstitials. This is what happens in practice.
6. Li- and Na-based batteries are compared in the table below.

Property	Li-based battery	Na-based battery
Mass	low	higher
Availability	limited	freely available
Operating temperature	room temperature	high
Recharging ability	good for current batteries	low for sodium-ion batteries but good for ZEBRA

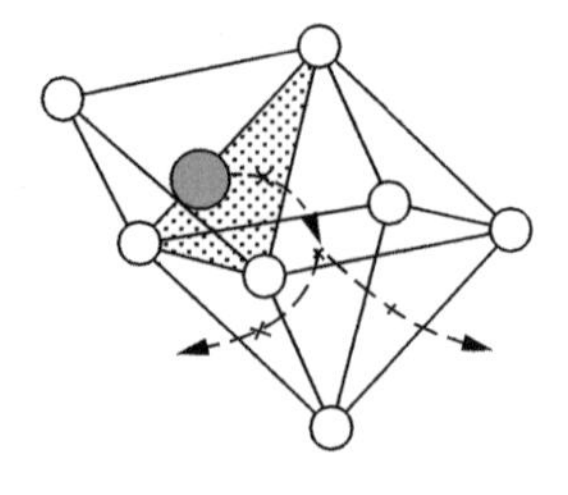

FIGURE A.7 Path taken by an anion moving from one tetrahedral site to another.

ANSWERS FOR CHAPTER 7

1. The chart in Figure 2.34 shows that the most likely Si environment for zeolite A is $Si(OAl)_3(OSi)$. However, we know that the Si/Al ratio is 1, so this coordination is not possible without having an Al–O–Al linkage. It was the spectroscopic work on this structure (and on zeolite ZK-4) that eventually confirmed the $Si(OAl)_4$ structure of zeolite A with strict alternation of Si and Al, and led to the extended ranges shown in the chart.
2. Catalytic activity tends to parallel the increase in acidity of the active sites as the Si ratio increases. Beyond a certain point, however, the number of acidic sites has decreased so much that catalytic activity also decreases.
3. This system demonstrates reactant shape-selective catalysis. The branched hydrocarbons are too bulky to pass through the pore openings in the catalyst.
4. We see here product shape-selective catalysis in operation: both reactant molecules are small enough to diffuse into the pores and be hydrogenated, but the slightly larger propane molecule cannot leave.
5. Zeolite A (Ca) shows reactant selectivity. The straight-chain *n*-hexane can pass through the windows and undergo reaction, but the branched-chain 3-methylpentane is excluded. The selective cracking of straight-chain hydrocarbons in the presence of branched chains is an important industrial process known as **selectoforming**, which improves the octane number of the fuel.
6. There are two possible reasons for the increased percentage of *para*-xylene. First, the increased crystal size increases the distance travelled by the diffusing molecules along the channels, giving more time for isomerisation to the *para*-form. Second, the shape-selective reactions take place in the internal pores of the zeolites, rather than on the surface. Larger crystals have a higher ratio of internal to external sites, and so we would expect selectivity to improve.
7. As the Si/Al ratio increases, the –OH stretching frequency falls. This indicates a decrease in the covalent bond strength, making ionisation of H^+ easier, that is, an increase in acid strength.
8. The property mentioned in the text is the length of the organic linker.
9. The removal of water produces a Lewis acid site.
10. Many MOFs are unstable with respect to high temperature, pH and/or the presence of water. They are also more expensive to produce than zeolites.
11. ZIFs have the same topologies as zeolites with the Si atoms replaced by metal ions or clusters.
12. The structure of zeolites is held together by covalent O-Si-O bonds. That of MOFs is bonded by dative bonds between the organic linkers (acting as ligands) and the metal centres. COFs have covalent bonds between light elements (C, H, N, O, B), 2D COFs are covalently bonded within layers but have π-bonding between layers.
13. MOFs, COFs, aluminosilicate materials such as MCM-41, organosilicas.

ANSWERS FOR CHAPTER 8

1. In Mn^{2+}, an electron can only go from one 3d level to another if it changes its spin. Transitions in which an electron changes its spin are forbidden and so give rise to weak spectral lines.
2. Radiation from a lamp excites electrons from the ground state, G, to states B, C, and D. Electrons in states C and D undergo nonradiative transitions to state B. Radiative transitions from B to lower states are forbidden, and so a large population of electrons in state B builds up. Eventually, a photon is emitted and an electron goes to state A. This photon induces further emission from B to A, eventually producing a beam of laser light.
   Transitions from B to both A and G must be forbidden for a large population to build up in B, but the transition B↔A must be less forbidden than the transition B↔G as it is B↔A that gives rise to laser action, and hence it must be the more likely to occur.
3. Blue light is absorbed in a transition from 4f to the higher 5d energy levels. A non-radiative transition occurs resulting in population of the lower 5d energy levels. Rather than build up a large population in these levels as in a laser, the electrons return to the 4f levels emitting green and yellow light as they do so.
4. The solid has an indirect band gap because the lowest energy interband transition (that is, from the highest energy level in the valence band to the lowest energy level in the conduction band) does not correspond to $\Delta k=0$ and is therefore forbidden.
5. This is an H-centre as two I atoms occupy one anion site.
6. A crystal containing F-centres contains anion vacancies. We would expect, therefore, that the density would be lower than that of the colourless crystal.
7. Silicon is an indirect band gap solid with an available nonradiative pathway from the conduction band to the valence band. In photovoltaic cells, electrons are promoted from the valence band to the conduction band and are then used to do electrical work. The promoted electrons do not return directly to the valence band either by emitting energy or by a nonradiative pathway. In LEDs, it is the return of the electrons to the valence band by emitting light that is important. This return has low probability because of the indirect band gap and the electrons use the nonradiative pathway instead.
   Promotion to the conduction band in the solar cell will also be of low probability, but there is no competing nonradiative route.
8. Polyphenylenevinylene has a π system delocalised over the benzene ring. It is likely that this is also delocalised over the conjugated double bond and hence to the benzene ring of the next unit. Thus, like polyacetylene, this polymer will have a delocalised π system, but it will include the π ring orbitals.
9. a. Rb would lose an electron to the polyacetylene, forming Rb^+, and give *n*-type polymer with a partially full conduction band.
   b. H_2SO_4 would act as an electron acceptor and give *p*-type polyacetylene.

10. $8(CH)_n + 9\delta nHClO_4 = 8\left[(CH)^{\delta+}(ClO_4^-)\delta\right]_n + \delta nHCl + 4\delta nH_2O$
11. If the polymer appears red, it is absorbing light in the blue/green region of the visible spectrum, suggesting that the band gap energy lies in the blue/green region. When acting as an LED, the energy of the emitted photons would correspond to the band gap energy or greater, thus the emitted light would be blue/green rather than red.
12. Sodium chloride has a cubic structure. Although the empirical formula of calomel is HgCl; its formula is usually written Hg_2Cl_2. It has a tetragonal crystal structure containing Hg_2 units. Thus, calomel crystals are asymmetric, a prerequisite for birefringence. Hg will be more polarisable than Na, giving a higher refractive index.
13. These oxides (at least formally) contain small highly charged ions. These have low polarisability and hence impart high refractive indices.
14. The photonic band gap wavelength varies with the periodicity and hence the colour of the reflected light that is cancelled out also varies.

ANSWERS FOR CHAPTER 9

1. Fe^{2+} has six 3d electrons and Fe^{3+} five 3d electrons.
   a) The maximum number of unpaired 3d electrons for Fe^{2+} is 4. The spin-only magnetic moment is thus 4.90 μ_B.
   b) The maximum number of unpaired 3d electrons for Fe^{3+} is 5. The spin-only magnetic moment is thus 5.92 μ_B.
   c) The minimum number of unpaired 3d electrons for Fe^{2+} is 0. The spin-only magnetic moment is thus 0 μ_B.
   d) The minimum number of unpaired 3d electrons for Fe^{3+} is 1. The spin-only magnetic moment is thus 1.73 μ_B.
2. For Ho^{3+}, $S=2$, $L=6$, and $J=8$.
   Thus $g=1.25$ and the magnetic moment $=10.60\ \mu_B$ to 2 dec. places.
3. Because the Mn atoms are further apart, the overlap of the 3d orbitals will be less. The 3d band will therefore be narrower than in manganese metal. With a narrower band, there is a larger interelectronic repulsion and a state with a number of unpaired spins comparable to the number of atoms becomes favourable. The alloy is thus ferromagnetic.
4. The magnetically ordered unit cell is identical to the high-temperature (paramagnetic) unit cell, so that all the layers of europium ions must be aligned with their spins parallel, giving a ferromagnetic compound.
5. The Zn^{2+} and half the V^{3+} ions are on octahedral sites with spins aligned, and the remaining V^{3+} ions are on tetrahedral sites aligned antiparallel. The net moment of the V^{3+} ions is zero. As all the electron spins are paired in Zn^{2+} ions, there is no overall magnetic moment and the compound is antiferromagnetic.
6. In MnS_2, the 3d electrons are localised on the evidence given. Above the Néel temperature, the solid is an insulator with a paramagnetic susceptibility corresponding to 5 unpaired electrons per Mn and this could be explained on the basis of filled bands below the 3d level and localised 3d

electrons. Below the Neél temperature, the localised electrons on different Mn ions interact to cancel out magnetic moments, possibly through a super-exchange mechanism involving the disulfide ions.

The properties of FeS_2 suggest that all spins are paired either in a band or localised on the iron ions. There are six 3d electrons per Fe ion, and this is just enough to fill the lower t_{2g} band. The semiconducting properties suggest that there is an empty band not much higher in energy. This is probably the e_g band. In CoS_2, there are enough electrons to partly occupy the e_g band, and thus this compound is metallic. The band is narrow and so gives rise to ferromagnetism.

7. The effect of deuterium substitution suggests that the hydrogen atoms are displaced when the ferroelectric phase is formed.
8. Li^+ is smaller than K^+ and so is easily displaced.
9. The ferroelectricity will derive from displacement of ions due to the lone pair on Pb^{2+}. The magnetism is due to the V^{4+} ions. Thus, the ferroelectricity and ferromagnetism arise from different ions and are unlikely to be strongly coupled (see $BiFeO_3$).

ANSWERS FOR CHAPTER 10

1. Hydrogen is a light element and the frequencies of vibrations involving hydrogen will be high.
2. The net oxidation state on the cations other than Cu is (2×0.8)(Hg) + (3×0.2)(Tl) + (2×2)Ba) + (2×2)(Ca) = 10.2. There are 8.33 oxygen atoms with oxidation state −2 giving −16.66. To achieve charge balance, the net oxidation state on the copper atoms is thus 16.66 − 10.2 = 6.46. There are three copper atoms, and so the average oxidation state is 6.46/3 = 2.153.
3. The Sm contributes +3. As as As^3 contributes −3, O contributes $(-2\times0.5) = -1$, F contributes $(-1\times0.5) = -0.5$. The net anion charge is thus −4.5. To achieve charge balance, the net oxidation state on Fe is (4.5 − 3) = +1. 5. Fe is usually found in oxidation states +2 and +3, so the oxidation state here is unusually low. It is possible the Fe_2As_2 layer is more alloy-like than ionic.
4. Both types have layered structures. In the cuprates, the superconducting layer is Cu-O_2 in square coordination. In the iron-based superconductors, the conducting layer is a square mesh of Fe with the anions (As^{3+}, Se^{2-}, Te^{2-}) above and below the Fe plane. The superconducting layers in both are interspersed with nonsuperconducting layers.
5.

Material	*Symmetry of order parameter*	*Spin state*
Metals and alloys	s-wave	singlet
Cuprates	d-wave	singlet
Iron-based superconductors	s-wave	singlet
Ferromagnetic superconductors	p-wave	triplet
Hydrides	s-wave	singlet

6. Here are some advantages and disadvantages. You may have thought of others.
   Advantages: Do not require cooling to very low temperatures. The usual coolant for superconductors is liquid helium. This is expensive. Supplies of helium are under threat of running out in the next 100 years.
   Disadvantages: It is difficult to make high-T_C cuprate superconductors into wires or otherwise manipulate them as they are brittle. The superconductivity is anisotropic, occurring in directions parallel to the Cu layers only.

ANSWERS FOR CHAPTER 11

1. The surface area of a sphere of radius r is given by $4\pi r^2$. The volume is given by $(4/3)\pi r^3$. The surface area to volume ratio is thus $3/r$.
   For the 2-nm-diameter sphere, this is 3 nm^1. For the 10-nm-diameter sphere this is 3/5 nm^1.
2. Because the particles are single domain, they are resistant to demagnetisation and so the bacteria can rely on them always lining up to indicate in which direction the bottom of the pond lies.
3. Quantum dots emit over a narrow band of frequencies and thus produce pure colors.
4. Nanoparticles of ZnO and TiO_2 are less efficient at scattering visible light than larger particles.
5. Figure A.8 shows the band structure of graphite intercalated with (a) an electron donor and (b) an electron acceptor.

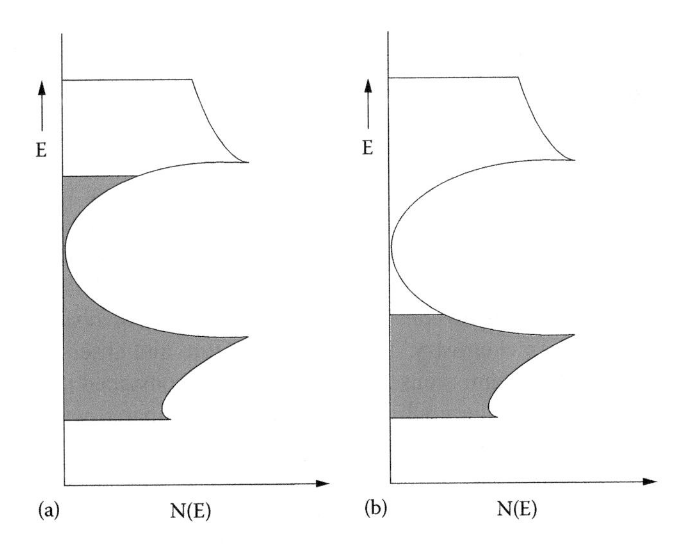

FIGURE A.8 Band structure of graphite doped with (a) an electron donor and (b) an electron acceptor.

6. The graphene was made from the bulk structure graphite so this could be considered a top-down method.
7. The smallest would be that with no hexagons and just twelve pentagons. This is a regular dodecahedron that has 20 vertices and the fullerene would have the formula C_{20}.
   In practice, pentagons are more strained than hexagons, and some hexagons are needed for stability.

ANSWERS FOR CHAPTER 12

1. Carbon is an element that is very abundant, worldwide, with no geopolitical boundaries to its use. By comparison, many of the elements required for electronic devices are relatively scarce (rare earths) or toxic (Co, Cd, etc.). In addition, the energy to process carbon-based materials and the consequent GHG emissions and water use can be considerably less than for many metals.
2. Rare-earth materials are not very abundant, but their electronic configurations give unique magnetic properties (see Chapter 8) that cannot be accomplished with other elements.
3. In 1 m^2, the volume of Si is 2×10^{-4} m^3. This corresponds to a mass of 460 g, which can provide a power of 46 W. To recover the embodied energy would require (2.8×10^8 J)/(46 J/s) which is 6×10^7 s, or about 700 days of fulltime use. At 8 hours per day of use, the payback time would be about 5.8 years.
4. Using Equation 12.1 and the data provided, and assuming that for the formation of each of the compounds $\Delta G \cong \Delta H$, then the exergies for $LiCoO_2$ and $LiFePO_4$ are 30 kJ mol^{-1} and 1485 kJ mol^{-1}, respectively. On this basis, it would take much less energy to produce $LiCoO_2$ than $LiFePO_4$, largely because of the highly negative (favorable) enthalpy of formation of the former.
5. Mechanochemical methods are generally carried out at ambient temperature, and with no solvents. In addition, there is almost no loss of material, as generally the reactants are mechanically mixed, in stoichiometric quantities, with products in very high yield. Compared with traditional solvent-based synthesis, mechanochemical methods prevent waste, exhibit atom economy, can be less hazardous (no solvents), use less solvent and no auxiliaries such as protecting agents, can be more energy efficient (since there are no high-temperature processes such as refluxing, but note that the mechanical processes do consume energy), and do not involve derivatives, all goals of green chemistry. The low temperature and absence of solvent would reduce GHG emissions. No water would be consumed in the synthesis itself, unless a water cooling jacket is used. Compared with high-temperature synthesis of solid-state materials, mechanochemical methods can use less material (containment is reusable), can be less hazardous (sealed tubes in furnaces can explode), and certainly use less energy, again goals of green chemistry. The reduction in energy requirements of mechanochemical approaches leads to reduced GHG emissions, and also lower water consumption because of the water use in production of heat furnaces.

Further Reading

This is a fast-moving field, and to keep up with current developments the reader is referred to review magazines and journals such as ***Chemistry World*** (http://www.rsc.org/chemistryworld/), ***Education in Chemistry*** (https://edu.rsc.org/eic), ***New Scientist, Scientific American, Angewandte Chemie, Science, Physics Today, Materials Today***, and ***Physics World*** (http://physicsweb.org).

GENERAL

Bruce, D.W., O'Hare, D. and Walton, R.I. (2010) *Molecular Materials*, John Wiley-Blackwell, Chichester.

Rao, C.N.R. (2012) *Modern Aspects of Solid-State Chemistry*, Springer.

Segal, D. (2017) *Materials for the 21st Century*, Oxford University Press, Oxford.

Tilley, R. (2013) *Understanding Solids: The Science of Materials*, 2nd edition, Wiley, Chichester.

West, A.R. (2014) *Solid-State Chemistry and Its Applications*, 2nd edition, Wiley, New York.

White, M.A. (2019) *Physical Properties of Materials*, 3rd edition, Taylor & Francis, CRC Press, Boca Raton.

CHAPTER 1

Cotton, F.A. (1990) *Chemical Applications of Group Theory*, 3rd edition, Wiley, New York.

Hahn, Th. (Ed.) (2005) *International Tables for Crystallography*, vol. A, 5th edition, *Space Group Symmetry: Brief Teaching Edition* (print); and (2016) 2nd online edition, International Union of Crystallography (IUCr), Chester. ISBN: 978-0-470-97423-0.

Hammond, C. (2015) *The Basics of Crystallography and Diffraction*, International Union of Crystallography (IUCr)/Oxford Science Publications. ISBN: 978-0-19-873868-8.

Ladd, M. (1998) *Symmetry and Group Theory in Chemistry*, Woodhead Publishing Ltd. ISBN: 978-1-898563-39-6.

Li, W.-K., Zhou, G.-D. and Mak, T. (2008) *Advanced Structural Inorganic Chemistry*, International Union of Crystallography (IUCr), Chester.

CHAPTER 2

Beis, K. and Evans, G. (Eds.) (2018) *Protein Crystallography: Challenges and Practical Solutions*, Royal Society of Chemistry.

Cullity, B.D. (1978) *Elements of X-ray Diffraction*, 2nd edition, Addison-Wesley Publishing Company. ISBN: 0201011743.

Dinnebier, R.E. and Bilinge, S.J.L. (2008) *Powder Diffraction: Theory and Practice*, RSC Publishing. ISBN: 978-0-85404-231-9.

Frank, J. (2018) *Single-Particle Cryo-Electron Microscopy: The Path Toward Atomic Resolution*. https://doi.org/10.1142/10844 | June 2018.

Martz, H.E., Logan, C.M., Schneberk, Daniel M.J. and Shull, P.J. (2016) *X-Ray Imaging: Fundamentals, Industrial Techniques and Applications*, 1st edition, CRC Press. ISBN: 9780849397721.

Russo, P. (2018) *Handbook of X-ray Imaging: Physics and Technology*, 1st edition, CRC Press. ISBN: 9781498741521.

Willmott, P. (2019) *An Introduction to Synchrotron Radiation: Techniques and Applications*, 2nd edition, Wiley. ISBN: 978-1-119-28039-2.

CHAPTER 3

Ashokkumar, M. (2016) *Ultrasonic Synthesis of Functional Materials*, SpringerBriefs in Molecular Science.

Biswas, K. and Rao, C.N.R. (2015) *Essentials of Inorganic Materials Synthesis*, John Wiley and Sons Inc, New York. ISBN: 978-1-118-89266-4.

Lalena, J.N., Cleary, D.A, Carpenter, E. and Dean, N.F. (2008) *Inorganic Materials Synthesis and Fabrication*, Wiley-Blackwell, New York.

Wright, J.D. and Sommerdijk, N.A.J.M. (2000) *Sol–Gel Materials: Chemistry and Applications*, CRC Press, Boca Raton, FL.

CHAPTER 4

From a more chemical perspective:

Canadell, E., Doublet, M.-L. and Iung, C. (2016) *Orbital Approach to the Electronic Structure of Solids*, OUP.

Cox, P.A. (1987) *The Electronic Structure and Chemistry of Solids*, OUP.

Hoffmann, R. (1988) *Solids and Surfaces: A Chemist's View of Extended Bonding*, VCH.

From a more physics perspective:

Kittel, C. (2004) *Introduction to Solid State Physics*, 8th edition, John Wiley.

Sutton, A.P. (1995) *Electronic Structure of Materials*, OUP.

Just one of the many codes for calculating the electronic structure of solids is CRYSTAL17. See the website (www.crystal.unito.it/index.php); there are a helpful set of tutorials. Other popular codes include VASP (www.vasp.at) and CASTEP (www.castep.org).

CHAPTER 5

Kumar, P., and Viswanath, B. (2016). *Effect of Sulfur Evaporation Rate on Screw Dislocation Driven Growth of M_oS_2 with High Atomic Step Density*. Crystal Growth & Design, 2016, 7145–7154.

Tilley, R.J.D. (2008) *Defects in Solids*, Wiley-Blackwell.

CHAPTER 6

Rajagopalan, R. and Zhang, L. (2019) *Advanced Materials for Sodium-Ion Storage*, CRC Press.

CHAPTER 7

Barrer, R.M. (1982) *Hydrothermal Chemistry of Zeolites*, Academic Press, New York.

Breck, D.W. (1974) *Zeolite Molecular Sieves*, John Wiley, New York.

Catlow, C.R.A., Smit, B. and van Santen, R.A. (Eds.) (2004) *Computer Modelling of Microporous Materials*, Academic Press, London.

Catlow, C.R.A., Smit, B. and van Santen, R.A. (Eds.) (2004) *Computer Modelling of Microporous Materials*, Academic Press, London.
Cejka, J., Corma, A. and Zones, S. (Eds.) (2010) *Zeolites and Catalysis: Synthesis, Reactions and Applications*, Wiley-VCH, Weinheim.
Kuznicki, S.M. and UOP (2012) *Zeolite Molecular Sieves: Structure Chemistry and Use*, 2nd edition, Wiley-Blackwell.
Long, J. and Yaghi, O. (Eds.) (2009) *Metal organic frameworks, Chemical Society Reviews*, **5**, 1201–1508.
Rothenburg, G. (2017) *Catalysis and Green Applications*, 2nd edition, Wiley, Chichester.
Tompsett, G.A., Conner, W.C. and Yngvesson, K.S. (2006) Microwave synthesis of nanoporous materials, *Chemical Physics Chemistry*, **7**, 296–319.
Xu, R., Pang, W., Yu, J., Huo, Q. and Chen, J. (2007) *Chemistry of Zeolites and Related Porous Materials: Synthesis and Structure*, Wiley-Blackwell.
Zhu, G. and Ren, H (2015) *Porous Organic Frameworks*, Springer Briefs in Molecular Science.

CHAPTER 8

Ball, P. (2018) Blueprints for real world invisibility, *Frontiers of Physics*, **13**, 134102.
Duffy, J.A. (1990) *Bonding, Energy Levels and Bands in Inorganic Solids*, Longman, London.
Heeger, A.J. (2010) Semiconducting polymers: The third generation, *Chemical Society Reviews*, **39**, 2354–2371.
Higgins, S.J., Eccleston, W., Sedgi, N. and Raja, M. (2003) Plastic electronics, *Education in Chemistry*, May, 70–73.
Padilla, W.J., Basov, D.N. and Smith, D.R. (2006) Negative refractive index materials, *Materials Today*, **9**, 28–35.
Pawadwe, V.B. and Dhobie, S.J. (2018) *Phosphors for Energy Saving and Conversion Technology*, CRC Press.
Pendry, J. (2009) Taking the wraps off cloaking, *Physics*, **2**, 95.
Stafford, N. (2010) LEDs to light up the world, *Chemistry World*, **7**, 42–45
Yablonovitch, E. (2001) Photonic crystals, *Scientific American*, **285**(6), 47–55.
Zhang, X. and Liu, Z. (2008) Superlenses to overcome the diffraction limit, *Nature Materials*, **7**, 435–441.

CHAPTER 9

Awschalom, D.W., Flatté, M.E. and Samarth, N. (2002) Spintronics, *Scientific American* (June), 68–73.
Coey, J.M.D. (2009) *Magnetism and Magnetic Materials*, Cambridge University Press, Cambridge, UK and New York.
Guinier, A. and Julien, R. (1989) *The Solid State from Superconductors to Superalloys*, Oxford University Press, New York.
Khomskii, D. (2009) Classifying multiferroics: Mechanisms and effects, *Physics*, **2**, 20.
Scott, J.F. and Gardner, J. (2018) Ferroelectrics, multiferroics and artifacts: Lozenge-shaped hysteresis and things that go bump in the night, *Materials Today*, **21**, 553–562.
Spaldin, N. (2011) *Magnetic Materials, Fundamentals and Applications*, 2nd edition, Cambridge University Press, Cambridge, UK and New York.
Tokura, Y. and Seki, S. (2010) Multiferroics with spiral spin orders, *Advanced Materials*, **22**, 1554–1565.
Vijay, M.S. (2017) *Piezoelectric Materials and Devices: Applications in Engineering and Medical Science*, CRC Press.
Vopson, M.M. (2015) Fundamentals of multiferroic materials and their possible applications, *Critical Reviews in Solid State and Materials Sciences*, **40**(4), 223–250.

CHAPTER 10

Buchel, W. (2004) *Superconductivity – Fundamentals and Applications*, Wiley-VCH.
Chubokov, A. and Hirschfeld, P.J. (2015) Iron-based superconductors, seven years later, *Physics Today*, **68**, 46.
Flouquet, J. and Buzdin, A. (2002) Ferromagnetic superconductors, *Physics World*, January. http://physicsweb.org.
Grant, P. (1987) Do-it-yourself superconductors, *New Scientist*, **115**, 36–39.
Hazen, R.M. (1990) Perovskites, *Scientific American*, June, 52–61.
King, A. (2019) World first as wind turbine upgraded with high temperature superconductor, *Chemistry World*, 22 November 2018. Accessed January 2019, https://www.chemistryworld.com/news/world-first-as-wind-turbine-upgraded-with-high-temperature-superconductor/3009780.article.
Norman, M.R. (2011) The challenge of unconventional superconductivity, *Science*, **332**, 196–200.

CHAPTER 11

Altavilla, C. and Ciliberto, E. (Eds.) (2010) *Inorganic Nanoparticles: Synthesis, Applications, and Perspectives*, CRC Press, Boca Raton, FL.
Broadwith, P. (2010) Bend me, shape me anyway you want me, *Chemistry World*, **7**, 23. (Structures made from carbon nanotubes.)
Cao, G. and Wang, Y. (2011) *Nanostructures and Nanomaterials: Synthesis, Properties and Applications*, 2nd edition, World Scientific.
Ozin, G.A., Arsenault, A. and Cademartiri, L. (2009) *Nanochemistry: A Chemical Approach to Nanomaterials*, Royal Society of Chemistry, Cambridge.

CHAPTER 12

Ashby, M.F. (2013) *Materials and the Environment*, 2nd edition, Elsevier.
Cindro, N., Tireli, M., Karadeniz, B., Mrla, T. and Užarević, K. (2019) Investigations of thermally controlled mechanochemical milling reactions, *ACS Sustainable Chemical Engineering*, **7**, 16301–16309.
Ercan, M., Malmodin, J., Bergmark, P. Kimfalk, E. and Nilsson, E. (2016) Life cycle assessment of a smartphone, *4th International Conference on ICT for Sustainability (ICT4S)*, Amsterdam, the Netherlands, 124–133.
Gaultois, M.W., Sparks, T.D., Borg, C.H.K., Seshadri, R., Bonificio, W.D. and Clarke, D.R. (2013) Data-driven review of thermoelectric materials: Performance and resource considerations, *Chemistry of Materials*, **25**, 2911–2920.
Graedel, T.E. (2011) On the future availability of energy metals, *Annual Review of Materials Research*, **41**, 323–335.
McDonough, W. and Braungart, M. (2002) *Cradle to Cradle*, North Point Press, New York.
MRS Bulletin, April 2012. Special Issue on Materials for Sustainable Development.
Rivero, R. and Garfias, M. (2006) Standard chemical exergy of elements updated, *Energy*, **31**, 3310–3326.
White, M.A. (2019) *Physical Properties of Materials*, 3rd edition, CRC Press, Boca Raton, Florida.

Index

D

F

G

H

Q

R

9780367135720